Ecological Studies, Vol. 215

Analysis and Synthesis

Ecological Studies

Further volumes can be found at springer.com

Ulrich Lüttge · Erwin Beck · Dorothea Bartels
Editors

Plant Desiccation Tolerance

Editors
Professor Dr. Ulrich Lüttge
Institute of Botany
Department of Biology
Technical University of Darmstadt
Schnittspahnstraße 3–5
64287 Darmstadt, Germany
luettge@bio.tu-darmstadt.de

Professor Dr. Erwin Beck
Department of Plant Physiology
University of Bayreuth
Universitätsstraße 30
95440 Bayreuth, Germany
erwin.beck@uni-bayreuth.de

Professor Dr. Dorothea Bartels
Institute of Molecular Plant Physiology
and Plant Biotechnology
University of Bonn
Kirschallee 1
53115-Bonn, Germany
dbartels@uni-bonn.de

ISSN 0070-8356
ISBN 978-3-642-19105-3 e-ISBN 978-3-642-19106-0
DOI 10.1007/978-3-642-19106-0
Springer Heidelberg Dordrecht London New York

Library of Congress Control Number: 2011931204

Cover design: deblik Berlin, Germany

Printed on acid-free paper

Springer is part of Springer Science+Business Media (www.springer.com)

Foreword

'Dark brown shrivelled dead-looking leaves' 'so dry they can be crushed to a powder between one's fingers' 'the plant cost its own weight in gold' 'surprisingly, the rehydrating leaves became green again' 'the leaves expanded to ten times their area when they were dry' – these phrases display the astonishment evoked by the novel vision of a desiccation tolerant 'resurrection plant' passing from a moribund dry state to a healthy active life as it re-hydrates! Such amazement has now been matched by rapidly advancing scientific understanding.

Researchers from countries widely spread from Hungary to New Zealand set out current knowledge on desiccation tolerant plants. It is appropriate that this book includes contributions of many eminent plant scientists from Germany – for German botanists played a substantial role in the initial reports of angiosperm species with desiccation tolerant foliage and in the subsequent research into the mechanisms involved in the survival and recovery of air-dry leaves. The first reports, by the taxonomist Kurt Dinter, of four desiccation-tolerant angiosperm species consisted of mere asides in his descriptions of species in the flora of southwest Africa. Fuller comments on one of these species, *Chamaegigas intrepidus*, were published by H. Heil in 1924. Four decades later P. Hoffman, G.H. Vieweg and H. Ziegler demonstrated renewed photosynthesis in re-hydrated shoots of the African 'resurrection bush' *Myrothamnus flabellifolia*, one of the species Dinter recognised to be desiccation-tolerant.

Focused exploration for desiccation-tolerant plants has extended our knowledge of the floristic spread and the geographic range greatly. Even a cursory perusal of the chapter topics shows that the phenomenon is found in the full range of phyla of chlorophyll-containing species from prokaryotes and cryptogams to angiosperms. Relatively few species in any one phylum have received intensive study. Among the angiosperms, Dinter's species *Craterostigma plantagineum, Chamaegigas intrepidus*, *Myrothamnus flabellifolia* and *Xerophyta viscosa* have all received considerable but by no means exhaustive scientific attention, as has also *Sporobolus stapfianus* that was recognised as a desiccation-tolerant grass in 1970.

The present tome also displays how the scope of desiccation-tolerant plant studies has expanded to embrace ecological, evolutionary, physiological, biochemical and molecular biological areas. In the last two instances, the increasing knowledge of desiccation tolerance is being driven by the explosive growth in the technology and understanding in these fields. The first investigations of gene expression of drying and rehydrating resurrection plants were conducted in the Max Planck Institute at Köln by Professor Dorothea Bartels and her colleagues. I am indebted to Professor Bartels for guiding my first steps in this important aspect of desiccation tolerance. The rapid growth of this field has given us insights into the complex changes in mRNA complements and the proteome that support the survival of drying leaves and the revival of rehydrating plants, not only in the foliage but also in the pollen and seed of most spermatophytes. The investigations of a widening number of researchers active in this area have elucidated much about the compounds and processes implementing desiccation tolerance. Much remains to be discovered on the mechanisms of regulating the implementation of desiccation tolerance. The visual drama of desolate air-dry plants re-imbibing water, re-expanding and reviving is matched by the intellectual fascination of the enabling molecular machinery. I hold the hope that a full comprehension of the regulatory processes will lead to genetic transformation of crop and pasture species to enable them to express throughout the full vegetative plant the desiccation tolerance of their seed and pollen – and so bring a full knowledge of the phenomenon of desiccation tolerance to its fullest practical yield.

Melbourne, Australia
January 2011

Donald F. Gaff

Contents

Part III The Cell Biological Level

Contributors

Dorothea Bartels Institute of Molecular Plant Physiology and Plant Biotechnology, University of Bonn, Kirschallee 1, 53115 Bonn, Germany, dbartels@uni-bonn.de

Erwin Beck Department of Plant Physiology, University of Bayreuth, Universitätsstr. 30, 95440 Bayreuth, Germany, erwin.beck@uni-bayreuth.de

Hans J. Bohnert Department of Plant Biology, Department of Crop Sciences, Center for Comparative and Functional Genomics, and Institute for Genomic Biology, University of Illinois at Urbana-Champaign, 1201 W. Gregory Drive, Urbana, IL 61801, USA

Burkhard Büdel Department of Biology, Botany, University of Kaiserslautern, 67663 Kaiserslautern, Germany, buedel@rhrk.uni-kl.de

John C. Cushman Department of Biochemistry and Molecular Biology, MS200, University of Nevada, Reno, NV 89557-0200, USA, jcushman@unr.edu

Sylvia K. Eriksson Department of Biochemistry and Biophysics, Arrhenius Laboratories, Stockholm University, Stockholm, Sweden

T.G. Allan Green Vegetal II, Farmacia Facultad, Universidad Complutense, 28040 Madrid, Spain; Biological Sciences, Waikato University, Hamilton, New Zealand, greentga@waikato.ac.nz

Ruth Grene Department of Plant Pathology, Physiology, and Weed Science, Virginia Tech, 435 Old Glade Road, Blacksburg, VA 24061-0330, USA, grene@vt.edu

Pia Harryson Department of Biochemistry and Biophysics, Arrhenius Laboratories, Stockholm University, Stockholm, Sweden, pia.harryson@dbb.su.se

Wolfram Hartung Julius-von-Sachs-Institut für Biowissenschaften, Lehrstuhl Botanik I, Universität Würzburg, Julius-von-Sachs-Platz 2, 97082, Würzburg, Germany, hartung@botanik.uni-wuerzburg.de

Ulrich Heber Julius-von-Sachs-Institute of Biosciences, University of Würzburg, Julius-von-Sachs-Platz 2, D-97082 Würzburg, Germany, heber@botanik.uni-wuerzburg.de

Hermann Heilmeier Interdisziplinäres Ökologisches Zentrum, TU Bergakademie Freiberg, Leipziger Str. 29, 09599 Freiberg, Germany, heilmei@ioez.tu-freiberg.de

Syed Sarfraz Hussain Institute of Molecular Plant Physiology and Plant Biotechnology, University of Bonn, Kirschallee 1, 53115 Bonn, Germany

Michael Lakatos Experimental Ecology, Department of Biology, University of Kaiserslautern, P.O. Box 3045, 67653 Kaiserslautern, Germany, lakatos@rhrk.uni-kl.de

Hartmut K. Lichtenthaler Botanical Institute (Molecular Biology and Biochemistry of Plants), Karlsruhe Institute of Technology, University Division, Kaiserstr. 12, 76133 Karlsruhe, Germany, hartmut.lichtenthaler@kit.de

Ulrich Lüttge Institute of Botany, Department of Biology, Technical University of Darmstadt, Schnittspahnstr. 3–5, 64287 Darmstadt, Germany, luettge@bio.tu-darmstadt.de

Melvin J. Oliver USDA-ARS Plant Genetics Research Unit, 205 Curtis Hall, University of Missouri, Columbia, MO 65211, USA

Ana Pintado Vegetal II, Farmacia Facultad, Universidad Complutense, 28040 Madrid, Spain

Stefan Porembski Department of Botany, Universität Rostock, Institute of Biosciences, Wismarsche Straße 8, 18051 Rostock, Germany, stefan.porembski@uni-rostock.de

Leopoldo G. Sancho Vegetal II, Farmacia Facultad, Universidad Complutense, 28040 Madrid, Spain

Renate Scheibe Department of Plant Physiology, University of Osnabrueck, Barbarastr. 11, 49076 Osnabrueck, Germany

Ernst Steudle Department of Plant Ecology, University of Bayreuth, Bayreuth, Germany, ernst.steudle@uni-bayreuth.de

Zoltán Tuba Plant Ecology Research Group of Hungarian Academy of Sciences, Godollo, Hungary; Department of Botany and Plant Physiology, Faculty of Agricultural and Environmental Sciences, Szent István University Gödöllő, Páter K. u.1, 2103 Godollo, Hungary

Cecilia Vasquez-Robinet Department of Plant Pathology, Physiology, and Weed Science, Virginia Tech, 435 Old Glade Road, Blacksburg, VA 24061-0330, USA; Department Biologie I – Botanik, Ludwig-Maximilians-Universität München, Großhadernerstr. 2-4, 82152 Planegg-Martinsried, Germany

Part I
Introduction

Chapter 1
Introduction

Dorothea Bartels, Ulrich Lüttge, and Erwin Beck

Evolution of life on earth began in aqueous environments. The oldest fossil records of green photosynthesizing organisms are the stromatolithes of cyanobacteria-like organisms about 3.5×10^9 years old. One of the major problems organisms were facing when leaving the water and conquering the land 400×10^6 years ago in the Devonian was exposure to a dry atmosphere. Among the present land plants, we observe a wealth of structural and functional adaptations suitable for shaping the water relations appropriate for life under such conditions. However, even plants in the aqueous habitats may have been subject to dry periods given by tidal rhythms or temporary drying out of their aqueous habitat. As primary water plants and not having evolutionary adaptations, these organisms needed to acclimate to dehydration conditions, the most extreme one of which is survival of desiccation, i.e. the loss of most of the cellular water.

Organisms that tolerate desiccation by dormancy and resume metabolic activity upon re-wetting have been termed poikilohydric. Their water content varies since they respond to the humidity of their environment like physical systems by shrinking and swelling. Unlike the non-desiccation-tolerant so-called homoiohydrous organisms, they are not differentiated into organs for absorption of water and structures that prevent loss of water (Schulze et al. 2005).

To date, we find many desiccation-tolerant forms among extant prokaryotic cyanobacteria (Chap. 2) and eukaryotic green algae (division Chlorobionta), such as the Chlorococcales in the class Chlorophyceae and species of the classes Trebouxiophyceae and Trentepohliophyceae, as well as species of *Porphyridium* among the red algae (class Rhodophyceae) (Chap. 4). Hence, desiccation tolerance must have evolved early and polyphyletically. However, in these algal taxa it was a primary step in evolution. Therefore, we consider the desiccation-tolerant cyanobacteria and algae as well as basic cryptogamic land plants such as bryophytes, lichens and fungi (Chaps. 5–7) termed as "primary poikilohydric" species. Desiccation tolerance can be expressed in somatic cells but particularly in special survival units, i.e. cysts, spores and zygotes, often surrounded by thick cell walls. Evolution has maintained this in vascular plants where spores of pteridophytes, pollen grains and most seeds of gymnosperms and angiosperms are highly desiccation tolerant. In some cases, seeds can survive dryness for years, up to one or two centuries and in the famous case of lotus (*Nelumbo nucifera*) even 1,200 years

U. Lüttge et al. (eds.), *Plant Desiccation Tolerance*, Ecological Studies 215,
DOI 10.1007/978-3-642-19106-0_1, © Springer-Verlag Berlin Heidelberg 2011

(Shen-Miller et al. 1995). While the germination success of the more than 10,000 years old Pyramid-wheat could not be substantiated, other seeds, e.g. from a date palm, which germinated and grew, have been dated more than 2,000 years BP (Sallon et al. 2008).

For the vegetative bodies of vascular plants, desiccation tolerance has evolved regressively or secondarily; these plants have been termed "secondary poikilohydric" (Chaps. 8 and 9). Secondary poikilohydric vascular plants are much fewer than the primary ones. They are predominantly found among the pteridophytes, the ferns and fern allies (700–1,000 species), but are rare in the angiosperms. Only a total of not more than 350 vascular poikilohydric angiosperm species have been described up to now (Chaps. 8 and 9). An impressive example of a secondary water plant, which is secondary poikilohydric, is the angiosperm *Chamaegigas intrepidus* (Chap. 12). Evidently, secondary poikilohydry has also evolved polyphyletically, but these were rare events. Development of desiccation avoidance in the evolution of vascular land plants was the more effective strategy than of desiccation tolerance. The latter requires complex dehydration and rehydration machineries.

There is a trade-off between desiccation tolerance and the size of plants as most desiccation-tolerant plants do not grow into tall plants. Poikilohydric thallophyte tissues usually consist of smaller cells, and the water potential of their protoplasts and organelles is in equilibrium with that of their immediate environment. Terrestrial poikilohydric thallophytes may have stomata-like structures, e.g. some liverworts and hornworts; however, these are immovable vents and are not able to control water loss to the atmosphere. Shrinkage upon dehydration is less dramatic as with vascular plants, and as a consequence, the compartment of the thallophytes differs fundamentally from that of the desiccation-tolerant vascular plants when losing and regaining water during a desiccation/rehydration cycle (Chaps. 6 and 10). One of the tallest and best-studied desiccation-tolerant angiosperms is the small dicotyledonous shrub *Myrothamnus flabellifolia*. The tree habit is also reached by some desiccation-tolerant monocotyledons (Sect. 8.2.2).

For the desiccation tolerance of taller vascular plants, hydraulic architecture is an important aspect. Cavitation and the replacement of water by air (embolism; Chap. 10) in the conducting elements of the xylem are outstanding implications of drought and the more so upon desiccation in vascular plants. The consequences for resurrection during rehydration are intriguing and the mechanisms of refilling are unknown as of yet. Little work has been performed on resurrection plants. However, the transition from temporal tolerance of tight water relations to drought resistance and further to the tolerance of desiccation is gradual. From savanna trees, we know daily courses where water-stress-related midday depression of hydraulic conductivity is followed in the afternoon by cavitations and embolisms in roots and leaves, which are refilled during the night (Bucci et al. 2003; Domec et al. 2006). There are also annual courses: For example, in the fern *Mohria caffrorum* poikilohydry is developed seasonally, i.e. plants are desiccation tolerant in the dry season but not during the rainy season (Farrant et al. 2008). Thus, Chap. 10 evaluates the structures and functions relevant for water flow, from the cellular to the organ and whole-plant level as an essential basis for any experimental approaches towards understanding their functional contribution to desiccation tolerance of vascular plants. One of the key points is indeed refilling of the conducting elements with water upon re-watering and during the process of resurrection. Often in nature extreme cases prove to be the best examples for understanding basic problems and Chap. 10 evidently ends up with the message that desiccation-tolerant plants are such an example challenging new research.

The poikilohydric cryptogams and among the vascular plants the majority of the poikilohydric ferns and dicotyledonous species retain their chlorophyll and much of the photosynthetic machinery during desiccation. They are termed "homoiochlorophyllous". Among the monocotyledonous plants, we find both homoiochlorophyllous and poikilochlorophyllous species. The latter degrade their chlorophyll molecules as well as the thylakoid membranes during desiccation. In evolution, homoiochlorophylly was primary and poikilochlorophylly was secondary. Interestingly, in this respect, some plants are only partially homoiochlorophyllous like *Ramonda serbica* (Degl'Innocenti et al. 2008).

Light is the most critical stress factor during dehydration, in the desiccated state and upon rehydration (Chaps. 3 and 7). The problem of homoiochlorophyllous plants is that they are under severe stress of photodestruction by maintaining light absorbing pigment complexes, but their advantage is that they recover photosynthetic activity rapidly upon rehydration. The advantage of the poikilochlorophyllous plants is that by dismantling their photosynthetic apparatus they avoid photodestruction. Their problem is that upon rehydration there is a substantial lag phase before they are able to resume photosynthesis (Chap. 9). Although homoiochlorophylly is considered to be a basic evolutionary trait, it does need a highly sophisticated machinery of photoprotection as it is described in several chapters (Chaps. 3, 7 and 11), while the more advanced trait of poikilochlorophylly requires a complex set of molecular and biochemical mechanisms (Chaps. 9, 13–16). Good examples to this are the needles of the winter hardy evergreen conifers, which, upon crystallization of tissue water in the intercellular spaces, may lose more than 90% of their liquid cellular water. These plants degrade a major part of their antenna pigments in the course of frost hardening still before the onset of frost (Beck et al. 2004). For a homoiohydric plant, this extreme degree of dehydration is only tolerable at subfreezing temperatures when biochemical reactions are greatly slowing down or even cease.

Functional diversity is profoundly determined by the homoiochlorophyllous or poikilochlorophyllous nature of desiccation-tolerant plants. Poikilochlorophylly determines the ecological niche acquisition by the respective species given by the extensions of dry periods (Chap. 9).

Thus, we face a large diversity in desiccation tolerance. This is covered in the various chapters of this book at different levels. At the phytogeographic level, we arrive at the diversity of habitats as an important facet (Chaps. 2, 4, 5 and 8). Ecological constraints of habitats determine selection of species. At the organismic level, we then consider the diversity of the organizational status of cyanobacteria, algae, bryophytes, lichens (Chaps. 2, 4 and 5) and vascular plants (Chap. 8). Functional diversity is seen in a variety of mechanisms of evolutionary adaptation as well as more short-term ecophysiological acclimation.

We realize that understanding of desiccation tolerance at the organismic level and in an ecological context has been continuously advanced (Part II). At the cell biological level, we distinguish biophysical mechanisms and biochemical processes starting from gene expression to the activity of proteins and the accumulation or disappearance of metabolites, unravelled by the various components of the so-called "omics" that provide the information basis for systems biology. Advanced methodology for highly sophisticated analyses of the biophysical processes of excitation of the photosynthetic apparatus and the dissipation of the energy of the excitons produced fosters understanding of principal problems and their potential solutions of green desiccation-tolerant organisms (Chaps. 3 and 7). This has impact at the level of the organisms (Part II) but also forms a link to the cellular level (Part III). There, it is re-considered from a biochemical viewpoint addressing oxidative stress and its function in cell biology under water deficit (Chap. 11) and the apparently paradoxical special case of an aquatic poikilohydric angiosperm (Chap. 12).

The major section (Chaps. 13–16) of the cell biological Part III fathoms the relevance of the enormous progress of molecular biology and genetics for the understanding of desiccation tolerance. We must recall that in terms not only of adaptation during evolution but also of acclimation to recurrent or arrhythmic environmental changes responses to water shortage and pronounced drought with an eventual coronation by desiccation tolerance are gradual. Therefore, just like for hydraulic architecture (Chap. 10), we must realize at the level of cell biology that drought tolerance in many aspects appears as a prelude to desiccation tolerance. Therefore, although this volume focuses on desiccation tolerance, certain aspects of responses to drought must also be included. Many defence strategies, e.g. against damage from radicals, are similarly involved in both drought and desiccation tolerance, and responses to drought and desiccation are, therefore, often quite similar. Desiccation tolerance especially of vascular plants is considered as a more advanced adaptation to severe and temporal shortage of water than drought tolerance. *Sensu stricto* desiccation tolerance involves the survival of losing the major fraction of tissue water under exposure to dry conditions, and showing recovery of full physiological competence after rehydration. At the molecular level, mechanisms providing for drought and desiccation tolerance are shared with respect to the genetic management of input of stress signals and of downstream processes of damage, repair, tolerance and avoidance. This raises the strong demand of a new comprehensive treatment considering genomics, transcriptomics, proteomics and metabolomics moving on from drought-tolerant to desiccation-tolerant plant systems.

Thus, Chap. 13 sets the scene by delineating the basic concepts of functional genomics, epigenomics, genetics, molecular biology and the sensing and signalling networks of systems biology, which we need when we consider stress physiology in general and with particular focus on tight water relations. A specific component of the complement is highlighted in Chap. 14, namely the dehydrin proteins. They have multiple general functions as chaperones modulating and protecting macromolecular cell structures and biomembranes. They occur in all seed plants and have been associated with the acquisition of desiccation tolerance of seeds. They are the best-characterized group of the so-called LEA proteins. LEA means "late embryogenesis abundant", i.e. they abound in seeds that are normally desiccation tolerant. Is there an evolutionary link to desiccation tolerance of somatic tissues? At least for one *bona fide* resurrection plant, *Craterostigma plantagineum*, some evidence for the involvement of dehydrins is available (Sect. 14.10). Overexpression of a dehydrin from barley in rice has been shown to increase tolerance of specific water-deficit stresses (Xu et al. 1996). We certainly must have an eye on dehydrins when further fathoming the mechanisms of desiccation tolerance, and Chap. 16 picks up the LEAs again.

The desiccation-tolerant moss *Physcomitrella patens* is fully sequenced. However, we do not have complete genome sequences of desiccation-tolerant higher plants. When this advances, comparisons with the genomes of other model plants such as *Arabidopsis thaliana* (Chap. 13) will turn out to be highly profitable. It is remarkable, however, how Chap. 15 can already advance from the conceptual basis of Chap. 13 towards revealing constituents of systems biology of desiccation tolerance using genomics, proteomics, metabolomics and fluxomics. A wealth of relevant genes from resurrection plants is identified, and the involvement of their gene products can be described. This is already much pertinent information and generates knowledge. It gives the basis and shows the direction towards understanding.

As far as it is possible at this stage of the progress of research Chap. 16 then reassembles many of the putative constituents of the desiccation-tolerance complement linking molecular biology with physiology. The challenge for further endeavours of investigation is obvious. The reward these endeavours will give for understanding plants, habitats, natural ecosystems as well as agro- and forest-ecosystems and biomes where water is one of the most essential ingredients, is similarly obvious.

References

Beck E, Heim R, Hansen J (2004) Plant resistance to cold stress: mechanisms and environmental signals triggering frost hardening and dehardening. J Biosci 29:449–459

Bucci SJ, Scholz FG, Goldstein G, Meinzer FC, da SL SL (2003) Dynamic changes in hydraulic conductivity in petioles of two savanna tree species: factors and mechanisms contributing to the refilling of embolized vessels. Plant Cell Environ 26:1633–1645

Degl'Innocenti E, Guidi L, Stevanovic B, Navari F (2008) CO_2 fixation and chlorophyll *a* fluorescence in leaves of *Ramonda serbica* during a dehydration–rehydration cycle. J Plant Physiol 165:723–733

Domec J-C, Scholz FG, Bucci SJ, Meinzer FC, Goldstein G, Villalobos-Vega R (2006) Diurnal and seasonal variation in root xylem embolism in neotropical savanna woody species: impact on stomatal control of plant water status. Plant Cell Environ 29:26–35

Farrant JM, Lehner A, Cooper K, Wiswedel S (2008) Desiccation tolerance in the vegetative tissues of the fern *Mohria cafforum* is seasonally regulated. Plant J 57:65–79

Sallon S, Solowey E, Cohen Y, Korchinsky R, Egli M, Woodhatch I, Simchoni O, Kislev M (2008) Germination, genetics, amd growth of an ancient date seed. Science 320:1464

Schulze E-D, Beck E, Müller-Hohenstein K (2005) Plant ecology. Springer, Berlin, pp 278–279

Shen-Miller J, Mudgett MB, Schopf JW, Clarke S, Berger R (1995) Exceptional seed longevity and robust growth. Ancient Sacred Lotus from China. Am J Bot 82:1367–1380

Xu D, Duan X, Wang B, Hong B, Ho T-HD WuR (1996) Expression of late embryogenesis abundant protein gene, HVA1, from barley confers tolerance to water deficit and salt stress, in transgenic Rice. Plant Physiol 110:249–257

Part II
The Organismic Level

Chapter 2
Cyanobacteria: Habitats and Species

Burkhard Büdel

2.1 Introduction

Cyanobacteria were most probably the first group of organisms performing an oxygen releasing photosynthesis. Their possible fossil origin ("look-alikes") from Apex chert of north-western West Australia dates back to about 3.46 billion years (Schopf 2000). However, the oldest unambiguous fossil cyanobacteria were found in tidal-flat sedimentary rocks and are about 2 billion years old (Hofmann 1976). With the onset of oxygenic photosynthesis between 2.45 and 2.32 billion years ago (Rasmussen et al. 2008), the ancient Earth's oxygen-free atmosphere experienced a deep impact with the sharp rise of oxygen. Before the evolution of respiration, the oxygen was highly toxic to life, and as a consequence, the first global catastrophe for most of the organisms living on earth to that date followed. Today, cyanobacteria are found in almost all habitats and biomes present on earth (Whitton and Potts 2000). However, to successfully colonize terrestrial habitats does also mean to be able to resist extreme desiccation. Air-drying does severely harm membrane structure, proteins, and nucleic acids and is lethal to the majority of organisms on Earth (Billi and Potts 2002). During their long evolutionary history, cyanobacteria developed the ability of their cells to undergo nearly absolute dehydration during air-drying without being killed, a phenomenon known as *anhydrobiosis*. This is also referred to as "desiccation tolerance" and is one mechanism of drought tolerance (Alpert 2005). Consequently, cyanobacteria colonized more and more of the available terrestrial habitats. Dehydration in air can lead to a removal of all but 0.1 g water/g dry weight (Billi and Potts 2000).

2.2 Cyanobacterial Anhydrobiosis and Resistance to Complete Desiccation

Desiccation-tolerant cyanobacteria must either protect cellular structures from damage and/or repair them upon rewetting. Dried aggregates of *Chroococcidiopsis* include live and dead cells, thus suggesting that desiccation resistance is not a

U. Lüttge et al. (eds.), *Plant Desiccation Tolerance*, Ecological Studies 215,
DOI 10.1007/978-3-642-19106-0_2,

simple process. In a recent study, Billi (2009) demonstrated that the desiccation surviving cells were avoiding and/or limiting genome fragmentation, preserve intact plasma membranes and phycobiliprotein autofluorescence, and exhibit spatially reduced reactive oxygen species accumulation and dehydrogenase activity upon rewetting. The percentage of cells avoiding subcellular damage was between 10 and 28% of dried aggregate. In the lichenized state, however, all cells of symbiotic *Chroococcidiopsis* species seem to survive desiccation, as there is no depression of photosynthetic CO_2 fixation rates of the same lichen thalli (genus *Peltula*, Lichinomycetes) before and after dry periods (Büdel et al. unpublished). This seems to be a general feature of lichenized cyanobacteria (Fig. 2.2g). The fungal host (mycobiont) apparently provides more than only a three-dimensional structure for optimized CO_2 and nutrient uptake, but also an environment for optimal (damage free) drying of cyanobacterial cells.

The common soil inhabiting cyanobacterium *Nostoc commune* Vaucher ex Bornet & Flahault developed protecting mechanisms to avoid genome fragmentation after prolonged cell desiccation (Shirkey et al. 2003). Changes at the ultrastructural level of *Chroococcidiopsis* cells have been demonstrated by Grilli Caiola et al. (1993). These authors could conclusively show that the thylakoid structure is changed in dry cells. The thylakoid double membrane opens between the single membranes, forming open spaces between them (Fig. 2.1a, b).

The filamentous, colony-forming cyanobacterium *N. commune* can tolerate simultaneous stresses of desiccation, UV irradiation, and oxidation. For protection, the acidic water stress protein A (wspA) and a highly stable and active

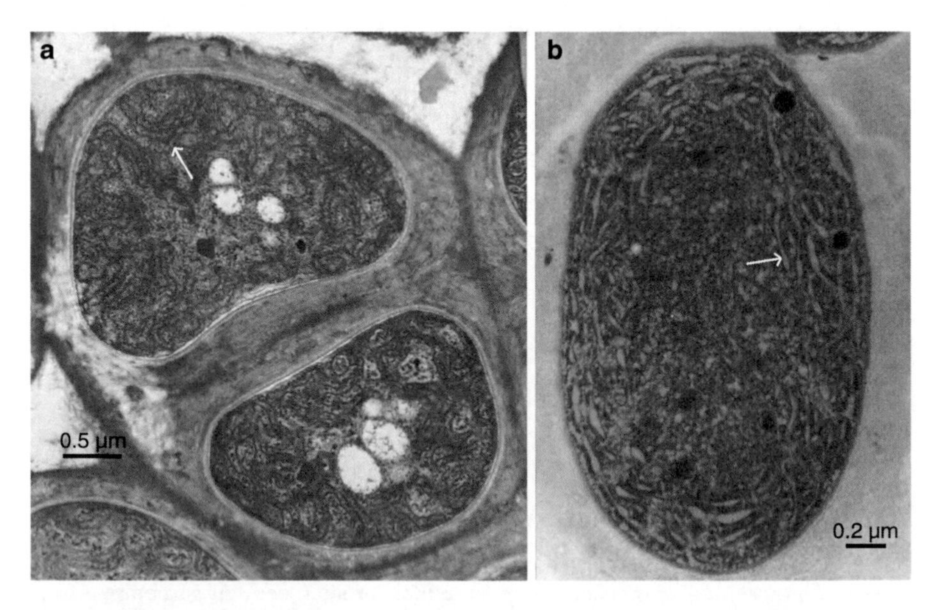

Fig. 2.1 Ultrastructure of wet and dry cells of the genus *Chroococcidiopsis*. (**a**) Wet cells, note the intact (closed double membrane layer) thylakoids (*arrow*). (**b**) Dry cell, freeze substituted over several months prior to ultrathin sectioning and microscopy in order to avoid preparation artifacts. Note the widened thylakoids (opening of the double membrane layer; *arrow*)

superoxidedismutase (sodF) were found to be secreted to the three-dimensional extracellular matrix. Transcription of wspA and sodF and synthesis and secretion of wspA were induced upon desiccation or UV-A/B irradiation of *N. commune* cells (Wright et al. 2005). The authors hypothesize that wspA plays a central role in the stress response of *N. commune* by modulation of structure and function of the three-dimensional extracellular matrix.

2.3 Habitats and Species

Once the importance of the microclimate for life conditions of small organisms was discovered, it soon became obvious that in terms of temperature, sunlight-exposed soils or rock surfaces are among the most extreme terrestrial habitats for photoautotrophic organisms (Kraus 1911; Jaag 1945). This led to the observation that two different exposures (SE versus NW in the northern hemisphere) on one large rock boulder can even result in a microclimate that reflects arctic alpine and sub-Mediterranean climates on one single rock (Schade 1917). As a result of high temperatures during insulation, lack of water and thus fast desiccation occurs in the colonizing organisms. Jaag (1945) showed that different rock types reach different surface temperatures under the same circumstances. At an air temperature of 16.5°C at 14:40 h in the Swiss capital of Zürich, he found surface temperatures of 35.7°C for diabas, 27.1°C for granite, and 24.4°C for marble during standardized measurements.

Cyanobacteria occur on almost all exposed rock surfaces on earth. There is hardly any rock surface to find without the presence of either epilithic (on the rock surface; Fig. 2.2d, e) or endolithic (inside the photic zone of the rock; Fig. 2.2a, c) growth of pro- or eukaryotic algae. Biofilms (thin smooth layers of pro- and eukaryotic algae only) or biological crusts (thick, uneven layers, often including lichens and sometimes bryophytes) occur on rocks of hot deserts (Büdel 1999; Büdel and Wessels 1991; Friedmann et al. 1967), polar deserts (Friedmann 1980; Broady 1981; Omelon et al. 2006; Büdel et al. 2008), inselbergs (Fig. 2.2c, d) in savannas and rain forests (Büdel et al. 2000), temperate regions (Boison et al. 2004), alpine zone (Horath and Bachofen 2009), and can even be found on the man-made surfaces of any kind of stone buildings (Eggert et al. 2006; Karsten et al. 2007).

A very conspicuous phenomenon of rock surfaces are the so-called Tintenstriche (German: Tinte = ink, Strich = stripe; Fig. 2.2e). They have been first described from the alpine habitat of the temperate region, but also occur in other biomes of the world (Lüttge 1997). Ink-stripes are bluish-black crusts on steep to more or less vertical rock surfaces such as dolomite (Diels 1914), granite (Golubic 1967; Büdel et al. 1994), sandstone (Wessels and Büdel 1995), and other rock types (Jaag 1945). The ink-stripe communities are dominated by cyanobacteria, sometimes accompanied by eukaryotic algae, fungal hyphae, and lichens, even mosses, and vascular plants can occur (Jaag 1945; Wessels and Büdel 1995). When studied under the light microscope, many cyanobacterial species expose very colorful sheaths (Fig. 2.3c, d, f, g). The bright yellow and red color mainly originates from scytonemin, an indol-alkaloid serving as light and UV protection (Buckley and Hougthon 1976; Garcia-Pichel and Castenholz 1991). When exposed to frequent desiccation

Fig. 2.2 Frequently dry habitats. (**a**) Upper Taylor Valley, from Rhone Bench, McMurdo Dry Valleys, Antarctica. (**b**) Quartz gravel pavement of the Knersvlakte semi desert, Western Cape Province, South Africa. (**c**) Uluru (Ayers Rock), an Inselberg consisting of sandstone arkose, Northern Territory, Australia. (**d**) Inselberg in the humid savanna along the Orinoco River, Venezuela. (**e**) "Tintenstriche" (ink-stripes) created by cyanobacterial growth on the rock surface, Milford Sound, South Island, New Zealand. (**f**) Cyanobacterial soil crust, moistened by early morning dewfall, semidesert at Soebatsfontein, Northern Cape province, South Africa. (**g**) The cyanolichen *Lichina pygmaea* at the rocky seashore of Île de Bréhat, Atlantic coast, France

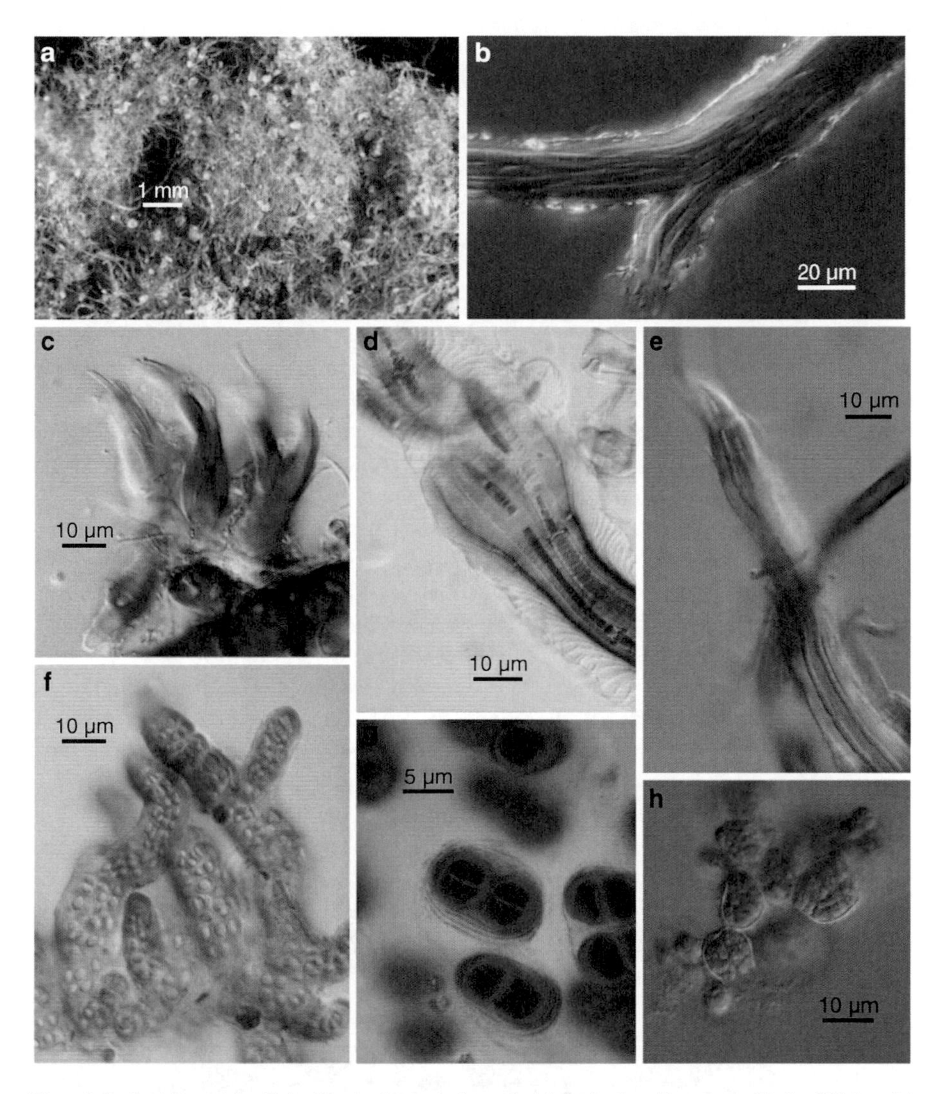

Fig. 2.3 Species. (**a**) *Coleofasciculus vaginatus*, biological soil crust, Utah, USA. (**b**) *Coleofasciculus vaginatus*, filament with numerous trichomes embedded in a thick, gelatinous sheath; isolated from a Sebkah, Tunisia. (**c**) *Calothrix parietina*, biofilm at an inselberg in the lowland rain forest at Les Nouragues, French Guiana. (**d**) *Petalonema alata*, biofilm on rock at the Nufenen Pass, European Alps, Switzerland. (**e**) *Schizothrix telephoroides*, biological soil crusts in *Trachypogon*-grass savanna, upper Orinoco River, Venezuela. (**f**) *Stigonema mammilosum*, biofilm of an inselberg in the humid savanna along the Orinocco River, Venezuela. (**g**) *Gloeocapsa sanguinea*, from an epilithic biofilm, Alps, Austria. (**h**) *Chroococcidiopsis* sp., hypolithic, from underneath translucent quartz rocks, Namib Desert, Namibia

events, scytonemin synthesis is significantly increased in the soil and rock inhabiting cyanobacteria *Nostoc punctiforme* and *Chroococcidiopsis* sp. compared to continuously hydrated samples (Fleming and Castenholz 2007).

Herbarium specimens of *N. commune* are known to maintain their viability for decades of storage (Whitton and Potts 2000). Desert inhabiting *Chroococcidiopsis* species (Fig. 2.3h) are able to enter a dormancy state at the desiccation onset and resume metabolic activities when water becomes available (Grilli Caiola and Billi 2007). Viable cells of *Chroococcidiopsis* spp. were recovered from rocks from the Negev Desert after storage for 30 years (Grilli Caiola et al. 1996). In Antarctica for example, where cyanobacteria are freeze dried when in equilibrium with air, their water content can drop down to less than 2% of dry weight at 5°C and a relative air humidity of 18%. *N. commune* populations from Antarctica, growing in water ponds (Fig. 2.4a) or on the bare soil (Fig. 2.4b), experience different modes of desiccation. While the population in the pond freezes in fully hydrated state and then stays dormant inside the ice clump, the soil population freezes either dry, or when freezing wet, it undergoes freeze drying thereafter. Although still belonging to the same species, they show some genetic differences (Novis and Smissen 2006).

No terrestrial photoautotrophic biofilm or biocrust is so well characterized by a single cyanobacterial genus as those in the endolithic habitat. There, the unicellular genus *Chroococcidiopsis*, an apparently old genus (Fewer et al. 2002) comprising about 14 species (Komárek and Anagnostidis 1998), is the most common cyanobacterium, extending from the polar desert (Friedmann and Ocampo-Friedmann 1976; Friedmann et al. 1988; Büdel et al. 2008; Büdel 2011) to the tropical rainforest (Sarthou et al. 1995; Büdel 1999). Other important cyanobacteria found in the endolithic habitat are *Cyanothece aeruginosa* (Büdel et al. 2008), *Gloeocapsa* spp. (Fig. 2.3g), *Hormathonema* sp. (Friedmann et al. 1988), *Gloeothece* sp., *Synechococcus* sp., and *Nostochopsis lobatus* (Weber et al. 1996).

Even rain forest trees can serve as substrata for epiphytic cyanobacteria. In West-Java for example, algal biofilms were found on the tree bark (=corticolous) composed of the filamentous cyanobacteria *Leptolyngbya* spp., *Nostoc punctiforme*, *Nostoc* spp., *Scytonema ocellatum*, *Scytonema* sp., and a large number of unicellular and filamentous green algae (Neustupa and Škaloud 2008). Freiberg (1998) reported cyanobacterial biofilms from leaves (=epiphyllous) of rain forest trees in Costa

Fig. 2.4 *Nostoc commune*, wet and dry. (**a**) Wet *N. commune* colonies in snow melt water pond, Observation Hill, Ross Island, Antarctica. (**b**) Freeze-dried *N. commune* colonies in channels between soil polygons, Taylor Valley floor, Ross Dependency, Antarctica

Rica. She found several species of the genus *Scytonema* and a species of *Stigonema* living on the leave surface, where they successfully performed di-nitrogen fixation.

Biological soil crusts (BSCs) are a widespread phenomenon of savannas, deserts (Fig. 2.2b, c, f), temperate zones, and polar tundra and desert (Fig. 2.2a) and are mainly formed by cyanobacteria (Belnap and Lange 2001). Like biofilms on rocks or trees, they are regularly exposed to short dry and wet cycles. These BSCs often comprise high numbers of cyanobacterial species, as for example 58 species in BSCs of south-western Africa (Büdel et al. 2009) or 33 in BSCs from the Bhubaneswar region in India (Tirkey and Adhikary 2005). With its long and multilayered trichomes (=naked cyanobacterial filaments) embedded in a sticky sheath, the cyanobacterial genus *Coleofasciculus* (compare Table 2.1; Fig. 2.3a, b) is the

Table 2.1 Cyanobacterial genera with anhydrobiotic species

Genus	Habitats
Calothrix (Nostocales and Rivulariaceae)	Rock (epilithic), soil, fresh water, marine, and lichenized[a]
Coleofasciculus[b] (Oscillatoriales and Phormidiaceae)	Soil (biological soil crusts, estuaries, marine, and freshwater)
Chroococcus (Chroococcales and Chroococcaceae)	Freshwater (bog pools), rock (epilithic), and lichenized[b]
Chroococcidiopsis [traditionally placed in the order Chroococcales, family Xenococcaceae, but position critical, forming natural sister group of Nostocales according to Fewer et al. (2002)]	Rock (epilithic and endolithic), rarely soil, and lichenized[a]
Cyanothece (Chroococcales and Cyanobacteriaceae)	Rock (endolithic), soil, and freshwater
Dichothrix (Nostocales and Rivulariaceae)	Rock (epilithic) and lichenized[a]
Gloeocapsa (Chroococcales and Microcystaceae)	Rock (epilithic and endolithic) and lichenized[a]
Hyphomorpha (Nostocales and Loriellaceae)	Epiphytic on leaves of liverworts and tree barks and lichenized[c]
Nostoc (Nostocales and Nostocaceae)	Rock (epilithic), soil, fresh water, rarely in brakish or marine environment, and lichenized[a]
Nostochopsis (Nostocales and Hapalosiphonaceae)	Rock (epi- and endolithic) and freshwater
Myxosarcina (Chroococcales, Xenococcaceae)	Rock (epilithic and endolithic) and lichenized[c]
Petalonema (Nostocales and Microchaetaceae)	Rock (epilithic)
Rivularia (Nostocales and Rivulariaceae)	Rock (epilithic), freshwater, lichenized[a]
Schizothrix (Pseudanabaenales and Schizotrichaceae)	Soil, rarely rock
Scytonema (Nostocales and Scytonemataceae)	Rock (epilithic), soil, fresh water, epiphytic on mosses and tree bark, lichenized[a]
Stigonema (Nostocales and Stigonemataceae)	Rock (epilithic), fresh water, lichenized[a]
Tolypothrix (Nostocales and Microchaetaceae)	Freshwater, soil, rock (epilithic), tree bark

[a]Regular lichen photobiont
[b]Formerly *Microcoleus*; see Siegesmund et al. (2008)
[c]Not safely verified or single observations only

main crust forming organism in tropical, subtropical, and sometimes also temperate regions (Belnap 1993; Belnap and Lange 2001), while the genus *Schizothrix* (Fig. 2.3e) seems to play a major role in biological soil crusts of neotropical savannas (Büdel et al. 1994). In boreal and polar regions, cyanobacteria of the genera *Nostoc*, *Scytonema*, or *Stigonema* (Fig. 2.3f) seem to take over this ecological niche (own observations, not yet properly confirmed).

2.4 Conclusion

The development of desiccation tolerance in cyanobacteria during the evolution of life opened new niches for them to colonize and to spread into almost all terrestrial habitats available until today. Nevertheless, reactivation of net photosynthesis after drought periods of either free living or symbiotic cyanobacteria heavily depends on the presence of water in its liquid form (Lange et al. 1986). Only after exposure to high relative air humidity (99%) for ecologically nonrealistic times (more than 24 h) could activate the net photosynthesis at low levels of the desert inhabiting cyanobacterium *Trichocoleus sociatus* (Lange et al. 1994). Net photosynthesis of many, if not most eukaryotic, cryptogamic plants can be reactivated with air humidity alone (Büdel and Lange 1991; Green and Lange 1995; Lange et al. 1986). Under moderate climatic conditions, events of high air humidity are frequent, and prolonged drought periods are rare. Here, eukaryotic, poikilohydrous, cryptogamic plants have a clear advantage as they can activate net photosynthesis more frequently than cyanobacteria and thus dominate over cyanobacterial competitors. This is most probably a major reason why cyanobacteria are less competitive under moderate climatic conditions and dominate in so-called extreme habitats.

Acknowledgments I would like to express my sincere thanks to U. Lüttge and E. Beck for offering me the opportunity to write this review and for many inspiring discussions. The German Research Foundation (DFG) is thanked for continuous financial support of my work. R. Honegger (Zürich) is thanked for preparation, thin sectioning and electron microscopy of dry *Chroococcidiopsis* cells.

References

Alpert P (2005) The limits and frontiers of desiccation-tolerant life. Integr Comp Biol 45:685–695

Belnap J (1993) Recovery rates of cryptobiotic crusts: inoculant use and assessment methods. Great Basin Nat 53:89–95

Belnap J, Lange OL (2001) Biological soil crusts: structure, function, and management. Springer, Berlin, pp 1–503

Billi D (2009) Subcellular integrities in *Chroococcidiopsis* sp. CCMEE 029 survivors after prolonged desiccation revealed by molecular probes and genome stability assays. Extremophiles 13:49–57

Billi D, Potts M (2000) Life without water: responses of prokaryotes to desiccation. In: Storey KB, Storey JM (eds) Environmental stressors and gene responses. Elsevier, Amsterdam, pp 181–192
Billi D, Potts M (2002) Life and death of dried prokaryotes. Res Microbiol 153:7–12
Boison G, Mergel A, Jolkver H, Bothe H (2004) Bacterial life and dinitrogen fixation at a gypsum rock. Appl Environ Microbiol 70:7070–7077
Broady PA (1981) The ecology of chasmolithic algae at coastal locations of Antarctica. Phycologia 20:259–272
Buckley CE, Hougthon JA (1976) A study of the effects of near UV radiation on the pigmentation of the blue-green alga *Gloeocapsa alpicola*. Arch Microbiol 107:93–97
Büdel B (1999) Ecology and diversity of rock-inhabiting cyanobacteria in tropical regions. Eur J Phycol 34:361–370
Büdel B (2011) *Chroococcidiopsis*. In: Reitner J, Thiel V (eds) Encyclopedia of geobiology. Springer, Heidelberg
Büdel B, Lange OL (1991) Water status of green and blue-green phycobionts in lichen thalli after hydration by water vapor uptake: do they become turgid? Bot Acta 104:361–366
Büdel B, Wessels DCJ (1991) Rock inhabiting blue-green algae/cyanobacteria from hot arid regions. Algol Stud 64:385–398
Büdel B, Lüttge U, Stelzer R, Huber O, Medina E (1994) Cyanobacteria of rocks and soils of the Orinoco lowlands and the Guayana uplands, Venezuela. Bot Acta 107:422–431
Büdel B, Becker U, Follmann G, Sterflinger K (2000) Algae, fungi, and lichens on Inselbergs. In: Porembski S, Barthlott W (eds) Inselbergs: biotic diversity of isolated rock outcrops in tropical and temperate regions. Springer, Berlin, pp 68–90
Büdel B, Bendix J, Bicker FR, Green TGA (2008) Dewfall as a water source frequently activates the endoltihic cyanobacteria communities in the granites of Taylor Valley, Antarctica. J Phycol 44:1415–1424
Büdel B, Darienko T, Deutschewitz K, Dojani S, Friedl T, Mohr KI, Salisch M, Reisser W, Weber B (2009) Southern african biological soil crusts are ubiquitous and highly diverse in drylands, being restricted by rainfall frequency. Microb Ecol 57:229–247
Diels L (1914) Die Algen-Vegetation der Südtiroler Dolomitriffe. Ein Beitrag zur Ökologie der Lithophyten. Ber Dtsch Bot Ges 32:502–526
Eggert A, Häubner N, Klausch S, Karsten U, Schumann R (2006) Quantification of algal biofilms colonising building material: chlorophyll *a* measured by PAM-fluorometry as a biomass parameter. Biofouling 22:79–90
Fewer D, Friedl T, Büdel B (2002) *Chroococcidiopsis* and heterocyst-differentiating cyanobacteria are each others's closest living relatives. Mol Phylogenet Evol 41:498–506
Fleming ED, Castenholz RW (2007) Effects of periodic desiccation on the synthesis of the UV-screening compound, scytonemin, in cyanobacteria. Environ Microbiol 9:1448–1455
Freiberg E (1998) Microclimatic parameters influencing nitrogen fixation in the phyllosphere in a Costa Rican premontane rain forest. Oecologia 17:9–18
Friedmann EI (1980) Endolithic microbial life in hot and cold deserts. Orig Life Evol Biosph 10:223–235
Friedmann EI, Ocampo-Friedmann R (1976) Endolithic blue-green algae in the dry valleys: primary producers in the antarctic desert ecosystem. Science 193:1257–1249
Friedmann EI, Lipkin Y, Ocampo-Paus R (1967) Desert algae of the Negev (Israel). Phycologia 6:185–200
Friedmann EI, Hua MS, Ocampo-Friedmann R (1988) Cryptoendolithic lichen and cyanobacterial communities of the Ross Desert, Antarctica. Ber Polarforsch 58:251–259
Garcia-Pichel F, Castenholz RW (1991) Characterization and biological implications of scytonemin, a cyanobacterial sheath pigment. J Phycol 27:395–409
Golubic S (1967) Die Algenvegetation an Sandsteinfelsen Ost-Venezuelas (Cumaná). Int Rev Hydrobiol 52:5–693

Green TGA, Lange OL (1995) Photosynthesis in poikilohydric plants: a comparison of lichens and bryophytes. In: Schultze ED, Caldwell MM (eds) Ecophysiology of photosynthesis. Springer, Berlin, pp 319–341

Grilli Caiola M, Billi D (2007) *Chroococcidiopsis* from desert to Mars. Cellular origin, life in extreme habitats and astrobiology, vol 11. Springer, Berlin, pp 555–568

Grilli Caiola M, Ocampo-Friedmann R, Friedmann EI (1993) Cytology of long-term desiccation in the desert cyanobacterium *Chroococcidiopsis* (Chroococcales). Phycologia 32:315–322

Grilli Caiola M, Billi D, Friedmann EI (1996) Effect of desiccation on envelopes of the cyanobacterium *Chroococcidiopsis* sp. (Chroococcales). Eur J Phycol 31:97–105

Hofmann HJ (1976) Precambrian microflora, Belcher Islands, Canada: significance and systematics. J Palaeontol 50:1040–1073

Horath T, Bachofen R (2009) Moelcular characterization of an endolithic microbial community in dolomite rock in the Central Alps (Switzerland). Microb Ecol. doi:10.1007/s00248-008-9483-7:

Jaag O (1945) Untersuchungen über die Vegetation und Biologie der Algen des nackten Gesteins in den Alpen, im Jura und im schweizerischen Mittelland. Beiträge zur Kryptogamenflora der Schweiz 9. Kommissionsverlag Buchdruckerei Büchler & Co., Bern, pp 1–560 + 21 plates

Karsten U, Schumann R, Mostaert AS (2007) Aeroterrestrial algae growing on man-made surfaces: what are the secrets of their ecological success? In: Seckbach J (ed) Algae and cyanobacteria in extreme environments. Springer, Dordrecht, pp 585–597

Komárek J, Anagnostidis K (1998) Cyanoprokaryota 1. Teil Chroococcales. Gustav Fischer Verlag, Jena, pp 1–548

Kraus G (1911) Boden und Klima auf kleinstem Raum Versuch einer exakten Behandlung des Standorts auf dem Wellenkalk. Gustav Fischer Verlag, Jena, pp 1–184

Lange OL, Kilian E, Ziegler H (1986) Water vapor uptake and photosynthesis of lichens: performance differences in species with green and blue-green algae as phycobionts. Oecologia 71:104–110

Lange OL, Meyer A, Büdel B (1994) Net-photosynthesis of a desiccated cyanobacterium without liquid water in high air humidity alone. Experiments with *Microcoelus sociatus* isolated from a desert soil crust. Funct Ecol 8:52–57

Lüttge U (1997) Cyanobacterial Tintenstrich communities and their ecology. Naturwissenschaften 84:526–534

Neustupa J, Škaloud P (2008) Diversity of subaerial algae and cyanobacteria on tree bark in tropical mountain habitats. Biologia 63:806–812

Novis PM, Smissen RD (2006) Two genetic and ecological groups of *Nostoc commune* in Victoria Land, Antarctia, revealed by AFLP analysis. Antarct Sci 18:573–581

Omelon CR, Pollard WH, Ferris FG (2006) Chemical and ultrastructural characterization of high Arctic cryptoendolithic habitats. Geomicrobiol J 23:189–200

Rasmussen B, Fletcher IR, Brocks JJ, Kilburn MR (2008) Reassessing the first appearance of eukaryotes and cyanobacteria. Nature 455:1101–1104

Sarthou C, Thérézien Y, Couté A (1995) Cyanophycées de l'inselberg des Nouragues (Guyane française). Nova Hedwig 61:85–109

Schade A (1917) Über den jährlichen Wärmegenuss von *Webera nutans* (Schreb.) Hedw. und *Leptoscyphus Taylori* (Hook.) Mitt. im Elbsandsteingebirge. Ber Deutsch Bot Ges 35:490–505

Schopf JW (2000) The fossil record: tracing the roots of the cyanobacterial lineage. In: Whitton BA, Potts M (eds) The ecology of cyanobacteria. Kluwer Academic Publishers, Dordrecht, pp 13–65

Shirkey B, McMaster NJ, Smith SC, Wright DJ, Rodriguez H, Jaruga P, Birincioglu M, Helm RF, Potts M (2003) Genomic DNA of *Nostoc commune* (Cyanobacteria) becomes covalently modified during long term (decades) desiccation but is protected from oxidative damage and degradation. Nucleic Acids Res 31:2995–3005

Siegesmund MA, Johansen JR, Karsten U, Friedl T (2008) *Coleofasciculus* gen. nov. (Cyanobacteria): morphological and molecular criteria for revision of the genus *Microcoleus* Gomont. J Phycol 44:1572–1585

Tirkey J, Adhikary SP (2005) Cyanobacteria in biological soil crusts of India. Curr Sci 89:515–521
Weber B, Wessels DCJ, Büdel B (1996) Biology and ecology of cryptoendolithic cyanobacteria of a sandstone outcrop in the Northern Province, South Africa. Algol Stud 83:565–579
Wessels DCJ, Büdel B (1995) Epilithic and cryptoendolithic cyanobacteria of Clarens sandstone cliffs in the Golden Gate Highlands National Park, South Africa. Bot Acta 108:220–226
Whitton BA, Potts M (2000) The ecology of cyanobacteria: their diversity in time and space. Kluwer Academic Publishers, Dordrecht, pp 1–704
Wright DJ, Smith SC, Jordar V, Scherer S, Jervis J, Warren A, Helm RF, Potts M (2005) UV Irradiation and desiccation modulate the three-dimensional extracellular matrix of *Nostoc commune* (Cyanobacteria). J Biol Chem 280:40271–40281

Chapter 3
Cyanobacteria: Multiple Stresses, Desiccation-Tolerant Photosynthesis and Di-nitrogen Fixation

Ulrich Lüttge

3.1 Multiple Stresses and Desiccation-Tolerant Cyanobacteria

Cyanobacteria are gram-negative photoautotrophic prokaryotes. Their traditional name, i.e., blue green algae, alludes to an aquatic life. Indeed, many cyanobacteria are bound to a submerged life. However, cyanobacteria are ubiquitous on our planet (Whitton and Potts 2000), and as shown in Chap. 2 cyanobacteria also have very successfully conquered habitats outside the water all over the world ranging from the occupation of all kinds of surfaces including rocks of high elevation mountains, rocks in the tropics as well as the Antarctic, soil crusts in deserts and man-made structures from concrete buildings to plastic garbage bins. Terrestrial cyanobacteria are also symbionts in lichens (Chaps. 5–7). In their terrestrial habitats, cyanobacteria are subject to a multitude of stress factors or stressors, such as high light intensities including ultraviolet radiation, high and low temperatures including freezing, osmotic stress, salinity and drought including desiccation (Allakhverdiev et al. 2000; Singh et al. 2002; Lin et al. 2004; Potts et al. 2005; Büdel et al. 2008). In-depth studies of ecophysiological adaptations of cyanobacteria to a plethora of stresses are increasingly facilitated by the accumulation of sequence data, available e.g., for *Prochlorococcus, Nostoc punctiforme, Gloeobacter* and *Synechococcus*. Complete genomic sequences have been obtained for the unicellular cyanobacterium *Synechocystis* sp. strain PCC 6803 and the filamentous, heterocyst-forming *Anabaena* sp. strain PCC 7120 (see Introduction of Singh et al. (2002) with references).

Most of the stressors challenging terrestrial cyanobacteria cause drying stress affecting water relations of the cells. This is quite evident for temperature, radiation, osmotic stress, salinity and drought. The extreme response of cyanobacteria to these stresses is such a large loss of cellular water that we can speak of desiccation *sensu stricto*. Some bacteria may survive in the air-dried state only for seconds. However, in principle we can find prokaryotic organisms that are able to survive desiccation and to recover after highly extended periods of time, e.g., some bacteria for thousands, perhaps millions, of years (Potts 1994) and cyanobacteria for at least 100 years (Cameron 1962; Hirai et al. 2004; Fukuda et al. 2008). Major genera of cyanobacteria with species that can tolerate the state of desiccation for prolonged

U. Lüttge et al. (eds.), *Plant Desiccation Tolerance*, Ecological Studies 215,
DOI 10.1007/978-3-642-19106-0_3, © Springer-Verlag Berlin Heidelberg 2011

periods are *Microcoleus, Lyngbya, Tolypothrix* and *Nostoc* (Potts 1999). *Nostoc commune* is one of the most intensely studied species.

3.2 Cell Physiological Responses of Cyanobacteria to Stress of Drying Leading the Path to Desiccation

As mentioned above, a variety of stressors are causing drying stress, which may be more or less strongly pronounced. Desiccation is the extreme response. However, a consideration of biochemical and physiological mechanisms allowing tolerance of milder forms of drying appears important as a prelude to evaluating resistance to desiccation as basically similar functions are involved. Loss of water from cells is accompanied by the destruction of water structures stabilizing macromolecules such as proteins and nucleotides and the lipoprotein arrangement of membranes. As in other drought- and desiccation-tolerant photosynthesizing organisms, a complement of mechanisms have evolved in cyanobacteria. Particular solutes, often called compatible solutes, are synthesized and accumulated to replace and mimic water structures. Special water stress proteins are accumulated. Protectants are produced to minimize destruction. Membrane lipids are adapted. Repair mechanisms are enhanced to overcome destruction.

3.2.1 Compatible Solutes

Compatible solutes (see Chap. 16) stabilize macromolecules and membranes. This is brought about by hydroxyl groups, e.g., in sugars and sugar alcohols, allowing the formation of hydrogen bonds, and by buried electrical charges such as the positive charge of a quaternary nitrogen in various betains, where H atoms of amino acids are replaced by methyl groups surrounding the N^+.

The major compatible solutes found in cyanobacteria are as follows:

- Sugars, often sucrose and especially trehalose (Hershkovitz et al. 1991; Potts 1994; Singh et al. 2002; Hincha and Hagemann 2004).
- Sugar derivatives such as 2-*O*-glucopyranosyl-glycerol (Allakhverdiev et al. 2001; Hincha and Hagemann 2004).
- Betains, such as glycine-betain and glutamate-betain (Allakhverdiev et al. 2001; Hincha and Hagemann 2004).

In dehydrated *Anabaena* cells, the expression of genes involved in trehalose synthesis is upregulated (Katoh et al. 2004; Higo et al. 2006).

A cytological consequence of the accumulation of the sugars is the suppression of crystallization of protoplastic constituents by the promotion of glass formation or vitrification controlling metabolism in the desiccated state at low water content

(Crowe et al. 1998; Potts 2001, Chap. 16). Sucrose and trehalose increase the stability of native proteins and assist in refolding of unfolded polypeptides so that we may call them chemical chaperones (Hottinger et al. 1994, Singer and Lindquist 1998; Higo et al. 2006). Sucrose and trehalose stabilize membranes by establishing hydrogen bonds to the P=O bond in the

$$-\overset{\overset{\displaystyle O}{\|}}{C}-O-\underset{\underset{\displaystyle O^-}{|}}{\overset{\overset{\displaystyle O}{\|}}{P}}-O^-$$

configuration of phospholipids. In this way, protection is obtained by a decrease of the temperature of transition from the gel phase to the liquid-crystalline lipid phase.

3.2.2 Heat Shock and Water Stress Proteins

Heat shock proteins (see also Chap. 16) were detected in the terrestrial epilithic cyanobacterium *Tolypothrix byssoidea* (Adhikary 2003). In the desiccation-tolerant cyanobacterium *N. commune,* during desiccation/rehydration cycles, acidic water stress proteins are secreted and accumulated in the glycan extracellular matrix and probably have a protective function at the structural level (Scherer and Potts 1989; Hill et al. 1994; Potts et al. 2005). These proteins are stable for decades during storage.

3.2.3 Sun Protectants

Desiccation of cyanobacteria at their sites in the field is naturally highly correlated with high solar irradiance. In extremely sun-exposed habitats, UV protection is provided by the production of effective sunblocking pigments, such as mycosporine-like amino acids and the indol-alkaloid scytonemin (Garcia-Pichel and Castenholz 1991; Büdel et al. 1997b; Sinha et al. 1998, 1999a, b, 2001; Büdel 1999; Singh et al. 2002; see also Chap. 2). In exposed epilithic cyanobacteria of Venezuela and French Guiana, scytonemin, with its absorption maximum at 380 nm, is found in such high concentrations that irradiance up to 500 nm is reduced inside the cells.

3.2.4 Membrane Lipids

The role of lipids in stress tolerance of cyanobacteria was reviewed by Singh et al. (2002). Lipids are particularly important in the phospholipid bilayers of membranes. The preservation of the structural integrity of membranes is highly important during the dynamics of dehydration, desiccation and rehydration to control the many membrane functions of transport and energy metabolism in respiration and photosynthesis (Sect. 3.3). Membrane lipids are particularly vulnerable to oxidation during various forms of stress where reactive oxygen species are formed because they are rich in olefinic bonds, which are a primary target for oxidative reactions. Hydrogen atoms at olefinic bonds are highly susceptible to oxidative attack. Phospholipids can be stabilized by sugars (Sect. 3.2.1). Responses to dehydration stress in cyanobacteria have been mainly studied with respect to osmotic stress and salinity. It appears that unsaturation of the fatty acids of lipids counteracts stress effects (Gombos et al. 1997). For photosynthesis of desiccation-tolerant cyanobacteria, it is interesting to note that with unsaturation particularly the photosynthetic machinery is protected, the repair of damaged photosystems I and II (PS I, PS II) is enhanced (Gombos et al. 1994; Allakhverdiev et al. 2001; Singh et al. 2002) and the important turnover of the D1 protein of PS II is sustained. However, it appears that the effects are not directly on the photosynthetic machinery in the membranes but more indirect by supporting effective control of the cytoplasmic environment controlled by mechanisms of water and ion transport at the plasma membrane (Allakhverdiev et al. 2001).

3.2.5 Polynucleotide Stability and Repair

Polynucleotides of cyanobacteria (*N. commune*), such as DNA (Wright 2004) and ribosomal RNA, rRNA (Han and Hu 2007), have been found to be remarkably stable in the state of desiccation for very many years. 16S rRNA was more stable in the desiccation-tolerant *N. commune* than in planktonic species of the Nostocaceae. For recovery from DNA destruction experimentally caused by radiation and possibly also occurring during desiccation, the synthesis of DdrA, a protein with an affinity for the 3′ ends of single-stranded DNA that protects those ends from nuclease digestion, is important. Among cyanobacteria, only the genome of *Gloeobacter violaceus* was found to contain a gene (*glr0712*), the product of which shows the significant similarity of 40% with DdrA. This cyanobacterium grows on exposed limestone in the Alps, where it is subject to solar radiation stress and desiccation/rehydration cycles (Potts et al. 2005).

3.3 Photosynthesis

3.3.1 Special Features of Cyanobacterial Photosynthesis

Compared with eukaryotic green photosynthesizing organisms, the light harvesting and photosynthetic electron transport machinery of cyanobacteria has a number of peculiarities (Papageorgiou 1996; review: Campbell et al. 1998):

- The major difference is that cyanobacteria as prokaryotes have no organelles, mitochondria and chloroplasts, for respiration and photosynthesis. Thus, the chains of electron transport carriers of respiration and photosynthesis are localized in the same membrane and are in fact overlapping. Respiratory and photosynthetic electron transport chains intersect at the level of the plastoquinone pool (Meunier et al. 1997). Through shared electron transport carriers respiration and photosynthesis directly influence each other's regulatory redox status.

- Another outstanding difference is that cyanobacteria have special super antennae for light harvesting, the so-called phycobilisomes, i.e., proteinaceous structures, which are their principal light harvesting complexes. Phycobilisomes are built up of the proteins phycoerythrin with the chromophore pigment phycoerythrobilin and phycocyanin and allophycocyanin with the pigment phycocyanobilin. The pigments are open tetrapyrrole systems. Phycobilisomes are localized on the thylakoid membrane in the vicinity of PS II to which they are connected via the allophycocyanin. The super antennae allow capturing photosynthetically active radiation from low photon flux densities and also from the green part of the spectrum.
- In photosynthesizing eukaryotes, the ratio of variable to maximum fluorescence, F_v/F_m, of chlorophyll a of PS II of dark adapted samples is a measure of potential quantum yield of PS II. In cyanobacteria, the absolute value of F_v/F_m is not a reliable indicator of PS II function because the phycobilisomes also contribute to fluorescence that overlaps with the spectrum of chlorophyll fluorescence and also because of the shared electron transport carriers with respiration. Hence, an F_v/F_m value of close to 0.8, which is considered to be the maximum potential quantum yield of PS II in non inhibited eukaryotes (Björkman and Demmig 1987), is never attained in cyanobacteria (see below, Figs. 3.1–3.5). Nevertheless, F_v/F_m in cyanobacteria can be regarded to correspond simply to the amounts of active PS II reaction centre (Satoh et al. 2002).
- $\Delta F/F'_m$, i.e., (variable fluorescence, F_v, minus maximum fluorescence, F'_m, in the light adapted state), divided by F'_m in eukaryotes is effective quantum yield of PS II and in the cyanobacterium *Synechoccus* was found to correlate with photosynthetic O_2 evolution (Campbell et al. 1998).
- The photochemical chlorophyll fluorescence-quenching coefficient q_p is a general index for the balance between energy capture and consumption.

3.3.2 Desiccation and Photoinhibition

As photosynthesizing organisms cyanobacteria are also subject to photoinhibition (Samuelsson et al. 1985, 1987; Raven and Samuelsson 1986). Desiccation-tolerant cyanobacteria of exposed habitats have light saturation of photosynthesis at an irradiance of 2,000 μmol photons m^{-2} s^{-1} or even above (Lüttge et al. 1995; Rascher et al. 2003). Hence, they perform like highly sun adapted photosynthetic organisms. Nevertheless, photoinhibition occurs in the hydrated state as shown in Fig. 3.1 by various samples obtained from mats on the dark floor of a glasshouse, rocks in an alpine valley and sun-exposed rock outcrops in the tropics when

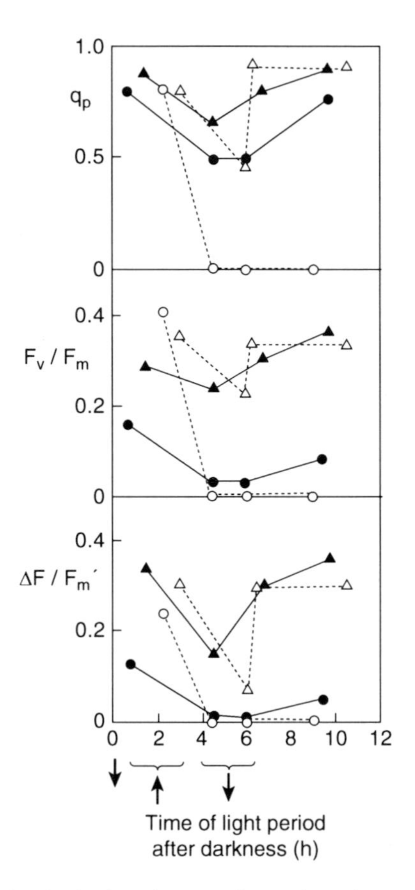

Fig. 3.1 Photoinhibition in the hydrated state of samples of cyanobacterial mats collected from the floor of a glasshouse in the Botanical Garden of the Technical University of Darmstadt (*closed circles*), the Schwarzwasser Valley in the Alps at Hirschegg, Austria (*open circles*), an inselberg near Seguéla, Ivory Coast (*closed triangles*) and limestone rock outcrops on the Paraguana Peninsula, Venezuela (*open triangles*). The light period of 12 h shown started with an irradiance of 240 μmol photons m^{-2} s^{-1} (*downward pointing arrow*). At the times of the upward pointing arrow for the various samples irradiance was experimentally increased to 1,280 μmol photons m^{-2} s^{-1}. Rapid reduction of q_p, F_v/F_m and $\Delta F/F'_m$ indicates photoinhibition, which was reversible upon return to 240 μmol photons m^{-2} s^{-1} (*downward pointing arrow*) except in the sample from the Alps. [Drawn after the data of Lüttge et al. (1995)]

irradiance was raised from 240 to 1,280 μmol photons m^{-2} s^{-1}. Except in one of the samples, photoinhibition was reversible when irradiance was reduced again to 240 μmol photons m^{-2} s^{-1}.

During desiccation and in the desiccated state, photoinhibition is a still much larger problem. Desiccation-tolerant poikilohydrous cyanobacteria are homoiochlorophyllous, i.e., they maintain their chlorophyll in the state of desiccation. In this state, at many of their natural sites they are subject to extraordinarily high solar radiation (Chap. 2). This is an extremely dangerous situation for them because excitation of chlorophyll continues while the energy of the excitons cannot be dissipated via photochemical work because the functioning of the enzymatic machinery of photosynthesis stalls in the state of desiccation. The site of initial damage can be the acceptor or the donor side of PS II. In acceptor-side photoinhibition damage is caused by singlet activated oxygen generated from the triplet state of chlorophyll, which is formed when other pathways of dissipation of exciton energy are insufficient. Donor-side photoinhibition occurs when electron donation into the reaction centre of PS II by water splitting does not match the rate of oxidation at the core pigment P_{680} of PS II. Oxidizing species on the donor side then become relatively long lived and can cause damage where the water-splitting Mn_4-cluster itself may be the initial target (Nixon et al. 2010).

To avoid destruction of the photosystems and other structures, the energy must be dissipated extremely rapidly given the half-lives of the first and second singlet excited states of chlorophyll of 10^{-11} to 10^{-9} and 10^{-15} to 10^{-13} s, respectively, which is well studied in homoiochlorophyllous poikilohydrous lichens and mosses (Chap. 7).

In eukaryotes, a mechanism of harmless energy dissipation as heat is the xanthophyll cycle with epoxide formation of zeaxanthin giving antheraxanthin and violaxanthin, which are reduced again to zeaxanthin. This cycle is missing in cyanobacteria (Fukuda et al. 2008). However, since energy dissipation not only during photoinhibition in the hydrated state (Fig. 3.1) but especially during drying, in the desiccated state and during rehydration, is highly important, cyanobacteria have other available mechanisms.

- Although the xanthophyll cycle is absent, carotenoids such as zeaxanthin (Demmig-Adams et al. 1990) and canthaxanthin (Albrecht et al. 2001; Lakatos et al. 2001) are present. They are mainly localized in the plasma membrane and not much in the thylakoids and may play protective roles in light shielding and preventing oxidative stress and photodamage (Masamoto and Furukawa 1997; Masamoto et al. 1999).
- An important protective mechanism is the inactivation of photosystems (Allakhverdiev et al. 2000; Satoh et al. 2002; Hirai et al. 2004; Fukuda et al. 2008). A state transition mechanism (Meunier et al. 1997) – also called "spill over" in eukaryotic photosynthesizing organisms – which regulates the distribution of excitation energy between PS II and PS I is involved. Non-photochemical quenching of chlorophyll fluorescence in cyanobacteria largely reflects changes in the PS II fluorescence yield as a result of such state transition. Deactivation of the PS II reaction centre activity and cessation of energy transfer from phycobilisomes to PS II occurs in the early stage of desiccation, and quenching of light energy absorbed by pigment–protein complexes is very important for desiccation tolerance in *N. commune* (Hirai et al. 2004).
- The D1 protein in the reaction centre of PS II is under a constant turnover of destruction and resynthesis. In *Synechocystis*, it was found to turn over rapidly with a half-live of about 30 min (Kanervo et al. 1993). When breakdown dominates synthesis PS II gets inactivated (Hirai et al. 2004, Fukuda et al. 2008). The role of lipid unsaturation in D1-protein turnover is discussed above

(Sect. 3.2.4). Particular proteases are involved in the degradation of damaged D1 protein (Adam and Clarke 2002).
- Cyanobacteria have evolved an inorganic carbon concentrating mechanism or a CO_2/HCO_3^- pump. The active uptake of CO_2 driven by redox systems and H^+-transporting ATPases in the plasma membrane and sustained by carbonic anhydrase catalysing the CO_2/HCO_3^--equilibrium concentrates CO_2 in special carboxysomes inside the cells (Skleryk et al. 1997; Sültemeyer et al. 1997). This facilitates the photochemical work of CO_2 assimilation, and, thus, controls photoinhibition albeit only in the hydrated stage when the biochemical reactions of the Calvin cycle are operative.

3.3.3 Recovery of Photosynthesis During Rewetting After Desiccation

If the protective mechanisms described in Sect. 3.2 do their job effectively, this should be reflected in rapid recovery of photosynthesis upon rehydration after desiccation. In fact, homoiochlorophylly of desiccation-tolerant cyanobacteria, in contrast to a poikilochlorophylly where chlorophyll is degraded during desiccation, should be a major advantage bestowing the capacity to quickly restore photosynthetic activity when water is available again. All authors having studied this observed recovery in desiccation/rehydration cycles within time scales of only a few minutes up to a few hours at the most after rewetting and the recovery of PS I is faster than that of PS II (Satoh et al. 2002; Harel et al. 2004). Short-term drying and rehydration cycles are shown in Fig. 3.2 where after the complete loss of overt photosynthetic activity given by measurements of q_P, F_v/F_m and $\Delta F/F'_m$ recovery of all parameters occurred within less than 2 h.

During desiccation, the ground fluorescence, F_o, is reduced to a fraction of that observed in hydrated samples (Figs. 3.3 and 3.4). In a dry cyanobacterial desert soil crust with primarily *Microcoleus* sp. kept in the dry state for at least 1 month, upon rewatering, F_o increased rapidly and reached a maximal value of six times that of the dry crust within about 3 min followed by a gradual decline to a constant value after about 50 min (Fig. 3.3a). The kinetics of rehydration depend on the duration of the dormant state of desiccation (Lange 2001; Figs. 3.3–3.5), and frequent drying and wetting cycles maintain stability (Scherer and Zhong 1991). In rapid recovery protein synthesis is not involved (*Microcoleus* sp., Harel et al. 2004). However, when recovery is slower protein synthesis, presumably for the repair of D1 protein, is required as has been found in *Nostoc flagelliforme* (Qiu et al. 2004). With the cyanobacterium *Synechocystis*, it was shown that chlorophyll does not turn over at the same rate as the D1 protein (Vavilin et al. 2005). Thus, chlorophyll associated with the D1 protein must be temporarily stored when in the D1 turnover damaged D1 protein is degraded by proteases and replaced by newly synthesized copies. It has been suggested that the phytol tail of chlorophyll is removed during PS II repair and the tetrapyrrole ring is reused (Vavilin and

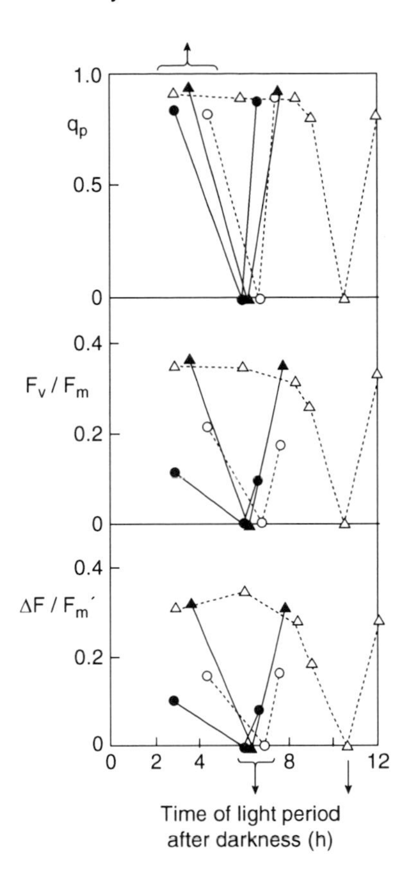

Fig. 3.2 Short-term drying (*upward facing arrow*) and rehydration cycles (water additions at the times of the *downward facing arrows*) of samples of cyanobacterial mats collected from the floor of a glasshouse in the Botanical Garden of the Technical University of Darmstadt (*closed circles*), the Schwarzwasser Valley in the Alps at Hirschegg, Austria (*open circles*), an inselberg near Seguéla, Ivory Coast (*closed triangles*) and limestone rock outcrops on the Paraguana Peninsula, Venezuela (*open triangles*). During drying q_p, F_v/F_m and $\Delta F/F'_m$ were completely lost but recovered in less than 2 h when water was added in the state of zero overt photosynthetic activity. [Drawn after the data of Lüttge et al. (1995)]

Vermaas 2007; Nixon et al. 2010). With respect to homoiochlorophylly during desiccation, this is an intriguing possibility deserving experimental exploration.

Cyanobacterial mats also occur on leaves of plants in moist tropical forests and may be subject to desiccation in these epiphyllous habitats (Chap. 2). Since dehydration/rehydration studies with them are rare, for the purpose of this review for comparison with the samples from notoriously dry sites some measurements were made on an epiphyllous mat collected in the warm and moist tropical glass house of the Botanical Garden of the Technical University of Darmstadt mainly composed of *Symploca* cf. *muscorum* Gomont ex Gomont with a few unicellular green algae of the Trebouxiophyceae in between the cyanobacteria. The sample was divided into pieces that were kept in dry air for up to 80 days and then rewetted. Recovery of F_o showed similar kinetics as in the desert crust. It was even faster and reached maximal values mostly in less than 30 s (Fig. 3.3b). Thereafter, it also declined, but after more extended time after rewetting it increased again. Thus, even cyanobacteria acclimated to a frequently moist or wet habitat show good recovery after desiccation.

In the desert crust, the amount of active PS II centres as given by F_v/F_m also increased within 50 min after rewetting to a constant level (Fig. 3.4a). A detailed analysis showed that activation of the photosynthetic apparatus did not require

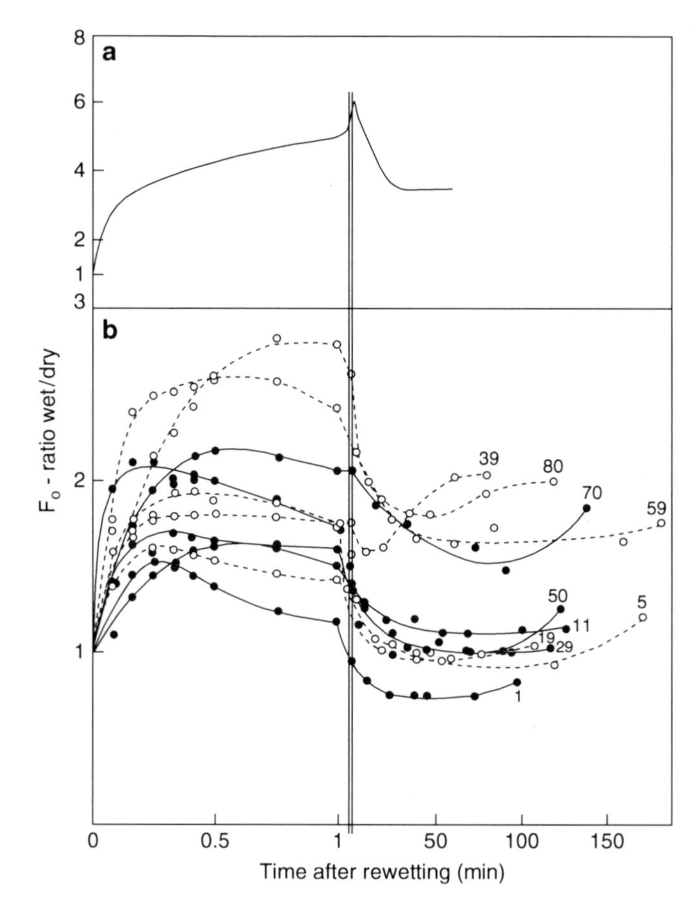

Fig. 3.3 Restoration of ground fluorescence, F_o, during rehydration as given by the ratio of F_o in the rehydrated wet to the desiccated dry state. (**a**) Cyanobacterial desert crust dominated by *Microcoleus* sp., rewetted after a desiccation time of 1 month. Curve drawn after data of Harel et al. (2004). (**b**) Epiphyllous mat of *Symploca* cf. *muscorum* Gomont ex Gomont with a few unicellular green algae of the Trebouxiophyceae in between the cyanobacteria collected in the warm and moist tropical glass house of the Botanical Garden of the Technical University of Darmstadt. The mat was cut into small pieces that were kept dry in darkness at room temperature for up to 80 days (numbers at the curves give days of the duration of desiccation). For measurements of chlorophyll fluorescence pieces of the mat were placed in the leaf clip holder of the Mini-PAM fluorometer of Heinz Walz (Effeltrich, Germany) and rehydrated and kept wet by adding droplets of tap water at regular intervals. The samples were kept in darkness during rehydration, and chlorophyll fluorescence measurements were made with saturating light pulses of 600 ms duration. (So far unpublished results of the author)

protein synthesis. Over 50% of the PS II activity, assembled phycobilisomes and PS I antennae were detected within less than 5 min of rehydration. Energy transfer to PS II and PS I by the respective antennae was fully restored within 10–20 min (Harel et al. 2004). In colonies of *N. commune*, PS I complexes recovered almost completely within 1 min, while recovery of PS II had a lag time of 5 min and

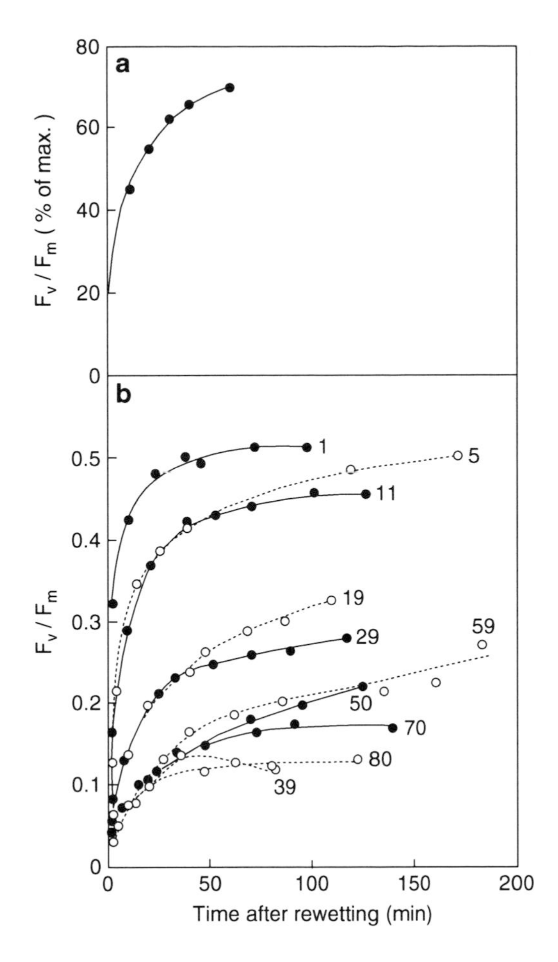

Fig. 3.4 Restoration of F_v/F_m after rewetting of (**a**) a cyanobacterial desert crust and (**b**) an epiphyllous cyanobacterial mat. Same samples and studies and all further details as in Fig. 3.3. F_v/F_m in (**a**) is given in per cent of the maximal value measured in (**b**) in absolute values. (b: So far unpublished results of the author)

recovered in two phases with half-times of about 20 min and 2 h. Photosynthetic CO_2-fixation was restored in parallel with the first recovery phase of PS II. Restoration of electron transport between PS I and PS II began at around 8 min (Satoh et al. 2002). In the epiphyllous sample from the moist glasshouse F_v/F_m within about 60 min increased to high values after up to 11 days of desiccation; medium values were still reached after up to 30 days, but after longer periods of desiccation the amount of active PS II centres as indicated by F_v/F_m remained rather low. After about 60 min of rewetting, the increase of F_v/F_m slowed down or even a constant value was reached (Fig. 3.4b).

The measurements of the epiphyllous mat from the moist glasshouse show that F_v/F_m measured after 1 and 60 min, respectively, of rehydration declined with increasing time of desiccation (Fig. 3.4b, Fig. 3.5a), while F_o increased (Fig. 3.5b) so that there was clear decline of the ratio of (F_v/F_m) to F_o (Fig. 3.5c). This inverse relation between (F_v/F_m) and F_o suggests that after rehydration of samples having

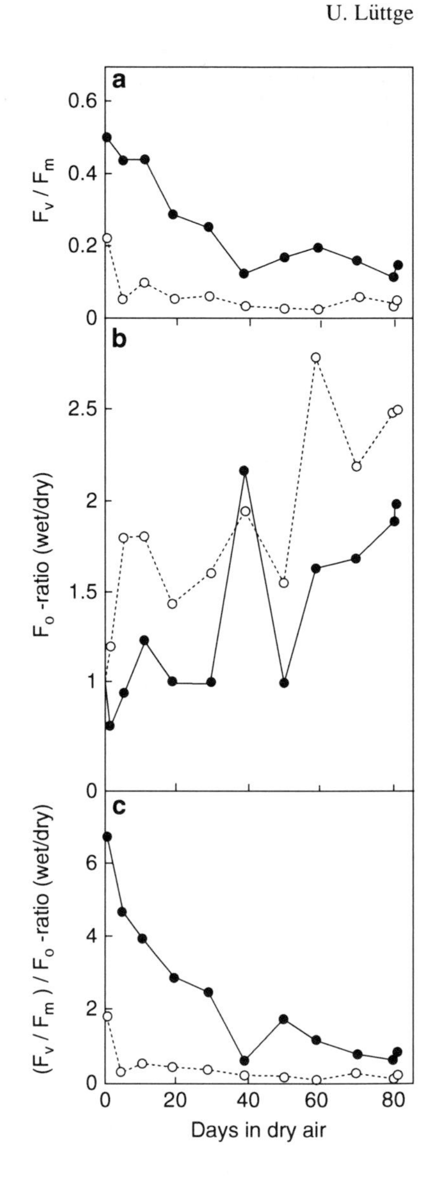

Fig. 3.5 Rehydration in the dark of the epiphyllous cyanobacterial mat of Figs. 3.3 and 3.4 kept desiccated in the dark for up to 81 days. All values are for 1 min (*open circles*) and 60 min (*closed circles*) of rewetting. (**a**) F_v/F_m, (**b**) ratio of F_o in the rehydrated (wet) to the desiccated (dry) state, (**c**) ratio of (F_v/F_m) to (F_o-wet/F_o-dry). (So far unpublished results of the author; for experimental details, see caption of Fig. 3.3)

suffered increasingly long times of desiccation electron transfer is restored to an increasingly lesser extent.

3.3.4 Physiological Ecology of Desiccation on the Rock Outcrops of Tropical Inselbergs

Among the most exposed sites of terrestrial cyanobacteria, we find the rock outcrops of tropical inselbergs, where the rock surface temperature may rise to 60°C.

Fig. 3.6 Run-off furrow coming down from a small vegetation island on the inselberg Pedra Grande at Atibaia, SP, Brazil

They are densely covered with epilithic cyanobacteria, which can only survive by tolerance of desiccation (Chap. 2). Strong rainfall events carve run-off furrows into the rock. The centre of the furrows after rainfall may keep wet for longer periods of time when water continues to flow down from small vegetation islands on the inselbergs (Fig. 3.6).

Over transects across the furrows on a small scale in space, we may encounter a gradation of cyanobacterial ecosystems. The cyanobacterial mini-ecosystems across the furrows are characterized by different dominating species, e.g., compact forms of colonies of *Gloeocapsa sanguinea* plus short branching *Stigonema mammilosum* able to resist the shearing forces of strong run-off currents after rainfall in the centre, long branching filamentous *Stigonema ocellatum* at the lateral slopes and filamentous *Scytonema myochrous* on the horizontal rocks on an inselberg in French Guiana (Rascher et al. 2003; Chap. 2).

Functioning of these systems is determined by the different times they keep moist after rainfall. A good indicator of that are δ^{13}C-values of the cyanobacterial biomass. Because after rainfall before drying again the mats of cyanobacteria in the centre of the furrows and at the lateral slopes are covered for longer times by water and films of water than the mats on the horizontal rocks outside the furrows, their photosynthesis is more pronouncedly limited by diffusion of CO_2 and HCO_3^- in the liquid phase. The inorganic carbon concentrating mechanism of cyanobacteria (Sect. 3.3.2) helps to counterbalance the liquid diffusion limited carbon supply, but diffusion limitation is still reflected in the stable isotope ratios δ^{13}C of the cyanobacteria. They are performing C_3-photosynthesis and the enzyme of primary CO_2-fixation ribulose-bis-phosphate carboxylase/oxygenase (RubisCO) has a ^{13}C-discrimination of +27‰. The δ^{13}C-values of cyanobacteria of inselbergs, however, are much less negative than −27‰, which would be obtained if RubisCO were mainly determining ^{13}C-discrimination during photosynthesis (Ziegler and Lüttge 1998; Fig. 3.7). This is due to the lower ^{13}C-discrimination of dissolution of CO_2 (−0.9‰) and diffusion of CO_2 and HCO_3^- in water (0.0‰) determining CO_2 delivery to RubisCO. Thus, as expected from these relations, there is a gradient from less negative to more negative δ^{13}C-values from the centre via the lateral slopes to the open rock surfaces in the inselbergs (Fig. 3.7). Maximum photosynthetic electron transport rates observed in the wet state tend to be somewhat higher (although not significantly statistically) outside the furrows (Fig. 3.7).

After rewetting desiccated inselberg cyanobacteria, effective quantum yield of PS II, $\Delta F/F'_m$, recovers as rapid within about 15 min (Figs. 3.8 and 3.9) as F_v/F_m and F_o shown in Sect. 3.3.3 (Figs. 3.3–3.5). Cycles of recovery and desiccation on the inselbergs are quite fast. Drying out depends on the position in the transect and the species composition and may take up to 30 min (Fig. 3.9)

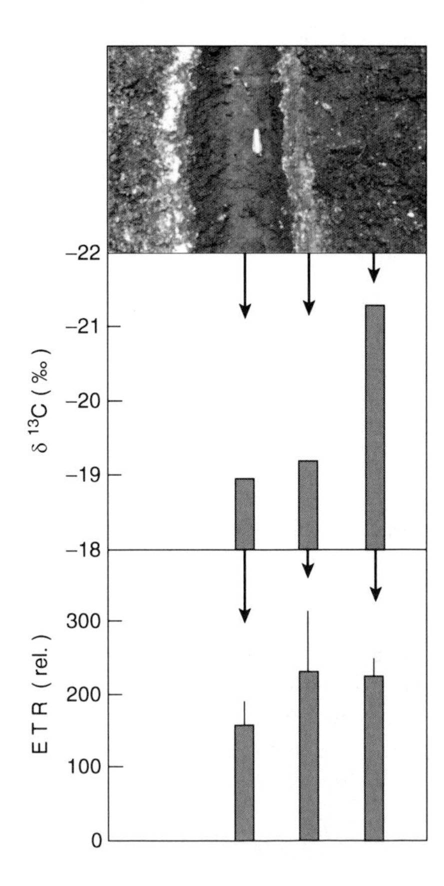

Fig. 3.7 Transects across run-off furrows with δ^{13}C-values (‰) and maximum apparent photosynthetic electron transport rates (ETR, relative units) measured at an irradiance of >1,750 μmol m^{-2} s^{-1}. *Top*: a run-off furrow on the inselberg at Galipero, near Puerto Ayacucho, Venezuela, and *centre*: the corresponding δ^{13}C-values (Ziegler and Lüttge 1998). *Bottom*: relative ETR values across a run-off furrow of the inselberg at Les Nouragues, French Guiana (Rascher et al. 2003)

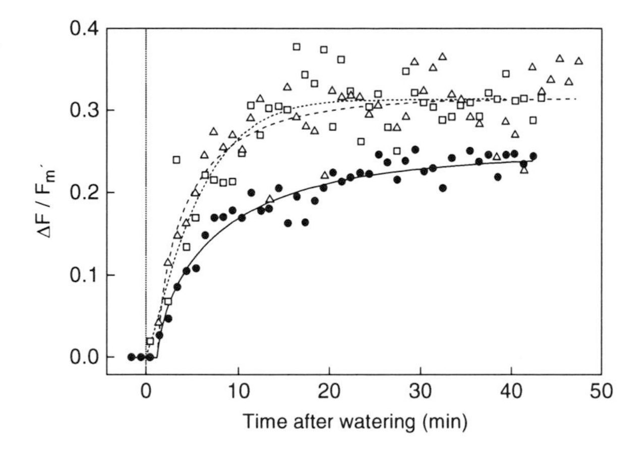

Fig. 3.8 Activation of effective electron transport of PS II, $\Delta F/F'_m$, by rehydration adding rain water at time zero to the cyanobacteria of a run-off furrow of the inselberg at Les Nouragues, French Guiana. *Closed circles* centre of the furrow, *open squares* lateral slopes, *open triangles* horizontal rocks (Rascher et al. 2003)

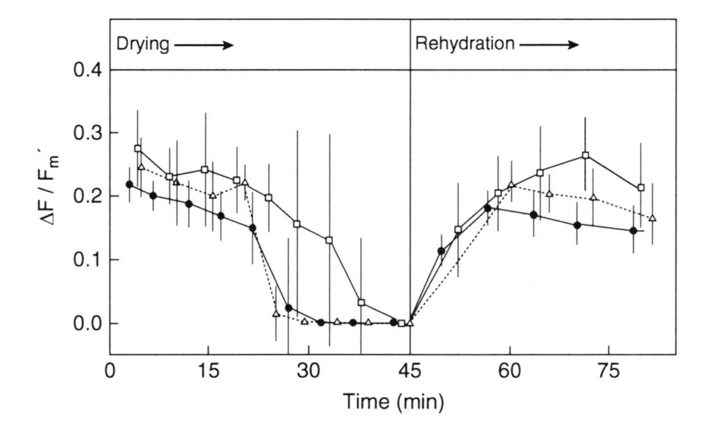

Fig. 3.9 Rehydration–desiccation cycle of $\Delta F/F'_m$ of cyanobacteria of a run-off furrow of the inselberg at Les Nouragues, French Guiana. *Closed circles* centre of the furrow, *open squares* lateral slopes, *open triangles* horizontal rocks (Rascher et al. 2003)

3.4 Biological Fixation of Di-nitrogen (N_2)

3.4.1 N_2-Fixation and Input into Ecosystems

Fixation of N_2 (Stewart 1980) is mediated by the enzyme nitrogenase. It is restricted to prokaryotes. As the nitrogenase is O_2-sensitive, it is often localized in special cells in the photosynthesizing filamentous cyanobacteria, i.e., the heterocytes, which have thick cell walls limiting O_2 diffusion into these cells, and lack PS II, and hence, photosynthetic O_2 evolution. From the possession of heterocytes, it can be concluded that most of the cyanobacteria of dry soil crusts and on the rock outcrops of inselbergs are N_2-fixers. Some taxa of cyanobacteria can also fix N_2 without having heterocytes (Belnap 2001).

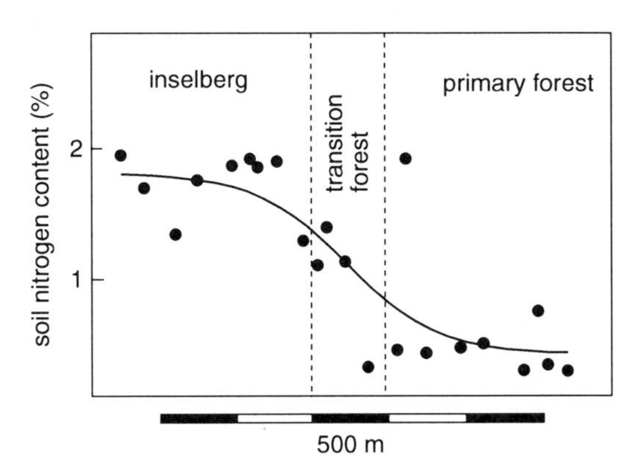

Fig. 3.10 Transect of soil nitrogen contents of the inselberg at Les Nouragues, French Guiana, from the inselberg itself down to the forest below (Dojani et al. 2007)

Immediately after rewetting of dry cyanobacterial biofilms, extracellular N is quickly released in the form of peptides, amino acids, amides and also nitrate and ammonium (Stewart 1963; Stewart et al. 1977; Belnap 2001; Russow et al. 2005). Release is caused by leaching from the cytoplasm, disruption, and abrasion of extracellular compounds including scytonemin and demolition and transfer of single cells or biofilm particles (Dojani et al. 2007). Thus, N_2-fixation by desiccation-tolerant cyanobacteria can make a considerable contribution to the N-budget of ecosystems with soil crusts and inselbergs (Evans and Ehleringer 1993; Belnap 2001), for example

- 3.5 kg ha^{-1} $year^{-1}$ by soil crusts, in the southern Sahel, Western Niger (Issa et al. 2001),
- 18 kg ha^{-1} $year^{-1}$ by inselbergs along the river Orinoco, Venezuela (Lüttge 2008),
- 130–235 kg ha^{-1} $year^{-1}$ by *Scytonema* on the inselberg at Les Nouragues, French Guiana (Dojani et al. 2007).

In the savanna soil at the foot of the Venezuelan inselberg at a distance of 10 m from the inselberg rock, soil content was 1.40 g N kg^{-1} and at distances of 30 and 50 m 0.45 g N kg^{-1} soil was measured (Büdel 1997a, b). For the inselberg at Les Nouragues, French Guiana, Fig. 3.10 shows the gradient of soil nitrogen from the inselberg itself to a transition forest and a primary forest below the inselberg. Soil of vegetation islands on the inselberg is more enriched in N; the values of about $\leq$0.5% found in the forest below are close to those of $\approx$0.14% at the bottom of the Venezuelan inselberg (Dojani et al. 2007).

3.4.2 *Recovery of N_2-Fixation During Rewetting After Desiccation*

As photosynthesis (Sect. 3.3) recovery of N_2-fixation during rewetting after desiccation depends on various conditions especially, however, on the duration of dormancy in the state of desiccation. Belnap (2001) has listed examples from the

Table 3.1 Time required for restoration of N_2-fixation in rewetted samples of cyanobacteria after various times in the state of desiccation. Compiled from the review of Belnap (2001) with references therein

Genera and species	Time in the state of desiccation	Time required for recovery
Nostoc sp.	A few days	Minutes to 3 h
Nostoc flagelliforme	2 days	30–50 h
Nostoc commune		
Nostoc sp.	Several months	8–12 h
Microcoleus	6 months	4–6 days
Scytonema		
Nostoc flagelliforme	2 years	120–150 h
Nostoc commune		

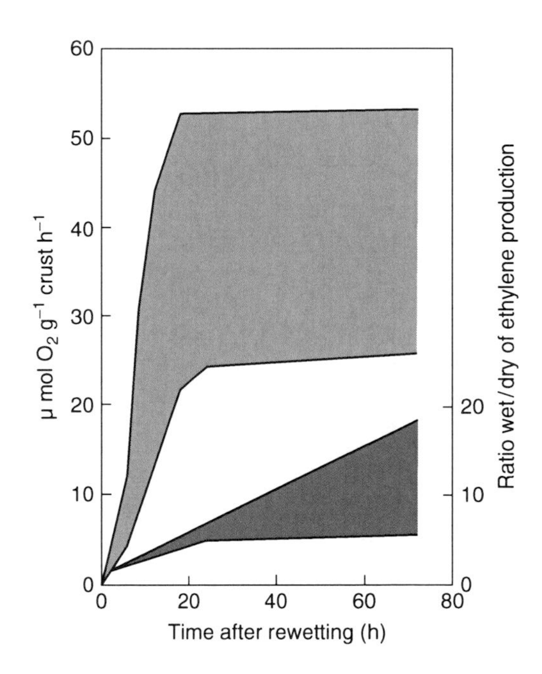

Fig. 3.11 Comparative kinetics of the reactivation after rewetting of photosynthesis (O_2-evolution, left ordinate, *light grey* range) and nitrogen fixation capacity (ratio of acetylene production in the dry to the rehydrated state, right ordinate, *dark grey* range) in dry soil crust samples collected at various locations in India. The *light grey* marked range covers lowest and highest values obtained in samples from four sites. The *dark grey* marked range covers the highest values obtained by rehydration in the light and the lowest values obtained by incubation in the dark (Tirkey and Adhikary 2005)

literature, which are summarized in Table 3.1. In all soil crust samples obtained by Issa et al. (2001) in the southern Sahel, which were desiccated for 3 years, N_2-fixation upon rewetting was restored within 2 h. N_2-fixation as measured by an acetylene-reduction test is shown to recover from desiccation more slowly than photosynthesis (Issa et al. 2001; Tirkey and Adhikary 2005; Fig. 3.11).

3.5 Conclusions

Ubiquitous terrestrial cyanobacteria constitute one of the most widely distributed life forms on earth, in total covering globally an enormous surface. Most of these sites, in particular but not only desert soil crusts and sun-exposed rocks, are highly stressful due to extreme insolation, high temperatures and only scarce or sporadic supply of water. Therefore, terrestrial cyanobacteria can only achieve their outstanding occupation of all kinds of surfaces by tolerance of desiccation with the protective mechanisms they have evolved for withstanding the physiological problems arising from the dynamics of drying and rehydration and of resting in the desiccation state for extended periods. The communities of desiccation-tolerant terrestrial cyanobacteria do not make conspicuous appearance and are therefore often overlooked. However, their photosynthesis makes a substantial contribution to global CO_2 sinks and their fixation of atmospheric di-nitrogen affords an important nitrogen input to their environment.

Acknowledgement I thank Burkhard Büdel for cooperation with much stimulating exchange and the identification of cyanobacteria samples.

References

Adam Z, Clarke AK (2002) Cutting edge of chloroplast proteolysis. Trends Plant Sci 7:451–456

Adhikary SP (2003) Heat shock proteins in the terrestrial epilithic cyanobacterium *Tolypothrix byssoidea*. Biol Plant 47:125–128

Albrecht M, Steiger S, Sandmann G (2001) Expression of a ketolase gene mediates the synthesis of canthaxanthin in *Synechococcus* leading to tolerance against photoinhibition, pigment degradation and UV-B sensitivity of photosynthesis. Photochem Photobiol 73:551–555

Allakhverdiev SI, Sakamoto A, Nishiyama Y, Inaba M, Murata N (2000) Ionic and osmotic effects of NaCl-induced inactivation of photosystems I and II in *Synechococcus* sp. Plant Physiol 123:1047–1056

Allakhverdiev SI, Kinoshita M, Inaba M, Suzuki I, Murata N (2001) Unsaturated fatty acids in membrane lipids protect the photosynthetic machinery against salt-induced damage in *Synechococcus*. Plant Physiol 125:1842–1853

Belnap J (2001) Factors influencing nitrogen fixation and nitrogen release in biological soil crusts. In: Belnap J, Lange OL (eds) Biological soil crusts: structure, function, and management. Ecol Stud 150. Springer, Berlin, Heidelberg, pp 241–261

Björkman O, Demmig B (1987) Photon yield of O_2 evolution and chlorophyll fluorescence characteristics at 77 K among vascular plants of diverse origins. Planta 170:489–504

Büdel B (1999) Ecology and diversity of rock-inhabiting cyanobacteria in tropical regions. Eur J Phycol 34:361–370

Büdel B, Becker U, Porembski S, Barthlott W (1997a) Cyanobacteria and cyanobacteria lichens from inselbergs of the Ivory Coast, Africa. Bot Acta 110:458–465

Büdel B, Karsten U, Garcia-Pichel F (1997b) Ultraviolet-absorbing scytonemin and mycosporine-like amino acid derivatives in exposed, rock inhabiting cyanobacterial lichens. Oecologia 112:165–172

Büdel B, Bendix J, Bicker FR, Green TGA (2008) Dewfall as a water source frequently activates the endolithic cyanobacterial communities in the granites of Taylor Valley, Antarctica. J Phycol 44:1415–1424

Cameron RE (1962) Species of *Nostoc* vaucher occurring in the Sonoran desert in Arizona. Trans Am Microsc Soc 81:379–384

Campbell D, Hurry V, Clarke AK, Gustafsson P, Öquist G (1998) Chlorophyll fluorescence analysis of cyanobacterial photosynthesis and acclimation. Microbiol Mol Biol Rev 62: 667–683

Crowe JH, Carpenter JF, Crowe LM (1998) The role of vitrification in anhydrobiosis. Annu Rev Physiol 60:73–103

Demmig-Adams B, Adams WW, Czygan F-C, Schreiber U, Lange OL (1990) Differences in the capacity for radiation less energy dissipation in the photochemical apparatus of green and blue-green algal lichens associated with differences in carotenoid composition. Planta 180:582–589

Dojani S, Lakatos M, Rascher U, Wanek W, Lüttge U, Büdel B (2007) Nitrogen input by cyanobacterial biofilms of an inselberg into a tropical rainforest in French Guiana. Flora 202:521–529

Evans RD, Ehleringer JR (1993) A break in the nitrogen cycle in arid lands? Evidence from $\delta^{15}N$ of soils. Oecologia 94:314–317

Fukuda S-Y, Yamakawa R, Hirai M, Khashino Y, Koike H, Satoh K (2008) Mechanisms to avoid photoinhibition in a desiccation-tolerant cyanobacterium, *Nostoc commune*. Plant Cell Physiol 49:488–492

Garcia-Pichel F, Castenholz RW (1991) Characterization and biological implications of scytonemin, a cyanobacterial sheath pigment. J Phycol 27:395–409

Gombos Z, Wada H, Murata N (1994) The recovery of photosynthesis from low-temperature photoinhibition is accelerated by the unsaturation of membrane lipids: a mechanism of chilling tolerance. Proc Natl Acad Sci USA 91:8787–8791

Gombos Z, Kanervo E, Tsvetkova N, Sakamoto T, Aro E-M, Murata N (1997) Genetic enhancement of the ability to tolerate photoinhibition by introduction of unsaturated bonds into membrane glycerolipids. Plant Physiol 115:551–559

Han D, Hu Z (2007) Mutations stabilize small subunit ribosomal RNA in desiccation-tolerant cyanobacteria *Nostoc*. Curr Microb 54:254–259

Harel Y, Ohad I, Kaplan A (2004) Activation of photosynthesis and resistance to photoinhibition in cyanobacteria within biological desert crusts. Plant Physiol 136:3070–3097

Hershkovitz N, Oren A, Cohen Y (1991) Accumulation of trehalose and sucrose in cyanobacteria exposed to matric water stress. Appl Environ Microbiol 57:645–648

Higo A, Katoh H, Ohmori K, Ikeuchi M, Ohmori M (2006) The role of a gene cluster for trehalose metabolism in dehydration tolerance of the filamentous cyanobacterium *Anabaena* sp. PCC 7120. Microbiology 152:979–987

Hill DR, Hladun SL, Scherer S, Potts M (1994) Water stress proteins of *Nostoc commune. DRH-1.* J Bacteriol 182:974–982

Hincha DK, Hagemann M (2004) Stabilization of model membranes during drying by compatible solutes involved in the stress tolerance of plants and microorganisms. Biochem J 383: 277–283

Hirai MH, Yamakawa R, Nishio J, Yamaji T, Kashino Y, Koike H, Satoh K (2004) Deactivation of photosynthetic activities is triggered by loss of a small amount of water in a desiccation-tolerant cyanobacterium, *Nostoc commune*. Plant Cell Physiol 45:872–878

Hottinger T, de Virgilio C, Hall MN, Boller T, Wiemken A (1994) The role of trehalose synthesis for the acquisition of thermotolerance in yeast. II. Physiological concentrations of trehalose increase the thermal stability of proteins in vitro. Eur J Biochem 219:187–193

Issa OM, Stal LJ, Défarge C, Couté A, Trichet J (2001) Nitrogen fixation by microbial crusts from desiccated Sahelian soils (Niger). Soil Biol Biochem 33:1425–1428

Kanervo E, Mäenpää P, Aro E-M (1993) D1 protein degradation and *psbA* transcript levels in *Synechocystis* PCC 6803 during photoinhibition in vivo. J Plant Physiol 142:669–675

Katoh H, Asthana RK, Ohmori M (2004) Gene expression in the cyanobacterium *Anabaena* sp. PCC 7120 under desiccation. Microb Ecol 47:164–174

Lakatos M, Bilger W, Büdel B (2001) Carotenoid composition of terrestrial cyanobacteria: response to natural light conditions in open rock habitats in Venezuela. Eur J Phycol 36:367–375

Lange OL (2001) Photosynthesis of soil-crust biota as dependent on environmental factors. In: Belnap J, Lange OL (eds) Biological soil crusts: structure, function, and management. Ecol Stud 150. Springer, Berlin, Heidelberg, pp 217–240

Lin Y, Hirai M, Kashino Y, Koike H, Tuzi S, Satoh K (2004) Tolerance to freezing stress in cyanobacteria, *Nostoc commune* and some cyanobacteria with various tolerances to drying stress. Polar Biosci 17:56–68

Lüttge U (2008) Physiological ecology of tropical plants. Springer, Berlin, Heidelberg

Lüttge U, Büdel B, Ball E, Strube F, Weber P (1995) Photosynthesis of terrestrial cyanobacteria under light and desiccation stress as expressed by chlorophyll fluorescence and gas exchange. J Exp Bot 46:309–319

Masamoto K, Furukawa K-I (1997) Accumulation of zeaxanthin in cells of the cyanobacterium *Synechococcus* sp. strain PCC 7942 grown under high irradiance. J Plant Physiol 151:257–261

Masamoto K, Zsiros O, Gombos Z (1999) Accumulation of zeaxanthin in cytoplasmic membranes of the cyanobacterium *Synechococcus* sp. strain PCC 7942 grown under high light condition. J Plant Physiol 155:136–138

Meunier PC, Colón-López MS, Sherman LA (1997) Temporal changes in state transitions and photosystem organization in the unicellular diazotrophic cyanobacterium *Cyanothece* sp. ATCC 51142. Plant Physiol 115:991–1000

Nixon PJ, Michoux F, Jianfeng Y, Boehm M, Komenda J (2010) Recent advances in understanding the assembly and repair of photosystem II. Ann Bot 106:1–16

Papageorgiou GC (1996) The photosynthesis of cyanobacteria (blue bacteria) from the perspective of signal analysis of chlorophyll *a* fluorescence. J Sci Ind Res 55:596–617

Potts M (1994) Desiccation tolerance of prokaryote. Microbiol Mol Biol Rev 58:755–805

Potts M (1999) Mechanisms of desiccation tolerance in cyanobacteria. Eur J Phycol 34:319–328

Potts M (2001) Desiccation tolerance: a simple process? Trends Microbiol 9:553–559

Potts M, Slaughter SM, Hunneke F-U, Garst JF, Helm RF (2005) Desiccation tolerance of prokaryotes: application of principles to human cells. Integr Comp Biol 45:800–809

Qiu BS, Zhang AH, Liu ZL, Gao KS (2004) Studies on the photosynthesis of the terrestrial cyanobacterium *Nostoc flagelliforme* subjected to desiccation and subsequent rehydration. Phycologia 43:521–528

Rascher U, Lakatos M, Büdel B, Lüttge U (2003) Photosynthetic field capacity of cyanobacteria of a tropical inselberg of the Guiana Highlands. Eur J Phycol 38:247–256

Raven JA, Samuelsson G (1986) Repair of photoinhibitory damage in *Anacystis nidulans* 625 (*Synechococcus* 6301): relation to catalytic capacity for, and energy supply to, protein synthesis, and implications for μ_{max} and the efficiency for light-limited growth. New Phytol 103:625–643

Russow R, Veste M, Böhme F (2005) A natural ^{15}N approach to determine the biological fixation of atmospheric nitrogen by biological soil crusts of the Negev desert. Rapid Commun Mass Spectrom 19:3451–3456

Samuelsson G, Lönneborg A, Rosenquist E, Gustafsson P, Öquist G (1985) Photoinhibition and reactivation of photosynthesis in the cyanobacterium *Anacystis nidulans*. Plant Physiol 79:992–995

Samuelsson G, Lönneborg A, Gustafsson P, Öquist G (1987) The susceptibility of photosynthesis to photoinhibition and the capacity of recovery in high and low light grown cyanobacteria, *Anacystis nidulans*. Plant Physiol 83:438–441

Satoh K, Hirai M, Nishio J, Yamaji T, Kashino Y, Koike H (2002) Recovery of photosynthetic systems during rewetting is quite rapid in a terrestrial cyanobacterium, *Nostoc commune*. Plant Cell Physiol 43:170–176

Scherer S, Potts M (1989) Novel water stress protein from a desiccation-tolerant cyanobacterium. Purification and partial characterization. J Biol Chem 264:12546–12553

Scherer S, Zhong ZP (1991) Desiccation independence of terrestrial *Nostoc commune* ecotypes (cyanobacteria). Microb Ecol 22:271–283

Singer MA, Lindquist S (1998) Multiple effects of trehalose on protein folding *in vitro* and *in vivo*. Mol Cell 1:639–648

Singh SC, Sinha RP, Häder D-P (2002) Role of lipids and fatty acids in stress tolerance in cyanobacteria. Acta Protozool 41:297–308

Sinha RP, Klisch M, Gröninger A, Häder D-P (1998) Ultraviolet absorbing/screening substances in cyanobacteria, phytoplankton and macroalgae. J Photochem Photobiol B 47:83–94

Sinha RP, Klisch M, Häder D-P (1999a) Induction of a mycosporine-like amino acid (MMA) in the rice-field cyanobacterium *Anabaena* sp. by UV radiation. J Photochem Photobiol B 52: 59–64

Sinha RP, Klisch M, Vaishampayan A, Häder D-P (1999b) Biochemical and spectroscopic characterization of the cyanobacterium *Lyngbya* sp. inhabiting mango (*Mangifera indica*) trees: presence of an ultraviolet-absorbing pigment, scytonemin. Acta Protozool 38:291–298

Sinha RP, Klisch M, Hebling EW, Häder D-P (2001) Induction of mycosporine-like amino acids (MMAs) in cyanobacteria by solar ultraviolet-B radiation. J Photochem Photobiol B 60: 129–135

Skleryk RS, Tyrell PN, Espie GS (1997) Photosynthesis and inorganic carbon acquisition in the cyanobacterium *Chlorogloeopsis* sp. ATCC 27193. Physiol Plant 99:81–88

Stewart WDP (1963) Liberation of extracellular nitrogen by two nitrogen-fixing blue-green algae. Nature 200:1020–1021

Stewart WDP (1980) Some aspects of structure and function in N_2-fixing cyanobacteria. Annu Rev Microbiol 34:497–536

Stewart WDP, Sampaio MJ, Isichei AO, Sylvester Bradley R (1977) Nitrogen fixation by soil algae of temperate and tropical soils. In: Döbereiner J, Burris RH, Hollaender H (eds) Limitations and potentialities for biological nitrogen fixation in the tropics. Plenum, New York, pp 41–63

Sültemeyer D, Klughammer B, Ludwig M, Badger MR, Price GD (1997) Random mutagenesis used in the generation of mutants of the marine cyanobacterium *Synechococcus* sp. strain PCC 7002 with an impaired CO_2 concentrating mechanism. Aust J Plant Physiol 24:317–327

Tirkey J, Adhikary SP (2005) Cyanobacteria in biological soil crusts of India. Curr Sci 89:515–521

Vavilin D, Vermaas W (2007) Continuous chlorophyll degradation accompanied by chlorophyllide and phytol reutilization for chlorophyll synthesis in *Synechocystis* sp PCC 6803. Biochim Biophys Acta 1767:920–929

Vavilin D, Brune DC, Vermaas W (2005) N-15-labeling to determine chlorophyll synthesis and degradation in *Synechocystis* sp PCC 6803 strains lacking one of both photosystems. Biochim Biophys Acta 1708:91–101

Whitton BA, Potts M (2000) The ecology of cyanobacteria: their diversity in time and space. Kluwer Academic, Dordrecht

Wright DJ (2004) Molecular biology of desiccation tolerance in the cyanobacterium *Nostoc commune*. Thesis Master of Science, Virginia Tech, Blacksburg, Virginia

Ziegler H, Lüttge U (1998) Carbon isotope discrimination in cyanobacteria of rocks of inselbergs and soils of savannas in the neotropics. Bot Acta 111:212–215

Chapter 4
Eukaryotic Algae

Burkhard Büdel

4.1 Introduction

Life on the land surface of the earth is impossible without the presence of water. Even simply organized, early prokaryotic organisms need to keep their cytoplasm hydrated for metabolic activity. The ability of early photosynthetic organisms to survive desiccation was one of the most important achievements for terrestrial life outside water. Desiccation tolerance must have evolved at least two times independently, first, in the prokaryotic algae (=cyanobacteria, Chap. 2) and, second, in the newly evolved eukaryotic algal lineages originating from either primary (green and red algae) or secondary endosymbiosis (brown algae). Desiccation-tolerant algae are found among the three major groups of the green land plants (Chlorobionta), the Chlorophyta, the Prasionophyta, and the Charophyta. Other desiccation-tolerant algae are found in the red algae (Rhodophyta) and the polyphyletic group of algae with heterokont flagellae, including the brown algae (Phaeophyceae).

The lowest limiting values for the state of hydration of algae in terms of water potential, at which the first irreversible effects or necrotic injuries appear, are employed as the measure of desiccation tolerance (sublethal at 5–10% injury, 50% is desiccation lethally, DL_{50}) (Larcher 2001). Marine littoral algae are among the most sensitive algae that can survive desiccation without injury at −1.4 to −4 MPa (99–97% relative air humidity, lower sublittoral), −7 to −20 MPa (95–86%, sublittoral), and −20 to −25 MPa (86–83%, eulittoral; Fig. 4.1). Algae of exposed rocks or the bark of trees may even tolerate water potentials as low as −140 MPa to indefinite for certain periods of time, depending on the functional group they belong to.

The term "desiccation tolerance" is often used to mean tolerance of partial desiccation, as is the case for example with most of the intertidal algae. In this text, it is distinguished between "partial", as mentioned above, and "complete" desiccation. Complete desiccation is defined as drying to equilibrium with moderately to very dry air or to 10% water content or less (Alpert 2006). Consequently, it is understood that desiccation tolerance is just one mechanism of drought tolerance, where drought is low water availability in the environment (Alpert 2005).

U. Lüttge et al. (eds.), *Plant Desiccation Tolerance*, Ecological Studies 215,
DOI 10.1007/978-3-642-19106-0_4,

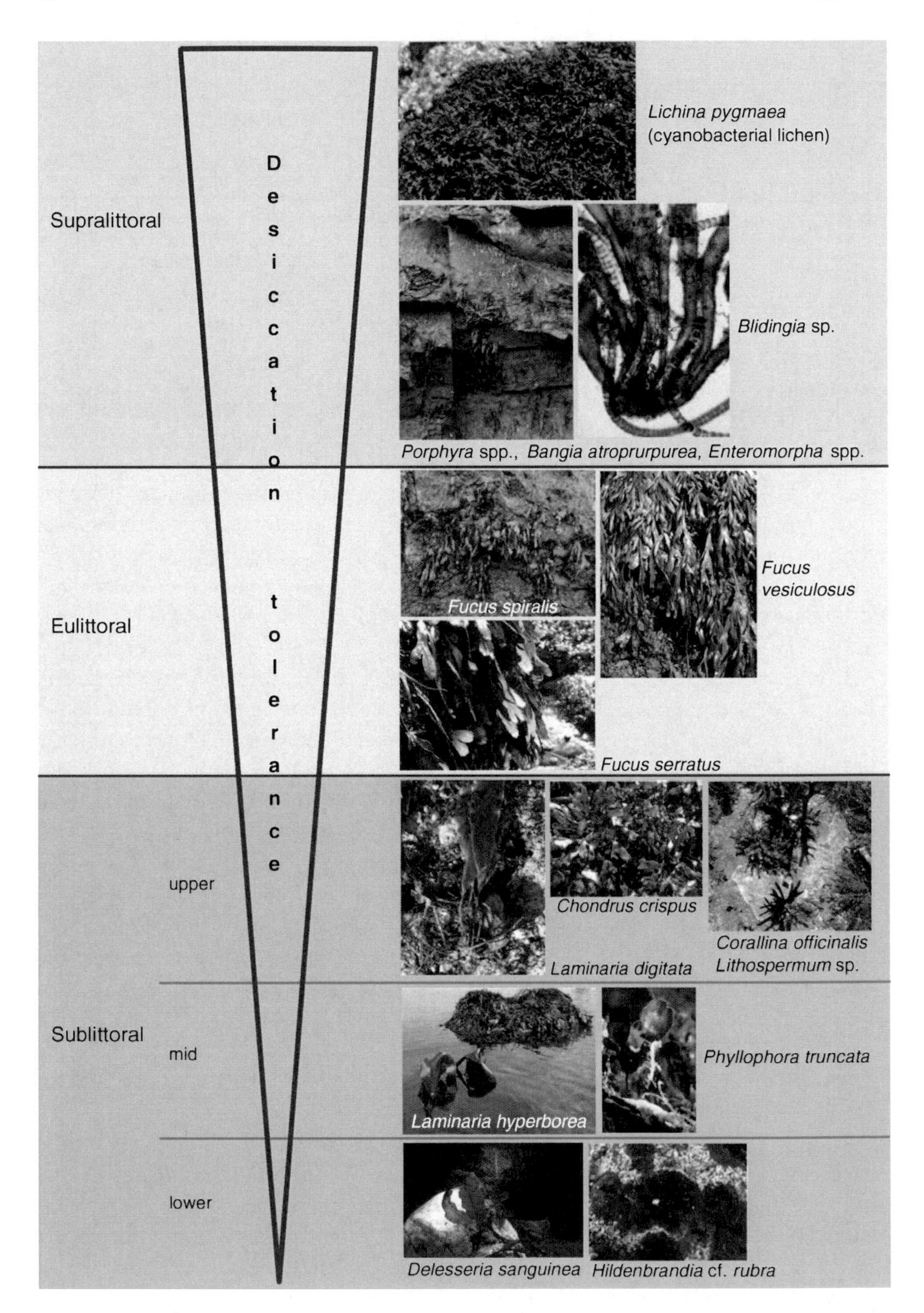

Fig. 4.1 Zonation of desiccation-tolerant algae of Northwest Atlantic coasts. With increasing time algae spend emerged in the upper littoral zones, desiccation tolerance also increases

4.2 Habitats and Species

Like cyanobacteria, eukaryotic algae occur in virtually every terrestrial habitat on earth. In contrast to prokaryotic algae, most eukaryotic algae can be reactivated by water vapor alone after desiccation (Lange et al. 1986). As a result of this general difference between the two basic algal types, eukaryotic algae often can outcompete cyanobacteria under moderate, less extreme climatic, or microclimatic conditions.

4.2.1 Marine Coastal Algae

Marine macroalgae (seaweeds) comprise about 20,000 species of which a large number can be found in the intertidal zone (Einav and Israel 2007). The intertidal shoreline zone of the oceans is a habitat of dramatic changes (Fig. 4.2a–d). Its upper regions are the beginning of terrestrial vegetation and are often sparsely inhabited by algae. There, scattered or abundant lichens occur and between them, patchy populations of macroscopic algae can be interspersed. In this upper intertidal zone and below, algae are found exposed on bare rock and not in rock pools. In terms of physiological constraints, this zone is the most stressful encountered by organisms in general, and by marine algae in particular. A recent review on this topic is that of Garbary (2007) for example. For marine intertidal algae, the benefits of drying out are often overshadowed by the stresses (e.g., light and salinity) involved (Hunt and Denny 2008). Algae of the littoral zone do not only experience partial desiccation during low tide. At the same time when falling dry during low tide, the algae become subject to high light exposure, drastic temperature de- or increase related with freezing–thawing circles at the higher latitudes, and extremely high changes in salinity. During incoming and outgoing tide and storms, strong mechanical forces destroy much of the algal fronds due to heavy swell (Fig. 4.2d). The role and importance of desiccation tolerance of marine coastal algae and the resulting zonation patterns, from the low tide zone with less expressed desiccation tolerance to the upper tide zone with a much more pronounced desiccation tolerance (Fig. 4.1), has been treated several times and is common knowledge in biological and ecological textbooks (Lüning 1985). Therefore, I will not treat this topic in general here, but rather try to summarize results of recent years.

4.2.1.1 Habitat Influence

Much of the desiccation resistance is determined by the ratio of surface area to volume, but not cell wall thickness or mucilage content (Dromgoole 1980). For a better comparison between different species, Rands and Davis (1997) developed

Fig. 4.2 Marine algae in coastal habitats: (**a**) extreme low tide, Wadden sea, Helgoland (Germany, North Sea; photo: kind permission of Daniel Weckbecker), main algae are *Laminaria hyperborea* (*brown*) and *Chondrus crispus* (*red*). (**b**) The green alga *Codium fragile* from rock pool, Roscoff, Bretagne, France. (**c**) The red alga *Chondrus crispus* exposed during low tide. (**d**) The brown alga *Durvillaea antarctica* at incoming high tide, west coast South Island, New Zealand, Pacific Ocean

two methods to determine (1) a relative measure of the activation energy of electrical conductance for the resistance of the protoplasm to desiccation; (2) they defined an initial evaporation rate constant for fresh, fully hydrated algae as an indicator of desiccation resistance of algal tissue. They used the subtidal species *Caulerpa prolifera* (P. Forsskal) Lamouroux (Chlorophyta), *Anadyomene stellata* (Wulfen) C. Agardh (Chlorophyta), *Sargassum fillipendula* C. Agardh (Phaeophyceae), and *Ulva lactuca* L. (Chlorophyta) from Cedar Key and Marineland, FL, USA, and compared them with species from Bay of Fundy, Nova Scotia, Canada. The comparison revealed that the species growing in the Bay of Fundy, subjected to extremes of tidal and drying conditions, displayed higher values of activation energy of conductance and lower desiccation rates than the species from the more moderate environment of Florida. These results conclusively showed that species subjected to extremes of tidal and drying conditions (Bay of Fundy, Nova Scotia) displayed higher values of activation energy of conductance and lower desiccation rates than the species from the more moderate environment of Florida. Their method provides quantitatively very good parameters to describe this phenomenon.

4.2.1.2 Invasive Species

The Japan native green alga *Codium fragile* ssp. *tomentosoides* (van Goor) P.C. Silva (Fig. 4.2b) is considered to be a high profile invader along the northwest Atlantic and southern Australian coasts (Occhipinti-Ambrogi and Savini 2003). It was conclusively shown that even small thallus pieces of this species can survive periods of emersion as long 90 days when kept under high relative air humidity (e.g., shipboard transportation). After 2 days of emersion and 4 days of submersed rehydration, thalli recover to their initial rates of net photosynthesis (Schaffelke and Deane 2005).

4.2.1.3 Mediterranean Sea

The intertidal zone of the Mediterranean Sea is rich in both algal biomass and the variety of seaweed species. Rocky beaches are generally richer in algal species than sandy beaches. In the eastern Mediterranean, intertidal algal communities are abundant with high standing stocks developing on abrasion platforms during the high growing seasons spring and fall. There, the algae are periodically exposed to air during low tides. Tidal fluctuations are limited to about 30 cm, but seaweeds become exposed to severe conditions. As a result of severe conditions and the input of nutrients by incoming waves within a generally oligotrophic sea, many species can grow there that are not found in the deeper waters of the benthic zone (Einav and Israel 2007).

4.2.1.4 Marine Versus Brackish Ecotypes

The brown alga *Fucus vesiculosus* L. is mainly an intertidal species found on the coasts of the North Sea, the western Baltic Sea, and the Atlantic and Pacific Oceans. In the Norwegian Sea, this species grows at 34 psu (practical salinity units, marine ecotype), whereas in the Bothnian Sea (part of Baltic Sea) it is found in brackish water with 5 psu (brackish ecotype). The brackish ecotype showed a decrease in the maximum quantum yield of photosystem II (a relative measure for potential photosynthetic activity) compared to the marine ecotype. This decrease occurred at all temperatures, but was most pronounced at 20°C. The investigation showed that the marine ecotype had higher emersion stress tolerance compared to the brackish type (Gylle et al. 2009). For *F. vesiculosus*, it can be concluded that growth in higher salinity environments results in a better osmotic adjustment compared to specimens from lower salinity environments.

4.2.1.5 Costs and Benefits

In a study on a seaweed from Pacific Grove, CA, USA, the costs and benefits of desiccation were determined for the intertidal turf alga *Endocladia muricata* (Postels and Ruprecht) J. Agardh (Rhodophyta). From laboratory experiments, the authors were able to show that during desiccation photosynthesis stops, but thermotolerance increases to a point, and that the alga is protected from head-induced mortality (Hunt and Denny 2008). They also found that this seaweed spends about 30% of the total time available for photosynthesis in the "drying-out" state. During these periods, the rate of drying determines how much time is spent hydrated and potentially for photosynthesis, but also vulnerable to high temperatures. Turf algae such as *E. muricata* dry from the edge of a clumb inward. Thus, the center needs more time to become dry than the edge and is thus subject to damage heat, resulting in a "fairy ring"-like growth pattern of the alga (Hunt and Denny 2008).

4.2.2 Terrestrial Algae

Algae are commonly known as freshwater or marine organisms, where they, of course, developed a high diversity and abundance, and are the main primary producers. At terrestrial sites, algae occur in almost all habitats one can think of, including the most hostile environments such as extreme arid, hot, and cold deserts. Also there, they are often the only primary producers. However, eukaryotic algae very often can be found in more moderate habitats such as, for example, tree bark, rock, or man-made stone walls, where they often compete with mosses, lichens, and sometimes even higher plants (Fig. 4.3a–h). Also in these habitats, desiccation tolerance is the key feature for a successful colonization.

4.2.2.1 Lithophytic Algae

Rock inhabiting algae live on or within rock substrates, expanding to a few millimeters underneath the rock surface (Hoffmann 1989). Golubic et al. (1981)

Fig. 4.3 Terrestrial habitats: (**a**) wall of the Fort Zeelandia, Paramaribo, Suriname, is covered by an assemblage of several green algae and cyanobacteria. (**b**) *Trentepohlia* sp. growth on the wall of a temple, Bali, Indonesia. (**c**) The green algal genus *Trentepohlia* (*yellow areas*), common on humid sandstone rock surfaces, entrance wall of a railway tunnel, Germany. (**d**) Wooden mill-wheel covered by tufts of the green alga *Chladophora* sp., Germany. (**e**) *Trentpohlia umbrina* on the bark of an apple tree, Germany. (**f**) *Cephaleuros/Phycopeltis* sp., on leaf of tea plant (*Camellia sinensis* cv.). (**g**) Green bark of a beech-tree (*Fagus sylvatica*), made of an assemblage of green algae like *Desmococcus olivaceus*, *Apatococcus lobatus*, and others, Germany. (**h**) Early successional type of a biological soil crust, dominated by the green alga *Zygogonium ericetorum*, Germany

distinguished several lithophytic habitats. *Epilithic* algae colonize the exposed rock surface, *endoliths* the interior of rocks. The latter group can be divided into three different subgroups, namely the *chasmoendoliths* that colonize fissures and cracks open to the rock surface, *cryptoendoliths* that colonize structural cavities within porous rocks, and the *euendoliths* that actively penetrate into the interior of the rock. *Hypolithic* algae live underneath the surface of translucent rocks that are partly embedded in the soil.

Epilithic

Green algae commonly colonize man-made surfaces such as walls of buildings, irrespective of whether they are plastered or not, consisting of natural rock, bricks, or concrete. So far, 150 unicellular species have been described from such habitats (Karsten et al. 2007). The green algal genera *Desmococcus* and *Apatococcus* (mostly the species *Desmococcus olivaceus* (Persoon ex Acharius) J.R. Laundon and *Apatococcus lobatus* (Chodat) J.B. Petersen; Fig. 4.4g, h) are among the most common taxa on these substrates, where they form bright green coatings. At the base of walls, where usually more moisture is available (e.g., spray water and capillary water from rain), an assemblage of species of the filamentous genus

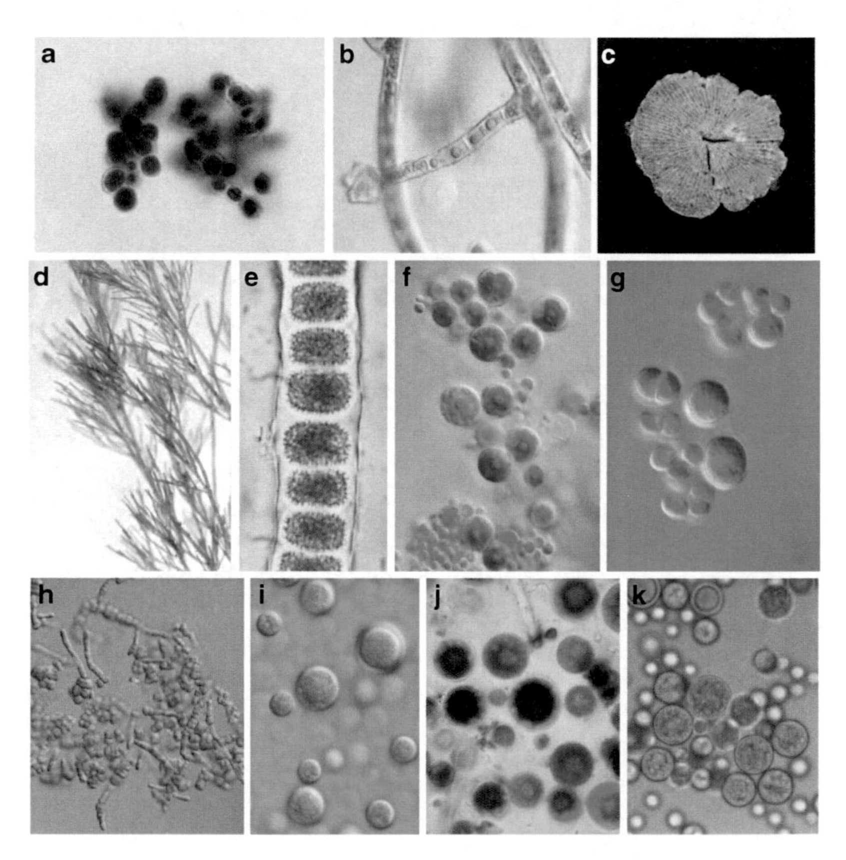

Fig. 4.4 Terrestrial algae: (**a**) *Trentepohlia umbrina*, bark of apple tree, Germany; (**b**) *Trentepohlia* cf. *aurea*, sandstone wall, Germany; (**c**) *Cephaleuros*/*Phycopeltis* sp., from leaf of tea plant (*Camellia sinensis* cv.); (**d**) *Cladophora* sp., from a wooden millwheel, Germany; (**e**) *Zygogonium ericetorum*, from soil crust, Germany; (**f**) *Trebouxia arboricola*, vegetative cells and autospores, lichen symbiont (courtesy of Thomas Friedl); (**g**) *Apatococcus lobatus*, from tree bark (courtesy of Thomas Friedl); (**h**) *Desmococcus olivaceus*, from tree bark (courtesy of Thomas Friedl); (**i**) *Macrochloris multinucleata*, from soil crust, Israel; (**j**) *Haematococcus pluvialis*, from roof gutter, Germany; (**k**) *Neochloris* sp., from soil crust, South Africa

Klebsormidium is dominant (Rindi 2007; Rindi et al. 2008). In humid climates or where a humid microclimate is realized, large yellow to orange patches often can be observed on the walls of buildings (e.g., Ireland, Singapore, Indonesia, and New Zealand; Fig. 4.3a–c). These patches are often produced by biofilm-forming filamentous green algae of the genus *Trentepohlia* (Rindi and López-Bautista 2007; Rindi and Guiry 2002). *Trentepohlia* species (Fig. 4.4b) are also common inhabitants of sandstones and calcareous rocks in the natural habitat, where they also occur as photobionts of lichens. A well-known biofilm producing species on depressions of horizontal parts of walls and roofs as well as in rain gutters, but also horizontal parts of natural rocks in temperate zones, is the unicellular green alga *Haematococcus pluvialis* Flotow (Fig. 4.4j).

From granite outcrops of a canyon at the Teteriv River in the forest zone of Ukraine, 67 species of eukaryotic algae were reported recently, 49 of them belonging to the Chlorophyta, 11 to the Streptophyta, 2 to the Xanthophyta, 4 to the Bacillariophyta, and 1 to the Eustigmatophyta (Mikhailyuk 2008). This unexpected high diversity leads to the conclusion that closer investigations of eukaryotic algal biofilms on rocks might reveal a much higher diversity than presently known.

Endolithic

Green algae as for example *Hemichloris antarctica* Tschermak-Woess and Friedmann and a species of the genus *Heterococcus* are known to occur in the cryptoendolithic habitat of Antarctic rocks (Friedmann and Ocampo-Friedmann 1984; Tschermak-Woess and Friedmann 1984; Friedmann 1982), where they can thrive in the desiccated state, often combined with freezing, for months. In a more recent study, a number of green algae were found growing endolithic in dolomitic limestone of an ancient cliff ecosystem of the Niagara Escarpment, Canada (Gerrath et al. 2000). In these rocks, the species diversity of eukaryotic algae matched that of cyanobacteria. In total, ten chlorophytes (*Chlorella*, *Coccobotrys*, *Muriella*, *Pseudendoclonium*, *Stichococcus*, and *Trebouxia*) and three xanthophytes (*Chloridella*, *Ellipsoidion*, and *Heterococcus*) were reported. So far, no green algae were detected living in the euendolithic habitat (Hoffmann 1989). From the few reports available on eukaryotic endolithic algae of arid climates, one might conclude that the endolithic habitat of arid environments seems not to be very suitable for eukaryotic algae.

Hypolithic

The hypolithic community was named "hypolithon" (Schlesinger et al. 2003) and occurs along a zone of sufficient light for positive net photosynthesis. Although the hypolithon is mainly composed by cyanobacteria, eukaryotic algae occur frequently. From the Namib Desert, even diatoms were reported from the hypolithic habitat (Rummrich et al. 1989). Vogel (1955) was the first who gave a species list of four green algae and one diatom identified from the hypolithon of South Africa. In a later investigation, Friedmann et al. (1967) determined 11 hypolithic green algal species from the Negev Desert.

4.2.2.2 Soil

There are a large number of green algae and diatoms living in the soil, either free living and not related to certain structures, or in association with other organisms forming biological soil crusts (BSCs). The closer these algae grow to the soil surface, the more they become exposed to frequent drying events. For a successful establishment of algae in the photic zone of soils, desiccation tolerance is a prerequisite. Using $^{13}C/^{12}C$ ratio, it has been found that a terrestrial biomass existed that was large enough to transfer photosynthetic C into coastal ground waters. These early terrestrial photosynthesizers were probably composed of algae, mosses, fungi, liverworts (Knauth and Kennedy 2009) at least 1 billion years ago, and lichens 600 million years ago (Yuan et al. 2005).

Some BSCs can be very rich in algal species and examples for high species diversity BSC have been reported from southwestern Africa (Fig. 4.4k), where along a 2,000 km transect more than 29 green algae were found (Büdel et al. 2009). Other algal rich BSC habitats were reported with 37 eukaryotic algae from Baja California, the Chihuahuan, and the Mojave Desert (Flechtner et al. 1998, 2008; Flechtner 2007), and 11 species in BSC from western North America (Lewis and Flechtner 2002). But also BSCs from the Gurbantunggut Desert, China, include with 26 species a large number of green algae (Zhao et al. 2008). Characteristic green algal genera of BSC in the European temperate region are *Klebsormidium* and *Zygogonium* (Figs. 4.3h and 4.4e), representing early successional types of BSC (Büdel 2001a; Hoppert et al. 2004; Langhans et al. 2009). Eukaryotic algae have recently been found to be extraordinarily diverse in desert-BSC, spanning five algal classes and encompassing many taxa new to science (Cardon et al. 2008). In this study, also the possible contribution and role of green algae to BSC is discussed in general. A list of eukaryotic algae associated with soil crusts is presented in Büdel (2001b).

4.2.2.3 Epiphytic Algae

The green algal species *D. olivaceus* and *A. lobatus* are not only common on rock surfaces and walls, but also the most common taxa on the bark of trees in the temperate zone (Fig. 4.3g). Together with other green algae such as *Trebouxia* spp., *Coccomyxa* spp., *Diplosphaera* sp., *Neochloris* spp., and the so-called "*Chlorella trebouxioides*" type alga, they form dense green biofilms on most forest trees in temperate Europe (Gärtner 1994). During the last 20–30 years, a number of *Trentepohlia* (e.g., *Trentepohlia umbrina* (Kützing) Bornet; Figs. 4.3e and 4.4a) species have been observed to form orange colored biofilms on the bark of orchard trees but also on poplar

(*Populus* spp.) and other forest trees (Lüttge and Büdel 2010). In northwestern Germany, the filamentous green algal species *Klebsormidium crenulatum* (Kützing) Lockhorst was found to form large green layers on tree bark (Frahm 1999).

From tropical regions of South-East Asia, a number of green algae of the families Chlorophyceae (four species), the Trebouxiophyceae (ten species), and the Ulvophyceae (six species) have been reported forming biofilms on tree bark (Neustupa and Škaloud 2008). The authors report a significant higher species diversity in open localities than in closed forest areas. In open localities with higher light intensities, Trentepohliales and cyanobacteria dominated, whereas closed forest areas were dominated by trebouxiophyceaen coccal green algae. Four species of the genus *Trentepohlia*, including two new species, were reported growing on trees in French Guiana. The new species *Trentepohlia chapmanii* Rindi and López-Bautista formed orange coatings on bamboo, and *Trentepohlia diffracta* var. *colorata* A. B. Cribb on tree bark in the rain forest of French Guiana (Rindi and López-Bautista 2007). The trentepohlioid genera *Cephaleuros* and *Phycopeltis* (Figs. 4.3f and 4.4c) are reported to grow subcuticular (often hemiparasitic) or epicuticular on the leaves of tea-shrubs and other shrubby species in tropical and subtropical regions (Chapman 1976; Chapman and Good 1976; Neustupa 2003; Sanders 2002). However, there is no information available yet on their capability to withstand drought periods. At least for the subcuticular species, it seems plausible that they might profit from the protected environment and probably never really fall dry.

Some 60 years ago, when millwheels still have been a common feature of European cultural landscapes (and probably elsewhere as well), otherwise amphibious freshwater algae often grew on such wooden constructions. Although not truly epiphytic, green algal species of the genus *Cladophora* and the filamentous red alga *Bangia atropurpurea* (Roth) Greville that were often found luxuriantly growing on millwheels on rivers (Figs. 4.3d and 4.4d) are listed here. There, these algal species often experienced long drought periods during low water levels and continued to grow after the water level rose again.

4.2.2.4 Airborne Algae

Viable microorganisms, including algae, enter the air naked, on small plant pieces, embedded in soil debris or from sea spray. They are finally deposited on surfaces by either impaction or gravitational settling. Dust storms can serve huge natural long-distance transport and contribute together with many more local natural sources to the so-called background microbial population (Andersen et al. 2009). Griffin et al. (2002) estimated that a quintillion (10^{18}) sediment-borne bacteria is moving around the planet Earth each year – enough to form a microbial bridge between Earth and Jupiter. It is critical that viable microorganisms survive after being transported in the atmosphere. This can be achieved by either resting cells (spores) or the ability to recover from complete desiccation of the vegetative cells themselves. Some of the most well-known deleterious conditions are solar UV radiation, relative humidity, temperature, and sometimes air pollutants. Dissolved substances as for example trehalose are also known to protect airborne cells (Andersen et al. 2009).

Newly exposed surfaces, such as abrasion plateaus, land slide areas, or even complete new islands due to volcanic activity, are subject to new colonization by autotrophic organisms. Algae are important colonizers of such isolated land areas (Wynn-Williams 1990). So far, 59 species of the Chlorophyta, 6 species of Tribophyceae, and 8 species of Bacillariophyceae have been identified from the atmosphere (Table 4.1). Tropical regions have a higher diversity and abundance of airborne algae than other regions (Schlichting 1969; Sharma et al. 2007), a fact that is attributed by the authors to the higher abundance of soil algae and other terrestrial algae.

Table 4.1 Airborne algae (Sharma et al. 2007, with additions by the author)

Cyanobacteria	Chlorophyta	Tribophyceae	Bacillariophyceae
Anabaena sp.	*Actinastrum* sp.	*Heterococcus* sp.	*Chaetoceros* sp.
Anacystis sp.	*Ankistrodesmus* sp.	*Heterothrix* sp.	*Coscinodiscus*-like
Aphanocapsa sp.	*Asterococcus*-like	*Heteropedia* sp.	*Cyclotella* sp.
Aphanothece sp.	*Borodinella* sp.	*Tribonema* sp.	*Hantschia* sp.
Arthrospira sp.	*Bracteacoccus*-like	*Botrydiopsis* sp.	*Melosira* sp.
Calothrix sp.	*Chlamydomonas* sp.	*Vaucheria* sp.	*Navicula* sp.
Chroococcus sp.	*Chlorella* spp.		*Nitzschia* sp.
Chroococcidiopsis sp.	*Chlorococcum* sp.		*Pinnularia* sp.
Cylindrospermum sp.	*Chlorhormidium* sp.		
Entophysalis sp.	*Chlorosphaera* sp.		
Fischerella sp.	*Chlorosphaeropsis* sp.		
Fremyella sp.	*Chlorosarcinopsis* sp.		
Gloeocapsa sp.	*Chlorosarcina* sp.		
Gloeothece sp.	*Chrysocapsa* sp.		
Hapalosiphon sp.	*Coelastrum* sp.		
Hydrocoleus sp.	*Cosmarium* sp.		
Lyngbya sp.	*Cylindrocystis* sp.		
Merismopedia sp.	*Dictyochloris* sp.		
Microchaete sp.	*Dimorphococcus* sp.		
Microcystis sp.	*Friedmannia* sp.		
Microcoleus sp.	*Gloeocystis* sp.		
Myxosarcina sp.	*Hormidium* spp.		
Nostoc sp.	*Klebshormotilopsis* sp.		
Oscillatoria spp.	*Microspora* sp.		
Petalonema sp.	*Nannochloris* sp.		
Phormidium spp.	*Neochloris* sp.		
Plectonema sp.	*Oedogonium* sp.		
Schizothrix sp.	*Oocystis* sp.		
Spirulina sp.	*Ourococcus* sp.		
Stigonema sp.	*Palmella* sp.		
Synechococcus sp.	*Palmellococcus* sp.		
Scytonema sp.	*Pediastrum* sp.		
Tolypothrix sp.	*Planktosphaeria* sp.		
	Pleodorina sp.		

(*continued*)

Table 4.1 (continued)

Cyanobacteria	Chlorophyta	Tribophyceae	Bacillariophyceae
	Pleurococcus sp.		
	Pleurastrum sp.		
	Prasiola sp.		
	Protococcus sp.		
	Protosiphon sp.		
	Pseudoulvella sp.		
	Radiococcus sp.		
	Radiosphaera sp.		
	Rhopalocystis sp.		
	Rhizoclonium sp.		
	Roya sp.		
	Scenedesmus spp.		
	Selenastrum sp.		
	Sphaerocystis sp.		
	Spongiochloris sp.		
	Spongiococcum sp.		
	Stichococcus spp.		
	Tetracystis sp.		
	Tetraspora sp.		
	Tetraedron sp.		
	Trebouxia sp.		
	Ulothrix sp.		
	Westella sp.		
	Zygnema sp.		

4.3 Physiological Ecology

Organisms that live on more or less exposed surfaces are subject to several environmental challenges such as high irradiance, temperature extremes, extreme water stress from being completely dry to oversaturated, and sometimes air pollutants. Algae have evolved a number of protective mechanisms to cope with these environmental challenges. One advantage of desiccation tolerance is that it increases thermotolerance for most of the organisms involved. However, not always do desiccation-tolerant algae or all individuals of a population survive extreme desiccation. In some cases, death due to anhydrobiosis can occur and several mechanisms are known to be responsible for it: irreversible phase changes to lipids, proteins, and nucleic acids such as denaturation and structural breakage through Maillard reactions, and accumulation of reactive oxygen species (ROS) during drying, especially under solar radiation (Rothschild and Mancinelli 2001).

4.3.1 Photosynthetic Patterns of Marine Algae

Carbon dioxide diffusion in water is by magnitudes slower than in air (Cowan et al. 1992). From that, one would expect that falling dry during low tide of coastal marine algae might improve CO_2 availability for photosynthesis. At least during the drying process, when water films disappear from the thallus surface, high rates of net photosynthesis should be reached. This was tested for several species of intertidal marine algae by Johnson et al. (1974) in a comparison of photosynthetic CO_2 uptake of submerged and emerged algae, coming from lower intertidal (*Ulva expansa* (Setch) S. and G. and *Prionitis lanceolata* (Harvey) Harvey) and from the middle and upper littoral (*Iridaea flaccida* (S. and G.) Silva, *Porphyra perforata* J. G. Agardh, *Fucus distichus* L., and *E. muricata*). The algae from the lower intertidal exposed reduced photosynthetic capacity in air compared to submerged rates. In contrast, however, species from the middle and upper littoral reached maximum photosynthesis after some degree of drying. Photosynthetic rates were 1.6–6.6 times greater in air than in water (same illumination and temperature). Given that desiccation rates under natural conditions are slow enough these algae might be capable of continuing high rates of photosynthetic activity for longer periods while exposed. They may even fix the bulk of their carbon at this time. Their intertidal zonation seems to be correlated with their capacity for sustained photosynthesis in air. The authors suggest that these relationships may be partially responsible for the vertical distribution of intertidal marine algae.

4.3.2 Release of Dissolved Organic Carbon During Rehydration

When desiccated seaweeds become rehydrated, a release of dissolved organic carbon (DOC) was observed (Moebus et al. 1974). During 30 min rehydration, two seaweeds that had lost up to 70% of their water released about 2–10 mg C/100 g dry weight (*Ascophyllum nodosum* (L.) Le Jolis) and the second (*F. vesiculosus*) 10–50 times more. Water losses exceeding 70% resulted in an even higher water DOC loss more or less similar for both species, approximating 2 g C/100 g dry weight. In the DOC released, the portion of carbohydrate-C was 0–5% with *A. nodosum* and 2–47% with *F. vesiculosus*.

4.3.3 Drought Period and Resurrection

The exposed isolated and cultured lichen photobiont *Trebouxia erici* Ahmadjian showed that the drying rate has a profound effect on the recovery of photosynthetic activity after rehydration, i.e., greater than the effects of desiccation duration. Basal chlorophyll fluorescence values (F'_0) in desiccated algae were significantly higher after rapid dehydration, suggesting higher levels of nonphotochemical light energy dissipation in slowly dried algae. In contrast, higher values of PS II electron transport were recovered after rehydration of slowly dried *T. erici* compared to rapidly dried algae. The main component of nonphotochemical quenching after slow dehydration was energy dependent, whereas after fast dehydration it was photoinhibition (Gasulla et al. 2009).

In a recent comparative study, Gray et al. (2007) investigated the recovery of PS II (photosynthetic quantum yield) from green algae of aquatic and terrestrial origin. These authors found that the recovery of terrestrial types was clearly improved compared to that of aquatic origin. In addition, the mode of dehydration (with or without light, slow or fast) as well as the mode of rehydration (with or without light) played an important role for the recovery speed and rate of PS II. When rehydration took place in the light, recovery of PS II was well enhanced compared to those that were rehydrated in the dark.

For two different types (green versus orange biofilms) of desiccation-tolerant bark algae communities, Lüttge and Büdel (2010) reported a different resurrection behavior upon rewetting. Recovery (potential quantum yield of PS II, F_v/F_m) was better for the green biofilm samples (*Apatococcus*, *Desmococcus*, *Trebouxia*, and *Coccomyxa*) from the bark of forest trees (good recovery even after 80 days of desiccation; F_v/F_m = ca. 50% of initial value) than for the orange biofilm (*T. umbrina*) from orchard trees (30–40 days; F_v/F_m = 20–55% of initial value).

4.3.4 Antioxidants as a Protective Means

Radiation and oxidative damage have always been common on Earth (Rothschild and Mancinelli 2001). When cells are desiccated in the light, chlorophyll molecules continue to be excited, but the energy not used in carbon fixation will cause formation of singlet oxygen (Kranner et al. 2005). Photoprotective mechanisms can dissipate much of it. These mechanisms include dissipation as heat via carotenoids, photorespiration, and morphological features that minimize light absorption. ROS such as superoxide and singlet oxygen are produced in chloroplasts by photoreduction of oxygen and energy transfer from triplet excited chlorophyll to oxygen, respectively. In addition, hydrogen peroxide and hydroxyl radicals can form as a result of the reactions of superoxide. All these ROS are reactive and potentially damaging, causing lipid peroxidation and inactivation of enzymes (Smirnoff 1993). Free radical scavengers are the two water-soluble antioxidants reduced glutathione (GSH) and ascorbate (vitamine C) and the lipid-soluble membrane-bound ROS scavenger α-tocopherol (vitamin E) (Kranner et al. 2005).

The green alga *Trebouxia excentrica* Archibald, isolated from the lichen *Cladonia vulcani* Savicz, can withstand desiccation, but has a limited capacity for photoprotection. During desiccation, it oxidizes GSH, but the alga loses considerable amounts of its photosynthetic pigments and tocopherol. Three weeks of desiccation does not cause major effects, but after 9 weeks they recover only partially (Kranner et al. 2005).

4.3.5 Compatible Solutes

Compatible solutes or osmoprotectants are small molecules that act as osmolytes and help to survive extreme osmotic stress. Organisms accumulate large amounts of disaccharides such as sucrose or trehalose. These sugars stabilize membranes and proteins in the dry state in a glass-like structure (Crowe et al. 1998). Examples of compatible solutes of the different categories encountered in various organisms include the following (Yancey et al. 1982):

- Polyols: glycerol, glucosylglycerol, arabitol, mannitol, sorbitol, and trehalose.
- Amino acids: glutamate, proline, γ-amino butyric acid, glycine betaine, and ectoine.

4.3.6 Ultrastructure

Vacuoles and cytoplasmic portions of desiccated samples of *Zygogonium ericetorum* Kützing appeared destroyed, whereas their nucleus and plastids remained intact. The thylakoid membranes showed lumen dilatations and numerous plastoglobules (Holzinger et al. 2010). The fast recovery of chlorophyll fluorescence of PS II, as described by Gray et al. (2007) for desiccation-tolerant green algae, is attributed by Holzinger et al. (2010) to the intact nucleus and plastids with only occasionally dilated thylakoids during desiccation.

4.4 Conclusion

In a recent publication of Alpert (2006), the speculation was raised why desiccation-tolerant organisms are so small or rare. While the ability to tolerate complete desiccation is widespread among organisms smaller than 5 mm, this is the exception in larger and morphologically more complex organisms. Shrinkage is one of the most problematic structural changes related with complete desiccation. It seems that organisms composed of a number of different tissues and cell types would need different strategies related to tissue and cell type to recover. To avoid such deleterious environmentally triggered damages might be more effective than investment in strategies that allow them to colonize ecologically extreme environments.

Acknowledgments I would like to thank the editors for the invitation to contribute to that highly interesting volume of Ecological Studies. Parts of this work have been supported by the German Research Foundation (DFG).

References

Alpert P (2005) The limits and frontiers of desiccation-tolerant life. Integr Comp Biol 45:685–695
Alpert P (2006) Constraints of tolerance: why are desiccation-tolerant organisms so small or rare? J Exp Biol 209:1575–1584
Andersen GL, Frisch AS, Kellogg CA, Levetin E, Lighthart B, Paterno D (2009) Aeromicrobiology/air quality. In: Schaechter M (ed) Encyclopedia of microbiology. Academic, Oxford, pp 11–26
Büdel B (2001a) Biological soil crusts in European temperate and Mediterranean regions. In: Belnap J, Lange OL (eds) Ecological studies, vol 150. Springer, Heidelberg, pp 75–87
Büdel B (2001b) Synopsis: comparative biogeography of soil-crust biota. In: Belnap J, Lange OL (eds) Ecological studies, vol 150. Springer-Verlag, Berlin, pp 141–152
Büdel B, Darienko T, Deutschewitz K, Dojani S, Friedl T, Mohr KI, Salisch M, Reisser W, Weber B (2009) Southern african biological soil crusts are ubiquitous and highly diverse in drylands, being restricted by rainfall frequency. Microb Ecol 57:229–247
Cardon ZG, Gray DW, Lewis LA (2008) The green algal underground: evolutionary secrets of desert cells. Bioscience 58:114–122
Chapman RL (1976) Ultrastructural investigation on the foliicolous pyrenocarpous lichen *Strigula elegans* (Fee) Müll. Arg. Phycologia 15:191–196
Chapman RL, Good BH (1976) Observations on the morphology and taxonomy of *Phycopeltis hawaiiensis* King (Chroolepidaceae). Pac Sci 30:187–195
Cowan IR, Lange OL, Green TGA (1992) Carbon-dioxide exchange in lichens: determination of transport and carboxylation characteristic. Planta 187:282–294
Crowe JH, Carpenter JF, Crowe LM (1998) The role of vitrification in anhydrobiosis. Annu Rev Physiol 60:73–103
Dromgoole FI (1980) Desiccation resistance of intertidal and subtidal algae. Bot Mar 23:149–159
Einav R, Israel A (2007) Seaweeds on the abrasion platforms of the intertidal zone of eastern Mediterranean shores. In: Seckbach J (ed) Cellular origin, life in extreme habitats and astrobiology, vol 11. Springer, Dordrecht, pp 195–207
Flechtner VR (2007) North American desert microbiotic soil crust communities. In: Seckbach J (ed) Cellular origin, life in extreme habitats and astrobiology, vol 11. Springer, Dordrecht, pp 539–551
Flechtner VR, Johansen JR, Clark WH (1998) Algal composition of microbiotic crusts from the central desert of Baja California, Mexico. Great Basin Nat 58:295–311
Flechtner VR, Johansen JR, Belnap J (2008) The biological soil crusts of the San Nicolas Island: enigmatic algae from a geographically isolated ecosystem. West N Am Naturalist 68:405–436
Frahm JP (1999) Epiphytische Massenvorkommen der fädigen Grünalge *Klebsormidium crenulatum* (Kützing) Lokhorst im Rheinland. Decheniana 152:117–119
Friedmann EI (1982) Endolithic microorganisms in the Antarctic cold desert. Science 215: 1045–1053
Friedmann EI, Ocampo-Friedmann R (1984) Endolithic microorganisms in extreme dry environments: analysis of a lithobiontic microbial habitat. In: Klug MJ, Reddey CA (eds) Current perspectives in microbial ecology. American Society for Microbiology, Washington, DC, pp 177–185
Friedmann EI, Lipkin Y, Ocampo-Paus R (1967) Desert algae of the Negev (Israel). Phycologia 6: 185–200
Garbary DJ (2007) The margin of the sea: survival at the top of the tides. In: Seckbach J (ed) Cellular origin, life in extreme habitats and astrobiology, vol 11. Springer, Dordrecht, pp 175–191
Gärtner G (1994) Zur Taxonomie aerophiler grüner Algenanflüge an Baumrinden. Ber nat -med Verein Innsbruck 81:51–59
Gasulla F, Gómez de Nova P, Esteban-Carrasco A, Zapata JM, Barreno E, Guéra A (2009) Dehydration rate and time of desiccation affect recovery of the lichenic algae *Trebouxia erici*: alternative and classical protective mechanisms. Planta 231:195–208

Gerrath JF, Gerrath JA, Matthes U, Larson DW (2000) Endolithic algae and cyanobacteria from cliffs of the Niagara Escarpment, Ontario, Canada. Can J Bot 78:807–815

Golubic S, Friedmann I, Schneider J (1981) The lithobiontic ecological niche, with special reference to microorganisms. J Sediment Petrol 51:475–478

Gray DW, Lewis LA, Cardon ZG (2007) Photosynthetic recovery following desiccation of desert green algae (Chlorophyta) and their aquatic relatives. Plant Cell Environ 30:1240–1255

Griffin D, Kellogg C, Garrsion V, Shinn E (2002) The global transport of dust. Am Sci 90:230–237

Gylle AM, Nygård CA, Ekelund NGA (2009) Desiccation and salinity effects on marine and brackish *Fucus vesiculosus* L. (Phaeophyceae). Phycologia 48:156–164

Hoffmann L (1989) Algae of terrestrial habitats. Bot Rev 55:77–105

Holzinger A, Tschaikner A, Remias D (2010) Cytoarchitecture of the desiccation-tolerant green alga *Zygogonium ericetorum*. Protoplasma 243:15–24

Hoppert M, Reimer R, Kemmling A, Schröder A, Günzl B, Heinken T (2004) Structure and reactivity of a biological soil crust from a xeric sandy soil in Central Europe. Geomicrobiol J 21:183–191

Hunt LJH, Denny MW (2008) Desiccation protection and disruption: a trade-off for an intertidal marine alga. J Phycol 44:1164–1170

Johnson WS, Gigon A, Gulmon SL, Mooney HA (1974) Comparative photosynthetic capacities of intertidal algae under exposed and submerged conditions. Ecology 55:450–453

Karsten U, Schumann R, Mostaert AS (2007) Aeroterrestrial algae growing on man-made surfaces: what are the secrets of their ecological success? In: Seckbach J (ed) Algae and cyanobacteria in extreme environments. Springer, Dordrecht, pp 585–597

Knauth LP, Kennedy MJ (2009) The late precambrian greening of the Earth. Nature 460(7256): 728–732

Kranner I, Cram WJ, Zorn M, Wornik S, Yoshimura I, Stabentheiner E, Pfeifhofer HW (2005) Antioxidants and photoprotection in a lichen as compared with its isolated symbiotic partners. Proc Natl Acad Sci USA 102:3141–3146

Lange OL, Kilian E, Ziegler H (1986) Water vapor uptake and photosynthesis of lichens: performance differences in species with green and blue-green algae as phycobionts. Oecologia 71:104–110

Langhans TM, Storm C, Schwabe A (2009) Community assembly of biological soil crusts of different successional stages in a temperate sand ecosystem, as assessed by direct determination and enrichment techniques. Microb Ecol 58:394–407

Larcher W (2001) Physiological plant ecology – ecophysiology and stress physiology of functional groups. Springer-Verlag, Berlin, pp 1–513

Lewis LA, Flechtner VR (2002) Green algae (Chlorophyta) of desert microbiotic crusts: diversity of North American taxa. Taxon 51:443–451

Lüning K (1985) Meeresbotanik. Verbreitung, Ökophsiologie und Nutzung der marinen Makrooalgen. Georg Thieme Verlag, Stuttgart, pp 1–375

Lüttge U, Büdel B (2010) Resurrection kinetics of photosynthesis in desiccation-tolerant terrestrial green algae (Chlorophyta) on tree bark. Plant Biol 12:4371–4444

Mikhailyuk TI (2008) Terrestrial lithophilic algae in a granite canyon of the Teteriv River (Ukraine). Biologia 63:824–830

Moebus K, Johnson KM, Sieburth JM (1974) Rehydration of desiccated intertidal brown algae: release of dissolved organic carbon and water uptake. Mar Biol 26:127–134

Neustupa J (2003) The genus *Phycopeltis* (Trentepohliales, Chlorophyta) from tropical Southeast Asia. Nova Hedwig 76:487–505

Neustupa J, Škaloud P (2008) Diversity of subaerial algae and cyanobacteria on tree bark in tropical mountain habitats. Biologia 63:806–812

Occhipinti-Ambrogi A, Savini D (2003) Biological invasions as a component of global change in stressed marine ecosystems. Mar Pollut Bull 46:542–551

Rands DG, Davis JS (1997) Comparative study of activation energies of conductance and desiccation rates of some marine algae. Aquat Sci 59:275–281

Rindi F (2007) Diversity, distribution and ecology of green algae and cyanobacteria in urban habitats. In: Seckbach J (ed) Cellular origin, life in extreme habitats and astrobiology, vol 11. Springer, Dordrecht, pp 621–638
Rindi F, Guiry MD (2002) Diversity, life history, and ecology of *Trentepohlia* and *Printzia* (Trentepohliales, Chlorophyta) in urban habitats in western Ireland. J Phycol 38:39–54
Rindi F, López-Bautista JM (2007) New and interesting records of *Trentepohlia* (Trentepohliales, Chlorophyta) from French Guiana, including the description of two new species. Phycologia 46:698–708
Rindi F, Guiry MD, López-Bautista JM (2008) Distribution, morphology, and phylogeny of *Klebsormidium* (Klebsormidiales, Charophyceae) in urban environments in Europe. J Phycol 44:1529–1540
Rothschild LJ, Mancinelli RL (2001) Life in extreme environments. Nature 409:1092–1101
Rummrich U, Rummrich M, Lange-Bertalot H (1989) Diatomeen als "Fensteralgen" in der Namib-Wüste und anderen ariden Gebieten von SWA/Namibia. Dinteria 20:23–29
Sanders WB (2002) Reproductive strategies, relichenization and thallus development observed in situ in leaf-dwelling lichen communities. New Phytol 155:425–435
Schaffelke B, Deane D (2005) Desiccation tolerance of the introduced marine green alga *Codium fragile* ssp. *tomentosoides* – clues for likely transport vectors? Biol Invasions 7:557–565
Schlesinger WH, Pippen JS, Wallenstein MD, Hofmockel KS, Klepeis DM, Mahall BE (2003) Community composition and photosynthesis by photoautotrophs under quartz pebbles, Southern Mojave Desert. Ecology 84:3222–3231
Schlichting HE (1969) The importance of airborne algae and protozoa. J Air Pollut Control Assoc 19:946–951
Sharma NK, Rai AK, Singh S, Brown RM Jr (2007) Airborne algae: their present status and relevance. J Phycol 43:615–627
Smirnoff N (1993) The role of active oxygen in the response of plants water deficit and desiccation. New Phytol 125:27–58
Tschermak-Woess E, Friedmann EI (1984) *Hemichloris antarctica*, gen. et. sp, nov. (Chlorococcales, Chlorophyta), a cryptoendolithic alga from Antarctica. Phycologia 23:443–445
Vogel S (1955) Niedere "Fensterpflanzen" in der südafrikanischen Wüste. Eine ökologische Schilderung. Beitr Biol Pflanz 31:45–135
Wynn-Williams DD (1990) Microbial colonization processes in Antarctic fellfield soil – an experimental overview. Proc NIPR Symp Polar Biol 3:164–178
Yancey PH, Clark ME, Hand SC, Bowlus RD, Somero GN (1982) Living with water stress: evolution of osmolyte systems. Science 217:1214–1222
Yuan X, Xiao S, Taylor TN (2005) Lichen-like symbiosis 600 million years ago. Science 308: 1017–1020
Zhao J, Zhang B, Zhang Y (2008) Chlorophytes of biological soil crusts in Gurbantunggut Desert, Xinjiang Autonomous Region, China. Front Biol China 3:40–44

Chapter 5
Lichens and Bryophytes: Habitats and Species

Michael Lakatos

5.1 Characteristics of Lichens and Bryophytes

Poikilohydric desiccation tolerance enables lichens and bryophytes to survive long periods of water limitation and to recover quickly by rehydration. The evolutionary success of this strategy is reflected by the fact that cryptogams inhabit almost all terrestrial habitats from the tropics to cold and hot deserts. As ecosystem components, lichens and bryophytes may considerably impact the surrounding environment through frequent desiccation–rewetting cycles. What are the differences in mechanism and functioning to successfully compete with vascular plants in many micro-sites and habitats? This chapter reviews key issues of cryptogamic desiccation tolerance with particular emphasis on the following aspects: (1) Comparison of mechanisms and processes of water exchange. (2) Function and impacts of micro-scale fluxes to illustrate the effects of desiccation–rewetting cycles on the environment. (3) Global patterns of lichens and bryophytes as an indication for their ecological relevance.

Lichens and bryophytes belong to the first photoautotrophic multi-cellular eukaryotes colonising terrestrial habitats with limited irregular water supply and frequent desiccation events. The phylogeny of both organisms is largely distinct. Polyphyletic lichens are nutrient specialised fungi, while bryophytes are non-vascular plants *senso stricto*: they evolved the same strategy of poikilohydry by (1) reducing metabolism when water availability is decreasing, (2) tolerating the loss of virtually all free water without dying, and (3) resuming growth after rehydration by high humidity or liquid water. This desiccation tolerance differs substantially from that of vascular plants (Oliver et al. 2000; Proctor and Tuba 2002) (Chap. 9), and lichens and bryophytes are the major groups implementing desiccation tolerance as a common and successful alternative life strategy (Proctor and Tuba 2002). Since both life forms are historically treated so by botanists and lack true vascular systems, they are grouped as "non-vascular plants".

Lichens are generally regarded as symbiotic between a fungal "mycobiont" and a photosynthetic algal and/or cyanobacterial "photobiont" partner. Approximately 85% of the lichen mycobionts are symbiotic with green algae, 10% with cyanobacteria, and 3–4% are simultaneously associated with both green algae and

U. Lüttge et al. (eds.), *Plant Desiccation Tolerance*, Ecological Studies 215,
DOI 10.1007/978-3-642-19106-0_5, © Springer-Verlag Berlin Heidelberg 2011

cyanobacteria (Honegger 1998). The number of species varies from 13,500 (Hawksworth and Hill 1984) to 18,882 (Feuerer and Hawksworth 2007; Hale 1983). In general, the layers of lichens mainly consist of fungal hyphae and those mostly of Ascomycetes. The photobionts are located extracellularly within the lichen thallus, arranged either as a distinct algal layer in internally stratified (heteromerous) thalli or randomly distributed in non-stratified (homoeomerous) thalli. The morphology and anatomy of heteromerous thalli is highly diverse, but basically they are composed of three layers: outer cortex (conglutinate zones), algal layer (photobiont), and medulla (aerial hyphae). Bryophytes are embryophytes with plant tissues and enclosed reproductive systems. Because bryophyte conducting tissue lacks lignin, it is not considered true vascular tissue. The bryophytes form a paraphyletic group placed in three separate divisions: the Marchantiophyta (liverworts), Anthocerotophyta (hornworts), and Bryophyta (mosses) with diverse morphology from thalloid to leaf possessing structures (Goffinet 2000). Bryophytes comprise 15,000 (Gradstein et al. 2001) to 25,000 species (Crum 2001). Both life forms are distributed worldwide and occur in every location habitable by photosynthetic organisms, occupying various microhabitats such as leaves, bark, rocks, soil, or even anthropogenic substrates.

In the last decade, reviews have focused on the comparison of poikilohydry versus homoiohydry (Proctor and Tuba 2002), on physiological and ecological aspects of either bryophytes (Proctor et al. 2007; Turetsky 2003) or lichens (Honegger 2006; Kranner et al. 2008; Palmqvist 2000), on functioning aspects in biogeochemistry (Cornelissen et al. 2007), and on anthropogenic contexts such as medicine, agriculture, and global change (Alpert 2005; Zotz and Bader 2009). Some recently published books give overviews on current state of the art for bryophytes (Glime 2007) and lichens (Nash 2008). Literature comparing poikilohydric mechanisms and functioning of lichens and bryophytes is currently scarce (Beckett 1997; Green and Lange 1995); however, the field of study is becoming more and more attractive to a broad variety of scientists for the understanding and manipulation of drought and desiccation tolerance as well as in the context of land use change and global warming.

5.2 Mechanisms of Water Exchange in Lichens and Bryophytes Allowing Desiccation Tolerance

Which morphological and ecological mechanisms force water exchange and reduce desiccation? Homoiohydric vascular plants maintain more or less stabile water contents within the plant tissue by continuous soil water uptake via roots and by limitation of water loss by cuticulated surfaces and controlled stomata opening. In contrast, poikilohydric non-vascular pants lack continuous water supply as well as other features to actively control water deficit. Their water status varies passively with surrounding conditions and, consequentially, these cryptogams are subject to frequent desiccation. To activate assimilation, non-vascular plants are able to utilise a variety of different liquid water sources such as rain, dew, and fog. Even water vapour is highly relevant for green algal lichens (Büdel and Lange 1991; Lange et al. 1986, 1993b) and some bryophytes (Dilks and Proctor 1979; Lange 1969; Leon-Vargas et al. 2006). Nevertheless, most bryophytes (Lange 1969) and all cyanobacterial lichens require liquid water to become photosynthetically active (Bilger et al. 1989). The absorbed water resources support photosynthesis of cryptogams in a delicate equilibrium between water availability and CO_2 uptake. Many lichens and bryophytes are able to maintain photosynthesis despite wide changes in water contents (WC) such as 250–400% of dry weight (DW) in green algal lichens (Blum 1973; Rundel 1988), 600–2,000% DW in cyanolichens (Lange et al. 1993b), and as much as 2,500% DW in bryophytes. But above certain WCs, supersaturation by water can cause a limitation of photosynthesis due to high CO_2 diffusion resistance (Dilks and Proctor 1979; Lange et al. 1993a, 2001; Lange and Tenhunen 1981; Zotz et al. 1997). In contrast, carbon gain can profit from desiccation if suprasaturated WC decreases and disables the limitation of carbon fixation and if nocturnal dryness reduces respiration, and thus carbon loss, during night hours.

Due to the lack of real vascular tissues, the location and conduction of absorbed water by non-vascular plants can be divided into (1) symplast (osmotic) water within the cell, (2) apoplast water held in cell-wall capillary spaces, and (3) external capillary water (Beckett 1996, 1997; Dilks and Proctor 1979; Proctor et al. 1998; Proctor and Tuba 2002). In green algal lichens, external water constitutes around 22% of total water (Beckett 1997), but in many cyanobacterial lichens and bryophytes, external water is a considerable and variable component. For the exploitation and allocation of water as well as for the prevention of rapid desiccation during the daytime, various cryptogamic functional types developed different mechanisms and morphological features to increase boundary, surface, and internal resistances (e.g. Cowan et al. 1992; Green et al. 1994; Proctor and Tuba 2002). For example, this includes low surface-to-volume ratio, thick cortices, hydrophobic surfaces, tomentum, and hair points (Larson 1981). Also, water storage mechanisms support decelerated desiccation by thick hyphal cell walls of the medulla and gelatinous sheaths of photobiont cells (Harris 1976) as well as water sacs, concavities, and external capillary spaces for lichens and bryophytes, respectively. In bryophytes, some ecological life forms (Bates 1998; Mägdefrau 1982; Richards 1984) occur endohydric with mainly symplastic conduction (thalloid liverworts, many Polytrichaceae, and Mniaceae), while the majority are characterised as ectohydric, taking advantage of external capillary water (Buch 1945, 1947). Lichens built various life forms with crustose, foliose (leaf-like), and fruticose (shrub-like) morphologies. All lichens absorb external water by capillary diffusion, and fully water-saturated lichens may hold of their total WC about 22% extracellular, 33% apoplastic, and 44% symplastic (Beckett 1997).

The key parameter to characterising the cell water status during desiccation and rewetting, integrating physiological and morphological impacts, is the water potential or osmotic potential of the organism (Proctor et al. 1998). The determination of water potential as an estimation of water relation is studied by pressure–volume

(P–V) curves analysing the correlation of relative water content (RWC) and water potential driven by the hygroscopic force of the organism. The water potential helps to establish a suitable understanding of rehydration and dehydration processes for non-vascular plants. Knowing water content at full turgor is an essential value for physiological comparisons within non-vascular plants and between vascular and non-vascular plants, because a substantial portion (around 1/5–1/3) of the water present is extracellular (Beckett 1995, 1997; Proctor et al. 1998) and can be lost without affecting the water status and physiology of the cell. At full turgor, mosses and thalloid liverworts exhibit WC of around 200% DW and 1,000–2,000% DW corresponding with an osmotic potential of around −2 to −1 MPa and −0.4 to −0.9 MPa, respectively (Dilks and Proctor 1979; Patterson 1946; Proctor et al. 1998), while the lichen *Roccella hypomecha* exhibits around 80% DW at −2.1 to −2.6 MPa (Beckett 1997). Full turgor loss happens between 60 and 80% RWC for bryophytes (Proctor et al. 1998) and between 45 and 55% RWC for lichens (Beckett 1995, 1997), corresponding to a water potential around −70 to −30 MPa. However, full turgor is not necessarily essential for growth in free living fungi (Harold et al. 1995; Heintzeler 1939) and terrestrial green algae (Bertsch 1966a; Edlich 1936; Zeuch 1934). Moreover, in lichens even collapsed photobiont cells with negative turgor (−10 MPa water potential; at 92% RH) can exhibit 10–40% of maximal net photosynthesis (Scheidegger et al. 1995).

5.3 Processes at Intermittent Desiccation Between Activity and Inactivity

Which processes are induced during dehydration and rehydration? If vascular vegetative cells reach around 30–40% of their RWC (tissue water potential below about −7 MPa), the limit of drought tolerance is reached and cells start to unrecoverably die. This is a crucial distinction of vegetative cells of non-vascular plants. Their photosynthesis becomes inactive by desiccation at RWC around 15–20%. Moreover, they commonly recover after desiccating to RWC down to around 5–10% corresponding to water potential of −100 to −400 MPa (Proctor et al. 2007).

On the cell-physiological level, this difference is explainable. Vascular resurrection plants induce a cellular protection mechanism by accumulating abscisic acid (ABA) [e.g. (Bartels and Sunkar 2005)] triggering gene products, stress proteins, and osmotically active sugars that mediate cellular protection during desiccation (see Chap. 16; Bartels 2005). Desiccation tolerance in non-vascular plants in contrast involves a suite of interacting mechanisms not only protecting cellular integrity but also providing multifaceted machinery for recovery. Like vascular plants, bryophytes invest in a constitutive cellular protection by ABA (Beckett et al. 2000) and osmotically active substrates, but in addition they highly invest in a rehydration-induced recovery mechanism (Oliver et al. 2005) by accumulating cell protecting and stabilising substrates as well as rapid translation controlled responses [e.g. (Alpert and Oliver 2002; Oliver et al. 2005; Wood 2007)].

The main players in bryophytes are dehydrins (Chap. 14) and other members of this late embryogenesis abundant (LEA) protein group interacting with hydrophobic and hydrophilic cellular structures (Tunnacliffe and Wise 2007; Wood and Oliver 2004). In a recent proteomic study, the five functional categories of desiccation-responsive proteins for material and metabolism, defence, cytoskeleton, and signal transduction were up-regulated in the moss *Physcomitrella patens* (Wang et al. 2009), which supports earlier findings. Several reactions are buffering oxidative damages such as denaturation of proteins, pigment loss, photosystem damages, and lipid peroxidations (Weissman et al. 2005a, b). A complex array of enzymes and redox molecules including the glutathione and ascorbate systems act as antioxidants and support thermal energy dissipation (see also Chaps. 6 and 8). But for lichens, it is unclear if ascorbate is produced at all (Kranner et al. 2005), while the glutathione/ascorbate cycle analogue ascorbate-peroxidase (Mayaba and Beckett 2001) and ABA (Unal et al. 2008) have been detected. The accumulation of osmotically active sugars also plays an important role. For bryophytes, these are mainly sucrose (Buitink et al. 2002) reaching levels up to 40% of dry mass (Bianchi et al. 1991). Lichens induce polyols (e.g. ribitol, sorbitol, arabitol, and mainly mannitol), with contents between 2 and 10% of thallus dry weight, and polyamines (Rai 1988; Unal et al. 2008), all depending on species, season, and dehydration status (Farrar 1976; Lewis and Smith 1967; Unal et al. 2008). The osmotically active sugars reduce the degradation of macromolecules as well as stabilise membranes and cells. Nevertheless, non-vascular plants growing in moist micro-environments such as shaded ground or humid tropics are relatively desiccation sensitive (Green et al. 1991; Lakatos et al. 2006; Lange et al. 1994; Pardow et al. 2010; Zotz et al. 1998).

Desiccation patterns at relative humidities lower 80% vary widely, and the time to lose 50% of photosynthetic capacity ranges within hours for lichens and several days for bryophytes (Dilks and Proctor 1979). In contrast, reactivation by rehydration takes place within a range of minutes up to 2 h for lichens as well as for some highly DT bryophytes such as *Tortula ruralis*, but may take a little longer for mesic adapted bryophytes. Commonly, the reactivation process starts by activating metabolism in the sequence of protein synthesis (Cowan et al. 1979; Oliver 1991), respiration (at ~ -20 MPa water potential), and finally photosynthesis (at ~ -10 MPa water potential) (Dilks and Proctor 1979; Lange 1969). The recovery of the cytoskeleton, cell cycle, and nutrient transport requires around a day, increasing with the length and degree of desiccation. This metabolic reactivation during rehydration is an exciting and exceptional attribute exclusively performed by poikilohydric organisms. Moreover, maintaining photosynthesis at very low water contents is one of the most outstanding features of non-vascular plants. Green algal lichens and some bryophytes (Lange 1969) are able to reactivate photosynthesis even in equilibrium with water vapour. Since its pioneering discovery by Butin (1954), numerous laboratory and field studies lead to the consensus that water vapour above 85% RH can reactivate photosynthesis in green algal lichens (e.g. Büttner 1971; Lange and Bertsch 1965; Lange et al. 1986, 1989, 2001; Lange and Kappen 1972; Lange and Kilian 1985). But even at lower air humidity, lichens such as *Ramalina menziesii* (Nash et al. 1990), *Ramalina thrausta* (Bertsch 1966b),

Lepraria membranacea (Brock 1975), or *Dendrographa minor* (Nash et al. 1990) are able to reactivate net photosynthesis at 84, 83, 80, and 74% RH, respectively. To absorb water vapour or fog for example, a high surface-to-volume ratio or/and hydrophilic surfaces are advantageous. Highly branched fruticose (shrub-like) lichens (e.g. *Bryoria*, *Teloschistes*, *Ramalina*, and *Usnea*) as well as weft and long tail bryophytes (e.g. *Frullania* and *Orthostichopsis*) follow this strategy. Vapour equilibration experiments during desiccation at 60% RH and traced by ^{18}O demonstrated total isotope exchange of thallus water with that of vapour within only 2 h for the fruticose *Usnea filipendula* and within 3–4 h for other lichen growth forms (Hartard et al. 2009). This equilibrium reaction is very fast in comparison to epiphytic vascular plants such as the epiphyte *Tillandsia usneoides*, which needs days for total equilibrium and which only happens at RH over 95% (Helliker and Griffiths 2007). Due to this very fast equilibration, lichens may serve as prospective long-term proxies for different environmental water sources (Hartard et al. 2009; Lakatos et al. 2007). Whether or not bryophytes, which commonly equilibrate slower than lichens, can also be used as environmental water indicators, as shown in the case of paleo-CO_2 proxies (Fletcher et al. 2005), has not been addressed yet.

An actual and fascinating example for the interaction of cell-physiological and morphological features is the activity pattern of epiphytic bryophyte growth forms that are unusually high in species diversity at a newly proposed "Tropical Lowland *Cloud* Forest" (TLCF). The precipitation in tropical lowland rain forests is generally high (>2,000 mm), but the microclimate in the canopy of trees is usually considered xeric and leads to short periods of activity and frequent desiccation (e.g. Zotz and Winter 1994). As a result, the canopies of regular tropical lowland forests are commonly inhabited by desiccation adapted epiphytes such as Bromeliaceae, crustose lichens, as well as short-tailed and loose matt bryophytes. The canopies of a TLCF, in contrast, exhibit an unusual richness of macrolichen (Normann et al. 2010) and of bryophyte surpassing 3- and 1.5-fold (Cornelissen and Gradstein 1990; Gradstein 1995, 2006) that of tropical lowland forests in the Amazon (Suromoni) and humid submontane/montane rainforests of the tropical Andes (Acebey et al. 2003; Wolf 1993). Though it has been suggested that the daily occurrence of fog events in these forests may account for the high diversity of epiphytic bryophytes by exploitation of this water source and by decrease of desiccation events (Gradstein et al. 2010; Gradstein 2006). Indeed, photosynthetic studies of desiccated weft liverwort, long tail moss, short tail moss, and loose matt liverwort showed fastest reactivation by vapour and fog for weft and slowest for loose matt liverworts (Fig. 5.1). These reactivation patterns correlate with preferential distribution patterns from wetter to dryer micro-sites and decreasing surface-to-volume ratio (Pardow and Lakatos unpublished). To transfer the interrelation of cell physiology and morphology to a broader ecological perspective, fast reactivation by vapour or fog also includes morphological structures with high surface-to-volume ratio and low water-holding capacities and thus fast desiccation. The ecological distribution of these life forms is mostly combined with a specific micro-environment supporting photosynthetic activity during the morning hour (eventually also supported by dew), desiccation at noon, and reactivation during the late

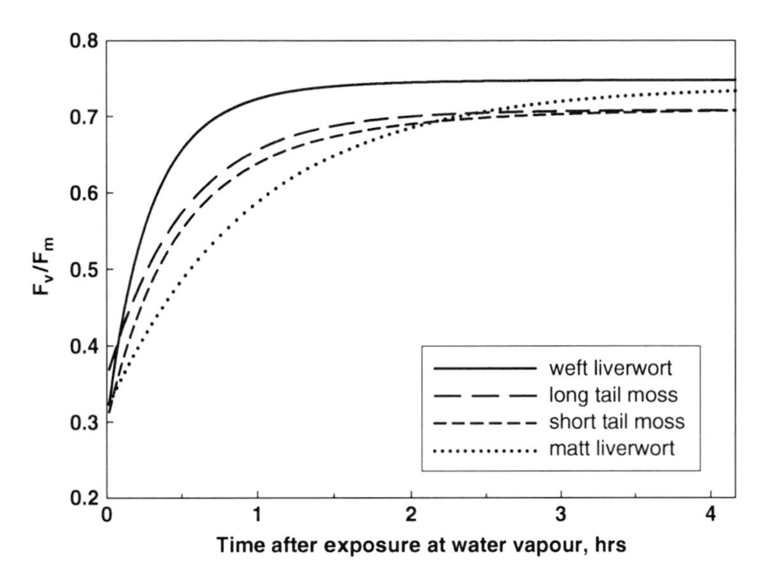

Fig. 5.1 Photosynthetic reactivation studies by chlorophyll a fluorescence measurements of desiccated weft liverwort (*solid line*), long tail moss (*long dashed line*), short tail moss (*short dashed line*), and loose matt liverwort (*dotted line*) after sunset in a putative tropical lowland cloud forest, French Guiana (relative humidity between 85 and 100%). The mean maximum quantum yield of photosystem II in the dark adapted state (F_v/F_m) revealed fastest reactivation by water vapour in situ for weft and slowest for loose matt liverworts (n = 3–5 per life form). These reactivation patterns correlate with preferential distribution patterns from wetter to dryer microsites and decreasing surface-to-volume ratio (Pardow and Lakatos unpublished)

afternoon as a response to frequent high humidity or fog. The strategy of retaining water commonly includes morphological structures that retard equilibrium with water vapour (thus combined with slow reactivation), but facilitate high water content capacities that decelerate desiccation. The ecological distribution of the latter is mostly combined with a micro-environment supporting frequent liquid water availability.

5.4 Functioning and Impacts of Non-vascular Plants at Microhabitats

5.4.1 Impacts of Non-vascular Plants at Microhabitats

What impact do plants with frequent desiccation–rewetting cycles have on the environment? Cryptogams are increasingly recognised for the important roles they play in the biogeochemical cycling (DeLucia et al. 2003; O'Connell et al. 2003a) and influencing vegetation–atmosphere exchanges in many ecosystems

(Lafleur and Rouse 1988; Shimoyama et al. 2004; Williams and Flanagan 1996). Recent studies have demonstrated the notable functioning of cryptogamic ground cover in water circulation, for both lichens (Hartard et al. 2008; Lakatos et al. 2007) and bryophytes (Douma et al. 2007; Flanagan et al. 1999). Non-vascular plants have impacts on hydrology (Pócs 1982; Pypker et al. 2006; Veneklaas et al. 1990) particularly in peatlands, but also in forest understories (e.g. Barbour et al. 2005). Despite these numerous impacts, non-vascular plants are less often considered in ecosystem processes (Cornelissen et al. 2007), although they obviously contribute. The difficulties seem to be the complex and strong interrelation of their adjacent local microclimate with some still not fully understood mechanisms, e.g. their micro-environmental water exchange and desiccation patterns. Therefore, the following will focus on some functional aspects of the interrelation between water availability, water exchange, and desiccation in cryptogamic cover on soils as well as on surfaces of plants.

5.4.2 Functioning of Non-vascular Soil Cover

Lichens and bryophytes are sometimes predominating components of ground cover vegetation. Their impact on the upper soil layer is hardly studied but shows significant influences on nutrients, moisture, and temperature. The knowledge about interaction between ground cover vegetation and soil is important especially in the context of land use changes (Pharo and Zartman 2007; Stofer et al. 2006), production of greenhouse gases (Bubier et al. 1995; Smith et al. 2004), and global warming (Bates et al. 2005; Bergamini et al. 2009; Dorrepaal et al. 2006; Frahm and Klaus 2001; Herk et al. 2002; Jägerbrand et al. 2009) since one of the main effects will be modification of evaporation processes, the latter being also one of the poorest studied (Beringer et al. 2001; Douma et al. 2007). For example, forest fragmentation and the disruption of soil vegetation cover push evaporation capacities of the pedo- and biosphere towards adverse water loss. Particularly in the context of global change, fewer but heavier rain events, long-lasting dry periods, and stronger storm events are predicted in Europe and North America (IPCC 2007), causing an adverse impact on water availability of soils if evaporation reducing vegetation is lacking. Thereby, germination, productivity, and biodiversity are negatively influenced. During the last 3 decades, studies have shown the important role of cryptogamic soil cover. Besides their effect on nutrient cycling (Turetsky 2003), mineralization due to leakages (Rieley et al. 1979; Vitt 2000), and soil stability (Belnap and Lange 2001; Evans and Johansen 1999), non-vascular plants seem to have high influences on thermal and hydrological as well as on carbon regimes of soils. In particular, their growth forms (e.g. cushions) generate diffusion resistances, which has most probably decisive functional effects on carbon and water cycles as a link between pedo-, bio-, and atmosphere (Betts 1999). First, because non-vascular plants exist at the interface of atmospheric boundary layer and soil, their function is also tightly coupled to variation in CO_2 partial pressure. Cryptogamic soil cover on the forest floor experiences elevated levels of CO_2 partial pressure that likely enhances photosynthesis. The CO_2 partial pressure immediately below leaves can vary from 48.6 Pa (DeLucia et al. 2003) up to 65 Pa (Tarnawski et al. 1994) for bryophytes on the floor of temperate rainforests, suggesting that the bryophyte community obtains most of its carbon from CO_2 derived from soil respiration. Observations within similar CO_2 partial pressures were made for terricolous lichen cushions, and the recycling of soil-respired $^{13}CO_2$ was supported by $\delta^{13}C$ composition of lichen bulk material (Lakatos et al. 2007, 2009). Second, the function of non-vascular soil cover is also tightly coupled with water content near the forest floor. The functioning of cryptogams as evaporation resistance and utilisers of vapour generates a surplus of water availability for soils (Betts 1999; Flanagan et al. 1999; Hartard et al. 2008). The depth to ground water affects soil respiration rates but also water content of the cryptogamic layer, greatly influencing their photosynthetic capacity (Johnson et al. 2005; Oechel et al. 1998). In Mediterranean forests, lichen cushions trap dew, fog, and water vapour leading to a permanent high humidity and buffered temperatures within the cushions (Hartard et al. 2008; Lakatos et al. 2007). This humidity and temperature buffer process reduces soil evaporation and increases soil water contents below these cushions. Interestingly, stable isotope studies on $\delta^{18}O$ of water indicate that these lichens profit more from the absorption of water vapour than from soil water uptake during the absence of rain (Hartard et al. 2008). Hence, the coupling of terricolous non-vascular plant cushions as interface between atmosphere and upper forest soil layer displays a complex soil-cushion-continuum with many interacting processes, e.g. recycling of soil-respired CO_2 by photosynthesis and utilisation of water vapour. This may greatly modify carbon storage, moisture content, and organic biomass of soils.

The desiccation process of non-vascular plants also has an impact on soil biodiversity. Lichens and bryophytes may influence biodiversity of associated soil organisms such as microbes, mycorrhiza, and fine roots by providing microhabitats with stable buffered temperature and higher soil moisture availability (Beringer et al. 2001). The latter, for example, exerts significant stimulations on mycorrhizal growth rates and branching density of very fine roots (Feil et al. 1988; Kottke and Agerer 1983). Besides their beneficial effect on mycorrhizal fungal growth due to carbohydrate leakage (Carleton and Read 1991), bryophytes and lichens also have antimicrobial properties. For example, homogenates of bryophytes have been shown to inhibit the decay of vascular plants (Sedia and Ehrenfeld 2006; Verhoeven and Toth 1995). Vascular germination is reduced by secondary metabolites of lichens (Bonan 1989; Cowles 1982; Sedia and Ehrenfeld 2005) and bryophytes (Basile et al. 2003), isolated flavanoids of mosses have antibacterial effects (Basile et al. 1999), and many bryophytes and lichens showed antibiotic activities (Banerjee and Sen 1979; Burkholder et al. 1944; Elo et al. 2007). Moreover, bryophyte tissue strongly chelates metals that are essential and may be limiting for microbes (Basiliko and Yavitt 2001). Thus, soil cover by lichens and bryophytes should enhance mycorrhiza and fine root biomass, but whether they are detrimental to that of microbes, in the context of beneficial carbohydrate leakage and antimicrobial effects, remains unclear. Hence, further knowledge regarding the impact of cryptogams on soil properties as well as water and carbon fluxes is in high demand by ecologists and atmosphere scientists.

5.4.3 *Functioning of Non-vascular Epiphytic Cover*

In tropical forests, non-vascular plants exhibit pronounced altitudinal distribution patterns in terms of diversity and biomass (e.g. Montfoort and Ek 1990; Sipman 1989; Sipman and Tan 1990; Wolf 1995). Species richness and biomass of lichens and bryophytes increase with altitude, reaching maximum values in tropical montane cloud forests (Fig. 5.2). But also in other ecosystems such as the boreal, temperate, and coastal (rain)forests, their biomass and impact are noteworthy. The proportion of non-vascular epiphytes in comparison to vascular species is around 75% for tropical and 60–90% for temperate rain forests (Affeld et al. 2008). Due to their poikilohydric nature, non-vascular plants are highly interactive with their surrounding abiotic conditions. These interactions are embedded in a complex system of hydrologic interrelations. Water, in the form of rain or fog, is intercepted by the entire tree canopy (made up of foliage, stem, branches, and epiphyte vegetation) and slowly drops or flows down the branches and stem (stem flow and throughfall water), where it is absorbed, retained, and evaporated by the epiphytes. The latter process can represent a substantially large amount. For example, in boreal ecosystems, epiphytic bryophytes were shown to intercept 15–60% of precipitation (compared to 23% of throughfall interception by terricolous bryophytes), most of which subsequently evaporates during desiccation (Price et al. 1997). In tropical ecosystems, canopy epiphytes intercepted 273 and 724 mm yr^{-1} in a submontane and in a cloud forest of Tanzania representing 10 and 18% of annual precipitation, respectively (Pocs 1980, 1982), whereas in a submontane rain forest of Uganda even 34% interception rates were recorded (Hopkins 1960). Most of these studies only included rainfall and did not account for cloud water deposition as water resource. The non-precipitating droplets of cloud water can deposit onto vegetation and can contribute from 2 to 61% of the total water at the central cordillera of Panamá (Cavelier et al. 1996) and up to 10 to 93% in elfin cloud forests of Venezuela and Colombia (Cavelier and Goldstein 1989). For pendant mosses and green algal lichens, interception of the equivalent of ca. 0.5 mm of cloud water droplets may be sufficient to recharge their water-holding capacity (Leon-Vargas et al. 2006).

In the context of water retention, cryptogams are not only able to exploit various water sources, but also release this water into the canopy in frequent cycles of wetting and drying. Generally, through evaporation, epiphytes contribute to maintaining high humidity within the canopy and in the understorey long after atmospheric inputs have stopped (e.g. Perry 1984; Veneklaas et al. 1990). It has been shown that evaporative drying was decreased by up to 20% at micro-sites in close proximity to vascular epiphytes compared to non-inhabited micro-sites (Stuntz et al. 2002). However, the contribution of non-vascular epiphytes in forests remains unknown. Considering their large water-holding capacities, cryptogams might be important in maintaining moist conditions within the canopy zone if their biomass is high. Cryptogams are capable of holding water as a community up to 500% of their dry weight, and as individuals, bryophyte or cyanobacterial lichen can even hold up to 1,400 and 2,000%, respectively (e.g. Pocs 1982). This provides potential for evaporative water vapour sources. Epiphytic bryophytes in a Costa Rican montane cloud forest were shown to evaporate up to 250% of their dry weight biomass in 3 days, exceeding evaporation of canopy humus by more than twofold (Köhler et al. 2007). The unique ability of cryptogams to take up and exploit water vapour underlines the importance of water vapour fluxes. However, fluxes of water vapour are among the least well understood of ecosystem processes (Worden et al. 2007). Moreover, dew deposition, humidity, and fog are important water sources in other biomes as well. In deserts, lichens achieve three times more positive cumulative carbon gain after rehydration from fog in comparison to dew (Lange et al. 1990), but dew formation alone may result in 53–63% of integrated carbon gain in other lichens (Lange et al. 2007). Estimation of the source of thallus hydration at coastal Mediterranean forests revealed that dew deposition was responsible for 15% and atmospheric water vapour uptake for 11% of calculated average annual gross CO_2 fixation (Matthes-Sears and Nash 1986).

5.5 Global Patterns of Desiccation-Tolerant Lichens and Bryophytes

5.5.1 Global Patterns as an Indication for the Ecological Relevance

How Successful Is the Desiccation Strategy of Non-vascular Plants in a Global Context? Approximately 8–30% of the terrestrial ecosystems are dominated or significantly affected by non-vascular plants (Larson 1987). In some ecosystems, they represent a substantial proportion of primary producers (Hofstede et al. 1993), affect nutrient fluxes (Clark et al. 2005; Forman 1975; Knops et al. 1996), and influence animal life (Pettersson et al. 1995; Richardson et al. 2000; van der Wal 2006). In the context of global change, this desiccation-tolerant plant group may profit by expanding its distribution due to desertification, while circum-polar warming might increase (Robinson et al. 1998; Rixen and Mulder 2005; Gordon et al. 2001; Washley et al. 2006) or decrease cryptogamic cover (e.g. Cornelissen et al. 2001; Walker et al. 2006; Wijk et al. 2004). Also, altitudinal range shifts of non-vascular plants are attributed to climate warming trends (Bergamini et al. 2009; Nadkarni and Solano 2002; Zotz and Bader 2009). Since non-vascular plants are poikilohydric organisms, their dominance correlates less with macroclimatic pattern (Lechowicz 1982) as we know from vascular plants regarding for example annual precipitation and diversity, leaf area index, and net primary production (NPP). A comparison of annual precipitation rates and around 58 studies on non-vascular biomass demonstrates no correlation (Fig. 5.2) – bearing the different study designs in mind – while a general trend exists that bryophyte distribution benefits from wetter habitats (Coley et al. 1993; Eldridge and Tozer 1997; Ellis and Coppins 2006; Frahm 2003). Thus, patterns of non-vascular plants depend rather on complex interactions with immediate microclimatic conditions (Green et al. 2007) and on alternative strategies

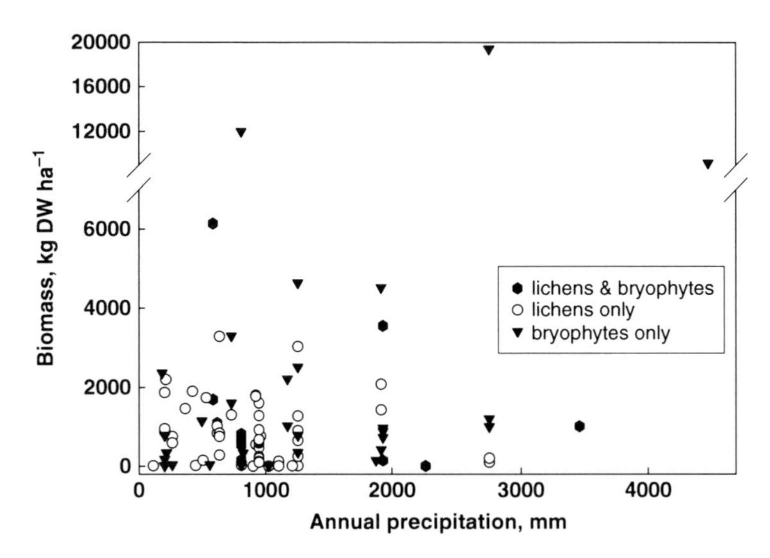

Fig. 5.2 Survey on biomass of epiphytic and terricol lichens (*open circles*) and bryophytes (*triangles*) modified from 58 publications and correlated with extracted annual precipitation from nearest weather station at the respective study site

to exploit different water sources. Due to the variable and complex effects of microclimate on cryptogamic physiology, up-scaling to ecosystem and global levels remains challenging. Recent extrapolations assume that the missing CO_2 sink in the global carbon cycle may partly be formed by arid cryptogams (Stone 2008; Wilske et al. 2009; Wohlfahrt et al. 2008), and all cryptogams worldwide produce around 6% of the estimated global net carbon uptake by terrestrial vegetation (Elbert et al. 2009). Yet, fully reliable estimates are not available, and the knowledge about general and specific impacts of cryptogamic ecosystem components is urgently demanded by ecologists and atmosphere scientists.

5.5.2 Impacts of Lichens and Bryophytes on the Carbon Cycle

Bryophytes, even though they may play a less obvious role in total global biomass, dominate the carbon and nitrogen cycling of many poorly drained terrestrial ecosystems (Turetsky 2003; Vitt 2000; Vitt et al. 2001), particularly in the boreal region, but also on a global scale (Longton 1992; O'Neill 2000). Such ecosystems, for example peatlands and to a lesser degree forested wetlands, cover at least 2.5×10^8 ha globally (Gorham 1991; Lugo et al. 1990; Matthews and Fung 1987), store approximately 455 Pg C, and sequester nearly a third of global soil C (Gorham 1991). Boreal forests, especially poorly drained areas, exhibit high rates of primary production (Camill et al. 2001; Vitt et al. 2001), slow decomposition rates, and bryophyte- and lichen-dominated successional pathways (O'Neill 2000; Turetsky 2003). However, non-vascular plants are increasingly recognised as playing notable roles in the biogeochemical cycling of well-drained boreal forests (DeLucia et al. 2003; O'Connell et al. 2003a, b) and as influencing vegetation–atmosphere exchanges in many ecosystems (Lafleur and Rouse 1988; Shimoyama et al. 2004; Williams and Flanagan 1996). The influence of bryophytes on hydrological (Barbour et al. 2005) and carbon regime seems to be relevant, but was largely overlooked until now. Depending on the studied ecosystem, NPP of bryophytes ranges from 22 in boreal forests (reviewed in Bisbee et al. 2001; Swanson and Flanagan 2001) to 350 g C m^{-2} yr^{-1} in the Antarctic (Fenton 1980). Rough estimations of modelled carbon capture of total forest floor CO_2 efflux by bryophytes range from approximately 10% in temperate rainforests (DeLucia et al. 2003) to 36% in an old black spruce forest (Swanson and Flanagan 2001). In arctic ecosystems, the moss contribution to ecosystem carbon uptake varied between 14 and even 96% (Douma et al. 2007). Approximations for lichen carbon capture are rare up to now but are often indicated as smaller compared with that of vascular plants and bryophytes (Bonan 1989; Elbert et al. 2009; Uchida et al. 2006).

5.5.3 *Impacts of Lichens and Bryophytes on the Hydrological Cycle*

Accurate evaporation estimates are required for a variety of climatological, hydrological, and ecological problems, including assessing the effects of changes in vegetation and climate on water loss and basin yield from catchments. These estimates are often calculated by the Penman–Monteith combination model specifying surface and aerodynamic resistances. For example, micrometeorological techniques, such as eddy covariance, enable ecosystem measurements of evaporation ("top down" approach), while whole-tree transpiration is commonly measured using sap flow sensors ("bottom-up" approach). Intensive studies on canopy transpiration, a component of the ecosystem water balance, have been conducted in a number of subtropical (Hutley et al. 1997), temperate (Berbigier et al. 1996; Granier et al. 2000; Köstner et al. 1992; Oren et al. 1998), and boreal forests (Cienciala et al. 1998; Kelliher et al. 1998; Zimmermann et al. 2000), but relatively little work has been done in low-productivity ecosystems such as circum-polar regions and old-growth forests (Barbour et al. 2005; Douma et al. 2007; Heijmans et al. 2004a). Yet, low-productivity forests reveal leaf area indices below 3 m^2 m^{-2}, and thus transpiration from understorey species and evaporation from the forest floor and wet surfaces become much larger components of total ecosystem evaporation (Heijmans et al. 2004b; Unsworth et al. 2004). Even though bryophytes, lichens, and standing dead wood are a feature of old-growth forests, their contribution to total evaporation has seldom been quantified (Unsworth et al. 2004). In a temperate coniferous rainforest, evaporation from the forest floor with a thick, nearly continuous moss layer and a high water table contributed up to 25% of ecosystem evaporation (Barbour et al. 2005), and values of 15–45% were estimated for boreal forests (Heijmans et al. 2004a). Nevertheless, these few case studies demonstrate that the impact of lichens and bryophytes on the hydrological cycle on whole ecosystems is challenging to quantify because of the complex interaction of precipitation, biomass, evaporation, and microclimate.

5.6 Conclusion

Land use change and climate change were identified as the two main drivers strongly affecting global biodiversity in this century (Sala et al. 2000). Two abiotic factors are assumed to change most dramatically: temperature and moisture – both of which intensify desiccation. Land use change increases fragmentation of landscapes. Particularly, forest clearances will change vegetation structure, local microclimate, and forest dynamics. For example, moist air is pulled out of the forests into adjacent clearings (Laurance et al. 2002), and moisture inputs by cloud formations are reduced in adjacent mountains (Lawton et al. 2001). Changes in precipitation are uncertain and are difficult to generalise (Sala et al. 2000), but moisture is likely to affect plant life by changes in precipitation regimes (Solomon et al. 2007), more frequent climatic extremes (Timmermann et al. 1999), and reduced input by cloud water in tropical montane forests (Still et al. 1999). Thus, hygrophilic cryptogams will seriously suffer from land use change and decreased moisture input for example in humid tropical forests representing a hot spot of cryptogamic diversity (Nadkarni and Solano 2002; Pardow et al. 2010; Zotz and Bader 2009). In contrast, biological soil crusts could be harmed by increase in precipitation (Belnap et al. 2004, 2008). These scenarios will form new assemblage of species, functional traits, and interactions, as already observed in Europe (Aptroot and van Herk 2007; Stofer et al. 2006).

Molecular biologists and medical scientists are realising the potential of desiccation-tolerant cryptogams in understanding gene function and cell protection. Ecosystem biologists are realising that non-vascular plants may play a major role in carbon, nutrient, and water cycling. Global climate modellers have become conscious that massive peatlands make substantial contributions to the modification of global greenhouse gases, temperatures, and water movement. Therefore, the scientific community is becoming increasingly aware that passively depending on environmental water availability is not simply a sign of primitiveness; rather it is an alternative strategy allowing non-vascular plants to successfully compete with vascular plants in many macro- and microhabitats.

References

Acebey A, Gradstein S, Krömer T (2003) Species richness and habitat diversification of bryophytes in submontane rain forest and fallows of Bolivia. J Trop Ecol 19:9–18

Affeld K, Sullivan J, Worner SP, Didham RK (2008) Can spatial variation in epiphyte diversity and community structure be predicted from sampling vascular epiphytes alone? J Biogeogr 35:2274–2288

Alpert P (2005) The limits and frontiers of desiccation-tolerant life. Integr Comp Biol 45:685–695

Alpert P, Oliver M (2002) Drying without dying. In: Black M, Pritchard H (eds) Desiccation and survival in plants: drying without dying. CABI, Wallingford, pp 3–43
Aptroot A, van Herk CM (2007) Further evidence of the effects of global warming on lichens, particularly those with *Trentepohlia* phycobionts. Environ Pollut 146:293–298
Banerjee R, Sen S (1979) Antibiotic activity of bryophytes. Bryologist 82:141–153
Barbour MM, Hunt JE, Walcroft AS, Rogers GND, McSeveny TM, Whitehead D (2005) Components of ecosystem evaporation in a temperate coniferous rainforest, with canopy transpiration scaled using sapwood density. New Phytol 165:549–558
Bartels D (2005) Desiccation tolerance studied in the resurrection plant *Craterostigma plantagineum*. Integr Comp Biol 45:696–701
Bartels D, Sunkar R (2005) Drought and salt tolerance in plants. Crit Rev Plant Sci 24:23–58
Basile A, Giordano S, López-Sáez J, Cobianchi R (1999) Antibacterial activity of pure flavonoids isolated from mosses. Phytochemistry 52:1479–1482
Basile A, Sorbo S, López-Sáez J, Castaldo Cobianchi R (2003) Effects of seven pure flavonoids from mosses on germination and growth of *Tortula muralis* HEDW. (Bryophyta) and *Raphanus sativus* L. (Magnoliophyta). Phytochemistry 62:1145–1151
Basiliko N, Yavitt J (2001) Influence of Ni, Co, Fe, and Na additions on methane production in Sphagnum-dominated Northern American peatlands. Biogeochemistry 52:133–153
Bates JW (1998) Is 'life-form' a useful concept in bryophyte ecology? Oikos 82:223–237
Bates JW, Thompson K, Grime JP (2005) Effects of simulated long-term climatic change on the bryophytes of a limestone grassland community. Glob Change Biol 11:757–769
Beckett RP (1995) Some aspects of the water relations of lichens from habitats of contrasting water status studied using thermocouple psychrometry. Ann Bot 76:211–217
Beckett RP (1996) Some aspects of the water relations of the coastal lichen *Xanthoria parietina* (L) TH FR. Acta Physiol Plant 18:229–234
Beckett RP (1997) Pressure-Volume analysis of a range of poikilohydric plants implies the existence of negative turgor in vegetative cells. Ann Bot 79:145–152
Beckett RP, Csintalan Z, Tuba Z (2000) ABA treatment increases both the desiccation tolerance of photosynthesis, and nonphotochemical quenching in the moss *Atrichum undulatum*. Plant Ecol 151:65–71
Belnap J, Lange O (2001) Biological soil crusts: structure, function, and management. Springer, Berlin
Belnap J, Phillips SL, Miller ME (2004) Response of desert biological soil crusts to alterations in precipitation frequency. Oecologia 141:306–316
Belnap J, Phillips SL, Flint S, Money J, Caldwell M (2008) Global change and biological soil crusts: effects of ultraviolet augmentation under altered precipitation regimes and nitrogen additions. Glob Change Biol 14:670–686
Berbigier P, Bonnefond JM, Loustau D, Ferreira MI, David JS, Pereira JS (1996) Transpiration of a 64-year old maritime pine stand in Portugal. Oecologia 107:43–52
Bergamini A, Ungricht S, Hofmann H (2009) An elevational shift of cryophilous bryophytes in the last century – an effect of climate warming? Divers Distrib 15:871–879
Beringer J, Lynch AH, Chapin FS, Mack M, Bonan GB (2001) The representation of arctic soils in the land surface model: the importance of mosses. Journal of Climate 14:3324–3335
Bertsch A (1966a) CO_2-Gaswechsel und Wasserhaushalt der aerophilen Grünalge *Apatococcus lobatus*. Planta 70:46–72
Bertsch A (1966b) Über den CO_2-Gaswechsel einiger Flechten nach Wasserdampfaufnahme. Planta 68:157–166
Betts A (1999) Controls on evaporation in a boreal spruce forest. J Climate 12:1601–1618
Bianchi G, Gamba A, Murelli C, Salamini F, Bartels D (1991) Novel carbohydrate metabolism in the resurrection plant *Craterostigma plantagineum*. Plant J 1:355–359
Bilger W, Rimke S, Schreiber U, Lange OL (1989) Inhibition of energy-transfer to photosystem II in lichens by dehydration: different properties of reversibility with green and blue-green phycobionts. J Plant Physiol 134:261–268

Bisbee K, Gower S, Norman J, Nordheim E (2001) Environmental controls on ground cover species composition and productivity in a boreal black spruce forest. Oecologia 129:261–270
Blum OB (1973) Water relation. In: Ahmadjian V, Hale ME (eds) The lichens, vol 2. Academic, New York, pp 381–400
Bonan G (1989) Environmental factors and ecological processes controlling vegetation patterns in boreal forests. Landscape Ecol 3:111–130
Brock TD (1975) Effect of water potential on a *Microcoleus* from a desert crust. J Phycol 11:316–320
Bubier J, Moore T, Juggins S (1995) Predicting methane emission from bryophyte distribution in northern Canadian peatlands. Ecology 76:677–693
Buch H (1945) Ueber die Wasser- und Mineralstoffversorgung der Moose I. Commentationes Biologicae 16:1–44
Buch H (1947) Über die Wasser- und Mineralstoffversorgung der Moose II. Commentationes Biologicae 20:1–61
Büdel B, Lange OL (1991) Water status of green and blue-green phycobionts in lichen thalli after hydration by water vapor uptake: do they become turgid? Bot Acta 104:361–366
Buitink J, Hoekstra F, Leprince O, Black M (2002) Biochemistry and biophysics of tolerance systems. In: Black M, Pritchard H (eds) Desiccation and survival in plants: drying without dying. CABI, Wallingford, pp 293–318
Burkholder PR, Evans AW, McVeigh I, Thornton HK (1944) Antibiotic activity of lichens. Proc Natl Acad Sci USA 30:250–255
Butin H (1954) Physiologisch-ökologische Untersuchungen über den Wasserhaushalt und die Photosynthese von Flechten. Biologisches Zentralblatt 73:459–502
Büttner R (1971) Untersuchungen zur Ökologie und Physiologie des Gasstoffwechsels bei einigen Strauchflechten. Flora 160:72–99
Camill P, Lynch JA, Clark JS, Adams JB, Jordan B (2001) Changes in biomass, aboveground net primary production, and peat accumulation following permafrost thaw in the boreal peatlands of Manitoba, Canada. Ecosystems 4:461–478
Carleton T, Read D (1991) Ectomycorrhizas and nutrient transfer in conifer- feather moss ecosystems. Can J Bot 69:778–785
Cavelier J, Goldstein G (1989) Mist and fog interception in elfin cloud forests in Colombia and Venezuela. J Trop Ecol 5:309–322
Cavelier J, Solis D, Jaramillo MA (1996) Fog interception in montane forest across the Central Cordillera of Panama. J Trop Ecol 12:357–369
Cienciala E, Running S, Lindroth A, Grelle A, Ryan M (1998) Analysis of carbon and water fluxes from the NOPEX boreal forest: comparison of measurements with FOREST-BGC simulations. J Hydrol 212:62–78
Clark KL, Nadkarni NM, Gholz HL (2005) Retention of inorganic nitrogen by epiphytic bryophytes in a tropical montane forest. Biotropica 37:328–336
Coley PD, Kursar TA, Machado J-L (1993) Colonization of tropical rain forest leaves by epiphylls: effects of site and host plant leaf lifetime. Ecology 74:619–623
Cornelissen JHC, Gradstein M (1990) On the occurence of bryophytes and macrolichens in different lowland rain forest types at Mabura Hill, Guyana. Trop Bryol 3:29–35
Cornelissen J et al (2001) Global change and Arctic ecosystems: is lichen decline a function of increases in vascular plant biomass. J Ecol 89:984–994
Cornelissen JHC, Lang SI, Soudzilovskaia NA, During HJ (2007) Comparative cryptogam ecology: a review of bryophyte and lichen traits that drive biogeochemistry. Ann Bot 99:987–1001
Cowan DA, Green TGA, Wilson AT (1979) Lichen metabolism. 1. The use of tritium labelled water in studies of anhydrobiotic metabolism in *Ramalina celastri* and *Peltigera polydactyla*. New Phytol 82:489–503
Cowan IR, Lange OL, Green TGA (1992) Carbon-dioxide exchange in lichens: determination of transport and carboxylation characteristics. Planta 187:282–294

Cowles S (1982) Preliminary results investigating the effect of lichen ground cover on the growth of black spruce. Naturaliste Canadien (Canada) 109:573–581

Crum H (2001) Structural diversity of bryophytes. University of Michigan, Herbarium

DeLucia EH et al (2003) The contribution of bryophytes to the carbon exchange for a temperate rainforest. Glob Change Biol 9:1158–1170

Dilks TJK, Proctor MCF (1979) Photosynthesis, respiration and water content in bryophytes. New Phytol 82:97–114

Dorrepaal E, Aerts R, Cornelissen JHC, van Logtestijn RSP, Callaghan TV (2006) Sphagnum modifies climate-change impacts on subarctic vascular bog plants. Funct Ecol 20:31–41

Douma JC, Van Wijk MT, Lang SI, Shaver GR (2007) The contribution of mosses to the carbon and water exchange of arctic ecosystems: quantification and relationships with system properties. Plant Cell Environ 30:1205–1215

Edlich F (1936) Einwirkung von Temperatur und Wasser auf aerophile Algen. Arch Microbiol 7:62–109

Elbert W, Weber B, Büdel B, Andreae MO, Pöschl U (2009) Microbiotic crusts on soil, rock and plants: neglected major players in the global cycles of carbon and nitrogen? Biogeosci Discuss 6:6983–7015

Eldridge DJ, Tozer ME (1997) Environmental factors relating to the distribution of terricolous bryophytes and lichens in semi-arid Eastern Australia. Bryologist 100:28–39

Ellis CJ, Coppins BJ (2006) Contrasting functional traits maintain lichen epiphyte diversity in response to climate and autogenic succession. J Biogeogr 33:1643–1656

Elo H, Matikainen J, Pelttari E (2007) Potent activity of the lichen antibiotic (+)-usnic acid against clinical isolates of vancomycin-resistant enterococci and methicillin-resistant Staphylococcus aureus. Naturwissenschaften 94:465–468

Evans RD, Johansen JR (1999) Microbiotic crusts and ecosystem processes. Crit Rev Plant Sci 18:183–225

Farrar JF (1976) Ecological physiology of the lichen Hypogymnia physodes. II. Effects of wetting and drying cycles and the concept of physiological buffering. New Phytol 77:105–113

Feil W, Kottke I, Oberwinkler F (1988) The effect of drought on mycorrhizal production and very fine root system development of Norway spruce under natural and experimental conditions. Plant Soil 108:221–231

Fenton J (1980) The rate of peat accumulation in Antarctic moss banks. J Ecol 68:211–228

Feuerer T, Hawksworth D (2007) Biodiversity of lichens, including a world-wide analysis of checklist data based on Takhtajan's floristic regions. Biodivers Conserv 16:85–98

Flanagan LB, Kubien DS, Ehleringer JR (1999) Spatial and temporal variation in the carbon and oxygen stable isotope ratio of respired CO_2 in a boreal forest ecosystem. Tellus B Chem Phys Meteorol 51:367–384

Fletcher BJ, Beerling DJ, Brentnall SJ, Royer DL (2005) Fossil bryophytes as recorders of ancient CO_2 levels: experimental evidence and a cretaceous case study. Global Biogeochemical Cycles 19:GB3012, Artn Gb3012

Forman RTT (1975) Canopy lichens with blue-green algae: a nitrogen source in a Colombian rain forest. Ecology 56:1176–1184

Frahm JP (2003) Climatic habitat differences of epiphytic lichens and bryophytes. Cryptogamic Bot 24:3–14

Frahm J, Klaus D (2001) Bryophytes as indicators of recent climate fluctuations in Central Europe. Lindbergia 26:97–104

Glime JM (2007) Bryophyte ecology. In: Physiological Ecology, vol. 1. sponsored by Michigan Technological University and the International Association of Bryologists

Goffinet B (2000) Origin and phylogenetic relationships of bryophytes. In: Shaw AJ, Goffinet B (eds) Bryophyte biology. Cambridge University Press, Cambridge, pp 124–149

Gordon C, Wynn JM, Woodin SJ (2001) Impacts of increased nitrogen supply on high arctic heath: the importance of bryophytes and phosphorus availability. New Phytologist 149

Gorham E (1991) Northern peatlands: role in the carbon cycle and probable responses to climatic warming. Ecol Appl 1:182–195
Gradstein S (1995) Bryophyte diversity of the tropical rainforest. Archives des sciences [Société de physique et d'histoire naturelle de Genève] 48:91–96
Gradstein SR (2006) The lowland cloud forest of French Guiana – a liverwort hotspot. Cryptogamie Bryologie 27:141–152
Gradstein S, Churchill S, Salazar-Allen N (2001) Guide to the bryophytes of tropical America. New York Botanical Garden Press, Bronx, NY
Gradstein S, Obregon A, Gehrig C, Bendix J (2010) Tropical lowland cloud forest – a neglected forest type. In: Bruijnzeel S, Scatena FN, Hamilton LS, Juvik J (eds) Mountains in the mist. Cambridge University Press, Cambridge
Granier A, Biron P, Lemoine D (2000) Water balance, transpiration and canopy conductance in two beech stands. Agric For Meteorol 100:291–308
Green TGA, Lange OL (1995) Photosynthesis in poikilohydric plants: a comparison of lichens and bryophytes. In: Schulz ED, Caldwell MM (eds) Ecophysiology of photosynthesis, vol chapter 16. Springer, Berlin, pp 319–341
Green TGA, Kilian E, Lange OL (1991) *Pseudocyphellaria dissimilis*: a desiccation-sensitive, highly shade-adapted lichen from New Zealand. Oecologa 85:498–503
Green TGA, Lange OL, Cowan IR (1994) Ecophysiology of lichen photosynthesis: the role of water status and thallus diffusion resistances. Cryptogamic Bot 4:166–178
Green T, Schroeter B, Sancho L (2007) Plant life in Antarctica. In: Pugnaire FI, Valladares F (eds) Handbook of functional plant ecology, 2nd edn. Marcel Dekker, New York, pp 389–433
Hale ME (1983) The biology of lichens, 3rd edn. Edward Arnold, London
Harold F, Harold R, Money N (1995) What forces drive cell wall expansion? Can J Bot 73:379–383
Harris GP (1976) Water and plant life: problems and modern approaches. Springer, Berlin
Hartard B, Máguas C, Lakatos M (2008) Delta O-18 characteristics of lichens and their effects on evaporative processes of the subjacent soil. Isot Environ Health Stud 44:111–125
Hartard B, Cuntz M, Máguas C, Lakatos M (2009) Water isotopes in desiccating lichens. Planta 231:179–193
Hawksworth DL, Hill DJ (1984) The lichen forming fungi. Blackie, Glasgow
Heijmans M, Arp WJ, Chapin FS III (2004a) Controls on moss evaporation in a boreal black spruce forest. Global Biogeochem Cycles 18:GB2004. doi:10.1029/2003GB002128
Heijmans M, Arp WJ, Chapin FS III (2004b) Carbon dioxide and water vapour exchange from understory species in boreal forest. Agric For Meteorol 123:135–147
Heintzeler I (1939) Das Wachstum der Schimmelpilze in Abhängigkeit von den Hydraturverhältnissen unter verschiedenen Außenbedingungen. Arch Microbiol 10:92–132
Helliker BR, Griffiths H (2007) Toward a plant-based proxy for the isotope ratio of atmospheric water vapor. Glob Change Biol 13:723–733
Herk CMv, Aptroot A, Dobben HFv (2002) Long-term monitoring in the Netherlands suggests that lichens respond to global warming. Lichenologist 34:141–154
Hofstede R, Wolf J, Benzing D (1993) Epiphytic biomass and nutrient status of a Colombian upper montane rain forest. Selbyana 14:37–45
Honegger R (1998) The lichen symbiosis – what is so spectacular about it? Lichenologist 30:193–212
Honegger R (2006) Water relations in lichens. In: Gadd G, Watkinson S, Dyer P (eds) Fungi in the environment. Cambridge University Press, Cambridge, pp 185–200
Hopkins B (1960) Observations on rainfall interception by a tropical forest in Uganda. East Afr Agric Forest J 25:255–258
Hutley LB, Doiey D, Yates DJ, Boonsaner A (1997) Water balance of an Australian subtropical rainforest at altitude: the ecological and physiological significance of intercepted cloud and fog. Aust J Bot 45:311–329

Jägerbrand A, Alatalo J, Chrimes D, Molau U (2009) Plant community responses to 5 years of simulated climate change in meadow and heath ecosystems at a subarctic-alpine site. Oecologia 161:601–610
Johnson SL, Budinoff CR, Belnap J, Garcia-Pichel F (2005) Relevance of ammonium oxidation within biological soil crust communities. Environ Microbiol 7:1–12
Intergovernmental Panel on Climate Change (IPCC) (2007) Climate Change 2007. The Fourth Assessment Report of the IPCC. Cambridge, New York: Cambridge University Press. Makers. Intergovernmental Panel on Climate Change, Bangkok
Kelliher F et al (1998) Evaporation from a central Siberian pine forest. J Hydrol 205:279–296
Knops JMH, Nash TH III, Schlesinger WH (1996) The influence of epiphytic lichens on the nutrient cycling of an oak woodland. Ecol Monogr 66:159–179
Köhler L, Tobon C, Frumau KFA, Bruijnzee LA (2007) Biomass and water storage dynamics of epiphytes in old-growth and secondary montane cloud forest stands in Costa Rica. Plant Ecol 193:171–184
Köstner BMM et al (1992) Transpiration and canopy conductance in a pristine broad-leaved forest of Nothofagus: an analysis of xylem sap flow and eddy correlation measurements. Oecologia 91:350–359
Kottke I, Agerer R (1983) Untersuchungen zur Bedeutung der Mykorrhiza in älteren Laub- und Nadelwaldbeständen des Südwestdeutschen Keuperberglandes. Mitteilungen des Vereins für Forstkundliche Standortskunde und Forstpflanzenzüchtung 30:30–39
Kranner I et al (2005) Antioxidants and photoprotection in a lichen as compared with its isolated symbiotic partners. Proc Natl Acad Sci USA 102:3141–3146
Kranner I, Beckett R, Hochman A, Nash TH (2008) Desiccation-tolerance in lichens: a review. Bryologist 111:576–593
Lafleur PM, Rouse WR (1988) The influence of surface cover and climate on energy partitioning and evaporation in a Subarctic wetland. Bound-Lay Meteorol 44:327–348
Lakatos M, Rascher U, Büdel B (2006) Functional characteristics of corticolous lichens in the understory of a tropical lowland rain forest. New Phytol 172:679–695
Lakatos M, Hartard B, Máguas C (2007) The stable isotopes $\delta^{13}C$ and $\delta^{18}O$ of lichens can be used as tracers of microenvironmental carbon and water sources. In: Dawson TE, Siegwolf RTW (eds) Stable isotopes as indicators of ecological change. Elsevier, Oxford, pp 73–88
Lakatos M, Hartard B, Máguas C (2009) Ökologie und Physiologie Borken bewohnender Flechten. In: Bayerische Akademie der Wissenschaften (ed) Rundgespräche der Kommission für Ökologie: Ökologische Rolle der Flechten, vol 36. Pfeil, Munich, pp 129–141
Lange OL (1969) CO_2-Gaswechsel von Moosen nach Wasserdampfaufnahme aus dem Luftraum. Planta 89:90–94
Lange OL, Bertsch A (1965) Photosynthese der Wüstenflechte *Ramalina maciformis* nach Wasserdampfaufnahme aus dem Luftraum. Naturwissenschaften 52:215–216
Lange OL, Kappen L (1972) Photosynthesis of Lichens from Antarctica. Antarct Res Ser 20:83–95
Lange OL, Kilian E (1985) Reaktivierung der Photosynthese trockener Flechten durch Wasserdampfaufnahme aus dem Luftraum: Artspezifisch unterschiedliches Verhalten. Flora 176:7–23
Lange OL, Tenhunen JD (1981) Moisture content and CO2 exchange of lichens. II. Depression of net photosynthesis in *Ramalina maciformis* at high water content is caused by increased thallus carbon dioxide diffusion resistance. Oecologia 51:426–429
Lange OL, Kilian E, Ziegler H (1986) Water vapor uptake and photosynthesis of lichens: performance differences in species with green and blue-green algae as phycobionts. Oecologia 71:104–110
Lange OL, Bilger W, Rimke S, Schreiber U (1989) Chlorophyll fluorescence of lichens containing green and blue green algae during hydration by water vapor uptake and by addition of liquid water. Bot Acta 102:306–313
Lange OL, Meyer A, Zellner H, Ulmann I, Wessels DCJ (1990) Eight days in the life of a desert lichen: water relations and photosynthesis of *Teloschistes capensis* in the coastal fog zone of the Namib Desert. Madoqua 17:17–30

Lange OL, Büdel B, Heber U, Meyer A, Zellner H, Green TGA (1993a) Temperate rainforest lichens in New Zealand: high thallus water content can severely limit photosynthetic CO_2 exchange. Oecologia 95:303–313
Lange OL, Büdel B, Meyer A, Kilian E (1993b) Further evidence that activation of net photosynthesis by dry cyanobacterial lichens requires liquid water. Lichenologist 25:175–189
Lange OL, Büdel B, Zellner H, Zotz G, Meyer A (1994) Field measurements of water relations and CO_2 exchange of the tropical, cyanobacterial basidiolichen *Dictyonema glabratum* in a panamanian rainforest. Bot Acta 107:279–290
Lange OL, Green TGA, Heber U (2001) Hydration-dependent photosynthetic production of lichens: what do laboratory studies tell us about field performance? J Exp Bot 52:2033–2042
Lange OL, Allan Green TG, Meyer A, Zellner H (2007) Water relations and carbon dioxide exchange of epiphytic lichens in the Namib fog desert. Flora 202:479–487
Larson D (1981) Differential wetting in some lichens and mosses: the role of morphology. Bryologist 84:1–15
Larson DW (1987) The absorption and release of water by lichens. Bibl Lichenol 25:351–360
Laurance W, Powell G, Hansen L (2002) A precarious future for Amazonia. Trends Ecol Evol 17:251–252
Lawton RO, Nair US, Pielke RA Sr, Welch RM (2001) Climatic impact of tropical lowland deforestation on nearby montane cloud forests. Science 294:584–587
Lechowicz MJ (1982) Ecological trends in lichen photosynthesis. Oecologia 53:330–336
Leon-Vargas Y, Engwald S, Proctor MCF (2006) Microclimate, light adaptation and desiccation tolerance of epiphytic bryophytes in two Venezuelan cloud forests. J Biogeogr 33:901–913
Lewis D, Smith D (1967) Sugar alcohols (polyols) in fungi and green plants. I. Distribution, physiology and metabolism. New Phytol 66:143–184
Longton R (1992) The role of bryophytes and lichens in terrestrial ecosystems. In: Bates J, Farmer A (eds) Bryophytes and lichens in a changing environment. Clarendon Press, Oxford, pp 32–76
Lugo A, Brinson M, Brown S (1990) Forested wetlands. Elsevier, New York, NY
Mägdefrau K (1982) Life-forms of bryophytes. In: Smith AJ (ed) Bryophyte ecology. Chapman and Hall, London, pp 45–58
Matthes-Sears U, Nash TH III (1986) The ecology of *Ramalina menziesii* V. Estimation of gross carbon gain and thallus hydration source from diurnal measurements and climatic data. Can J Bot 64:1698–1702
Matthews E, Fung I (1987) Methane emission from natural wetlands: global distribution, area, and environmental characteristics of sources. Global Biogeochem Cycles 1(1):61–86
Mayaba N, Beckett R (2001) The effect of desiccation on the activities of antioxidant enzymes in lichens from habitats of contrasting water status. Symbiosis 31:113–121
Montfoort D, Ek R (1990) Vertical distribution and ecology of epiphytic bryophytes and lichens in a lowland rain forest in French Guyana. In: vol. MSc. Utrecht, Holland, Utrecht, p 61
Nadkarni NM, Solano R (2002) Potential effects of climate change on canopy communities in a tropical cloud forest: an experimental approach. Oecologia 131:580–586
Nash TH III (2008) Lichen biology, 2nd edn. Cambridge University Press, Cambridge
Nash TH III, Reiner A, Demmig-Adams B, Kilian E, Kaiser WM, Lange OL (1990) The effect of atmospheric desiccation and osmotic water stress on photosynthesis and dark respiration of lichens. New Phytol 116:269–276
Normann F et al (2010) Diversity and vertical distribution of epiphytic macrolichens in lowland rain forest and lowland cloud forest of French Guiana. Ecol Indic 10:1111–1118
O'Connell KEB, Gower ST, Norman JM (2003a) Comparison of net primary production and light-use dynamics of two boreal black spruce forest communities. Ecosystems 6:236–247
O'Connell KEB, Gower ST, Norman JM (2003b) Net ecosystem production of two contrasting boreal black spruce forest communities. Ecosystems 6:248–260
O'Neill K (2000) Role of bryophyte-dominated ecosystems in the global carbon budget. In: Shaw AJ, Goffinet B (eds) Bryophyte biology. Cambridge University Press, Cambridge, pp 344–368

Oechel WC, Vourlitis GL, Hastings SJ Jr, RPA BP (1998) The effects of water table manipulation and elevated temperature on the net CO_2 flux of wet sedge tundra ecosystems. Glob Change Biol 4:77–90

Oliver MJ (1991) Influence of protoplasmic water loss on the control of protein synthesis in the desiccation-tolerant moss *Tortula ruralis*: ramifications for a repair-based mechanism of desiccation tolerance. Plant Physiol 97:1501–1511

Oliver MJ, Tuba Z, Mishler BD (2000) The evolution of vegetative desiccation tolerance in land plants. Plant Ecol 151:85–100

Oliver MJ, Velten J, Mishler BD (2005) Desiccation tolerance in bryophytes: a reflection of the primitive strategy for plant survival in dehydrating habitats? Integr Comp Biol 45:788–799

Oren R, Phillips N, Katul G, Ewers B, Pataki D (1998) Scaling xylem sap flux and soil water balance and calculating variance: a method for partitioning water flux in forests. Ann For Sci 55:191–216

Palmqvist K (2000) Carbon economy in lichens [review]. New Phytol 148:11–36

Pardow A, Hartard B, Lakatos M (2010) Morphological, photosynthetic and water relations traits underpin the contrasting success of two tropical lichen groups at the interior and edge of forest fragments. Ann Bot Plants plq004:10

Patterson P (1946) Osmotic values of bryophytes and problems presented by refractory types. Am J Bot 33:604–611

Perry DR (1984) The canopy of the tropical rain-forest. Sci Am 251:138–147

Pettersson R, Ball J, Renhorn K, Esseen P, Sjöberg K (1995) Invertebrate communities in boreal forest canopies as influenced by forestry and lichens with implications for passerine birds. Biol Conserv 74:57–63

Pharo EJ, Zartman CE (2007) Bryophytes in a changing landscape: the hierarchical effects of habitat fragmentation on ecological and evolutionary processes: the conservation ecology of cryptogams. Biological Conservation 135:315–325

Pocs T (1980) The epiphytic biomass and its effect on the water-balance of 2 rain-forest types in the Uluguru Mountains (Tanzania, East-Africa). Acta Botanica Academiae Scientiarum Hungaricae 26:143–167

Pocs T (1982) Tropical forest bryophytes. In: Smith AJE (ed) Bryophyte ecology. Chapmann & Hall, London, pp 59–104

Price AG, Dunham K, Carleton T, Band L (1997) Variability of water fluxes through the black spruce (*Picea mariana*) canopy and feather moss (*Pleurozium schreberi*) carpet in the boreal forest of Northern Manitoba. J Hydrol (Amsterdam) 196:310–323

Proctor MCF, Tuba Z (2002) Poikilohydry and homoihydry: antithesis or spectrum of possibilities? New Phytol 156:327–349

Proctor MC, Nagy Z, Csintalan Z, Takacs Z (1998) Water-content components in bryophytes: analysis of pressure-volume relationships. J Exp Bot 49:1845–1854

Proctor MCF et al (2007) Desiccation-tolerance in bryophytes: a review. Bryologist 110:595–621

Pypker TG, Unsworth MH, Bond BJ (2006) The role of epiphytes in rainfall interception by forests in the Pacific Northwest. II. Field measurements at the branch and canopy scale. Canadian J Forest Research-Revue Canadienne De Recherche Forestiere 36:819–832

Rai A (1988) Nitrogen metabolism. In: Galun M (ed) Handbook of lichenology, vol 1. CRC, Boca Raton, FL, pp 201–237

Richards PW (1984) The ecology of tropical forest bryophytes. In: Schuster RM (ed) New manual of bryology. The Hattori Botanical Laboratory, Nichinan, pp 1233–1270

Richardson B, Rogers C, Richardson M (2000) Nutrients, diversity, and community structure of two phytotelm systems in a lower montane forest, Puerto Rico. Ecol Entomol 25:348–356

Rieley J, Richards P, Bebbington A (1979) The ecological role of bryophytes in a North Wales woodland. J Ecol 67:497–527

Rixen C, Mulder C (2005) Improved water retention links high species richness with increased productivity in arctic tundra moss communities. Oecologia 146:287–299

Robinson CH, Wookey PA, Lee JA, Callaghan TV Press MC (1998) Plant community responses to simulated environmental change at a high arctic polar semi-desert. Ecology 79:856–866
Rundel PW (1988) Water relation. In: Galum M (ed) Handbook of lichenology II, vol VII.B.1, pp 17–36
Sala OE et al (2000) Global biodiversity scenarios for the year 2100. Science 287:1770–1774
Scheidegger C, Schroeter B, Frey B (1995) Structural and functional processes during water vapour uptake and desiccation in selected lichens with green algal photobionts. Planta 197:399–409
Sedia E, Ehrenfeld J (2005) Differential effects of lichens, mosses and grasses on respiration and nitrogen mineralization in soils of the New Jersey Pinelands. Oecologia 144:137–147
Sedia E, Ehrenfeld J (2006) Differential effects of lichens and mosses on soil enzyme activity and litter decomposition. Biol Fertil Soils 43:177–189
Shimoyama K, Hiyama T, Fukushima Y, Inoue G (2004) Controls on evapotranspiration in a west Siberian bog. J Geophys Res Atmos 109:D08111
Sipman HJM (1989) Lichen zonation in the parque Los Nevados transect. Cramer, Berlin-Stuttgart
Sipman HJM, Tan BC (1990) A field impression of the lichen and bryophyte zonation on Mount Kinabalu. Flora Malesiana Bulletin 10:241–244
Smith LC et al (2004) Siberian peatlands a net carbon sink and global methane source since the early Holocene. Science 303:353–356
Solomon S et al (2007) Climate change 2007: the physical science basis; Contribution of Working Group I to the fourth Assessment Report of the Intergovenmental Panel on Climate Change
Still C, Foster P, Schneider S (1999) Simulating the effects of climate change on tropical montane cloud forests. Nature 398:608–610
Stofer S et al (2006) Species richness of lichen functional groups in relation to land use intensity. Lichenologist 38:331–353
Stone R (2008) ECOSYSTEMS: have desert researchers discovered a hidden loop in the carbon cycle? Science 320:1409
Stuntz S, Simon U, Zotz G (2002) Rainforest air-conditioning: the moderating influence of epiphytes on the microclimate in tropical tree crowns. Int J Biometeorol 46:53–59
Swanson R, Flanagan L (2001) Environmental regulation of carbon dioxide exchange at the forest floor in a boreal black spruce ecosystem. Agric For Meteorol 108:165–181
Tarnawski MG, Green TGA, Büdel B, Meyer A, Zellner H, Lange OL (1994) Diel changes of atmospheric co_2 concentration within, and above, cryptogam stands in a new zealand temperate rainforest. NZ J Bot 32:329–336
Timmermann A, Oberhuber J, Bacher A, Esch M, Latif M, Roeckner E (1999) Increased El Nino frequency in a climate model forced by future greenhouse warming. Nature 398:694–697
Tunnacliffe A, Wise M (2007) The continuing conundrum of the LEA proteins. Naturwissenschaften 94:791–812
Turetsky MR (2003) The role of bryophytes in carbon and nitrogen cycling. Bryologist 106: 395–409
Uchida M, Nakatsubo T, Kanda H, Koizumi H (2006) Estimation of the annual primary production of the lichen *Cetrariella delisei* in a glacier foreland in the High Arctic, Ny-Ålesund, Svalbard. Polar Res 25:39–49
Unal D, Senkardesler A, Sukatar A (2008) Abscisic acid and polyamine contents in the lichens *Pseudevernia furfuracea* and *Ramalina farinacea*. Russ J Plant Physiol 55:115–118
Unsworth MH et al (2004) Components and controls of water flux in an old-growth douglas-fir-western hemlock ecosystem. Ecosystems 7:468–481
van der Wal R (2006) Do herbivores cause habitat degradation or vegetation state transition? Evidence from the tundra. Oikos 114:177
Veneklaas EJ, Zagt RJ, Leerdam A, Ek R, Broekhoven AJ, Genderen M (1990) Hydrological properties of the epiphyte mass of a montane tropical rain forest, Colombia. Plant Ecol 89:183–192

Verhoeven J, Toth E (1995) Decomposition of Carex and Sphagnum litter in fens: effect of litter quality and inhibition by living tissue homogenates. Soil Biol Biochem 27:271–275
Vitt D (2000) Peatlands: ecosystems dominated by bryophytes. In: Shaw A, Goffinet B (eds) Bryophyte biology. Cambridge University Press, Cambridge, pp 312–343
Vitt D, Halsey L, Campbell C, Bayley S, Thormann M (2001) Spatial patterning of net primary production in wetlands of continental western Canada. Ecoscience 8:499–505
Walker MD et al (2006) Plant community responses to experimental warming across the tundra biome. Proc Natl Acad Sci USA 103:1342–1346
Wang XQ et al (2009) Exploring the mechanism of *Physcomitrella patens* desiccation tolerance through a proteomic strategy. Plant Physiol 149:1739–1750
Wasley J, Robinson SA, Lovelock CE, Popp M (2006) Climate change manipulations show antarctic flora is more strongly affected by elevated nutrients than water. Global Change Biology 12:1800–1812
Weissman L, Garty J, Hochman A (2005a) Characterization of enzymatic antioxidants in the lichen *Ramalina lacera* and their response to rehydration. Appl Environ Microbiol 71:6508–6514
Weissman L, Garty J, Hochman A (2005b) Rehydration of the lichen *Ramalina lacera* results in production of reactive oxygen species and nitric oxide and a decrease in antioxidants. Appl Environ Microbiol 71:2121–2129
Wijk MTv et al (2004) Long-term ecosystem level experiments at Toolik Lake, Alaska, and at Abisko, Northern Sweden: generalizations and differences in ecosystem and plant type responses to global change. Glob Change Biol 10:105–123
Williams TG, Flanagan LB (1996) Effect of changes in water content on photosynthesis, transpiration and discrimination against (CO2)-^{13}C and (COO)-^{18}O-^{16}O in *Pleurozium* and *Sphagnum*. Oecologia 108:38–46
Wilske B et al (2009) Modeling the variability in annual carbon fluxes related to biological soil crusts in a Mediterranean shrubland. Biogeosci Discuss 6:7295–7324
Wohlfahrt G, Fenstermaker LF, Arnone JA III (2008) Large annual net ecosystem CO_2 uptake of a Mojave Desert ecosystem. Glob Change Biol 14:1475–1487
Wolf JHD (1993) Diversity patterns and biomass of epiphytic bryophytes and lichens along an altitudinal gradient in the northern Andes. Ann Mo Bot Gard 80:928–960
Wolf JHD (1995) Non-vascular epiphyte diversity patterns in the canopy of an upper montane rain forest (2550–3670 m), central Cordillera, Colombia. Selbyana 16:185–195
Wood AJ (2007) The nature and distribution of vegetative desiccation-tolerance in hornworts, liverworts and mosses. Bryologist 110:163–177
Wood A, Oliver M (2004) Molecular biology and genomics of the desiccation-tolerant moss *Tortula ruralis*. In: Wood A, Oliver M (eds) New frontiers in bryology: physiology, molecular biology and functional genomics. Kluwer Academic Publisher, Dodrecht, pp 71–90
Worden J, Noone D, Bowman K (2007) Importance of rain evaporation and continental convection in the tropical water cycle. Nature 445:528–532
Zeuch L (1934) Untersuchungen zum Wasserhaushalt von *Pleurococcus vulgaris*. Planta 22:614–643
Zimmermann R et al (2000) Canopy transpiration in a chronosequence of Central Siberian pine forests. Glob Change Biol 6:25–37
Zotz G, Bader MY (2009) Epiphytic plants in a changing world-global: change effects on vascular and non-vascular epiphytes. Progress in Botany 70:147–170
Zotz G, Winter K (1994) Photosynthesis and carbon gain of the lichen, *Leptogium azureum*, in a lowland tropical forest. Flora 189:179–186
Zotz G, Büdel B, Meyer A, Zellner H, Lange OL (1997) Water relations and CO_2 exchange of tropical bryophytes in a lower montane rain forest in Panama. Bot Acta 110:9–17
Zotz G, Büdel B, Meyer A, Zellner H, Lange OL (1998) In situ studies of water relations and CO_2 exchange of the tropical macrolichen, *Sticta tomentosa*. New Phytol 139:525–535

Chapter 6
Ecophysiology of Desiccation/Rehydration Cycles in Mosses and Lichens

T.G. Allan Green, Leopoldo G. Sancho, and Ana Pintado

6.1 Introduction

Lichens and bryophytes are all poikilohydric which is defined as meaning that their water content (WC, thallus water content) will tend to equilibrium with the water status of the environment. Under wet conditions they become hydrated and active, under dry conditions they dry out and become dormant. This relatively simple definition actually hides a complexity of ecological and physiological performance. Although occasionally still included in the definition, there is no generality in the abilities of these poikilohydric organisms to withstand drying. There are large differences in ability to survive both brief and long term desiccation. Species-specific differences also exist in rates of rehydration and recovery from desiccation and in response to rate of desiccation. Most important, lichens and bryophytes are in no way similar organisms. Although both groups were probably some of the first organisms to colonise the land they represent vastly different structural solutions to the problems that they faced. Bryophytes (Fig. 6.1) are true plants and members of the Plantae (in the six Kingdom classifications). Lichens (Fig. 6.2), in contrast, are a symbiosis composed of a heterotrophic fungus (Kingdom Fungi) with a photosynthetic partner that can be algal (Kingdom Protista) or cyanobacterial (Kingdom Eubacteria) (Fig. 6.2b). Some lichens can be made up of the members of three Kingdoms, a fungus, a photosynthetic green alga and a nitrogen fixing cyanobacterium. The morphological range is very broad (Figs. 6.1 and 6.2). Lichens are made up of predominantly unicellular, colonial or filamentous algae and cyanobacteria inside a hyphal fungus, whilst bryophytes are true multicellular organisms and at least two-dimensional and normally three-dimensional. With such a massive difference in structure and base physiology, should we expect that bryophytes and lichens will behave similarly with respect to changes in water content?

The poikilohydric lifestyle links the two groups and is also the key to their success, and they are very successful groups. They are most visible in polar, alpine and desert areas. About 8% of the terrestrial surface of the world is dominated by lichens (Larson 1987; Fig. 6.3) and bryophytes are major components of the extensive arctic tundra and bogs (Fig. 6.4). Recent estimates indicate an important role in the world carbon cycle for soil crusts (Fig. 6.3b), which are dominated by

U. Lüttge et al. (eds.), *Plant Desiccation Tolerance*, Ecological Studies 215,
DOI 10.1007/978-3-642-19106-0_6, © Springer-Verlag Berlin Heidelberg 2011

Fig. 6.1 Moss and liverwort structure and form: (**a**) Ectohydric moss: clump of *Bryum argenteum* in flush at Cape Hallett, Antarctica, showing closely packed shoots; *bar* is 1 cm. Photograph by Prof. Roman Türk, Salzburg, with permission. (**b**) Ectohydric moss: loose clump of *Tortula ruralis* showing the typical one cell thick leaves that are fully exposed because they are hydrated; *bar* is

lichens and bryophytes (Elbert et al. 2009). High biomasses are also reached in wet forests where they occupy the epiphytic habitat that is not available for higher plants.

In recent years there have been several excellent and extensive reviews of the poikilohydric lifestyle (Kappen and Valladares 2007; Proctor and Tuba 2002), of the ecophysiology of bryophytes with an emphasis on structure and water relations (Proctor 2008) and of the mechanisms of desiccation tolerance of bryophytes (Oliver 2008) and lichens (Beckett et al. 2008; Becket and Minibayeva 2008). In this chapter we will first briefly summarise the limits to desiccation tolerance and the underlying physiology. Then, we will consider photosynthesis and how it responds to thallus water content (WC) and will do this because of the substantial literature available, because photosynthesis must be linked to productivity and because the differences between the groups are very clear. We intend to not only consider the changes that occur during hydration and desiccation but also to show that the interaction between WC and photosynthesis is reflected in the ecology of the two groups with each tending to exploit different niches on the response curve. In view of the fact that the majority of the recent reviews are on bryophytes we will do this mainly from a lichen standpoint. Also, although light is a key stress factor during desiccation and rehydration, it is covered in Chap. 7 in this book and will not be dealt with in detail.

6.1.1 Desiccation Tolerance: The Limits

Desiccation tolerance in bryophytes and lichens has been defined as the ability to equilibrate their internal water potential with that of moderately dry air, and then resume normal function when rehydrated (Alpert 2000). This means equilibration with 50% relative humidity at 20°C which leads to a thallus water content (WC) around 10% and equilibration with a water potential of −100 MPa. This threshold of 10% WC seems to have physiological meaning and to correspond to the point at which there is no longer enough water to form a monolayer around macromolecules thus stopping enzymatic reactions and metabolism (Billi and Potts 2002). In the

Fig. 6.1 (Continued) 1 cm. Photograph by Prof. Paul Busselen, Katholieke Universiteit Leuven, Belgium, with permission. (**c**) Endohydric moss: male shoot of *Dawsonia superba* with linear leaves that are many cells thick and have lines of photosynthetic lamellae along the upper surface that are capped with wax; *bar* is 1 cm. (**d**) Endohydric moss: *Dendrilogotrichum squamosum* shoots amongst *Gunnera magellanica* at Seño Almirantazgo, Tierra del Fuego, Chile; *bar* is 2 cm. (**e**) Leafy liverwort: shoots of *Schistochila* sp. in rain forest, New Zealand, showing typical two ranked appearance of the leaves; *bar* is 1 cm. (**f**) Thalloid liverwort: *Anthoceros* sp., showing the typical sporophyte of Anthocerotae; from Beagle Channel, Tierra del Fuego, Chile; *bar* is 1 cm. (**g**) *Sphagnum* shoots showing typical leaves appressed to lateral branches which store water in large dead cells; Seno Almirantazgo, Tierra del Fuego, Chile; *bar* is 1 cm. (**h**) *Andreaea* sp. clump growing on mountain rock, Cerro Bandera, Navarino Island, Chile, note dark protective colouring; *bar* is 2 cm

Fig. 6.2 Lichen structure and form. (**a**) Foliose lichen: thallus of *Nephroma antarcticum* growing on a tree stump in *Nothofagus* forest on Cerro Bandera, Navarino Island, Chile; *bar* is 2 cm. (**b**) Foliose lichen: an example of a photosymbiodeme, a single thallus that has green algal (*green colour*) and cyanobacterial (*dark colour*) sectors from a *Nothofagus* evergreen rain forest,

laboratory much lower hydrations, less than 1%, can be achieved by drying over P_2O_5 and levels almost as low, around 2%, can occur in the field when dry lichens are heated by the sun to temperatures that can reach 70°C. Lichens and bryophytes are often given as examples of organisms that show desiccation tolerance. An excellent example of this ability, at least for lichens, is the survival with no apparent ill effects and with full recovery of their photosystems and photosynthesis of desiccated lichens exposed to full space conditions for 14 days (Sancho et al. 2007). However, only a small number of species of lichens and bryophytes have actually been tested for desiccation tolerance (Alpert 2000; Lange 1953, 1955) and data for extended desiccation tolerance are few and somewhat surprising (Table 6.1, p. 98).

It is clear that many groups, including animals such as rotifers and tardigrades, can withstand several years of desiccation. The surprise is the lichens which, although regarded as the classic organisms for survival in extreme habitats including dry habitats, show considerably less tolerance than all other groups. It is curious that they fail even to survive as long as a poikilohydrous pteridophytes or angiosperms and, in his review on the *Limits and frontiers of life* (Alpert 2005), they are not even mentioned in the section on inherent limits.

Possibly they have not been well researched with lichenologists making the assumption that lichens are desiccation tolerant. Alternatively, perhaps lichens as a group are adapted to short periods of desiccation which can be severe but do not survive if the desiccation is prolonged. This could reflect their occupation of habitats with irregular water availability but not long periods of dryness.

The relationship between the normal habitat conditions of a bryophyte and lichen and degree of survival following drying and desiccation is important and should be taken into consideration at all times. The survival times presented in Table 6.1 represent one extreme in desiccation tolerance whilst, at the other extreme, there are examples of species that have almost zero tolerance of dehydration, let alone desiccation. For lichens, *Pseudocyphellaria dissimilis* cannot survive even 20 h equilibration at 12% relative humidity (Green et al. 1991) whilst species of the moss order Hookerales and many leafy liverworts (Fig. 6.1e) are also

Fig. 6.2 (Continued) Waikaremoana, North Island, New Zealand. The *green* sectors have been shown to reactivate from high relative humidity almost every night whilst the cyanobacterial sectors require rainfall. The free-living forms of the sectors are *Pseudocyphellaria lividofusca* (green algal) and *P. knightii* (cyanobacterial). *Bar* is 2 cm. Photograph by Prof. Otto Lange, Würzburg, with permission. (**c**) Fruticose lichens: (*Usnea* sp.) hanging from tree branch in Omora Park, Navarino Island, Chile. (**d**) Fruticose lichen: *Ramalina maciformis* growing on flint stone in the Negev desert, Israel. *Bar* is 1 cm; photograph by Prof. Otto Lange, Würzburg, with permission. (**e**) Crustose lichen: *Buellia frigida* growing on granite, Mt. Falconer, Taylor Valley, Southern Victoria Land, continental Antarctica; *bar* is 1 cm. (**f**) Cladonia form: lichens with vertical podetia and basal squamules; *Cladonia* aff. *ustulata* from Beagle Channel area, Tierra del Fuego, Chile. (**g**) Endolithic lichen, growing inside Beacon sandstone from Dry Valleys, Southern Victoria Land, continental Antarctica; *bar* is 1 cm. Photograph by Prof. Imre Friedmann. (**h**) Cross section (scanning electron micrograph) of foliose lichen thallus (*Pseudocyphellaria* sp.); *bar* is 500 μm. The *red arrow* indicates a single algal cell in the photobiont layer below the upper cortex

Fig. 6.3 Habitats of lichens. (**a**) Antarctic cold desert: Linnaeus Terrace, Dry Valleys, continental Antarctica with rocks (Beacon sandstone) containing endolithic lichens (*brown coloured areas*). Photograph by Catherine Beard with permission. (**b**) Soil crust: lichen *Diploschistes diacapsis* in the Tabernas Desert, Almeria, Spain; the lichen has protected the soil under its thallus from erosion

desiccation intolerant (Proctor and Pence 2002). It is important to remember that most bryophytes and lichens fall between these extremes of desiccation tolerance.

The response to desiccation has been studied in much more depth in bryophytes than lichens. The ability to recover is influenced by the rate of desiccation, the depth of desiccation and the length of the recovery time and, very important, the bryophyte being studied. In a study on six mosses subjected to 60 days desiccation at water potentials from −40 to −400 MPa with recovery assessed using the chlorophyll fluorescence parameter F_v/F_m (optimal quantum efficiency of Photosystem II) measured at 20 min and 24 h after rehydration there is a large difference between the tolerant mosses, such as *Tortula ruralis* (Fig. 6.1b) and non-tolerant species, such as *Anomodon viticulosus* (Fig. 6.5; from Proctor and Pence (2002). The latter species showed, even after 24 h, no recovery when stored at −300 MPa and lower, and minimal recovery at −40 MPa. In contrast, *T. ruralis* achieved near optimal F_v/F_m after all treatments in 20 min, and full recovery after 24 h. Twenty minutes after treatment, all species except the very sensitive *A. viticulosus* showed a near-optimal recovery over the range of water potentials from −100 to −300 MPa treatment, and decreased recovery at lower and higher potentials. There is a relationship between the level of desiccation tolerance and the ecology of the species with the tolerant *T. ruralis* from open soil crusts and the sensitive *A. viticulosus* from shaded sites. Degree of drought tolerance is generally greatest in plants from dry habitats (Clausen 1952, 1964; Johnson and Kokila 1970; Dilks and Proctor 1974). Desiccation tolerance is also not a static property, there is clear evidence that mosses can change their sensitivity seasonally (Beckett and Hoddinott 1997) and with wetness of their habitat (Robinson et al. 2000). Also, both lichens and mosses can be prehardened before the main drying. A suitable light desiccation or even the addition of abscisic acid (ABA) (Beckett 1999), a plant growth substance that can cause both desiccation hardening and change in structural form, can be used for prehardening, however, ABA is still poorly understood in these groups (Dietz and Hartung 2008; Beckett et al. 2000). Studies exist showing that seasonal acclimation can occur in lichens as well, in this case of respiration to temperature (Lange and Green 2005). All of these results suggest that desiccation tolerance is potentially dynamic and changing and that physiological

Fig. 6.3 (Continued) by rain. Photographed area is approximately 10 × 10 cm. (**c**) Exposed rocks: ridges of "Armorican" quartzite in Despeñaperros Natural Park, Andalusia, Spain, covered with crustose lichens (*Acarospora* sp. is bright yellow). (**d**) Hot desert fog community: Namib fog desert with ground covered with the red fruticose lichen *Teloschistes capensis*. Activity of this lichen depends almost entirely on regular hydration by the fog. Photograph by Prof. Otto Lange, Würzburg, with permission. (**e**) Mountain rock surface: (Monfrague National Park, Extremadura, Spain) showing *Umbilicaria freyi* with other crustose and fruticose lichens; *bar* is 2 cm. (**f**) Epiphytic: lichens in forest (La Quesera pass, Spain) with the large foliose species (*Lobaria pulmonaria*) on a horizontal branch and hanging fruticose *Usnea* sp.; *bar* is 5 cm. (**g**) Ephemeral habitat: *Placopsis pycnotheca* growing on fresh glacial mud at Alemania Glacier, Tierra del Fuego, Chile. The mud surface is probably only a few months old; *bar* is 1 cm. (**h**) Ephemeral habitat: foliicolous lichens, foliose and fruticose, growing on a 2 year old *Camellia* leaf in a garden in Hamilton, North Island, New Zealand; *bar* is 1 cm

Fig. 6.4 Habitats of bryophytes: (**a**) Antarctic flush: bryophyte flush area at Botany Bay, Southern Victoria Land, continental Antarctica; substantial moss growth is confined to areas with regular water flow from snow melt; the moss clumps are strongly coloured to protect against the high light levels when the water is flowing; the mosses are *Ceratodon purpureus* (*brown*) and *Bryum argenteum*

studies on desiccation tolerance need to take the habitat conditions and even recent history of the studied organism into account when interpreting results. The vast majority of the data are from studies on bryophytes, and work on lichens is much needed. A new emphasis is, perhaps, also needed; rather than make the slightly negative statement that a moss or lichen is more desiccation sensitive than another, it might be better to say that they have a desiccation tolerance suitable for their particular (possibly consistently moist) environments. The very existence of desiccation sensitive poikilohydric bryophytes and lichens, together with seasonal changes in desiccation sensitivity, suggest that there is some form of metabolic cost to maintain tolerance which will be avoided if the habitat conditions allow.

6.1.2 Desiccation Tolerance: Physiology

The physiology underlying the ability to withstand desiccation has been much more extensively studied in bryophytes than lichens, probably because mosses and liverworts are single, rather than two or three, organisms and also true green plants so that any findings might be extended to higher plants. There have been several recent reviews, Oliver (2008), Buitink et al. (2002) for bryophytes, and Beckett et al. (2008), Becket and Minibayeva (2008), and Kranner et al. (2008) for lichens, Bartels and Sunkar (2005; Chaps. 9, 11 and 16) for plants in general. Bewley (1979) defined the properties desiccation-tolerant protoplasm must possess: it must (a) limit damage to a repairable level; (b) maintain integrity when in a dried state; and (c) upon rehydration, rapidly mobilise mechanisms that repair any damage that has been sustained as a result of both desiccation and rehydration. Detail has been added to these properties, such as the need for orderly shutdown of metabolism (Alpert and Oliver 2002) and that a critical level of physical cell order be maintained in the dry state so that there is effective packing of structures (Thompson and Platt 1997).

Oliver (2008) states that the effectiveness of cellular protection aspects of a particular mechanism for desiccation tolerance and, to some extent, the repair processes determine the intensity of desiccation that a particular species can survive.

Fig. 6.4 (Continued) (*green-yellow*); *bar* is 50 cm. (**b**) Rock surfaces: clumps of mosses, all ectohydric, on granite rock surface near Pia Glacier, Tierra del Fuego, Chile; clumps provide extra capillary water storage between the closely packed shoots; note strong *brown or golden-brown colours* due to protective pigments; *bar* is 5 cm. (**c**) Arctic tundra: Svalbard, which has extensive bryophyte and lichen cover. (**d**) Temperate rain forest: an environment rich in epiphytic and ground mosses and liverworts, here growing near a stream in a *Nothofagus* forest, Beagle Channel, Chile. (**e**) Epiphytic habitat: typical moss clump (*Leptostomum* sp.) on tree branch, clumps store water and extend the active time in this low light environment; *Nothofagus* forest, Beagle Channel, Chile; *bar* is 10 cm. (**f**) Tundra: a tundra area in Beagle Channel, Tierra del Fuego with the moss *Rhacomitrium lanuginosum* forming balls on the ground. (**g**) Rain forest: pendulous moss, *Weymouthia* sp. in a lowland rain forest, Mount Egmont, North Island, New Zealand; *bar* is 10 cm

Table 6.1 Selected data for survival of long periods of desiccation of various types of organisms (more data in Table 1 – Alpert 2000)

Species	Desiccation period survived	Reference
Lichens		
Ramalina maciformis	1 year 1% WC	Lange (1969)
Lasallia pustulata	>1 year air dry	Lange (1953)
Umbilicaria spp.	10 year air dry (at −20°C)	Larson (1989)
Unknown cyanobacterial	5 year air dry office table	Thomas (1921)
Cyanobacteria		
Nostoc commune	55 year air dry	Shirkey et al. (2003)
Fungus		
Schizophyllum commune	34 year at <0.01 mm Hg	Bisby (1945)
Moss		
Anoectangium compactum	19 year air dry	Malta (1921) cf Richardson (1981)
Tortula ruraliformis	14 year air dry	Bristol (1916) cf Mueller (1972)
Liverwort		
Riccia macrocarpa	23 year air dry	Breuil-Sée (1993)
Pteridophyte		
Pellaea atropurpurea	5 year air dry	Pickett (1931)
Selaginella densa	3.5 year air dry	Webster and Steeves (1964)
Angiosperms		
Chamaegigas intrepidus	4 year air dry	Hickel (1967)
Xerophyta squarrosa	5 year air dry	Gaff (1977)

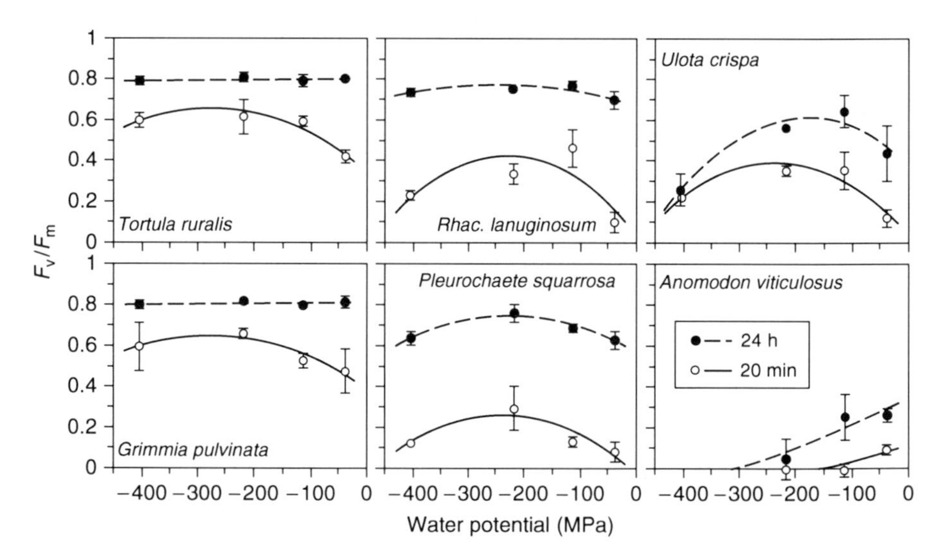

Fig. 6.5 The effect of 60 days desiccation at selected water potentials on 6 mosses from different habitats. Chlorophyll fluorescence (F_v/F_m, optimal quantum efficiency of Photosytem II) was used to determine recovery and was measured 20 min and 24 h after rewetting (from Proctor and Pence 2002)

We can restate this to make it ecologically relevant by saying that bryophytes and lichens will possess mechanisms sufficient to survive the desiccation stress of their habitat and that this protection level might well be dynamic, seasonally or spatially.

Such a statement is justified on resource grounds, organisms are unlikely to invest resources in excess of that needed to survive in a particular environment. Looking from this direction we would expect species to show a variety of responses to desiccation depth, length, drying rapidity and rehydration, and this indeed is what has been found with, once again, most studies being on bryophytes.

The highly desiccation-tolerant moss *T. ruralis* (Fig. 6.1b) can withstand rapid desiccation from hydrated to <-500 MPa in 30 min. Cellular protection in such a case must be constitutive as there is little time available for the synthesis of new protective compounds. Protein synthesis is sensitive to dehydration and stops rapidly indicating that the proteins required for desiccation protection are already present. *T. ruralis* also maintains membrane integrity when dehydrated (Platt et al. 1994). This situation can be contrasted with what occurs in the few desiccation-tolerant angiosperms where slow drying is required in order for new systems to be initiated and protective measures established (Kappen and Valladares 2007; Chaps. 9 and 16). The mechanisms of desiccation tolerance have really only been studied in depth in *T. ruralis* so there is little information about what happens in more sensitive bryophytes, and in lichens. However, it is reasonable to assume that they may have some similarities with the non-constitutive angiosperms and certainly there are examples already known of induced desiccation tolerance following a brief and light dehydration before full desiccation in lichens (Beckett 1999).

Whether in constitutive or induced modes it appears that several major components need to be in place to ensure survival (1) compatible solutes, in particular sugars; (2) protective proteins; (3) amphiphilic metabolites; (4) antioxidants (Hoekstra et al. 2001; Oliver 2008). It is possible that these occur and are functionally similar in bryophytes, angiosperms and lichens although very little detail is available for bryophytes and even less for lichens.

1. Compatible solutes: these are molecules that can be present in quantity but do not interfere with normal cell structure and function. There is a long list but notable are sucrose, oligosaccharides, fructans, polyols, trehalose (Hoekstra et al. 2001). *T. ruralis* contains about 10% sucrose (dw basis), a level that remains relatively constant whether hydrated or dry (Oliver 2008). Lichens are well known to have high polyol levels, *Xanthoria elegans* contains 5.4% arabitol, 4.3% mannitol and 1.7% ribitol (dw basis) again with little change when wet or dry (Aubert et al. 2007) and a role for these compounds in desiccation tolerance had been proposed for many years (Farrar 1976). The suggested protection mechanism during dehydration is preferential exclusion, sugars and polyols are all preferentially excluded from the surface of proteins, thus keeping the proteins preferentially hydrated. At very low water contents, <0.3 g H_2O g dw^{-1}, when no bulk water remains in the cytoplasm and preferential exclusion can no longer work, only sugars are effective at structurally and functionally preserving proteins and membranes by water molecule replacement thus stabilising hydrogen bonds at lower water contents (Hoekstra et al. 2001). Other compatible solutes can be effective only at higher water contents during dehydration. At even lower water contents, ~0.1 g H_2O g dw^{-1} at room temperature, sugars

can contribute to vitrification, biological glass formation that maintains structure and prevents crystallisation of other substances. The glasses formed have properties, in particular, the collapse temperature when the glass disappears and molecular mobility increases massively, which differ from experimentally produced sugar glasses, suggesting that other compounds, possibly proteins (see below), are also involved. Vitrification appears to occur only at very low water contents and has not yet been demonstrated in bryophytes and, possibly, it may not be important, as best survival is often at medium rather than higher or lower water potentials (Proctor 2008).

2. Protective proteins: LEAs (late embryogenesis abundant proteins) and HSPs (heat shock proteins) accumulate to high quantities during acquisition of desiccation tolerance in higher plants and are thought to play a primary role in desiccation tolerance (Buitink et al. 2002; Chaps. 13–15). Work with *Arabidopsis* has shown that mutants lacking many of the LEA proteins do not tolerate desiccation (Meurs et al. 1992). The situation in bryophytes is far from clear but dehydrins, a class of LEA (Chap. 14), have been identified in *T. ruralis* (Bewley et al. 1993) where they appear to be constitutively expressed and remain constant in quantity during desiccation and rehydration. It is certain that proteins will be involved in desiccation tolerance in bryophytes and lichens but the major difference to higher plants is that they appear to be constitutive, being always present. This is certainly true for desiccation-tolerant species like *T. ruralis* and also appears to occur in moderately tolerant species such as *Polytrichum formosum* as well (Proctor et al. 2007). There is also convincing evidence that protein synthesis is not necessary for protection during rehydration in some mosses (Proctor and Smirnoff 2000).
3. Amphiphilic metabolites: these partition from the cytoplasm into the lipid phase when their concentration in the cytoplasm increases upon water loss. Amphiphilic partitioning into membranes causes membrane disturbance and has positive aspects by possibly assisting the automatic insertion of antioxidants or phospholipase inhibitors into the membrane. It is possible that the membrane disturbance also leads to a reduction in some metabolic processes, such as respiration, and would assist with an orderly shut down during dehydration (Hoekstra et al. 2001).
4. Antioxidants: these have been reviewed for lichens by Kranner et al. (2008) and Beckett et al. (2008). It is suggested that during dehydration and rehydration (Weissman et al. 2005) there can be accumulation of ROS (reactive oxygen species) especially when absorbed light energy is in excess as photosynthesis declines with fall in water content. ROS are regarded as a potential major source of cellular damage through protein denaturation, pigment loss and lipid peroxidation in membranes (Smirnoff 1993; Kranner et al. 2008). Protection is by the presence of antioxidants such as glutathione, ascorbate, polyols, carbohydrates and many others. It has been shown that glutathione concentration correlates with desiccation tolerance in three lichens (Kranner 2002) but a fuller review shows that the presence and quantity of antioxidants does not show a simple linkage to desiccation tolerance (Kranner et al. 2008). An important point is that

the ever present, and in quantity, sugars and polyols are also effective antioxidants and this may be one of their important functions during desiccation. Certainly, during dehydration they can continue to function as enzymatic systems slow, or stop and, at very low water contents, they are the only substances that can continue to remove ROS. At present both the actual risk from ROS and the processes to scavenge ROS in lichens and bryophytes are not well understood. However, significant ROS production is possible during desiccation as the photosystems become uncoupled from carbon fixation and it is now becoming clear that systems exist even in desiccated lichens to minimise this risk (see Chap. 7).

6.2 Photosynthetic Response to Thallus Water Content

All bryophytes and lichens have a thallus water content (WC) that changes as the thallus tends to equilibrium with the environment. Changes in WC affect the rates of photosynthesis and respiration and bryophytes and lichens all show a similar pattern of response of net photosynthesis or respiration to WC. The response curve has important implications to the ecological performance of these organisms. Obviously, no carbon gain is possible unless they are hydrated, but there is also a complex interplay between water content, free water, water potential and dehydration and rehydration processes that contribute to bryophytes and lichens having very different ecologies. These complexities are explored in the following sections and then final comments are made on the need for physiological research to take the normal habitat into consideration when planning experiments and interpreting results.

6.2.1 Overall Structure of the Photosynthesis/Water Content Response

Figure 6.6 shows an idealised response of net photosynthesis (NP) to thallus water content (WC) derived from the lichen *Ramalina maciformis* (Lange et al. 2001; Fig. 6.2d). The main features of the curve are also applicable to bryophytes. WC is low to the left and it increases to the right. At very low WC, there is no detectable CO_2-exchange and metabolism is effectively halted. As WC increases NP starts and continues to rise with WC until a maximal value for NP is reached at WC_{opt}; further increases in WC can result in a steady maximal NP or some degree of decline in NP, so-called supraoptimal depression of NP (Lange et al. 1993).

The response curve can be divided into three sections: one is effectively a point, WC_{opt}, the WC at which maximal NP is first reached, which many studies have shown, occurs when the lichen or bryophyte reaches full turgor with a water

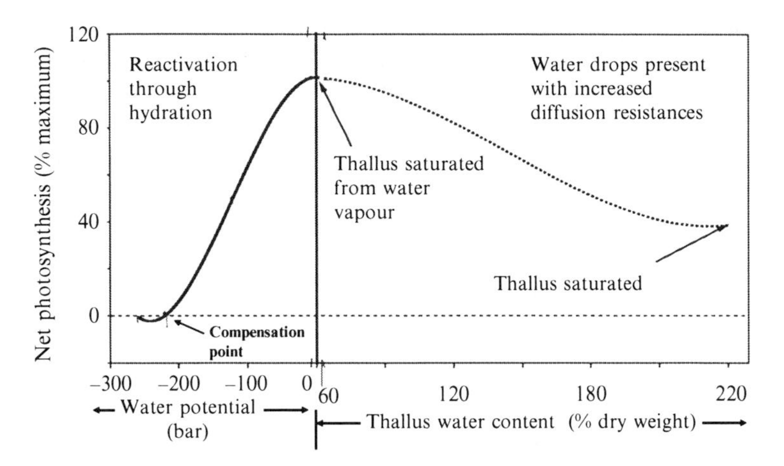

Fig. 6.6 Response of net photosynthesis (CO_2 exchange) to thallus water content; idealised curve from the lichen *Ramalina maciformis*. *Y*-axis is net photosynthesis as % of maximum; *X*-axis is thallus water content shown as water potential in bars below WC_{opt} and as % dry weight above WC_{opt}

potential close to zero (Proctor 2008). At low WC, to the left on the curve and lower than WC_{opt}, NP and the water potential steadily decline as WC falls and this can be called the *potential dominated* zone. At WC above WC_{opt} the additional water is free and external to the cells and cell walls and is in capillaries or spaces of various dimensions between cells, tissues or hyphae; this is the *external capillary water* zone of the response curve.

6.2.2 Thallus Water Content: The Limits

Thallus water content for bryophytes and lichens is normally related to dry weight:

$$WC\ (g\ H_2O\ g\ dw^{-1}) = [\ (fresh\ weight - dry\ weight)/dry\ weight\ ]$$

It is often expressed as a percentage so that a value of 100% indicates equal amounts of water and dry matter (1 g H_2O 1 g dw^{-1}).

Bryophytes typically have much higher maximal WC than lichens, values below 700% are not common and absolute maxima can be around 2,000% especially in *Sphagnum* species (Fig. 6.1g) (Dilks and Proctor 1979; Proctor 2008). Lichens show more variability depending on photobiont type and thallus structure (Fig. 6.2) as can be seen in the data from a study on New Zealand rain forest lichens (Lange et al. 1993, Fig. 6.7). The cyanobacterial lichen, *Collema laeve*, has a homoiomerous jelly thallus structure made up mainly of cyanobacterial filaments in gelatinous sheaths, and a maximal WC around 2,100% (value determined in laboratory). Heteromerous, foliose cyanobacterial lichens (Fig. 6.2b) have much lower maximal WC, from 417% for *P. dissimilis* to 1,175% for *Coccocarpia palmicola*. Heteromerous green algal lichens generally have a lower maximal WC, around 150%

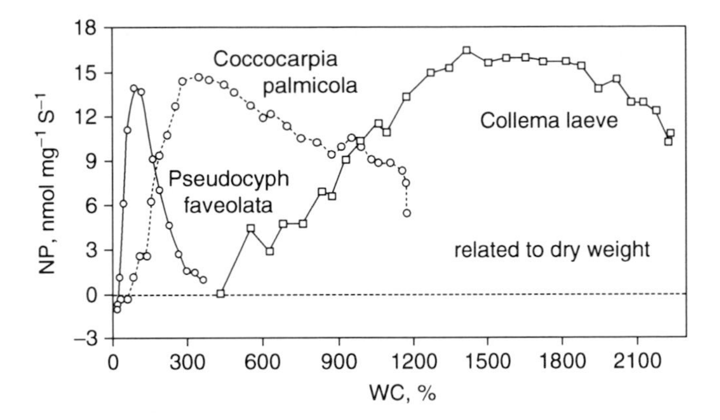

Fig. 6.7 Net photosynthesis (NP, CO_2 exchange) as a function of thallus water content for, from left to right, *Pseudocyphellaria faveolata*, a green algal lichen, *Coccocarpia palmicola*, cyanobacterial non-gelatinous lichen, and *Collema laeve*, a gelatinous homoiomerous lichen. From Lange et al. (1993)

(Richardson 1993) but also show a wide range from 189% for *P. colensoi* to 1,011% for *Sticta filix*. There also appears to be, although not yet conclusively demonstrated, a positive correlation in lichens between high maximal WC and a higher WC_{opt} (Fig. 6.7).

6.2.3 Water Content Response Curve: WC_{opt}

If we accept that WC_{opt} is approximately the lowest WC to achieve full turgor (Proctor 2008) then all water at that WC will be internal, in the sense that it is inside the plasmalemma or within the pores of the cell walls and adjacent gels but does not lie between cells or hyphae, and the wide range in WC_{opt} must represent structural differences in the cells or hyphae. For example, in bryophytes there will be a large difference in WC at full turgor between species with thick cell walls and those, like many thalloid liverworts (Fig. 6.1f), with large cells and very thin cell walls. Extracellular gels, like those around cyanobacteria in homoiomerous lichens, can support a large amount of water with little dry matter; a good laboratory example would be an 1% agarose gel which has a water content of around 9,900%.

6.2.4 Water Content Response Curve: The Ψ Dominated Zone

In both bryophytes and lichens, WC lower than WC_{opt} will be at negative water potentials and this has been confirmed many times as it is possible to construct this part of the response curve by equilibrating the thalli with the air of different relative humidities and therefore different water potentials (Cowan et al. 1979; Dilks and Proctor 1979; Schlensog et al. 2000). Any lichen or bryophyte in this part of the curve will have water relations dominated by the negative water potentials and we name this the *Ψ dominated zone*. Metabolism can continue to very low water potentials, metabolites that are involved in the tricarboxylic acid cycle were still involved in enzymatic reactions at 45% relative humidity, 20°C, approximately −100 MPa and WC <10% (Table 6.2; Cowan et al. 1979). Sugar alcohols were still being labelled at −56 MPa. These results act as a warning to physiologists studying air-dried lichens; at 50% relative humidity and 20°C, the lichen metabolism may still be active and not totally dormant as many expect.

There is a steady decline in net photosynthetic rate (NP, Fig. 6.6) with decline in WC below WC_{opt} until cessation at WC as low as 10–20% dw. This fall in NP with decrease in water potential occurs in all lichens and bryophytes but is rarely commented on. It is actually remarkable as the organisms seem able to lose and recover photosynthetic ability in the Ψ dominated WC range with no negative effects. No explanation for how NP is down-regulated seems to be available and it is worth considering this phenomenon in more detail. The key work by Nash et al. (1990) shows that the decline in NP occurs at the same rate whether the negative water potentials are generated by equilibration with humid air (the normal natural situation) or by osmotic solutions (NaCl or sorbitol). This suggests that water potential alone might be the controlling factor but does not completely remove the possibility that increased diffusion resistances brought about by increasing protoplasm viscosity within the cells could also be contributing.

Table 6.2 Tritium labelling of metabolites in *Ramalina celastri* (green algal, fruticose) at chosen atmospheric humidities (RH in %)

% RH								
Compound	100	86	75	66	54	45	33	15
Glu	***	***	**	*	**	***	t	–
Ala	***	***	***	***	**	**	?	–
GABA	**	**	**	**	t	t	–	–
Asp	*	*	t	t	t	t	–	–
Mal	*	*	*	t	t	t	–	–
Cit	*	**	*	t	t	t	–	–
Gln	***	***	***	t	t	t	–	–
U_1	**	**	**	**	*	t	–	–
Pentitol	***	***	**	**	–	–	–	–
Ser/Gly	***	**	t	t	–	–	–	–
Sucrose	***	t	–	–	–	–	–	–
Sugar-P	*	t	–	–	–	–	–	–

Labelled compounds were separated by two-dimensional paper chromatography, detected by scintillation autography, and identified by co-chromatography; from Cowan et al. (1979)
*** Heavily labelled, ** moderate label, * light label, t trace, ? uncertain, – none detected
Glu glutamate, *Ala* alanine, *GABA* γ-amino butyrate, *Asp* aspartate, *Mal* malate, *Cit* citrate, *Gln* glutamine, *Ser* serine, *Gly* glycine, *Sugar-P* sugar phosphates, U_1 compound tentatively identified as ethanolamine

Proctor (2008) showed that if water content is normalised to full turgor (relative water content = 1.0) then the NP decline with decreasing WC is identical for a moss, a liverwort and a higher plant (Fig. 6.8). This means a similar proportional depression of NP at the same water potential but different WC (these organisms have different values for WC_{opt}) which tends to not support a diffusion effect as the main control. However, it does suggest that the mechanism leading to decline in NP is identical in these very different photosynthetic organisms and this points strongly at Rubisco activity, an enzyme that they all possess, as the important factor influencing NP. Parry et al. (2002) showed in tobacco that decreased Rubisco activity under drought stress was due to the binding of a strong inhibitor to some of the enzyme molecules. Removal of the inhibitor immediately restored full Rubisco activity. Such a system of positive control would seem to be very applicable to lichens and bryophytes and would provide the required link between water potential and NP and also the possibility for the organism to rapidly change Rubisco activity. The search for a protein that can act as a water potential sensor has not yet been successful except for an example in fungi, the osmosensors SLN1 and SHO1 in yeast (Bartels and Sunkar 2005).

The above information suggests that lichens and bryophytes behave similarly in the Ψ dominated zone of the NP response curve. This is probably correct in terms of metabolism but, in terms of the amount of time spent under natural conditions at WC in the Ψ dominated zone, there appears to be a very large difference between the groups.

During dehydration both bryophytes and lichens are suggested to spend only brief periods in the Ψ dominated zone of the NP response curve to WC. The high maximal WC in bryophytes also means that, during dehydration, the plants spend only a small proportion of the time in the Ψ dominated zone. In bryophytes rehydration is referred to as being almost instantaneous creating in its own right a potentially stressful cellular event (Oliver 2008). Measurements by Dilks and Proctor (1979) show that bryophytes in damp, shady habitats stay wet for long

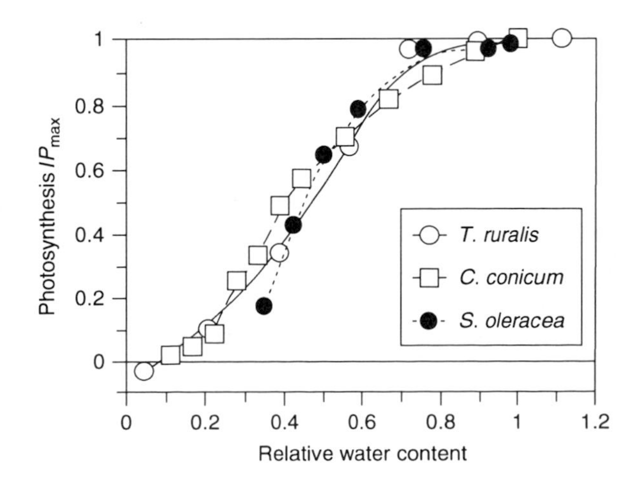

Fig. 6.8 Response of net photosynthesis (relative measure = actual NP/ maximal NP (P_{max}) to relative water content (maximal NP, P_{max}, is assumed to be at full turgor for each species at relative water content = 1). Results from Proctor (2008) for the moss *Tortula ruralis*, the thalloid liverwort *Conocephalum conicum* and the higher plant *Spinacia oleracea*

periods whilst those in more open sites show many periods of dryness. To bryophytes, a habitat is either wet, when they are active, or dry, when they are desiccated (Proctor 2008).

A position in the Ψ dominated zone on the response curve is assured if hydration is achieved by equilibration with humid air or if only a small amount of water is supplied (e.g. through dew fall, fog or mist). Rehydration in equilibrium with humid air can occur to such an extent that WC becomes high enough for respiration and photosynthesis to occur. This has been demonstrated for mosses, but only rarely. Lange (1969) surveyed several mosses and found all the pleurocarpous species to be capable of positive net photosynthesis after rehydration from humid air. In a field study on *T. ruralis* (Fig. 6.1b), hydration did occur to WC sufficient for positive NP but the overall net carbon balance was negative (Csintalan et al. 1999). The situation for lichens is completely different. It is now well demonstrated that most green algal lichens that have been tested can hydrate from water vapour alone to the extent that they become photosynthetically active (Lange and Bertsch 1965; Lange and Kilian 1985; Green et al. 2002; Pintado and Sancho 2002). In addition, early morning dew is sufficient to hydrate lichens to WC below WC_{opt} but still high enough for substantial photosynthesis to occur before dehydration by the morning sun. The classic papers are those on *R. maciformis* (Fig. 6.2d) in the Negev Desert in which the morning "gulp" of photosynthetic carbon fixation driven by overnight humidity and dew hydration is very clear (Lange et al. 1970). In these dew/humidity situations the lichens rarely reach WC_{opt} and almost all active time is spent in the Ψ dominated zone of the WC–response curve.

Cyanobacterial lichens, in contrast to the green algal species (Fig. 6.2b), do not show net photosynthesis when hydrated by water vapour alone but require liquid water (Lange et al. 1986, 1988). It has been suggested that liquid water needs to be present for energy transfer from phycobilins to the main photocentres in cyanobacteria (Bilger et al. 1989) but the exact explanation is still unclear.

It is important to understand that, although cyanobacterial and green algal lichens are very different in their ability to achieve positive NP, they differ little or not at all in the WC reached when equilibrated with air humidity as shown in Fig. 6.9 (Schlensog et al. 2000). The reason for the difference in activating NP appears to be simple, the WC required to achieve positive NP, i.e.: to exceed the water compensation point, is higher in cyanobacterial than in green algal lichens. Lange et al. (1988) compared green algal and cyanobacterial sectors from photosymbiodemes (Fig. 6.2b) and the WC for NP compensation was on average 20% (range 15–30%) for the five green algal lichens and 85 and 180% for the two cyanobacterial species. The latter never achieved positive NP in such experiments because the WC reached in equilibrium with 96.5% relative humidity at 15°C was 55%. Lichens and mosses with high maximal WC also tend to have high WC for NP compensation and, because bryophytes generally have much higher maximal WC than lichens, one would expect that humidity activation would be much less frequent in bryophytes. This seems to be the case as activation by humid air alone is rarely reported (Lange 1969; Csintalan et al. 1999).

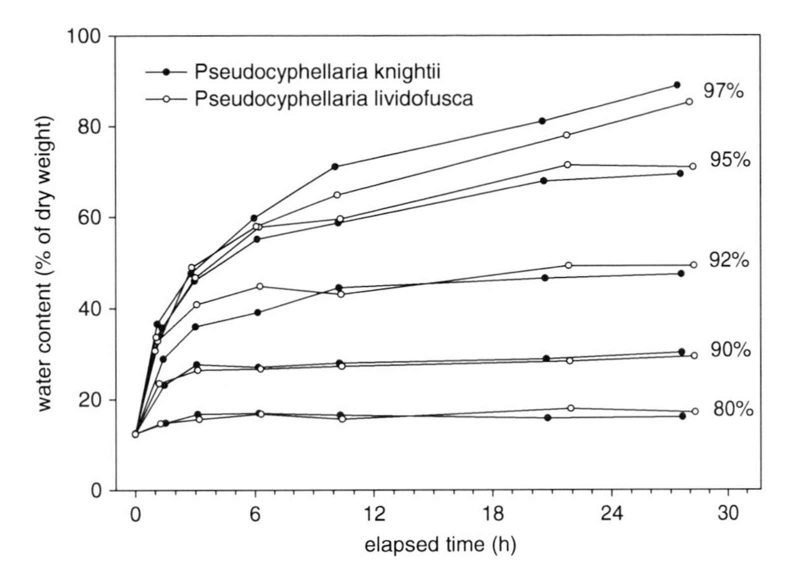

Fig. 6.9 Time course for change in thallus water content (% dw) for the green algal *Pseudocyphellaria lividiofusca* (*open symbols*) and the cyanobacterial *P. knightii* (*closed symbols*), two species that also occur as a combined photosymbiodeme. Thalli were exposed at 15°C to air of relative humidities 80, 90, 92, 95 and 97% for up to 30 h. At any selected humidity there is little difference between the water content increase in cyanobacterial and green algal lichens

The Ψ dominated zone of the response curve is, therefore, a niche that is occupied often by green algal lichens but not by bryophytes. The ecological implications of this ability of lichens are discussed in the later Sect. 6.4.1.

6.2.5 Water Content Response Curve: External Water Zone

Any water at WC above WC_{opt} is principally external water (Proctor 2008) and is held outside the cell walls or gels in spaces between cells, hyphae or formed by larger structures like moss and liverwort leaves. This water is at very high potential, at or close to zero, and can be lost through desiccation without affecting the water potential of the lichen or bryophyte. Water potential will only start to decline once WC falls below WC_{opt}. The external water zone is the part of the NP to water content response curve that does not exist in homoiohydric higher plants which have a water status that is similar to, or slightly lower than, WC_{opt} in lichens and bryophytes (Proctor 2008; Lange et al. 2001; Green and Lange 1994). Diffusion of small molecules like CO_2 is about 10,000 times slower in water than in air and, because of this, external water poses a potential problem to both lichens and bryophytes because it can block air diffusion pathways and cause decreased NP. The linkage between the depression in NP and increasing diffusion resistances within the lichen thallus has been demonstrated using Helox techniques (Cowan et al. 1992).

The problem is worse where CO_2 diffusion occurs through small pores as in the thalloid liverwort *Marchantia*; a 10-μm thick layer of water over the pores would almost completely eliminate NP (Green and Snelgar 1982). It is no surprise, therefore, that many lichens (Lange et al. 1993) and bryophytes (Dilks and Proctor 1979) show depressed NP at WC greater than WC_{opt} as shown for *R. maciformis* in Fig. 6.6 (see also Fig. 6.7). The depression in NP at high WC is extremely variable and appears to be species specific; few ideas exist as yet for the structural basis for the various responses of NP to high WC produced by the different lichens.

The ability to reconcile the potentially conflicting requirements of water storage and free gas exchange for photosynthesis is an area of major difference between bryophytes and lichens. There is no doubt that the depressed NP at WC greater than WC_{opt} is a much more common and much stronger phenomenon in lichens than in bryophytes (Green and Lange 1994). Some lichens show massive depressions in NP, *Pseudocyphellaria pubescens*, for example, has around 10% maximal NP at a WC of 200% and WC_{opt} at 100% (Lange et al. 1993). In contrast, *Sticta caliginosa* shows no depression at all even at 500% WC. Mosses and liverworts (Fig. 6.1) also often show some depression of NP at high WC but to a much lesser extent, about 20%, and at much higher WC than lichens, around 1,000–1,500% (Dilks and Proctor 1979). Some bryophyte species can reach very high WC with little sign of any effect on NP, 1,400% for *Hookeria lucens* (Dilks and Proctor 1979) and *Weymouthia mollis* (Plate 6.4g) (Snelgar et al. 1980).

The underlying reasons for the ability of bryophytes to handle much higher water contents without great effect on NP have been thoroughly analysed by Proctor (2008). Many bryophytes have simple leaves, often only one cell thick, and, by various means, separate water storage from gas exchange (Fig. 6.1b). One method is to simply have concave leaves with the outer i.e. lower surface of the leaf water repellent (often by wax deposits) and the water stored on the hydrophilic concave upper surface. This is a highly efficient structure with photosynthetic cells carrying out gas exchange on the convex side and accessing stored water on the other. This is taken further in many bryophyte species by arranging shoot systems with closely overlapping concave leaves, the inner surfaces functioning for storage and the outer for photosynthesis, with the added advantage that water movement along the stem is also possible by capillarity. Good examples of this are *Weymouthia* spp., *Pleurozium schreberi* and *Scleropodium* spp. (Proctor 2008). Rundel and Lange (1980) analysed NP and water content in the desert moss *Barbula aurea* and show the use of concave leaves for water storage at higher WC (Fig. 6.10). Some species have papillate or mammillate leaf surfaces on which the water is moved and stored between the papillae or mamillae and gas exchange occurs through the water repellent apices (Dilks and Proctor 1979).

This level of structural complexity is apparently not possible in lichens (Green and Lange 1994). The vast majority of lichens are heteromerous with the photobionts embedded within the main structure of the lichen which is composed of fungal hyphae (Fig. 6.2h). The possibilities for water storage do not appear to be extensive. Water can be stored within the cortex or medulla and this has been visualised with suitable electron microscope techniques (Scheidegger et al. 1995,

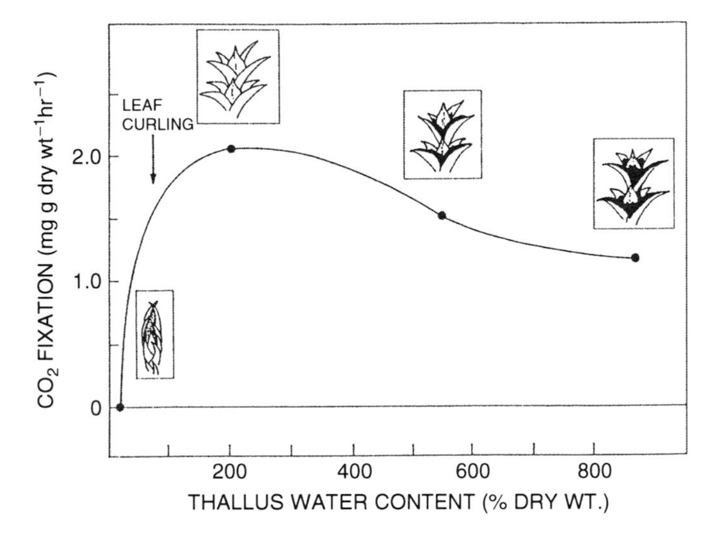

Fig. 6.10 Net photosynthesis (CO_2 exchange) as a function of thallus water content at a temperature of 20°C and saturating irradiance for the desert moss *Barbula aurea*. The *inset* pictures indicate the status of the thalli at low water content, curled leaves, at WC_{opt} with fully expanded leaves but no external water, and at higher WC with the external water in *black* on the concave side of the leaves leaving the outer convex side mainly free for gas exchange. From Rundel and Lange (1980)

1997; Honegger 1991, 2008; Souza-Egipsy et al. 2002). Experimental filling of the air spaces in the medulla of lichens leads to a major decrease in CO_2 diffusion and NP as shown by Cowan et al. (1992). The lowest diffusion resistance occurs when the lichen is hydrated by humidity so that no external water is present. This has been demonstrated to be important in the field by Pintado and Sancho (2002) looking at the photosynthetic ecology of *Ramalina capitata*. When hydrated by humid air alone this lichen has higher NP at any selected WC below WC_{opt} compared to thalli at the same WC but which have reached it after drying down from higher WC produced by hydration with liquid water (Fig. 6.11). This investigation also showed that dark respiration was much lower at WC below WC_{opt} after hydration by humid air compared to liquid water activation (Fig. 6.11). Because of this, photosynthetic efficiency [calculated following Palmqvist (2000)] was twice as high for the humidity activated thalli representing a much increased carbon gain. It appears that this gain in efficiency occurs because the mycobiont is proportionately less activated than the photobiont. This might be a strategy of more general occurrence. The vagrant lichen *Aspicilia fruticosa* shows a similar effect with maximal NP at a very low 60% whilst dark respiration is still increasing at its maximal WC of 120% (Fig. 6.12; Sancho et al. 2000).

The structure of a heteromerous lichen thallus, with larger voids in the medulla than in the cortices (Fig. 6.2g), prevents water entry into the medulla unless forced in under pressure. The presence of hydrophobins on medullar and photobiont surfaces also prevents medulla flooding (Honegger 2008). Should water ever enter the

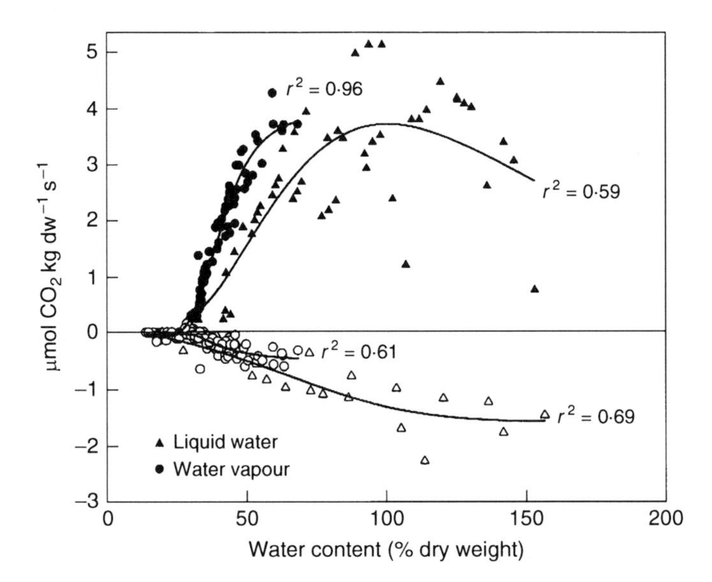

Fig. 6.11 Dependence of net photosynthesis (*solid symbols*, PPFD 250 μmol m^{-2} s^{-1}) and dark respiration (*open symbols*) on water content during desiccation of a hydrated thallus (*triangles*) and water uptake from water vapour (*circles*) of *Ramalina capitata* var. *protecta* at 10°C. From Pintado and Sancho (2002)

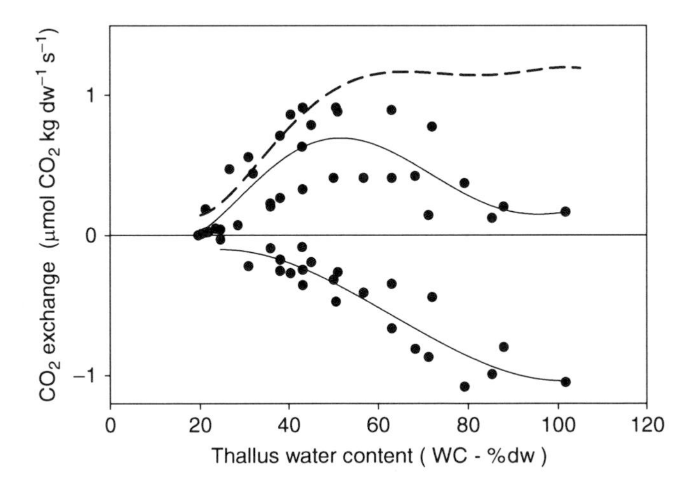

Fig. 6.12 Effect of thallus water content (WC – % dw) on net photosynthesis (NP – *upper solid line*), dark respiration (DR – *lower solid line*) and gross photosynthesis (GP = NP + DR – *dashed line*) related to dry weight for *Aspicilia fruticuloso-foliaceae* at 400 μmol m^{-2} s^{-1} and 15°C. From Sancho et al. (2000)

medullar space, then the hydrophobins by forcing the formation of small water droplets with a higher water potential encourage the medulla to dry out by water transfer through the vapour phase from the droplet to capillary spaces and cellular structures elsewhere in the thallus. The structural limitations of the lichen thallus, in particular the lack of ability to construct photosynthetic tissues in sheets that allow water storage on one side and gas exchange on the other, mean that the high water storage without diffusion pathway blockage achieved by bryophytes is not readily attainable by lichens. The nearest equivalents are the complex thalli of *Pseudocyphellaria* and *Sticta* species which carry out CO_2-exchange entirely through cyphellae or pseudocyphellae, special pore structures in their lower cortex, thus freeing much of the rest of the thallus surface for water storage (Green et al. 1985). We can now see that bryophytes and lichens, because of their structures tend to occupy two different photosynthetic niches; lichens occupy the Ψ dominated, low WC part of the WC response curve whilst bryophytes, with their high levels of water storage, occupy the external water zone.

There is, however, some room for caution as the different strategies may not always be clearly divided between lichens and bryophytes. In a rare study Lange et al. (1993) compared the maximal WC determined by standard laboratory procedure with values measured in the field for 21 lichen species growing in a New Zealand temperate rain forest (Fig. 6.13). Almost all lichens, the obvious exception were the ground fruticose species, reached WC in the field during rainfall that were several times higher than the laboratory value. The extreme was *Pseudocyphellaria rufovirescens* that had a maximal field value of 3,116%, about five times higher than the 613% measured by typical laboratory procedures. The difference was put down to the adhering water droplets in the field (which usually are removed by shaking in laboratory studies in order to receive reproducible amounts of water

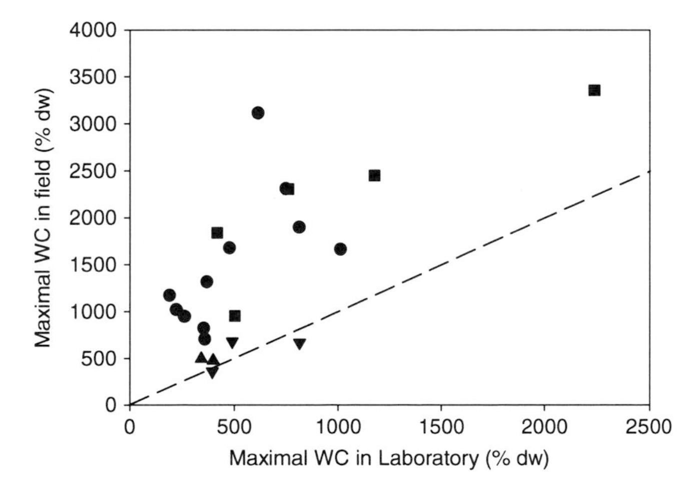

Fig. 6.13 A comparison of maximal thallus water content measured in the field (*Y*-axis) and in the laboratory (*X*-axis) for 21 lichen species from a New Zealand temperate rain forest; data from Lange et al. (1993). Explanation for symbols: *closed square* – foliose cyanobacterial lichens; *closed circle* – foliose green algal lichens; *closed square* – fruticose on trees; *downward closed square* – fruticose on ground. *Dotted line* indicates the 1:1 relationship, data points above the line have a higher maximal water content in the field than found in the laboratory

content) and the form of the lichens (e.g. faveolate) that provided water storage possibilities in the depressions in the upper cortex. Water storage at this site does not affect NP as CO_2-exchange occurs completely through the pseudocyphellae in the lower cortex (Green et al. 1981, 1985). It appears, in the evergreen rain forest in New Zealand, that lichens are able to adopt the external water storage strategy similar to bryophytes. A possible reason for this is the constant low PPFD level in these forests, usually below 4% of full sunlight, so that high light stress is unlikely to occur however long the lichens remain active. However, this is an area where more data is needed before any greater generalisations can be made.

6.3 Aligning Physiology with Habitat

Most studies on desiccation tolerance look at the effect of some variable e.g. depth of desiccation, on survival of the bryophyte or lichen. It is then possible to rank the tested organisms against the tested variable. In Fig. 6.14 a dehydration/rehydration event is shown schematically. Below the event line are the environmental variables that can influence survival of the organism. While desiccated the depth and length of the desiccation are the main variables, the speed of desiccation will also be important and also whether there has been any prehardening following an earlier

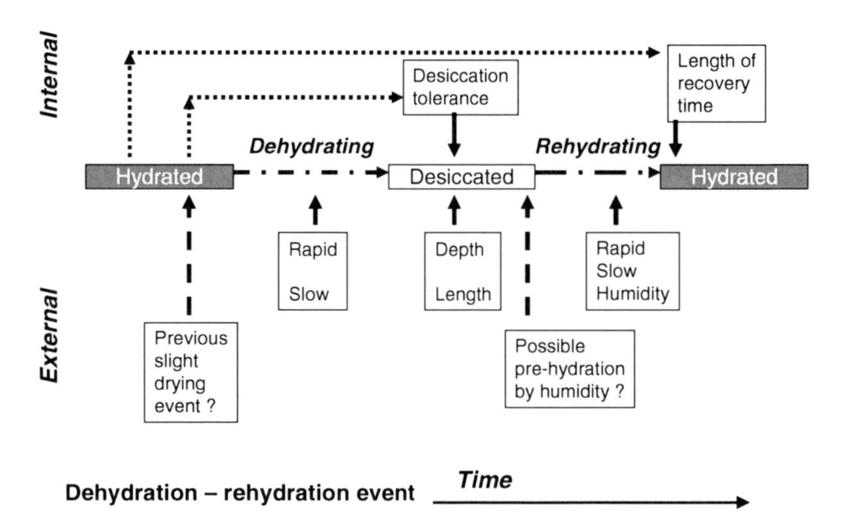

Fig. 6.14 Environmental and organism drivers of desiccation tolerance in lichens and bryophytes. A hydrated – desiccated – rehydrated time line is shown running left to right; below this event line are the various ways in which the environment can affect the survival of the event. *Solid black arrows* pointing up show impacts that will normally always occur e.g. slow or rapid drying; whilst *dashed arrows* pointing up are impacts that may not be consistent. Above the event line are the organism drivers that determine final recovery. *Dotted arrows* above the event line show that these organism drivers can be influenced by events in the previous hydrated stage e.g. season, desiccation event

desiccation event. Rehydration can also be slow or fast with prehydration by humidity being a possible additional factor. Above the event line are properties of the organism that can influence survival. These include its desiccation tolerance and recovery time after rewetting. Both of these properties are apparently under the control of the organism as previous events will initiate hardening, and the length and stability of the previous active period is inversely related to recovery speed (Schlensog et al. 2004).

The message from this schematic, and from much of the research that has been carried out, is that desiccation tolerance is dynamic, can be altered rapidly, and is almost certainly set at a level that is suitable for the organism to survive in that particular environment. Such acclimation, or selection over a longer period, would represent a strategy of optimal resource use.

We suggest that this should be kept in mind and, rather than just ranking organisms against a treatment factor, it might be better to consider exactly what the environment of the organism is and how this will set the desiccation tolerance need. The schematic shows that this might not be an easy task but it would be much more useful in gaining an understanding of the ecological performance of bryophytes and lichens.

6.4 Ecophysiological Implications of Hydration, Rehydration and the NP Response to WC

6.4.1 What Constrains the Bryophyte/Lichen Niche?

There seems little doubt that some forms of bryophytes and lichens were the first multicellular photosynthetic organisms to colonise the land at some time before the Silurian when the first vascular plants appeared (Proctor 2000). The strategy of poikilohydry was almost certainly primitive and evolved from the production of desiccation-tolerant spores (Oliver et al. 2000). Organisms employing poikilohydry are confined to a small overall size, in particular because of the limitations to water transport that relies on capillarity (Proctor and Tuba 2002). The evolution of homoiohydric higher plants with their greater size, ability to exploit soil water sources and larger productivity would have removed poikilohydric plants from many habitats but created the new epiphytic habitat that is now so important (Chap. 5; Figs. 6.3f, h and 6.4d, e, g).

It is possible to define broadly the constraints that determine the present day lichen and bryophyte niche (after Green et al. 2008; Chap. 5)

(a) They are excluded from habitats that favour higher plants.
(b) They are favoured, and higher plants excluded, from habitats with low water storage such as rock surfaces (Figs. 6.3c, e, h and 6.4b); these habitats can have high or low, $\pm$ erratic absolute water availability, but if the presence of the

water is transient then poikilohydry is preferred. Many epiphytic habitats (Figs. 6.3h and 6.4d, e, g) are physiologically dry (too low water storage for higher plants to exploit).

(c) They are favoured in habitats where water is not delivered by rain but rather in the form of fog, dew or high humidity (Figs. 6.3d and 6.4g). These sources are almost inaccessible to higher plants.

(d) They are favoured in habitats with low mean temperatures, examples are where trees are excluded from polar and alpine zones (Fig. 6.4a, c), and all higher plants from more extreme locations like Antarctica (Figs. 6.3a and 6.4a).

6.4.2 Lichens Versus Bryophytes: The Differences

The above constraints are applicable to both lichens and bryophytes but there are major differences between the two groups and these relate to their NP response to WC.

Lichens. green algal lichens (about 90% of lichens) exploit the Ψ dominated zone of the NP response curve and are able to spend a considerable part of their active time at WC below WC_{opt}. This is possible because lichens hydrate rapidly in equilibrium with humid air so that a WC of around 55% is reached in under 6 h at 95% RH (Fig. 6.9). Photosynthetic compensation is at around 20% WC so a WC of 55% is sufficient for reasonable rates of NP, between 38 and 69% of maximal NP depending on the species (Lange et al. 1988). Dew fall is also sufficient to allow photosynthesis by these lichens (Belnap and Lange 2003). All NP below WC_{opt} is, of course suboptimal but it is now becoming clear that lichens spend most of their active time under suboptimal conditions and NP rarely reaches maximal values (Lange 2003a, b; Green et al. 2008). In addition to suboptimal NP due to low WC, for many lichens there is also the time at high water contents when NP can also be depressed due to increased diffusion resistances (Lange 2003a, b). An additional contributor to suboptimal performance at all WC is low light; lichens dry out rapidly and are only rarely photosynthetic at light levels suitable to give maximal NP (Lange 2003a). Cyanobacterial lichens cannot achieve positive NP with humidity alone; however, Lange et al. (1998) showed that the soil crust lichen *Collema tenax* followed a different strategy so that, when hydrated by rainfall it achieved substantial carbon fixation at high light and temperatures, conditions avoided by green algal species at the same site. Cyanobacterial, gelatinous homoiomerous lichens seem to follow a similar strategy to bryophytes.

Bryophytes. although there are few data available it appears that most bryophytes gain little carbon at WC suboptimal for NP, i.e. at WC below WC_{opt}, but profit instead from their high maximal WC and ability to continue photosynthesis for longer periods. A good example of this is the continuous recordings of *Hennediella heimii* in continental Antarctica (Pannewitz et al. 2003). This moss, which grows in an area kept wet by meltwater from a glacier (Fig. 6.4a), was continuously hydrated and active for the entire recording period of just over 2 weeks with continuous light and maximal light levels reaching 75% of nominal full sunlight. The measurements by Dilks and Proctor (1979) also show mosses in shady habitats to remain wet and active for long periods (Fig. 6.4d). In an opposite strategy to most lichens, bryophytes can construct cushions and turfs in order to enhance water storage for longer activity (Figs. 6.1a, h and 6.4a, b, f). It should be noted that to benefit from this longer active period the bryophytes must be able to tolerate very high light levels when active, often full sunlight (Plates 6.1h and 6.4b), something that lichens rarely have to do.

6.5 Conclusions

Lichens and bryophytes are both poikilohydric and share many features related to this strategy. Both groups in general have the ability to withstand dehydration and sometimes exceptionally deep desiccation. However, in both groups many species are not highly desiccation tolerant or highly desiccation sensitive but occupy a middle ground between these extremes, and this suggests that there is strong organism control over this property.

The groups utilise different strategies with respect to thallus water content. Green algal lichens typically have lower maximal WC and a low WC for NP compensation. This allows these lichens to rehydrate with substantial photosynthesis by equilibrium with humid air or by dew fall of fog. In general, when these lichens are active the environmental conditions, especially light, are suboptimal for photosynthesis. Many bryophytes and cyanobacterial lichens seem to operate a different strategy that emphasises high WC and water storage. This is achieved by two different methods, the lichens employ gels to build water storage whilst the bryophytes are able to build structures that separate water storage and gas exchange. In many cases this will require that these species can tolerate very high light levels while active.

The two strategies given above are founded on use of different sectors of the NP to water content response curve so that the two groups are exploiting two different niches.

At the start of this chapter we noted that lichens, despite the commonly held view, seem not to tolerate long periods of desiccation (in terms of many months or years), even poilkilohydric ferns do better, and of course spores and seeds. The question is, is this reality or are we simply looking at an under-researched area. We suggest that this is reality and the cause is the ability of lichens to rapidly equilibrate with humid air, something that can happen overnight in even the driest deserts. Some lichens can withstand very severe desiccation to WC $<2\%$ dw, as occurs when heated to over 60°C in desert areas, but the length is usually a small number of months, rather than years.

References

Alpert P (2000) The discovery, scope, and puzzle of desiccation tolerance in plants. Plant Ecol 151:5–17

Alpert P (2005) The limits and frontiers of desiccation-tolerant life. Integr Comp Biol 45:685–695

Alpert P, Oliver MJ (2002) Drying without dying. In: Black M, Prichard HW (eds) Desiccation and survival in plants: drying without dying. CABI, Wallingford, UK, pp 3–43

Aubert S, Juge C, Boisson AM, Gout E, Bligny R (2007) Metabolic processes sustaining the reviviscence of lichen *Xanthoria elegans* in high mountain environments. Planta 226: 1287–1297

Bartels D, Sunkar R (2005) Drought and salt tolerance in plants. Crit Rev Plant Sci 24:23–58

Beckett RP (1999) Partial dehydration and ABA induce tolerance to desiccation-induced ion leakage in the moss *Atrichum androgynum*. S Afr J Bot 65:212–217

Beckett RP, Hoddinott N (1997) Seasonal variations in tolerance to ion leakage following desiccation in the moss *Atrichum androgynum* from a KwaZulu-Natal afromontane forest. S Afr J Bot 63:276–279

Becket RP, Minibayeva FV (2008) Desiccation tolerance in lichens. In: Jenks MA, Wood AJ (eds) Plant desiccation tolerance. Blackwell, Iowa, pp 91–114

Beckett RP, Csintalan V, Tuba Z (2000) ABA treatment increases both the desiccation tolerance of photosynthesis, and nonphotochemical quenching in the moss *Atrichum undulatum*. Plant Ecol 151:65–71

Beckett RP, Kranner I, Minibayeva FV (2008) Stress physiology and the symbiosis. In: Nash TH III (ed) Lichen biology, 2nd edn. Cambridge University Press, Cambridge, pp 134–151, viii + 486 pages

Belnap J, Lange OL (eds) (2003) Biological soil crusts: structure, function, and management. Springer, Berlin, p 503

Bewley JD (1979) Physiological aspects of desiccation tolerance. Annu Rev Plant Physiol 30:195–238

Bewley JD, Reynolds TL, Oliver MJ (1993) Evolving strategies in the adaptation to desiccation. In: Close TJ, Bray EA (eds) Plant responses to cellular dehydration during environmental stress, vol 10, Current topics in plant physiology: American society of plant physiologists series. American Society of Plant Physiologists, Rockville, MD, pp 193–201

Bilger W, Rimke S, Schreiber U, Lange OL (1989) Inhibition of energy-transfer to photosystem II in lichens by dehydration: different properties of reversibility with green and blue-green phycobionts. J Plant Physiol 134:261–268

Billi D, Potts M (2002) Life and death of dried prokaryotes. Res Microbiol 153:7–12

Bisby GR (1945) Longevity of *Schizophyllum commune*. Nature 155:732–733

Breuil-Sée A (1993) Recorded desiccation-survival times in bryophytes. J Bryol 17:679–684

Bristol BM (1916) On the remarkable retention of vitality in moss protonema. New Phytol 15:137–143

Buitink J, Hoekstra FA, Leprince O (2002) Biochemistry and biophysics of tolerance systems. In: Black M, Pritchard HW (eds) Desiccation and survival in plants: drying without dying. CABI, Wallingford, Oxon, pp 293–318

Clausen E (1952) Hepatics and humidity. A study of the occurrence of hepatics in a Danish tract and the influence of relative humidity on their distribution. Dansk Botanisk Arkiv 15:1–80

Clausen E (1964) The tolerance of hepatics to desiccation and temperature. Bryologist 67:411–417

Cowan DA, Green TGA, Wilson AT (1979) Lichen *metabolism*. 1. The use of tritium labelled water in studies of anhydrobiotic metabolism in *Ramalina celastri* and *Peltigera polydactyla*. New Phytol 82:489–503

Cowan IR, Lange OL, Green TGA (1992) Carbon-dioxide exchange in lichens: determination of transport and carboxylation characteristics. Planta 187(2):282–294

Csintalan Z, Proctor MCF, Tuba Z (1999) Chlorophyll fluorescence during drying and rehydration in the mosses *Rhytidiadelphus loreus* (Hedw.)Warnst., *Anomodon viticulosus* (Hedw.) Hook & Tayl. and *Grimmia pulvinata* (Hedw.) Sm. Ann Bot 84:235–244

Dietz S, Hartung W (2008) Abscisic acid in lichens: variation, water relations and metabolism. New Phytol 138:99–106

Dilks TJK, Proctor MCF (1974) The pattern of recovery of bryophytes after desiccation. J Bryol 8:97–115

Dilks TJK, Proctor MCF (1979) Photosynthesis, respiration and water content in bryophytes. New Phytol 82:97–114

Elbert W, Weber B, Büdel B, Andreae MO, Pöschl U (2009) Microbiotic crusts on soil, rock and plants: neglected major players in the global cycles of carbon and nitrogen. Biogeosci Discuss 6:6983–7015, www.biogeosciences-discuss.net/6/6983/2009/

Farrar JF (1976) Ecological physiology of the lichen *Hypogymnia physodes*. II. Effects of wetting and drying cycles and the concept of physiological buffering. New Phytol 77:105–113

Gaff DF (1977) Desiccation tolerant vascular plants of Southern Africa. Oecologia 31:95–109

Green TGA, Lange OL (1994) Photosynthesis in poikilohydric plants: a comparison of lichens and bryophytes. In: Schulze E-D, Caldwell MC (eds) Ecophysiology of photosynthesis. Ecological Studies. Springer, Berlin, pp 319–341

Green TGA, Snelgar WP (1982) A comparison of photosynthesis in two thalloid liverworts. Oecologia 54:275–280

Green TGA, Snelgar WP, Brown DH (1981) Carbon dioxide exchange in lichens: CO_2 exchange through the cyphellate lower cortex of *Sticta latifrons* Rich. New Phytol 88:421–426

Green TGA, Snelgar WP, Wilkins AL (1985) Photosynthesis, water relations and thallus structure of Stictaceae Lichens. In: Brown DH (ed) Lichen physiology. Plenum, New York

Green TGA, Kilian E, Lange OL (1991) *Pseudocyphellaria dissimilis*: a desiccation-sensitive, highly shade-adapted lichen from New Zealand. Oecologia 85:498–503

Green TGA, Schlensog M, Sancho L, Winkler J, Broom FD, Schroeter B (2002) The photobiont (cyanobacterial or green algal) determines the pattern of photosynthetic activity within a lichen photosymbiodeme: evidence obtained from *in situ* measurements of chlorophyll *a* fluorescence. Oecologia 130:191–198

Green TGA, Nash TH III, Lange OL OL (2008) Physiological ecology of carbon dioxide exchange. In: Nash TH III (ed) Lichen biology, 2nd edn. Cambridge University Press, Cambridge, pp 152–181, viii + 486 pages

Hickel B (1967) Contributions to the knowledge of a xerophilic water plant, *Chamaegigas intrepidus*. Int Rev Gesamten Hydrobiol 53:361–400

Hoekstra FA, Golovina EA, Buitink J (2001) Mechanisms of plant desiccation tolerance. Trends Plant Sci 6:431–438

Honegger R (1991) Functional aspects of the lichen symbiosis. Annu Rev Plant Physiol Plant Mol Biol 42:553–578

Honegger R (2008) Morphogenesis. In: Nash TF III (ed) Lichen biology, 2nd edn. Cambridge University Press, Cambridge, pp 69–93, viii + 486 pages

Johnson A, Kokila P (1970) The resistance to desiccation of ten species of tropical mosses. Bryologist 73:682–686

Kappen L, Valladares F (2007) Opportunistic growth and desiccation tolerance, the ecological success of the poikilohydrous strategy. In: Pugnaire F, Valladares F (eds) Functional plant ecology. Taylor & Francis, New York, pp 8–65

Kranner I (2002) Glutathione status correlates with different degrees of desiccation tolerance in three lichens. New Phytol 154:451–460

Kranner I, Beckett R, Hochman A, Nash TH III (2008) Desiccation-tolerance in lichens – A review. Bryologist 111:576–593

Lange OL (1953) Hitze- und Trockenresistenz der Flechten in Beziehung zu ihrer Verbreitung. Flora 140:39–97

Lange OL (1955) Untersuchungen über die Hitzresistenz der Moose in Beziehung zu ihrer Verbreitung. II. Die Resistenz stark ausgetrockneter Moose. Flora 142:381–399

Lange OL (1969) CO_2-Gaswechsel von Moosen nach Wasserdampfaufnahme aus dem Luftraum. Planta 89:90–94

Lange OL (2003a) Photosynthetic productivity of the epilithic lichen *Lecanora muralis*: long-term field monitoring of CO_2 exchange and its physiological interpretation. II. Diel and seasonal patterns of net photosynthesis and respiration. Flora 198:55–70

Lange OL (2003b) Photosynthetic productivity of the epilithic lichen *Lecanora muralis*: long-term field monitoring of CO_2 exchange and its physiological interpretation. III. Diel, seasonal, and annual carbon budgets. Flora 198:277–292

Lange OL, Bertsch A (1965) Photosynthese der Wustenflechte *Ramalina maciformis* nach Wasserdampfaufnahme aus dem Luftraum. Naturwissenschaften 52:215–216

Lange OL, Green TGA (2005) Lichens show that fungi can acclimate their respiration to seasonal changes in temperature. Oecologia 142:11–19

Lange OL, Kilian E (1985) Reaktivierung der Photosynthese trockener Flechten durch Wasserdampfaufnahme aus dem Luftraum: Artspezifisch unterschiedliches Verhalten. Flora 176:7–23

Lange OL, Schulze E-D, Koch W (1970) Experimentell-ökologische Untersuchungen an Flechten der Negev-Wueste. II- CO_2-Gaswechsel und Wasserhaushalt von *Ramalina maciformis* (Del.) Bory am natürlichen Standort während der sommerlichen Trockenperiode. Flora 159:38–62

Lange OL, Kilian E, Ziegler H (1986) Water vapour uptake and photosynthesis of lichens: performance differences in species with green and blue-green algae as phycobionts. Oecologia 71:104–110

Lange OL, Green TGA, Ziegler H (1988) Water status related photosynthesis and carbon isotope discrimination in species of the lichen genus *Pseudocyphellaria* with green or blue-green photobionts and in photosymbiodemes. Oecologia 75:494–501

Lange OL, Büdel B, Heber U, Meyer A, Zellner H, Green TGA (1993) Temperate rainforest lichens in New Zealand: high thallus water content can severely limit photosynthetic CO_2 exchange. Oecologia 95:303–313

Lange OL, Belnap J, Reichenberger H (1998) Photosynthesis of the cyanobacterial soil-crust lichen *Collema tenax* from arid lands in southern Utah, USA: role of water content on light and temperature responses of CO_2 exchange. Funct Ecol 12:195–202

Lange OL, Green TGA, Heber U (2001) Hydration-dependent photosynthetic production of lichens: what do laboratory studies tell us about field performance. J Exp Bot 52:2033–2042

Larson DW (1987) The absorption and release of water by lichens. In: Peveling E (ed) Progress and problems in lichenology in the eighties. Bibliotheca Lichenologica No. 25. J. Cramer, Berlin, pp 351–360

Larson DW (1989) The impact of ten years at $-20°C$ on gas exchange in five lichen species. Oecologia 78:87–92

Malta N (1921) Versuche über die Widerstandsfähigkeit der Moose gegen Austrocknung. Acta Univ Latviensis 1:125–129

Meurs C, Basra AS, Karssen CM, van Loon LC (1992) Role of abscisic acid in the induction of desiccation tolerance in developing seeds of *Arabidopsii thaliana*. Plant Physiol 98:1484–1493

Mueller DMJ (1972) Observations on the ultrastructure of *Buxbaumia* protonema. Bryologist 75:63–68

Nash TH III, Reiner A, Demmig-Adams B, Kilian E, Kaiser WM, Lange OL (1990) The effect of atmospheric desiccation and osmotic water stress on photosynthesis and dark respiration of lichens. New Phytol 116:269–276

Oliver MJ (2008) Biochemical and molecular mechanisms of desiccation tolerance in bryophytes. In: Goffinet B, Shaw AJ (eds) Bryophyte biology. Cambridge University Press, Cambridge, pp 269–298

Oliver MJ, Tuba Z, Mishler BD (2000) The evolution of vegetative desiccation tolerance in land plants. Plant Ecol 151:85–100

Palmqvist K (2000) Carbon economy in lichens. New Phytol 148:11–36

Pannewitz S, Green TGA, Scheidegger C, Schlensog M, Schroeter B (2003) Activity pattern of the moss *Hennediella heimii* (Hedw.) Zand. in the Dry Valleys, Southern Victoria Land, Antarctica during the mid-austral summer. Polar Biol 26:545–551

Parry MAJ, Andralojc PJ, Khan S, Lea PJ, Keys AJ (2002) Rubisco activity: effects of drought stress. Ann Bot 89:833–839

Pickett FL (1931) Notes on xerophytic ferns. Am Fern J 21:49–56

Pintado A, Sancho LG (2002) Ecological significance of net photosynthesis activation by water vapour uptake in *Ramalina capitata* from rain protected habitats in central Spain. Lichenologist 34:403–413

Platt KA, Oliver MJ, Thomson WW (1994) Membranes and organelles of dehydrated *Selaginella* and *Tortula* retain their normal configuration and structural integrity: freeze fracture evidence. Protoplasma 178:57–65

Proctor MCF (2000) The bryophyte paradox: tolerance of desiccation, evasion of drought. Plant Ecol 151:41–49

Proctor MCF (2008) Physiological ecology. In: Goffinet B, Shaw AJ (eds) Bryophyte biology. Cambridge University Press, Cambridge, pp 237–268

Proctor MCF, Pence VC (2002) Vegetative tissues: bryophytes, vascular resurrection plants and vegetative propagules. In: Black M, Prichard HW (eds) Desiccation and survival in plants: drying without dying. CAB International, Wallingford, UK, pp 3–43

Proctor MCF, Smirnoff N (2000) Rapid recovery of photosystems on rewetting desiccation-tolerant mosses: chlorophyll fluorescence and inhibitor experiments. J Exp Bot 51:1695–1704

Proctor MCF, Tuba Z (2002) Poikilohydry and homoihydry: antithesis or spectrum of possibilities? New Phytol 156:327–349

Proctor MCF, Duckett JG, Ligrone R (2007) Desiccation tolerance in the moss *Polytrichum formosum* Hedw.: physiological and fine-structural changes during desiccation and recovery. Ann Bot 99:75–93

Richardson DHS (1981) The biology of mosses. Blackwell Scientific, New York

Richardson DHS (1993) The physiology of drying and rewetting in lichens. In: Jennings DH (ed) Stress tolerance of fungi. Academic, London, pp 275–296

Robinson SA, Wasley J, Popp M, Lovelock CE (2000) Desiccation tolerance of three moss species from continental Antarctica. Aust J Plant Physiol 27:379–388

Rundel PW, Lange OL (1980) Water relations and photosynthetic response of a desert moss. Flora 169:329–335

Sancho LG, Schroeter B, Del-Prado R (2000) Ecophysiology and morphology of the globular erratic lichen *Aspicilia fruticulosa* (Eversm.) Flag. from central Spain. In: Schroeter B, Schlensog M, Green TGA (eds) New aspects in cryptogamic research. Contributions in honour of Ludger Kappen. Bibliotheca Lichenologica. J. Cramer, Berlin, pp 137–147

Sancho LG, de la Torre R, Horneck G, Ascaso C, de Los RA, Pintado A, Wierzchos J, Schuster M (2007) Lichens survive in space: results from the 2005 LICHENS experiment. Astrobiology 7:443–54

Scheidegger C, Schroeter B, Frey B (1995) Structural and functional processes during water vapour uptake and desiccation in selected lichens with green algal photobionts. Planta 197:399–409

Scheidegger C, Frey B, Schroeter B (1997) Cellular water uptake, translocation and PSII activation during rehydration of desiccated *Lobaria pulmonaria* and *Nephroma bellum*. Bibl Lichenol 67:105–117

Schlensog M, Schroeter B, Green TGA (2000) Water dependent photosynthetic activity of lichens from New Zealand: differences in the green algal and the cyanobacterial thallus parts of photosymbiodemes. In: Schroeter B, Schlensog M, Green TGA (eds) New aspects in cryptogamic research. Contributions in honour of Ludger Kappen. Bibliotheca Lichenologica. J. Cramer, Berlin, pp 149–160

Schlensog M, Pannewitz S, Green TGA, Schroeter B (2004) Metabolic recovery of continental antarctic cryptogams after winter. Polar Biol 27:399–408

Shirkey B, McMaster NJ, Smith SC, Wright DJ, Rodriguez H, Jaruga P, Birincioglu M, Helm RF, Potts M (2003) Genomic DNA of *Nostoc commune* (Cyanobacteria) becomes covalently modified during long-term (decades) desiccation but is protected from oxidative damage and degradation. Nucleic Acids Res 31:2995–3005

Smirnoff N (1993) The role of active oxygen in the response of plants to water deficit and desiccation. New Phytol 125:27–58

Snelgar WP, Brown DH, Green TGA (1980) A provisional survey of the interaction between net photosynthetic rate, respiratory rate, and thallus water content in some New Zealand cryptogams. NZ J Bot 18:247–56

Souza-Egipsy V, Ascaso C, Sancho LG (2002) Water distribution within terricolous lichens revealed by scanning electron microscopy and its relevance in soil crust ecology. Mycol Res 106:1367–1374

Thomas HH (1921) Some observations on plants in the Libyan Desert. J Ecol 9:75–89

Thomson WW, Platt KA (1997) Conservation of cell order in desiccation mesophyll of *Selaginella lepidophylla* (Hook. & Grev.) Spring. Ann Bot 79:439–447

Webster TR, Steeves TA (1964) Observations on drought resistance in *Selaginella densa* Rydb. Am Fern J 54:189–196

Weissman L, Garty J, Hochman A (2005) Rehydration of the lichen *Ramalina lacera* results in production of reactive oxygen species and nitric oxide and a decrease in antioxidants. Appl Environ Microbiol 71:2121–2129

Chapter 7
Lichens and Bryophytes: Light Stress and Photoinhibition in Desiccation/Rehydration Cycles – Mechanisms of Photoprotection

Ulrich Heber and Ulrich Lüttge

Abbreviations

Fm	Maximum modulated fluorescence under strong actinic illumination, Q_A reduced
F_o	Minimum modulated chlorophyll fluorescence of hydrated organisms, quinone acceptor Q_A in the reaction center of photosystem II oxidized
F_o'	Minimum modulated fluorescence in the absence of water, Q_A oxidized or reduced
$F_v/F_m = (F_m - F_o)/F_m$	Quantum efficiency of charge separation in PSII
PPFD	Photosynthetically active photon flux density
PS	Photosystem
RC	Reaction center

7.1 Introduction

Light is absorbed by photosynthetic pigments. Excitons migrate within the pigment bed of the photosynthetic apparatus, but only a few chlorophyll molecules are highly reactive and capable of undergoing fast photochemical reactions. They are localized in reaction centers (RCs) where light energy is trapped and charge separation results in the formation of a strong oxidant and a reductant within a few picoseconds (Zinth and Kaiser 1993; Holzwarth et al. 2006). This is the first step of photosynthetic energy conservation in hydrated photoautotrophs. Further reactions result in water oxidation and in the reduction of substrates such as carbon dioxide. This is the photochemical work of the production of new biomass. Indeed, high quantum yields of the photoreactions facilitate autotrophic growth of shade-adapted organisms even in very low light.

However, light absorbed in excess to that which can be used for photosynthesis threatens to cause damage to photosynthetically active organisms by oxidative

U. Lüttge et al. (eds.), *Plant Desiccation Tolerance*, Ecological Studies 215,
DOI 10.1007/978-3-642-19106-0_7, © Springer-Verlag Berlin Heidelberg 2011

reactions. This is particularly dangerous for poikilohydric photoautotrophs when they retain their chlorophyll during desiccation. They are termed homoiochlorophyllous. All poikilohydric cryptogams (cyanobacteria, Chap. 3; algae, Chap. 4; bryophytes and lichens, Chap. 6) belong to this functional group. In contrast, chlorophyll and chloroplast structures are degraded during desiccation in most of the few poikilohydric vascular plants, which are termed poikilochlorophyllous (Chap. 9). They are considered to be secondary poikilohydric. Poikilochlorophylly has the advantage of reducing irradiance stress under excess illumination. Its disadvantage is that upon rewetting, complex and slow reactions are required to restructure the photosynthetic apparatus. Homoiochlorophylly has the advantage of rapid resumption of photosynthetic carbon assimilation after rewetting, but its disadvantage is that chlorophyll continues to absorb photons after desiccation, while the energy of absorbed light cannot be used for and dissipated by photochemical work. To prevent photooxidative damage in the desiccated state, homoiochlorophyllous poikilohydric cryptogams are equipped with highly efficient mechanisms of energy dissipation. The first excited state of chlorophyll is thermally taken to the ground state before it can give rise to charge separation in functional reaction centers. Ultrafast energy dissipation protects the photosynthetic apparatus of desiccated homoiochlorophyllous photoautotrophs from photooxidation. Its mechanism is activated during desiccation and inactivated by hydration. In this way, it does not interfere with photosynthetic energy conservation. When this does not prevent excitons from reaching reactive RCs, RC function may be altered so as to engage RCs themselves in photoprotective energy dissipation.

7.2 Conservation Versus Thermal Dissipation of Absorbed Light Energy in Hydrated Poikilohydric Photoautotrophs

In photosynthesis, main light-harvesting pigments are chlorophylls and carotenoids. The latter support photosynthesis by capturing photons and transferring the absorbed light energy to the chlorophylls (Terashima et al. 2009), but they have also photoprotective functions. Mutants devoid of carotenoids, and plants, in which inhibitors block carotenoid synthesis, do not survive exposure to strong light. A special carotenoid, the xanthophyll zeaxanthin, protects against the oxidative activity of excess absorbed light energy but does not interfere with ongoing photosynthesis (Demmig-Adams 1990; Demmig-Adams et al. 1990a, b; Björkman and Demmig-Adams 1994). Together with antheraxanthin, zeaxanthin is formed in the so-called xanthophyll cycle in a light-dependent reaction from violaxanthin. The side-by-side occurrence of energy conservation in photosynthesis and of energy dissipation in competitive pathways requires adjustment of reaction rates.

After much investment of research by several international laboratories (Horton et al. 1996, 2005; Gilmore and Govindjee 1999; Niyogi 1999; Ma et al. 2003; Holt et al. 2004; Li et al. 2004; Pascal et al. 2005; Cogdell 2006; Avenson et al. 2008, and others), which include studies on poikilohydric homoiochlorophyllous photoautotrophs such as mosses and lichens (Demmig-Adams et al. 1990a, b; Bukhov et al. 2001; Heber et al. 2001, 2006a, b, 2010), a tentative picture of the molecular mechanism and of the regulation of zeaxanthin-dependent thermal energy dissipation has emerged. Necessary, but not sufficient, is the presence of zeaxanthin and of a thylakoid protein termed PsbS protein. Protonation of the PsbS protein is needed in addition for the activation of energy dissipation. As long as photosynthesis is not light-saturated, protons deposited during light-dependent electron transport into the intrathylakoid space are consumed during ATP synthesis. As they are not available for the protonation of the PsbS protein, energy dissipation remains inactive. Additional protons, presumably from cyclic electron transport and sufficient for protonation, become available only when light pressure increases so as to surpass the capacity of the photosynthetic apparatus for carbon assimilation (Heber 2002). The extent of protonation of the PsbS protein, and perhaps also of other pigment proteins, determines energy dissipation. It decides on the balance between energy conservation and energy dissipation. As proton availability is a function of light intensity, a reduction in light intensity permits deprotonation of the PsbS protein. This results in inactivation of thermal energy dissipation. In this way, light intensity regulates energy dissipation so that it is not faster than energy capture by RCs. Zeaxanthin-dependent energy dissipation does not drain light energy from the RCs that are engaged in energy conservation. In this situation, minor photooxidative damage to the RCs by rogue reactions appears to be unavoidable. Continuous metabolic repair and replacement of damaged protein ensures effective conversion of light energy into redox energy for photosynthesis (Aro et al. 1993).

The mechanism of zeaxanthin-dependent energy dissipation uses light as signal to regulate energy dissipation. Light flux is coupled to the protonation of the PsbS protein. However, protonation does not a priori need light. CO_2 is a potential acid with a pK of 5.6. It is capable of protonating weaker acids. When elevated concentrations of CO_2 were permitted to pass over a hydrated desiccation-tolerant moss in near darkness, thermal energy dissipation was activated as indicated by strong fluorescence quenching (Bukhov et al. 2001). Removal of CO_2 reversed fluorescence quenching. Only few molecules of zeaxanthin per PSII RC were sufficient for activation of energy dissipation by CO_2 in near darkness. No activation of energy dissipation by CO_2 was observed after the moss had been depleted of zeaxanthin, which proved to be essential for energy dissipation. Very similar observations are made with lichens that are associated with green algae. These algae possess a xanthophyll cycle that permits the light-dependent conversion of violaxanthin to zeaxanthin (Demmig-Adams et al. 1990a, b). Apparently, the weak acid CO_2 was able to replace light. It was able to protonate the PsbS protein. Importantly, CO_2 failed to activate energy dissipation in darkened leaves of higher plants. Perhaps, by having less capacity for photosynthetic energy conservation, shade-adapted poikilohydric photoautotrophs are particularly sensitive to strong light. For survival in the hydrated state, they depend on a highly sensitive system to activate photoprotective energy dissipation. This evidently distinguishes them from poikilohydric vascular plants.

Much less information than on the mechanism of zeaxanthin-dependent energy dissipation is available on photoprotective mechanisms of hydrated lichens that are not associated with green algae but use prokaryotic cyanobacteria as photobionts. Cyanobacterial lichens do not possess a xanthophyll cycle for light-dependent zeaxanthin synthesis (Demmig-Adams et al. 1990b). Photoprotection of these organisms is not governed by a light-dependent protonation reaction as in mosses and in lichens, which harbor green algae as photobionts. Like green plants, they use light intensity as signal for activation of energy dissipation, but the transduction pathway and the mechanism of energy dissipation are not yet known.

7.3 Changes in Conservation and Thermal Dissipation of Absorbed Light Energy During Slow Desiccation

With the beginning of desiccation, loss of water changes the balance between energy conservation and energy dissipation. Chlorophyll fluorescence and light-dependent charge separation in the reaction centers decrease. Thermal energy dissipation increases. Coexistence of energy conservation in functional RCs and of thermal energy dissipation as it exists in hydrated organisms can no longer ensure survival of poikilohydric homoiochlorophyllous organisms under strong light after water has been lost. It cannot prevent the formation of radicals in rogue reactions that are derived from the light-dependent initial charge separation in functional RCs. Metabolic repair of damage is impossible. In this situation, desiccation opens other pathways for photoprotection. Efficient photoprotection of desiccated poikilohydric photoautotrophs requires fast migration of absorbed light energy to dissipation centers, and fast trapping in these centers before trapping in functional RCs can give rise to charge separation and the formation of reactive radicals.

Emission of light as fluorescence, use of light energy for charge separation, and dissipation of absorbed light energy as heat are, according to the first law of thermodynamics, competitive processes. During slow drying of desiccation-tolerant photoautotrophs, fluorescence decreases and charge separation in PSII RCs is gradually lost. Figure 7.1 shows this for a moss, *Rhytidium rugosum*, for a chlorolichen, *Hypogymnia physodes*, and for a cyanolichen, *Peltigera neckeri*. It illustrates both similarities and (slight) differences in the response of fluorescence emission to hydration and desiccation of poikilohydric organisms as different as mosses and lichens in which either algae (*H. physodes*) or cyanobacteria (*P. neckeri*) serve as photobionts. When these photoautotrophs are desiccated, fluorescence emission is at a minimum level F_0'. Even very strong light pulses do not elicit easily detectable fluorescence responses. Owing to the competitive relationship between thermal energy dissipation and fluorescence emission, when photochemical energy use is negligible, loss of fluorescence indicates the activation

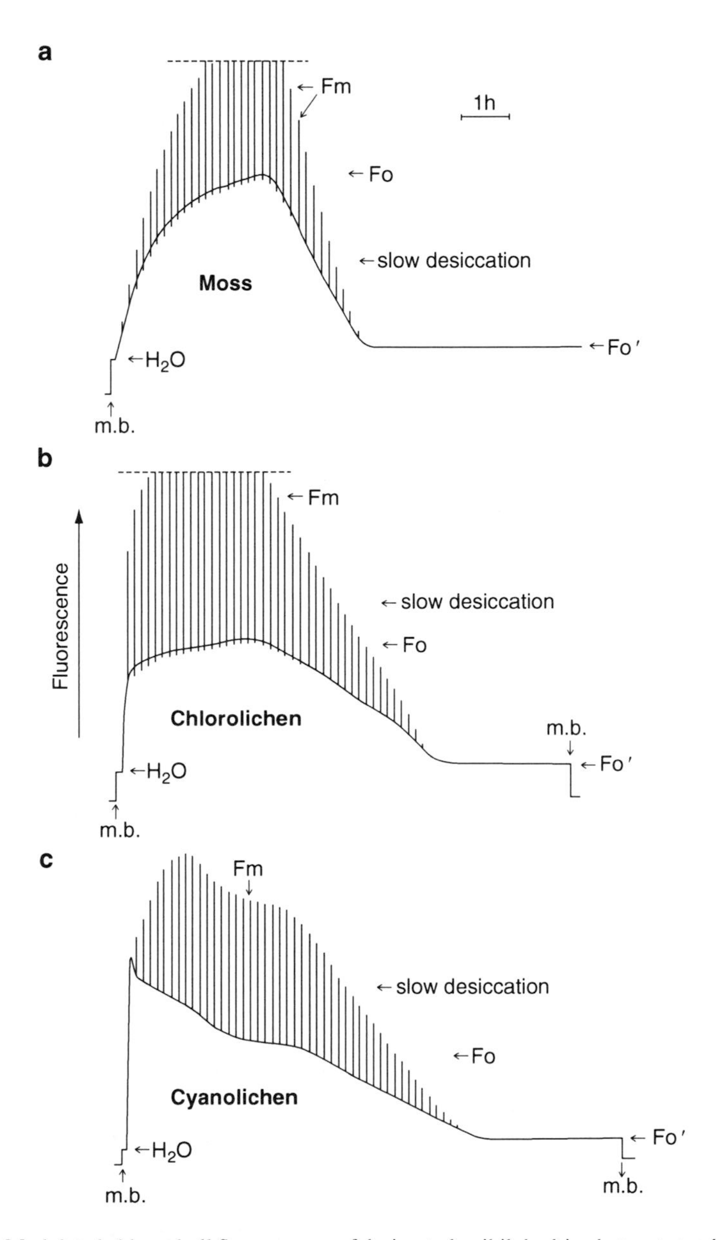

Fig. 7.1 Modulated chlorophyll fluorescence of desiccated poikilohydric photoautotrophs: effects of hydration (↑H_2O) and slow desiccation on fluorescence emission and on charge separation in RCs of PSII as indicated by rapid light-induced fluorescence responses. (**a**) *Rhytidium rugosum*; (**b**) *Hypogymnia physodes*; (**c**) *Peltigera neckeri*. PPFD of a modulated measuring beam was 2 μmol m^{-2} s^{-1}; 1 s actinic light pulses of PPFD 10,000 μmol m^{-2} s^{-1} were given every 500 s. *m.b.* measuring beam. For explanation, see text

of thermal energy dissipation. Hydration by a little water increases fluorescence. This is mainly the result of the inactivation of thermal energy dissipation. Only to a small part is it caused by increased light absorption, which is best seen in the cyanolichen experiment of Fig. 7.1c where it accounts for a larger initial increase in fluorescence emission than in the moss and chlorolichen experiments of Fig. 7.1a, b. Hydration changes the gray color of the desiccated cyanolichen to a bluish tinge. This effect disappears as desiccation progresses. Both the moss and the chlorolichen remain green during hydration.

Loss of fluorescence emission during desiccation and its recovery upon rehydration were seen not only under laboratory conditions but also in the field, e.g., in three poikilohydric bryophyte species (*Campylopus savannarum*, *Racocarpus fontinaloides*, and *Ptychomitrium vaginatum*) on the highly sun-exposed granite surface of an inselberg in Brazil (Lüttge et al. 2008). Fluorescence quenching is indicative of photoprotective acclimation during desiccation.

In hydrated photoautotrophs, the quantum efficiency of photosynthesis is maximal under strictly rate-limiting light. Charge separation in reaction centers approaches 100% efficiency. Even under these conditions, low fluorescence emission is observed. It is close to or at the F_o level (Fig. 7.1). Thermal energy conservation is not active under these conditions and fluorescence emission is in equilibrium with photosynthetic energy conservation. Charge separation in the RCs initiates energy conservation. It takes as little as a few picoseconds in PSII RCs (Zinth and Kaiser 1993; Holzwarth et al. 2006). Therefore, F_o fluorescence of hydrated desiccation-tolerant photoautotrophs as different as mosses and lichens (Fig. 7.1) is in equilibrium with and indicative of ongoing fast charge separation. In consequence, the decrease in F_o during desiccation not only shows the activation of a mechanism of thermal energy dissipation but also that excitons are now thermally deactivated faster than they can be trapped in the RCs. Thermal energy dissipation drains energy from the RCs before charge separation can take place. This explains the loss of charge separation in the RCs as desiccation-tolerant photoautotrophs lose water during slow desiccation. It also explains photoprotection of RCs. If this deduction is correct, then ratios of F_o/F_o' (Fig. 7.1) are indicators of the effectiveness of photoprotection. Large ratios mean that RCs of desiccated photoautotrophs have very little chance for charge separation even if they are functionally intact. They are effectively protected by the ultrafast thermal deactivation of excitons in dissipation centers before charge separation in the RCs is possible. Hydration results in the inactivation of the mechanism(s) of energy dissipation. Therefore, fluorescence increases and charge separation returns.

In the experiments of Fig. 7.1, recovery of fluorescence emission upon rewetting is rapid in the lichens and slower in the moss. In the cyanolichen, the initial increase in fluorescence results not only from increased quantum efficiency of fluorescence emission but also from increased light absorption while the lichen turns from gray to almost blue. Later on, preferential loss of water from the cortex increases light scattering. This reestablishes the gray appearance of the lichen, permits an estimation of the F_o level of fluorescence (see Fig. 7.1c), and explains initial differences in the kinetics of fluorescence responses to hydration between the cyanolichen and the green organisms. As water becomes available, charge separation in functional RCs in response to short light pulses becomes possible. It increases fluorescence reversibly to maximum values of F_m because electron transport reduces electron carriers during the duration of short light pulses. As water is slowly lost, both the minimum quantum yield of fluorescence emission in the hydrated state (F_o) and the pulse-induced charge separation in RCs decrease. F_o/F_o' ratios were comparable in the moss (4.8), the chlorolichen (4.7), and the cyanolichen (5.3). As the decrease of F_o results from the activation of thermal energy dissipation, it reveals strongly increased photoprotection. This occurs at the expense of both energy conservation and fluorescence. Losses of charge separation in PSII RCs and of fluorescence go hand in hand, while a large part of the water is lost. Later on inactivation of charge separation accelerates. Apparently, an additional factor becomes involved in RC inactivation during the final phases of water loss. This is illustrated for the lichen *Hypogymnia* in Fig. 7.2, which compares loss of fluorescence as depicted by a Stern–Volmer-type equation ($F_o/F_o' - 1$) with loss of charge separation as shown by $(F_m - F_o)/F_m = F_v/F_m$ (Paillotin 1976; Genty et al. 1989). Loss of RC activity is essentially complete even before water vapor equilibrium with dry air has been attained.

Figure 7.2 is also an example for very similar responses of fluorescence parameters of other desiccation-tolerant photoautotrophs to slow desiccation. Inactivity of potentially active RCs in the dry state is evidence that energy dissipation is faster than charge separation in functional RCs, whether or not reaction constants are altered by desiccation (Heber 2008). If reaction constants remain unaffected, it must be concluded that energy dissipation is faster than the few picoseconds needed for charge separation in the hydrated state (Zinth and Kaiser 1993; Holzwarth et al. 2006). As long as RCs remain principally active in

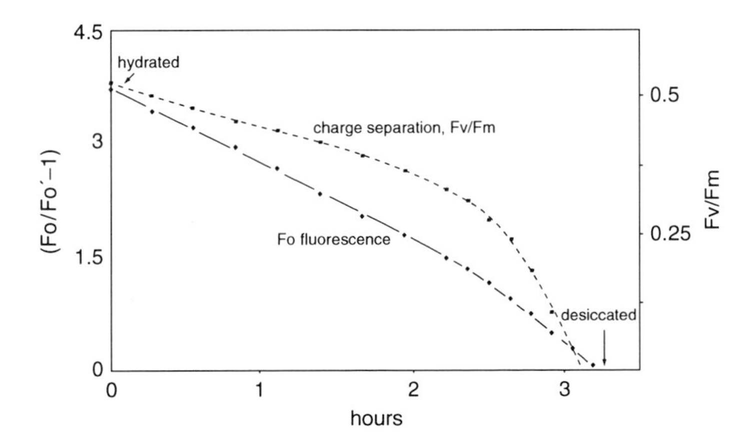

Fig. 7.2 Nonphotochemical fluorescence loss during slow desiccation of the hydrated chlorolichen *Hypogymnia physodes* as indicated by the Stern–Volmer relation ($F_o/F_o' - 1$) versus loss of pulse-induced charge separation in PSII RCs as indicated by F_v/F_m. *Abscissa*: time of desiccation; *left ordinate*: F_v/F_m; *right ordinate*: ($F_o/F_o' - 1$)

the absence of water, only ultrafast thermal energy dissipation can protect PSII RCs against photodamage.

Activation of energy dissipation during loss of water is a regulated process. It is inhibited by glutaraldehyde that possesses two reactive aldehyde groups (Heber et al. 2007, 2010; Heber 2008). Glutaraldehyde prevents the loss of fluorescence and of charge separation in PSII RCs during desiccation by reacting with proteins (Coughlan and Schreiber 1984). It does not react with the pigments of the photosynthetic apparatus. When a glutaraldehyde solution is used for the hydration of desiccated mosses or lichens, it also inhibits the inactivation of desiccation-induced energy dissipation. The glutaraldehyde experiments suggest that conformational changes of a pigment protein, which are inhibited by glutaraldehyde, play a central role in the mechanism that protects dessicated photoautotrophs against photodamage. Other experiments support this conclusion. Fast drying of desiccation-tolerant photoautotrophs leads to less loss of fluorescence and of light-dependent charge separation in PSII RCs than slow drying, suggesting that the putative conformational change of a thylakoid protein that leads to photoprotection is not fast (Heber et al. 2007). Involvement of peptide folding in the mechanism of energy dissipation is seen in heating experiments where subjecting desiccated lichens to temperatures of up to 80°C gradually reverses the loss of fluorescence suffered during desiccation. This suggests that thermal unfolding of protein structures destroys dissipation centers releasing pigments for fluorescence emission (Heber and Shuvalov 2005; Heber 2008). More fluorescence is quenched after slow than after fast drying, and more fluorescence is emitted by heating desiccated lichens after the lichens had been dried slowly than after fast drying.

It thus appears that conformational changes of a pigment protein complex are at the basis of achieving photoprotection during desiccation. Conformational changes may alter pigment–pigment interactions within the protein so as to facilitate thermal energy dissipation. This conclusion has been drawn by Horton et al. (2005) and by Pascal et al. (2005) for the photoprotection of higher plants. The LHCII complex of PSII is highly fluorescent when isolated, but no longer after crystallization. Loss of fluorescence is thought to result from conformational changes of the LHCII complex that alters the spatial organization of bound chlorophyll and carotenoids in relation to one another. Ruban et al. (2007) have reported a change in the configuration of bound neoxanthin on crystallizing the LHCII complex.

Related changes in protein configuration may facilitate desiccation-induced energy dissipation of mosses and lichens. However, it is worth noting that cyanolichens do not possess the LHCII complex of higher plants. They nevertheless activate thermal energy dissipation during desiccation as shown by the loss of fluorescence (see *Peltigera* experiment of Fig. 7.1c). Activation of thermal energy dissipation during desiccation is different from the activation of zeaxanthin-dependent energy dissipation in the xanthophyll cycle in that neither light nor protonation is required. Nevertheless, desiccation under illumination increases loss of fluorescence indicating increased photoprotection of chlorolichens that possess the xanthophyll cycle (Heber 2008; Stepigova et al. 2008; Heber et al. 2010). However, nigericin inhibits zeaxanthin synthesis in both the xanthophyll cycle and the protonation of the PsbS protein, but it fails to inhibit desiccation-induced fluorescence quenching. Therefore, and because desiccation in darkness is sufficient for the activation of considerable photoprotection, zeaxanthin is not an essential component of the mechanism of desiccation-induced photoprotection.

7.4 Desiccation-Induced Decreased Light Absorption and Shading of Photobionts as Auxiliary Mechanisms of Photoprotection

A multitude of lichen species produce pigments functioning as sun protectants. Frequently found pigments are calycin (orange to red), parietin (yellow to orange), sordidone (yellow), pulvinic acid and its lactone (yellow), vulpinic acid (yellow), and rhizocarpic acid (yellow) (Wirth 1995). Some lichens such as *Leptogium* or *Collema* species are almost black. Pigments are usually located in the cortex above the photobionts. In *Xanthomaculina convoluta*, the lower side of the thallus is black and the upper side green. On drying, the thallus rolls so as to shade the upper side. Depending on their absorption spectrum and their localization within the thallus, pigments decrease absorption of light by the algal or cyanobacterial photobionts. It is well known that pigmentation of the hyphal cortex of lichens provides protection to photobionts against excessive photosynthetically active and ultraviolet light (see Rikkinen 1995). After prolonged sun exposure thalli of *Lobaria pulmonaria* form a melanin pigment in the cortex (Solhaug et al. 2003; Gauslaa and Solhaug 2004; McEvory et al. 2007), while shaded thalli remain gray to green. The brown melanin pigment is known to have very strong absorption in the UV between 240 and 320 nm (Gauslaa and Solhaug 2001).

Desiccation also decreases light absorption by photobionts by increasing light scattering in the hyphal cortex. This extends the path of light (Butler 1962). Backscattering results in the escape of light from the cortex. Hydration of desiccated lichens has been shown to increase light absorption (Dietz et al. 2000; Gauslaa and Solhaug 2001; Heber et al. 2007). Reduction of light absorption by shading mechanisms, or by increased light scattering, reduces, but does not eliminate the danger of photooxidative damage. This is particularly important in the case of desiccated photoautotrophs where photooxidative damage accumulates with time of exposure to strong light while metabolic repair of damage is impossible (Gauslaa and Solhaug 1999; Gray et al. 2007). Full photoprotection in the desiccated state demands essentially complete avoidance of photooxidative reactions. This cannot be achieved by reducing light absorption. It requires the absence of RC activity.

7.5 Fast Thermal Energy Dissipation in Desiccated Poikilohydric Photoautotrophs as Central Mechanism of Photoprotection

After desiccation has activated thermal energy dissipation, photoreactions can no longer be readily observed because light energy is efficiently drained away from the RCs. It migrates to and is dissipated in dissipation centers outside the RCs. This is the main basis of photoprotection. Observation of photoreactions in desiccated photoautotrophs is possible only after saturating the dissipation centers with extremely strong light. Full sunlight in the summer rarely exceeds a PPFD of 1,800 μmol m^{-2} s^{-1}, although on very exposed sites such as the biofilms of soil crusts of deserts and sand dunes and the rocks of inselbergs on clear days it may attain levels close to the maximum possible solar irradiance of 2,500 μmol m^{-2} s^{-1}. These are the photon fluxes against which protection needs to be effective. Under illumination with PPFDs as high as 10,000 μmol m^{-2} s^{-1}, photochemistry becomes observable in desiccated mosses and lichens. Quantum efficiencies of these reactions are very low owing to competition with effective thermal energy dissipation. Transmission spectrophotometry is not possible with opaque lichen thalli, but in dried leaves reversible band bleaching was observed around 500 nm. It was attributed to the photooxidation of β-carotene in the RCs of PSII (Shuvalov and Heber 2003). Reflection photometry with desiccated lichens revealed reversible formation of absorption bands around 800 and close to 1,000 nm (Heber 2008; Heber et al. 2010). Chlorophyll radicals that include P700$^+$ in the RC of PSI absorb around 800 nm and oxidized carotenes around 1,000 nm. Decay kinetics of the 800 nm signal revealed more than one underlying photochemical reaction even after P700 oxidation had been saturated by background far-red light that oxidizes P700. Band formation was accompanied by reversible changes of fluorescence emission from the dry lichens. They consisted either in an increase or in a decrease of fluorescence. Figure 7.3 shows such responses for different species. In the moss *R. rugosum* (Fig. 7.3a) and the chlorolichen *P. neckeri* (Fig. 7.3d),

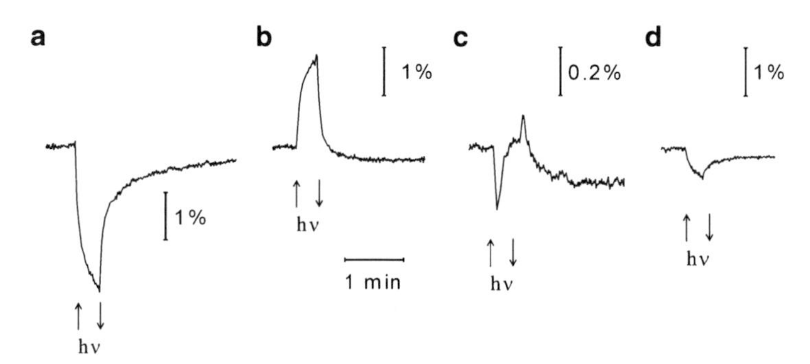

Fig. 7.3 Responses of chlorophyll fluorescence of desiccated photoautotrophs to a 20 s illumination period with PPFD = 7,000 μmol m^{-2} s^{-1} (see *arrows*; increased fluorescence ↑; decreased fluorescence ↓) in percent of residual fluorescence. (**a**) *Rhytidium rugosum*, a moss; (**b**) *Hypogymnia physodes*, a chlorolichen; (**c**) *Parmelia caperata*, a chlorolichen; (**d**) *Peltigera neckeri*, a cyanolichen. For explanation, see text

fluorescence was reversibly quenched; in the chlorolichen *H. physodes*, it was reversibly increased (Fig. 7.3b). The same had been observed for the chlorolichen *Parmelia sulcata* (Heber 2008). In the chlorolichen *Parmelia caperata*, a fast quenching response was followed by an increase in fluorescence (Fig. 7.3c). These changes were reversed on darkening in the same order. When the moss *Rhytidium* was collected during a rainy period and dried in darkness, strong illumination increased fluorescence reversibly in contrast to the response shown in Fig. 7.3a for the sun-dried moss. With the chlorolichen *H. physodes*, responses were occasionally as complex as those shown for *P. caperata* in Fig. 7.3c when thalli were collected from the western-exposed side of tree trunks. Apparently, fluorescence responses of desiccated mosses and thalli of chlorolichens depended on the extent of photoprotection of the thalli by mechanisms such as those described in the previous section. Only in the cyanolichen *P. neckeri* was reversible formation of the 800 nm absorption band invariably accompanied by small reversible fluorescence quenching.

Because chlorophylls of the RCs are chemically reactive, the different fluorescence responses of desiccated mosses and of lichens, which differ in their photosensitivity, are likely to originate from different reaction pathways within PSII RCs even though emission of fluorescence is delocalized. Increased fluorescence under strong light is interpreted to reflect normal activity of PSII RCs, i.e., charge separation followed by the reduction of the primary quinone acceptor Q_A in the RC (Duysens and Sweers 1963). Decreased fluorescence suggests altered RC activity, i.e., charge separation followed by oxidation of β-carotene (Shuvalov and Heber 2003). The carotene radical, in turn, can take an electron from a neighboring chlorophyll, ChlZD2, which is a fluorescence quencher when oxidized (Faller et al. 2006). This interpretation can explain opposite fluorescence responses, either a decrease or an increase of fluorescence in the light, or both in sequence in the same thallus (Fig. 7.3c), by electron flow in different directions within the RC. Even simply changing the temperature can change fluorescence responses. In desiccated *P. sulcata*, fluorescence increases reversibly under very strong illumination at temperatures above 25°C (Heber and Shuvalov 2005). Close to and below 0°C, fluorescence is reversibly quenched. The observations suggest convertibility of PSII RCs. By definition, quencher formation within an RC is photoprotective because, by opening a pathway for thermal energy dissipation, it competes with potentially harmful photochemistry (Heber et al. 2006a; Ivanov et al. 2008). Hydration would permit reduction of the quencher radical in the RC. It would reconvert energy-dissipating RCs into energy-conserving RCs and permit normal photosynthesis.

7.6 Changes in Conservation and Thermal Dissipation of Absorbed Light Energy upon Hydration

After slow desiccation has decreased fluorescence and largely eliminated charge separation in functional reactions centers, hydration restores fluorescence emission and activates charge separation in the RCs. However, patterns of restoration are not uniform. When in the chlorolichen *P. sulcata* the F_o/F_o'ratio had been large (≥5) suggesting strong photoprotection, charge separation (as indicated by $(F_m - F_o)/F_m$ ratios; Genty et al. 1989) recovered rapidly after hydration in close relation to the recovery of fluorescence emission. When desiccation had resulted in ratios of $F_o/F_o' < 5$, charge separation recovered after hydration usually more slowly than fluorescence emission. These observations need to be seen in relation to the biphasic loss of charge separation during slow desiccation (Fig. 7.2), which shows that an additional factor contributed to inactivation of charge separation as water loss approached water vapor equilibrium with dry air. Restoration of fluorescence emission after hydration results from the inactivation of desiccation-induced energy dissipation. When charge separation in PSII RCs is restored in direct relation to the recovery of fluorescence emission upon hydration, the RCs had evidently remained

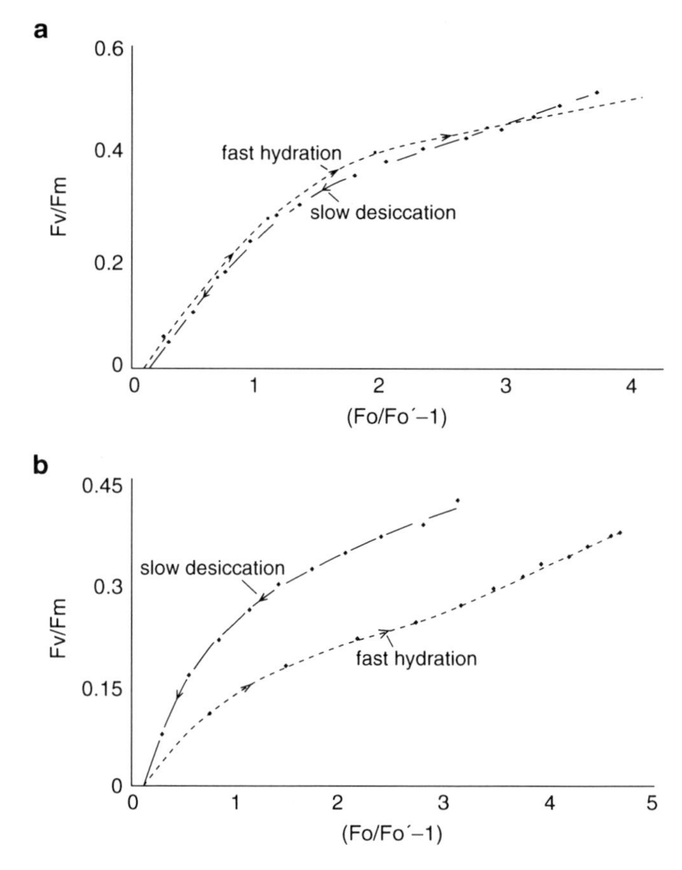

Fig. 7.4 Charge separation (F_v/F_m) in PSII RCs of the chlorolichen *Hypogymnia physodes* (**a**) and the moss *Rhytidium rugosum* (**b**) versus changes in fluorescence as expressed by $(F_o/F_o' - 1)$ during slow desiccation and fast hydration. During slow desiccation, the basal fluorescence level F_o decreases to its lowest value F_o'. After desiccation is complete, $(F_o/F_o' - 1) = 0$. Desiccation: *arrow* pointing towards $F_v/F_m = 0$; hydration: *arrow* pointing toward maximum F_v/F_m

unaltered during desiccation. If however, charge separation lags behind restoration of fluorescence emission, then the RCs had been in a state different from that which permits normal activity. They could not resume activity as soon as energy dissipation was inactivated by hydration. Recovery of normal activity is a process that requires more time than hydration-dependent inactivation of energy dissipation. This interpretation of the observations agrees with the fact described above that RC activity in desiccated thalli under extremely strong illumination is accompanied either by increased fluorescence, which indicates reduction of Q_A, or by decreased fluorescence, which reveals formation of a quencher in the RCs.

Figure 7.4 shows two experiments with a chlorolichen and a moss (see also Fig. 7.1). It compares loss of charge separation (decreasing ratios F_v/F_m) during desiccation (as shown by a Stern–Volmer-type equation ($F_o/F'_o - 1$)) with restoration of charge separation upon hydration. In the chlorolichen *Hypogymnia*, losses during desiccation and restoration upon hydration had the same pattern (Fig. 7.4a). The fluorescence experiment of Fig. 7.3b had suggested for *Hypogymnia* that RCs had remained essentially unaltered during desiccation. In the moss *Rhytidium*, on the other hand, restoration of charge separation on hydration was retarded compared to the loss of charge separation during desiccation (Fig. 7.4b). For this moss, the fluorescence quenching shown in the experiment of Fig. 7.3a had suggested alteration of PSII RCs so as to convert them to dissipation centers. Light-dependent quencher formation was also recorded for *P. neckeri* in Fig. 7.3d. The fast restoration of fluorescence emission and the slow recovery of charge separation shown in Fig. 7.1c reveal that in this cyanolichen recovery of charge separation was also much retarded compared with restoration of fluorescence emission. Apparently, conversion of the quencher shown in Fig. 7.3d to functional RCs lagged far behind recovery of fluorescence on hydration.

7.7 Vulnerability of PSII RCs to Photooxidative Damage

Charge separation of PSII RCs in desiccated photoautotrophs proceeds via the charge separated state $^1P680^+Pheo^-Q_A$ to the more stable state $^1P680^+PheoQ_A^-$. It results in increased fluorescence emission. Pheo and Q_A represent pheophytin and the primary quinone acceptor Q_A in PSII RCs, respectively. Spin conversion during charge recombination produces long-lived triplet $^3P680^*$, which reacts with oxygen to form $^1P680$ and the highly oxidative singlet oxygen $^1O_2$ (Krieger-Liszkay 2005; Krieger-Liszkay et al. 2008). The triplet chlorophyll cannot be deactivated by the β-carotene in the RC because of the considerable distance between P680 and the carotenes (Faller et al. 2006). Singlet oxygen appears to be the oxidant that is mainly responsible for oxidative damage to photosensitive desiccated photoautotrophs such as the shade-adapted lichen *L. pulmonaria* (Gauslaa and Solhaug 1999). To prevent its formation by preventing charge separation, thermal energy dissipation of desiccated photoautotrophs needs to be highly effective.

7.8 Molecular Mechanisms of Photoprotection

After desiccation of lichens and desiccation-tolerant mosses, preferential quenching of the main emission band at 685 nm is accompanied by far-red emission, part of which is attributable to PSII (Heber and Shuvalov 2005). This suggests involvement of a long-wavelength emitter in desiccation-induced fluorescence quenching. Strong support for this conclusion has been presented by picosecond fluorescence spectroscopy (Veerman et al. 2007; Komura et al. 2010). The excitation lifetime of the far-red emitter was dramatically shortened by desiccation of lichens, which makes it a strong competitor with reaction centers for the capture of excitation energy.

Different proposals exist to explain energy dissipation in higher plants. According to Holt et al. (2005), Ahn et al. (2008), and Avenson et al. (2008), charge transfer between excited chlorophyll and zeaxanthin reduces a chlorophyll molecule and oxidizes a zeaxanthin molecule. This is in principle similar to charge separation in PSII RCs, which produces an oxidized chlorophyll, $P680^{+}$, and reduced pheophytin $Pheo^{-}$. Charge transfer is followed by a dissipative recombination reaction between reduced chlorophyll and oxidized zeaxanthin.

On the other hand, Horton et al. (1996, 2005) suggest that aggregation of the light-harvesting complex LHCII activates energy dissipation. This complex is highly fluorescent when isolated, but no longer after crystallization (Pascal et al. 2005). Quenching is thought to result from conformational changes of the LHCII complex that alters the spatial organization of chlorophyll and carotenoids in the complex in relation to one another, thereby permitting the fast transfer of excitation energy from chlorophyll to low-lying optically forbidden excited states of carotenoids followed by thermal deexcitation. It is worth noting that fluorescence lifetime measurements and X-ray structure determinations have recently provided arguments against a direct participation LHCII in fast thermal energy dissipation in vivo (Barros et al. 2009). In all proposals under discussion, energy dissipation involves conformational changes of proteins, whether induced by protonation as in hydrated plants or by desiccation in cryptogams.

7.9 Conclusions

Primary poikilohydric cryptogams, bryophytes, lichens, and algae that retain their chlorophyll during desiccation are equipped with highly effective mechanisms of thermal energy dissipation, which protect them against photooxidative damage while they are desiccated. Activation of photoprotection is triggered by the loss of water. Light is not necessary for activation, but its presence increases photoprotection particularly in chlorolichens (Heber 2008; Stepigova et al. 2008). Desiccation-induced thermal energy dissipation is prevented by glutaraldehyde that reacts with proteins. This reveals the participation of a pigment protein complex in the mechanism of desiccation-induced energy dissipation, which is located outside the reaction centers. When in operation, dissipation centers active within the protein complex drain excitation energy from functional reaction centers, thereby preventing them from engaging in potentially destructive photoreactions. Excitation energy still reaching reaction centers can convert them to dissipation centers increasing thereby photoprotection. Hydration inactivates these mechanisms of energy dissipation.

While poikilohydric cryptogams are hydrated, other photoprotective mechanisms come into play. Organisms possessing the xanthophyll cycle form zeaxanthin from violaxanthin in a light-dependent reaction. They are protected against excess illumination by a zeaxanthin-dependent mechanism of energy dissipation, which is regulated by light so as not to interfere with photosynthetic energy conservation. Activation of this mechanism is triggered by a light-dependent protonation reaction. Little is known on molecular details of photoprotection of hydrated poikilohydric cyanobacteria and cyanobacterial lichens, which do not possess the xanthophyll cycle (Demmig-Adams et al. 1990b).

References

Ahn TK, Avenson TJ, Ballottari M, Cheng Y-C, Niyogi KK, Bassi R, Fleming GR (2008) Architecture of a charge-transfer state regulating light harvesting in a plant antenna protein. Science 320:794–797

Aro EM, Virgin I, Andersson B (1993) Photoinhibition of photosystem II. Inactivation, protein damage and turnover. Biochim Biophys Acta 1143:113–134

Avenson TJ, Ahn TK, Zigmantas D, Niyogi KK, Li Z, Ballottari M, Bassi R, Fleming GR (2008) Zeaxanthin radical formation in minor light-harvesting complexes of higher plant antenna. J Biol Chem 283:3550–3558

Barros T, Royant A, Standfuss J, Dreuw A, Kühlbrandt W (2009) Crystal structure of plant light-harvesting complex shows the active, energy-transmitting state. EMBO J 28:296–306

Björkman O, Demmig-Adams B (1994) Regulation of photosynthetic light energy capture, conversion and dissipation in leaves of higher plants. In: Schulze E-D, Caldwell MM (eds) Ecophysiology of photosynthesis, vol 100, Ecological studies. Springer, Heidelberg, Germany, pp 17–70

Bukhov NG, Kopecky J, Pfündel EE, Klughammer C, Heber U (2001) Few molecules of zeaxanthin per reaction centre of photosystem II permit effective thermal dissipation of light energy in a poikilohydric moss. Planta 212:739–748

Butler WL (1962) Absorption of light by turbid materials. J Opt Soc Am 52:292–299

Cogdell RJ (2006) The structural basis of non-photochemical quenching is revealed? Trends Plant Sci 11:59–60

Coughlan SJ, Schreiber U (1984) The differential effects of short-time glutaraldehyde treatments on light-induced thylakoid membrane conformational changes, proton pumping and electron transport properties. Biochim Biophys Acta 767:606–617

Demmig-Adams B (1990) Carotenoids and photoprotection of plants: a role for the xanthophyll zeaxanthin. Biochim Biophys Acta 1020:1–24

Demmig-Adams B, Máguas C, Adams WW III, Meyer A, Kilian E, Lange OL (1990a) Effect of high light on the efficiency of photochemical energy conversion in a variety of lichen species with green and blue–green phycobionts. Planta 180:400–409

Demmig-Adams B, Adams WW III, Czygan F-C, Schreiber U, Lange OL (1990b) Differences in the capacity for radiationless energy dissipation in the photochemical apparatus of green and blue-green algal lichens associated with differences in carotenoid composition. Planta 180:582–589

Dietz S, Büdel B, Lange OL, Bilger W (2000) Transmittance of light through the cortex of lichens from contrasting habitats. Bibl Lichenol 75:171–182

Duysens LMN, Sweers HE (1963) Mechanisms of photochemical reactions in algae as studied by means of fluorescence. In: Studies on microalgae and photosynthetic bacteria, special issue of plant and cell physiology. Tokyo University Press, pp 353–372

Faller P, Fufezan C, Rutherford AW (2006) Side path electron donors: cytochrome b559, chlorophyll Z and β-carotene, chapter 15. In: Wydrzynski T, Satoh K (eds) Photosystem II: the water/plastoquinone oxidoreductase in photosynthesis. Kluwer, Dordrecht, pp 347–365

Gauslaa Y, Solhaug KA (1999) High-light damage in air-dry thalli of the old forest lichen *Lobaria pulmonaria* – interactions of irradiance, exposure duration and high temperature. J Exp Bot 50:697–705

Gauslaa Y, Solhaug KA (2001) Fungal melanins as a sun screen for symbiotic green algae in the lichen *Lobaria pulmonaria*. Oecologia 126:462–471

Gauslaa Y, Solhaug KA (2004) Photoinhibition in lichens depends on cortical characteristics and hydration. Lichenologist 36:130–143

Genty B, Briantais J-M, Baker NR (1989) The relationship between the quantum yield of photosynthetic electron transport and quenching of chlorophyll fluorescence. Biochim Biophys Acta 990:87–92

Gilmore AM, Govindjee (1999) How higher plants respond to excess light: energy dissipation in photosystem II. In: Singhal GS, Renger G, Sopory SK, Irrgang K-D, Govindjee (eds) Concepts in photobiology: photosynthesis and photomorphogenesis. Narosa Publishing House, New Delhi, pp 513–548

Gray DW, Lewis LA, Cardon ZG (2007) Photosynthetic recovery following desiccation of desert green algae (Chlorophyta) and their aquatic relatives. Plant Cell Environ 30:1240–1255

Heber U (2002) Irrungen, Wirrungen? The Mehler reaction in relation to cyclic electron transport in C_3 plants. Photosynth Res 73:223–231

Heber U (2008) Photoprotection of green plants: a mechanism of ultra-fast thermal energy dissipation in desiccated lichens. Planta 228:641–650

Heber U, Shuvalov VA (2005) Photochemical reactions of chlorophyll in dehydrated photosystem II: two chlorophyll forms (680 and 700 nm). Photosynth Res 84:85–91

Heber U, Bukhov NG, Shuvalov VA, Kobayashi Y, Lange OL (2001) Protection of the photosynthetic apparatus against damage by excessive illumination in homoiohydric leaves and poikilohydric mosses and lichens. J Exp Bot 52:1999–2006

Heber U, Bilger W, Shuvalov VA (2006a) Thermal energy dissipation in reaction centers of photosystem II protects desiccated poikilohydric mosses against photooxidation. J Exp Bot 57:2993–3006

Heber U, Lange OL, Shuvalov VA (2006b) Conservation and dissipation of light energy by plants as complementary processes involved in sustaining plant life: homoiohydric and poikilohydric autotrophs. J Exp Bot 57:1211–1223

Heber U, Azarkovich M, Shuvalov VA (2007) Activation of mechanisms of photoprotection by desiccation and by light: poikilohydric photoautotrophs. J Exp Bot 58:2745–2759

Heber U, Bilger W, Türk R, Lange OL (2010) Photoprotection of reaction centres in photosynthetic organisms: mechanisms of thermal energy dissipation in desiccated thalli of the lichen *Lobaria pulmonaria*. New Phytol 185:459–470

Holt NE, Fleming GR, Niyogi KK (2004) Toward an understanding of the mechanism of nonphotochemical quenching in green plants. Biochemistry 43:8281–8289

Holt NE, Tigmantas D, Valkunas L, Li X-P, Niyogi KK, Fleming GR (2005) Carotenoid cation formation and the regulation of photosynthetic light harvesting. Science 307:433–436

Holzwarth AR, Muller MG, Reus M, Nowazyk M, Saner J, Rogner M (2006) Kinetics and mechanism of electron transfer in intact photosystem II and in the isolated reaction center: pheophytin is the primary electron acceptor. Proc Nat Acad Sci USA 103:6895–6900

Horton P, Ruban AV, Walters RG (1996) Regulation of light harvesting in green plants. Annu Rev Plant Physiol Plant Mol Biol 47:655–684

Horton P, Wentworth M, Ruban A (2005) Control of the light harvesting function of chloroplast membranes: the LHCII-aggregation model for non-photochemical quenching. FEBS Lett 579:4201–4206

Ivanov AG, Sane PV, Hurry V, Öquist G, Huner NPA (2008) Photosystem II reaction center quenching: mechanisms and physiological role. Photosynth Res 98:565–574

Komura M, Yamagishi A, Shibata Y, Iwasaki I, Itoh S (2010) Mechanism of strong quenching of photosystem II chlorophyll fluorescence under drought stress in a lichen, *Physciella melanchla*, studied by subpicosecond fluorescence spectroscopy. Biochim Biophys Acta 1797:331–338

Krieger-Liszkay A (2005) Singlet oxygen production in photosynthesis. J Exp Bot 56:337–346

Krieger-Liszkay A, Fufezan C, Trebst A (2008) Singlet oxygen production in photosystem II and related protection mechanism. Photosynth Res 98:551–564

Li X-P, Gilmore AM, Caffari S, Bassi R, Golan T, Kramer D, Niyogi KK (2004) Regulation of photosynthetic light harvesting involves intrathylakoid lumen pH sensing by the PsbS protein. J Biol Chem 279:22866–22874

Lüttge U, Meirelles ST, de Mattos EA (2008) Strong quenching of chlorophyll fluorescence in the desiccated state in three poikilohydric and homoiochlorophyllous moss species indicates photo-oxidative protection on highly light-exposed rocks of a tropical inselberg. J Plant Physiol 165:172–181

Ma Y-Z, Holt NE, Li X-P, Niyogi KK, Fleming GR (2003) Evidence for direct carotenoid involvement in the regulation of photosynthetic light harvesting. Proc Nat Acad Sci USA 100:4377–4382

McEvory M, Gauslaa Y, Solhaug KA (2007) Changes in pools of depsidones and melanins, and their function, during growth and acclimation under contrasting natural light in the lichen *Lobaria pulmonaria*. New Phytol 175:271–282

Niyogi KK (1999) Photoprotection revisited: genetic and molecular approaches. Annu Rev Plant Physiol Plant Mol Biol 50:333–359

Paillotin G (1976) Movement of excitations in the photosynthetic domains of photosystem II. J Theor Biol 58:237–252

Pascal AA, Liu Z, Broess K, van Oort B, van Amerongen H, Wang C, Horton P, Robert B, Chang W, Ruban A (2005) Molecular basis of photoprotection and control of photosynthetic light-harvesting. Nature 436:134–137

Rikkinen J (1995) What's behind the pretty colours? A study on the photobiology of lichens. Bryobrothera 4:1–239

Ruban AV, Berera R, Ilioaia C, van Stokkum IHM, Kennis JTM, Pascal AA, van Amerongen H, Robert B, Horton P, van Grondelle R (2007) Identification of a mechanism of photoprotective energy dissipation in higher plants. Nature 450:575–578

Shuvalov VA, Heber U (2003) Photochemical reactions in dehydrated photosynthetic organisms, leaves, chloroplasts, and photosystem II particles: reversible reduction of pheophytin and chlorophyll and oxidation of β-carotene. Chem Phys 294:227–237

Solhaug KA, Gauslaa Y, Nybakken L, Bilger W (2003) UV-induction of sun-screening pigments in lichens. New Phytol 158:91–100

Stepigova J, Gauslaa Y, Cempirkova-Vrablikova H, Solhaug KA (2008) Irradiance prior to and during desiccation improves the tolerance to excess irradiance in the desiccated state of the old forest lichen *Lobaria pulmonaria*. Photosynthetica 46:286–290

Terashima I, Fujita T, Inoue T, Chow WS, Oguchi R (2009) Green light drives photosynthesis more efficiently than red light in strong white light: revisiting the enigmatic question of why leaves are green. Plant Cell Physiol 50:684–697

Veerman J, Vasil'ev S, Paton GD, Ramanauskas J, Bruce D (2007) Photoprotection in the lichen *Parmelia sulcata*: the origins of desiccation-induced fluorescence quenching. Plant Physiol 145:997–1005

Wirth V (1995) Die Flechten. Verlag Eugen Ulmer, Stuttgart

Zinth W, Kaiser W (1993) Time-resolved spectroscopy of the primary electron transfer in reaction centres of *Rhodobacter sphaeroides* and *Rhodopseudomonas viridis*. In: Deisenhofer J, Norris JR (eds) The photosynthetic reaction centre. Academic, San Diego, CA, pp 71–88

Chapter 8
Evolution, Diversity, and Habitats of Poikilohydrous Vascular Plants

Stefan Porembski

8.1 Introduction

Only a few species of vascular plants are able to cope with extreme temporal variations of water availability. Most higher plants are homoiohydrous, i.e., their water content varies very little. Very exceptionally the water content of vascular plants follows fluctuations of humidity in their environment. Walter (1931) called plants whose water content closely follows fluctuations of humidity in their environment poikilohydrous. Desiccation tolerant vascular plants are able to survive cycles of dehydration and rehydration without losing viability. In the desiccated state they survive the loss of up to 95% of their cellular water. Detailed overviews of the ecological and physiological adaptations of resurrection plants were provided by Gaff (1981, 1989), Bewley (1995), Hartung et al. (1998), Tuba et al. (1998), Kluge and Brulfert (2000), Walters et al. (2002) and Kappen and Valladares (2007). A survey of the anatomy of desiccation tolerant vascular plants was given by Fahn and Cutler (1992). Recently, the molecular genetics of desiccation tolerance became objects of research (for surveys see Ingram and Bartels 1996; Phillips et al. 2002, see also Chaps. 13–17).

Desiccation tolerance is widespread among cryptogams but is very rare among higher plants. Early reports of desiccation tolerant angiosperms were provided by e.g., Dinter (1918) and Heil (1924). Knowledge about their natural growth sites became more detailed through the studies of e.g., Hambler (1961) and Gaff (1977) which already emphasized the role of rock outcrops as habitats for desiccation tolerant vascular plants. Since almost two decades own studies are devoted to plant ecological investigations on granitic and gneissic outcrops (inselbergs) and ferricretes over a broad geographic spectrum. Inselbergs are characterized by harsh microclimatic and edaphic conditions, and desiccation-tolerant vascular plants dominate in certain plant communities (e.g., monocotyledonous mats). Based on extensive fieldwork and laboratory experiments, the existing knowledge on the systematic position and the ecology of desiccation tolerant vascular plants was summarized by Porembski and Barthlott (2000). It was demonstrated clearly that rock outcrops such as inselbergs form centres of diversity for poikilohydrous vascular plants. Rather neglected hitherto was the fact that the canopy of forests

U. Lüttge et al. (eds.), *Plant Desiccation Tolerance*, Ecological Studies 215,
DOI 10.1007/978-3-642-19106-0_8,

(mostly tropical) harbours numerous desiccation tolerant ferns which outnumber poikilohydrous rock outcrop dwellers. Over the last decade, more information on the number and ecology of desiccation tolerant plants have accumulated and will here be reported in an updated survey in the following.

8.2 Systematic Distribution and Evolutionary Aspects

Among higher plants desiccation tolerant species occur within ferns and fern allies and angiosperms and are completely lacking within gymnosperms. The absence of poikilohydrous species among the mostly phanerophytic gymnosperms can be explained by the fact that certain ecophysiological constraints exclude trees from being desiccation tolerant.

Preceding surveys have estimated the number of desiccation tolerant vascular plants on rock outcrops to be around c. 300 species (Porembski and Barthlott 2000). No such account has been given on the number of desiccation tolerant epiphytic vascular species. Based on own calculations their number can be estimated to comprise between 700 and 1,000 species (almost exclusively ferns). This number contains a considerable percentage of the mainly epiphytic filmy ferns (Hymenophyllaceae) that are probably mainly desiccation tolerant. Consequently the number of desiccation tolerant vascular plant species could reach c. 1,300 if all members of the Hymenophyllaceae are desiccation tolerant which seems to be rather likely (Kornás 1977; Nitta 2006). A list of desiccation tolerant genera is given in Table 8.1.

8.2.1 "Ferns" and "Fern Allies"

Within the paraphyletic group of "ferns and fern allies" both lycophytes and monilophytes (ferns sensu Pryer et al. 2004) contain desiccation tolerant species. A considerable number of both terrestrial and epiphytic ferns and fern allies are notable for colonizing xeric habitats such as inselbergs and forest canopies. For many clades, however, only anecdotic evidence is available with regard to the number of desiccation tolerant species.

Selaginellaceae form an ancient group of lycopods and date back to the Carboniferous Period (330–350 million years ago). The genus *Selaginella* (Fig. 8.1) comprises desiccation tolerant species in the subgenera *Tetragonostachys* (moss-like species with small leaves that possess thick cuticles, distributed throughout the tropics) and *Stachygynandrum* (rosette forming species such as the "Rose of Jericho" *Selaginella lepidophylla*, distributed throughout the tropics) as well as in a clade that is still unnamed (Korall and Kenrick 2002). According to the latter authors, desiccation-tolerant species evolved at least three times in different clades of *Selaginella*. Based on own observations and on available literature it can be

Table 8.1 Genera of vascular plants containing desiccation tolerant taxa (see also Table 9.1)

Genus	Family	Distribution	Growth sites
*Acanthochlamys**	Velloziaceae	SW China	Rock outcrops
Actiniopteris	Pteridaceae	Paleotropics	Rock outcrops
Afrotrilepis	Cyperaceae	W. Africa	Rock outcrops
Anemia	Schizaeaceae	S. America	Rock outcrops
*Aponogeton**	Aponogetonaceae	Paleotropics	Rock outcrops
Asplenium	Aspleniaceae	Subcosmop.	Rock outcrops
Barbacenia	Velloziaceae	S. America	Rock outcrops
Barbaceniopsis	Velloziaceae	S. America	Rock outcrops
Boea	Gesneriaceae	Paleotropics	Rock outcrops
Borya	Boryaceae	Australia	Rock outcrops
Bulbostylis	Cyperaceae	S. America	Rock outcrops
Burlemarxia	Velloziaceae	S. America (only Brazil)	Rock outcrops
Cheilanthes	Pteridaceae	Subcosmop.	Rock outcrops
Coleochloa	Cyperaceae	W./E. Africa, Mad.	Rock outcrops Rarely epiphytic
*Corallodiscus**	Gesneriaceae	Trop. Asia	Rock outcrops
Craterostigma	Linderniaceae	W./E. Africa	Rock outcrops
Doryopteris	Pteridaceae	S. America	Rock outcrops
Drynaria	Polypodiaceae	Subcosmop.	Canopy
Eragrostiella	Poaceae	Australia	Rock outcrops
Eragrostis	Poaceae	E./S. Africa	Rock outcrops
Fimbristylis	Cyperaceae	Trop. Africa, Australia	Rock outcrops
*Guzmania**	Bromeliaceae	Neotropics	Canopy
Haberlea	Gesneriaceae	S. Europe	Rock outcrops
Hemionitis	Pteridaceae	S. America	Rock outcrops
*Henckelia**	Gesneriaceae	Trop. Asia	Rock outcrops
Hymenophyllum	Hymenophyllaceae	Subcosmop.	Canopy
Jancaea	Gesneriaceae	Greece	Rock outcrops
Limosella	Plantaginaceae	S. Africa	Rock outcrops
Lindernia	Linderniaceae	Trop. Africa	Rock outcrops
Micrairia	Poaceae	Australia	Rock outcrops
Microchloa	Poaceae	Mostly Paleotropics	Rock outcrops
Microdracoides	Cyperaceae	W. Africa	Rock outcrops
Myrothamnus	Myrothamnaceae	E./S. Africa, Mad.	Rock outcrops
Nanuza	Velloziaceae	Brazil	Rock outcrops
Notholaena	Pteridaceae	N./S. America	Rock outcrops
Oropetium	Poaceae	Paleotropics	Rock outcrops
*Paraboea**	Gesneriaceae	Trop. Asia	Rock outcrops
Pellaea	Pteridaceae	Trop. Africa	Rock outcrops
Phymatosorus	Polypodiaceae	Paleotropics	Canopy Rock outcrops
Platycerium	Polypodiaceae	Mainly Paleotropics	Canopy Rock outcrops
Pleopeltis	Polypodiaceae	Subcosmop.	Canopy
Pleurostima	Velloziaceae	Brazil	Rock outcrops
Polypodium	Polypodiaceae	Subcosmop.	Rock outcrops Canopy
Ramonda	Gesneriaceae	S. Europe	Rock outcrops
Satureja	Lamiaceae	S. America	Rock outcrops
Schizaea	Schizaeaceae	E. Africa, Seychelles	Rock outcrops
Selaginella	Selaginellaceae	Pantrop., N. America	Rock outcrops

(*continued*)

Table 8.1 (continued)

Genus	Family	Distribution	Growth sites
Sporobolus	Poaceae	Paleo-/Neotropics	Rock outcrops
Streptocarpus	Gesneriaceae	E. Africa/Mad.	Rock outcrops
Talbotia	Velloziaceae	S. Africa	Rock outcrops
Trichomanes	Hymenophyllaceae	Subcosmop.	Canopy
Trilepis	Poaceae	S. America	Rock outcrops
Tripogon	Poaceae	Subcosmop.	Rock outcrops
*Trisepalum**	Gesneriaceae	Trop. Asia	Rock outcrops
Vellozia	Velloziaceae	S. America (mainly Brazil)	Rock outcrops
Xerophyta	Velloziaceae	Trop. Africa, Mad.	Rock outcrops

Information on geographic distribution and growth sites is based on literature sources and own personal observations. For genera marked with * direct proof of desiccation tolerance is still lacking. Indications on growth sites refer to desiccation tolerant taxa within the respective genus

Fig. 8.1 Dry season aspect of *Selaginella* spec. on inselberg in southern India (Karnataka). Throughout the tropics species of the genus *Selaginella* are common on rock outcrops

estimated that more than 50 species of this genus are desiccation tolerant with the vast majority living on rock outcrops both in temperate and tropical regions.

Within the monilophytes desiccation tolerant species have evolved independently several times. Among orders that contain resurrection plants are Hymenophyllales, Polypodiales, and Schizaeales. On the family level desiccation tolerant species occur in Anemiaceae (*Anemia*), Aspleniaceae (*Asplenium* s.l.), Hymenophyllaceae (*Hymenophyllum* s.l., *Trichomanes* s.l., Fig. 8.2), Polypodiaceae (e.g., *Drynaria*, *Phymatosorus*, *Platycerium*, *Polypodium*, Fig. 8.3), Pteridaceae (e.g., *Actiniopteris*, *Cheilanthes*, *Doryopteris*, *Hemionitis*, *Notholaena*, *Pellaea*, Fig. 8.4), and Schizaeaceae (*Schizaea*). It has to be emphasized, however, that more experimental tests are needed in order to conclusively decide about the desiccation tolerance of a large number of ferns (e.g., within Hymenophyllaceae). Bearing in mind the lack of robust data it can only be speculated that the number of

Fig. 8.2 Despite their delicate appearance filmy ferns such as *Trichomanes reniforme* (New Zealand) are able to survive long dry spells in a desiccated state

Fig. 8.3 The staghorn fern *Platycerium stemaria* grows epiphytically and epilithcally in wetter parts of tropical Africa

desiccation tolerant ferns ranges between 200 and 1,200 species with the higher number being more probable.

8.2.2 Angiosperms

Among angiosperms, the desiccation tolerant monocotyledons outnumber the dicotyledons. Otherwise it is difficult to identify clear patterns in the systematic distribution of desiccation tolerant angiosperms. However, it is obvious that the basal lineages of angiosperms do not contain any resurrection plants. Desiccation

Fig. 8.4 Typical elements of rock outcrop vegetation in the Paleotropics are species of the fern genus *Actiniopteris*

Fig. 8.5 Velloziaceae are the largest family of desiccation tolerant angiosperms. *Nanuza plicata* frequently occurs on inselbergs in the Brazilian Mata Atlantica-region

tolerance evolved several times independently within angiosperms and mostly within rather herbaceous lineages. Within the monocotyledons, resurrection plants have evolved independently within Alismatales (Aponogetonaceae, *Aponogeton desertorum* from Namibia seems to be poikilohydric), Asparagales (Boryaceae), Pandanales (Velloziaceae), and Poales (Cyperaceae, Poaceae and possibly Bromeliaceae). Among the monocotyledons Velloziaceae comprise most desiccation tolerant species (more than 200, Fig. 8.5) whereas Poaceae (e.g., within the genera *Microchloa*, *Tripogon*), Cyperaceae (e.g., *Afrotrilepis*, *Coleochloa*, *Microdracoides*, Fig. 8.6), and Boryaceae (*Borya*, Fig. 8.7) are by far less speciose. In addition, there are hints that Bromeliaceae too contain desiccation tolerant species.

Fig. 8.6 The stem-forming Cyperaceae *Microdracoides squamosus* is endemic on rock outcrops in West Africa

Fig. 8.7 In particular on rock outcrops in Western Australia several desiccation tolerant species of *Borya* occur

According to Zotz and Andrade (1998), the epiphytic *Guzmania monostachya* can lose more than 90% of the water present in full turgor and shows a typical response of desiccation tolerant plants. Moreover, the genus *Tillandsia* might include desiccation tolerant epiphytic and lithophytic representatives but detailed data are not available yet. Remarkable is the acquisition of the tree habit by desiccation tolerant arborescent monocotyledons (within Boryaceae, Cyperaceae, Velloziaceae, see Porembski 2006). These have mostly developed in the tropics with *Borya* being a temperate outlier.

Fig. 8.8 *Lindernia welwitschii* is a typical element of shallow depressions on rock outcrops in southern Africa

Among the dicotyledons, desiccation tolerant representatives occur within Gunnerales (Myrothamnaceae) and Lamiales (Gesneriaceae, Linderniaceae, Plantaginaceae: *Limosella*). Here Linderniaceae (*Craterostigma*, *Lindernia*, incl. *Chamaegigas*), Fig. 8.8) and Gesneriaceae (e.g., *Boea*, *Streptocarpus*, possibly also *Corallodiscus*, *Henckelia*, *Paraboea*, *Trisepalum*, see Weber 2004) each contain more than a dozen species whereas Myrothamnaceae comprise only two species (*Myrothamnus flabellifolius* in tropical Africa, *Myrothamnus moschatus* in Madagascar, Fig. 8.9).

In total c. 300 desiccation tolerant species occur within angiosperms but it should be emphasized again that more information is needed about the behaviour of certain taxa (e.g., *Tillandsia*) that might include resurrection plants too.

8.3 Habitats and Geographic Distribution

In contrast to poikilohydrous cryptogams that are conspicuous in arid ecosystems throughout the world (incl. hot deserts and Antarctica), poikilohydrous vascular plants are not centred in arid and semi-arid regions. The majority of desiccation tolerant vascular plants occur on zonal growth sites, which are characterized by very harsh environmental conditions (for details see Szarzyinski 2000) but where precipitation is higher than in deserts. It is only under these conditions that resurrection plants are not outcompeted by homoiohydrous plants and they are thus usually not found in zonal ecosystems. A further hitherto rather neglected hotspot of diversity for vascular resurrection plants is the canopy of tropical and temperate forests where poikilohydrous epiphytes (almost exclusively ferns) can occur in great profusion.

Fig. 8.9 The shrub *Myrothamnus moschatus* is endemic to Malagasy rock outcrops

Most prominent terrestrial habitats are rock outcrops such as inselbergs. The latter form centres of diversity for resurrection plants where mat-forming monocotyledons can form extensive stands. Monocotyledonous mat-formers prefer freely exposed rocky slopes, ferns rather occur in shaded places and Linderniaceae (e.g., *Craterostigma*) show a preference for shallow depressions and rock pools. Among fern allies, several species of *Selaginella* grow sun-exposed on rock outcrops in temperate and tropical regions.

The most extensive stands globally of desiccation tolerant vascular plants are found in the mountain range of the Western Ghats that in the western parts of India runs for more than 1,500 km in north–south direction. Here nearly vertical rocky slopes are characteristic where desiccation tolerant grasses (*Tripogon* spp.) cover large parts of the rocky surface (Fig. 8.10).

Moreover, lateritic plateaus (e.g., ferricretes) that are characterized by sharp contrasts between flooding in the rainy season and drought in the dry season form growth sites for desiccation tolerant plants (Fig. 8.11) in seasonal parts of the tropics. Here, Poaceae are most important with the genera *Microchloa* and *Oropetium* being prominent in the Paleotropics. In addition, ferns (e.g., *Actiniopteris* spp., *Polypodium* spp.) and Velloziaceae (in the Neotropics) are common on these flat outcrops.

Ferns represent the largest number of desiccation tolerant epiphytes with Hymenophyllaceae (only a minority of filmy ferns grows terrestrially) being particularly prominent. Based on observations by Kornás (1977) and Nitta (2006) who

Fig. 8.10 Steep rocky cliffs of the Indian Western Ghats are colonized by resurrection plants, such as grasses (*Tripogon* spp.)

Fig. 8.11 In many parts of the tropics lateritic plateaus occur with desiccation tolerant Poaceae (e.g., *Oropetium* spp., *Tripogon* spp.) being particularly important

confirmed desiccation tolerance for a considerable number of filmy fern species it could be assumed that Hymenophyllaceae (comprising c. 700 spp.) are the largest family of vascular resurrection plants. Likewise members of the Polypodiaceae (e.g., *Phymatodes*, *Platycerium*, *Polypodium*) occur with desiccation tolerant epiphytic species in tropical but also in temperate regions. Very rarely desiccation tolerant Poaceae (*Tripogon* spp., in India) and Cyperaceae (*Coleochloa* spp., tropical Africa) can be found growing epiphytically.

Reports on desiccation tolerant epiphytic angiosperms are rare with *G. monostachya* being the only example (Zotz and Andrade 1998) hitherto known. A closer examination

Fig. 8.12 The fern *Asplenium ceterach* is a common colonizer of walls in southern and maritime Europe

of other epiphytic bromeliads (particularly within *Tillandsia*) might reveal that their number is higher than hitherto expected.

A small number of resurrection plants have managed to colonize appropriate growth sites in human settlements. In Europe different species of *Asplenium* (e.g., *A. ceterach*, *A. trichomanes*, Fig. 8.12) can be found in crevices of buildings. Here they are characteristic elements of clearly circumscribed plant communities (e.g., *Asplenietea trichomanis*) that are characterized by prolonged droughts. In wetter parts of tropical Africa the staghorn fern *Platycerium stemaria* is common on roofs and walls of buildings where it can withstand periods of desiccation.

The vast majority of resurrection plants occur as epiphytes or lithophytes in the tropics. Both the tropical parts of Africa and South America are rich in desiccation tolerant monocotyledons in contrast to tropical Asia where this group is less species rich. However, our knowledge about resurrection plants from tropical Asia is comparatively poor and thus species numbers might increase for this region depending on future explorations. It can, however, be stated that southern Asia forms a centre of diversity for desiccation tolerant Gesneriaceae (e.g., *Boea*, *Corallodiscus*, *Paraboea*) where some representatives occur in altitudes above 4,500 m (*Corallodiscus kingianus*, experimental proof of desiccation tolerance is still missing). In this context it would be interesting to get more information about the desiccation tolerance of *Acanthochlamys bracteata* (high altitude areas in SW China) which is the only Asiatic representative of the largely poikilohydrous Velloziaceae.

In tropical Africa the Sudano-Zambezian Region and Madagascar form centres of diversity for poikilohydrous vascular plants. Particularly rich in species are Linderniaceae (*Craterostigma*, *Lindernia*) and Velloziaceae (*Xerophyta*) whereas

Cyperaceae (*Afrotrilepis*, *Coleochloa*, *Microdracoides*) and Poaceae (*Microchloa*, *Oropetium*, *Tripogon*) are less speciose but can become dominant on e.g., inselbergs. Some poikilohydrous members of *Lindernia* are colonizers of seasonally water-filled rock pools (e.g., the famous *L. intrepidus*, endemic to Namibia, see Heil 1924, Hartung et al. 1998; Chap. 12). Most important mat-forming taxa are *Afrotrilepis pilosa* (West Africa, see Porembski et al. 1996), *Coleochloa setifera* (East Africa, Madagascar), *Microdracoides squamosus* (West Africa), and *Xerophyta* spp. (mainly in the Zambezian Region). Endemic to tropical Africa and Madagascar are the shrubby Myrothamnaceae *M. flabellifolia* (Zambezian Region) and *M. moschatus* (Madagascar) that typically occur in monocot-mats and crevices. In the same region *Streptocarpus* spp. (Gesneriaceae) can be found on rock outcrops. Among ferns and fern allies, both lithophytes (e.g., *Asplenium*, *Pellaea*, *Selaginella*) and epiphytes (e.g., *Hymenophyllum*, *Platycerium*) occur widespread in tropical Africa and Madagascar.

In North and Central America resurrection plants are exclusively represented by mostly rock-colonizing ferns and grasses (*Sporobolus atrovirens*) that can be found from the warm temperate to the arctic regions. Most typical is the genus *Selaginella* with *Cheilanthes*, *Notholaena*, *Pellaea*, and *Polypodium* (with the lithophytic/epiphytic *P. polypodioides* and *Polypodium virginianum*) being less important (cf. Iturriaga et al. 2000). Of particular interest is the so-called Rose of Jericho *S. lepidophylla*, a characteristic colonizer of open sites in the Chihuahuan desert.

In South America rock outcrops in Brazil form centres of diversity for resurrection plants with both inselbergs and quartzitic outcrops harbouring many species (Porembski et al. 1998). The so-called campo rupestre vegetation (widespread in Minas Gerais and Bahia) is also rich in resurrection plants. Of particular importance are Velloziaceae which can be found from inselbergs at sea level up to the high altitude zone of the Itatiaia Mts. (above 2,200 m). The shrubby to tree-like Velloziaceae have their centre of diversity in the quartzitic Serra do Espinhaço where numerous local endemics (mostly *Vellozia* spp.) occur. Nearly all species grow on sun-exposed rocks and form mat-like communities or grow in crevices. Ferns and fern allies (e.g., *Doryopteris*, *Selaginella*) are the second largest taxonomic group whereas Cyperaceae (*Bulbostylis*, *Trilepis*) are of less importance and desiccation tolerant dicotyledons are almost absent. The only exception is the tiny Cactaceae *Blossfeldia liliputana* (northern Argentina, southern Bolivia). Informations on poikilohydrous epiphytic ferns in the Neotropics are sparse but it can be assumed that species of *Pleopeltis* and *Trichomanes* belong to this group (cf. Hietz and Briones 1998). Moreover, it should be tested of whether certain lithophytic and epiphytic Bromeliaceae are poikilohydrous.

In Europe ferns dominate among resurrection plants with mainly limestone colonizing species of e.g., *Asplenium*, *Cheilanthes*, *Hymenophyllum* and *Polypodium* being most prominent. These genera link tropical and temperate zones where they occur with numerous desiccation tolerant species. Only a few desiccation tolerant angiosperms occur in temperate regions such as the relictual gesneriads *Haberlea*, *Jancaea*, and *Ramonda* (Fig. 8.13) that are endemic in certain mountain areas (e.g., the Balkans) around the Mediterranean Sea.

Fig. 8.13 The gesneriad *Ramonda serbica* occurs as a paleoendemic in the Balkans

Endemic to Australia is the genus *Borya* that is common on rock outcrops where it is particularly typical in the temperate parts of this continent. In addition, ferns (e.g., *Cheilanthes*), grasses (*Micrairia* spp.), and the gesneriad *Boea hygroscopica* are present (Gaff and Latz 1978).

8.4 Adaptive Traits

Within certain groups of poikilohydrous vascular plants specific key adaptive traits occur which obviously have been evolved several times independently in close connection with their particular way of life. In the following, a concise survey is given about their most important adaptive traits from a morphological–anatomical viewpoint.

There are differences in the ability to survive periods of drought between individual desiccation tolerant species. It is conceivable that factors such as speed of tissue desiccation, duration of desiccation, and temperature influence the desiccation tolerance of poikilohydrous plants. However, the data hitherto available are still too fragmentary, and it is thus not possible to draw conclusions on the relationships between the degree of desiccation tolerance and taxonomic or ecological characteristics. Nevertheless one can possibly outline some tendencies. Based on own observations it can be stated that both in temperate and tropical regions certain species of *Selaginella* seem to be among the most desiccation tolerant colonizers of rock outcrops. According to Gaff (1981) the most hardy resurrection plants (e.g., *Borya nitida*) survive the loss of over 94% of their water content at full turgor. Dehydration of leaves of poikilohydrous species is often accompanied by a change in leaf colour. With regard to this aspect, one can distinguish between poikilochlorophyllous (i.e., losing most or all of their chlorophyll) and homoiochlorophyllous (i.e., species that preserve their chlorophyll content) plants. Most poikilohydrous monocotyledons are poikilochlorophyllous whereas most desiccation tolerant dicots and ferns are homoiochlorophyllous. The latter are usually more rapid in recovering their water content and photosynthetic activity.

As far as their reproductive traits are concerned, vascular resurrection plants have not developed unique strategies. Among desiccation tolerant angiosperms, both pollination by wind (Cyperaceae, Myrothamnaceae, Poaceae) and animals (Gesneriaceae, Linderniaceae, Velloziaceae) can be found. Among the latter, entomophily seems to dominate but for certain Velloziaceae pollination by bats and birds has been reported too (e.g., Sazima and Sazima 1990). Dispersal of diaspores over larger distances seems to be almost exclusively by wind but on a smaller scale water dispersal could play a role too. Moreover, accidental transport of diaspores by birds cannot be ruled out. The lack of fleshy fruits is probably a consequence of the vagaries of water availability that makes the production of berries or drupes for desiccation tolerant plants too risky. Most poikilohydrous angiosperms have hermaphrodite flowers but dioecy (*Microdracoides*, *Myrothamnus*) and monoecy (Cyperaceae) are likewise present.

Leaf size of poikilohydrous vascular plants varies widely from tiny leaflets (e.g., *Selaginella*) up to the large fronds (e.g., *P. stemaria*) of certain ferns. With regard to leaf shape it can be stated that all poikilohydrous angiosperms possess undivided leaves whereas other types of leaves are completely lacking. This is in contrast to ferns and fern allies where both undivided and divided leaves occur. Very common in all groups of vascular resurrection plants is the curling of leaves during the process of desiccation what appears to be a mechanism for avoiding photoinhibitory damage.

Desiccation tolerant arborescent monocotyledons form a remarkable example of convergent evolution. This type of arborescent monocotyledons occurs in both tropical and temperate regions and is found within Boryaceae (*Borya*), Cyperaceae (*Afrotrilepis*, *Bulbostylis*, *Coleochloa*, *Microdracoides*), and Velloziaceae (e.g., *Vellozia*, *Xerophyta*). They possess a number of ecophysiological and morphoanatomical adaptations (e.g., roots with velamen radicum, Porembski and Barthlott 1995) that render them perfectly adapted for the colonization of rock outcrops. Besides their treelike habit, their ability to form clonal populations of considerable age (i.e., hundreds of years) by means of stolons or by basal branching allows for the long lasting occupation of suitable sites (for details see Porembski 2006). The trunks of arborescent Cyperaceae and Velloziaceae are regularly colonized by vascular epiphytes with certain orchids (e.g., *Polystachya* spp. in tropical Africa, *Pseudolaelia vellozicola* in Brazil) showing a remarkably high degree of phorophyte specificity (Porembski 2005).

A further remarkable example of a highly specialized resurrection plant is provided by *Lindernia* (*Chamaegigas*) *intrepidus* (Chap. 12) which is endemic to Namibia where it occurs in seasonally water-filled rock pools (Heil 1924; Heilmeier et al. 2005). This species has desiccation tolerant submerged leaves which are contractile and develops desiccation sensitive floating leaves after rainfall. The submerged leaves shrink by 75–80%, mainly due to contraction of xylem vessels that are characterized by extremely densely packed helical thickenings (Schiller et al. 1999).

8.5 Economic Importance

Resurrection plants have become important experimental models for understanding the physiological and molecular aspects of desiccation tolerance. In the future this knowledge might be of interest for the development of drought tolerant crop species (Vicré et al. 2004).

Whereas the vast majority of resurrection plants are currently not yet used economically a limited number of them is of economic relevance. One of the best known examples is the "Rose of Jericho" (*S. lepidophylla*, a Chihuahuan element, USA and Mexico) which is globally sold for ornamental purposes. Apart from this example, resurrection plants are rarely used as ornamentals. This is the case with *M. squamosus* in the surroundings of the Cameroonian capital Yaoundé. Here individuals of this showy species are collected on inselbergs and sold in markets. A further case of commercial use of poikilohydrous plants is the sale of the epiphytic fern *P. stemaria* as ornamental in West Africa (Porembski and Biedinger 2001).

An additional case is the "wonder bush" (*M. flabellifolia*) which is marketed in large amounts in parts of southern Africa. The species is commonly used as a medicinal plant (for its antimicrobial attributes see van Vuuren 2008) throughout its distributional range where local collecting activities sometimes take a heavy toll on local populations on inselbergs (own observations in Angola). Certain leaf ingredients provide components of skin moisturizing creams which are sold globally as is the case with individual twigs of this species that are marketed under the name "wonder bush". Own recent observation have shown that the Malagasy inselberg endemic *M. moschatus* too is used as a medicinal plant which is sold in local markets.

Alves (1994) reports on the sale of c. 50 cm long stem segments of *Vellozia* species in Brazil (Bahia) which are used as fire starters that burn even in torrential rains.

8.6 Conservation

Up to now almost no vascular resurrection plants are protected by law in their countries of origin. This is mainly due to the fact that most of them occur in rather inaccessible habitats and thus have escaped the attention of conservationists. However, there are numerous dangers to these species with the destruction of their habitats being the most relevant factor for the decrease in population numbers. The list of the driving forces behind the destruction of their habitats (forest canopies, rock outcrops) includes the conversion of forest in farmland, quarrying, human-lit fires, grazing, and off-road driving. In addition, the introduction of invasive weeds can also impose serious threats on desiccation tolerant plants. For example, this has been the case on certain inselbergs in the West African rainforest zone where introduced pineapples (*Ananas comosus*) have occasionally outcompeted *A. pilosa*. All over the world the most serious threats to the habitats of desiccation

Fig. 8.14 Burnt stems of *Microdracoides squamosus* due to human lit fires on an inselberg in Cameroon

tolerant rock outcrop colonizing plants are both quarrying and fire. Due to the ever increasing demand for rocks for e.g., construction works the quarrying of inselbergs has regionally reached dramatic extents as can be seen around Bangalore (southern India) where numerous inselbergs have completely disappeared during the last decades. Human lit fires are particularly widespread nowadays on inselbergs where mat communities made up by desiccation tolerant Cyperaceae and Velloziaceae are heavily influenced (Fig. 8.14). Whereas old specimens are relatively well protected against fire by their insulating sheaths of adventitious roots and old leaves, recruitment by juveniles becomes extremely limited.

Certain resurrection plants are endangered by the collection of huge amounts of individuals for ornamental purposes (s. above). One of the best known examples is *S. lepidophylla* that is currently imported in large numbers into the European Community as to warrant monitoring.

Acknowledgements Financial support is gratefully acknowledged for rock outcrop studies by the Deutsche Forschungsgemeinschaft. The author is deeply indebted for valuable discussions and remarks to W. Barthlott (Bonn), J.-P. Ghogue (Yaoundé), S. D. Hopper (Kew), N. Korte (Rostock), Z. Tuba (Gödöllö) and G. Zotz (Oldenburg).

References

Alves RJV (1994) Morphological age determination and longevity in some *Vellozia* populations in Brazil. Folia Geobot Phytotaxon 29:55–59

Bewley JD (1995) Physiological aspects of desiccation tolerance – a retrospect. Int J Plant Sci 156:393–403

Dinter K (1918) Botanische Reisen in Deutsch-Südwest-Afrika. Feddes Rep Beih 3:1–169

Fahn A, Cutler DF (1992) Xerophytes. In: Braun HJ, Carlquist S, Ozenda P, Roth I (eds) Handbuch der Pflanzenanatomie, vol 13, part 3, Spezieller Teil. Borntraeger, Berlin

Gaff DF (1977) Desiccation tolerant vascular plants of Southern Africa. Oecologia 31:95–109

Gaff DF (1981) The biology of resurrection plants. In: Pate JS, McComb AJ (eds) The biology of Australian plants. University of Western Australia Press, Perth, pp 114–146

Gaff DF (1989) Responses of desiccation tolerant "resurrection" plants to water stress. In: Kreeb KH, Richter H, Hinckley TM (eds) Structural and functional responses to environmental stresses: water shortages. SPB Academic Publishing, The Hague, pp 264–311

Gaff DF, Latz PK (1978) The occurrence of resurrection plants in the Australian flora. Aust J Bot 26:485–492

Hambler DJ (1961) A poikilohydrous, poikilochlorophyllous angiosperm from Africa. Nature 191:1415–1416

Hartung W, Schiller P, Dietz K-J (1998) The physiology of poikilohydric plants. Prog Bot 59:299–327

Heil H (1924) *Chamaegigas intrepidus* Dtr., eine neue Auferstehungspflanze. Beitr Bot Zentralbl 41:41–50

Heilmeier H, Durka W, Woitke M, Hartung W (2005) Ephemeral pools as stressful and isolated habitats for the endemic aquatic resurrection plant *Chamaegigas intrepidus*. Phytocoenologia 35:449–468

Hietz P, Briones O (1998) Correlation between water relations and within-canopy distribution of epiphytic ferns in a Mexican cloud forest. Oecologia 114:305–316

Ingram J, Bartels D (1996) The molecular basis of dehydration tolerance in plants. Annu Rev Plant Physiol Plant Mol Biol 47:377–403

Iturriaga G, Gaff DF, Zentella R (2000) New desiccation-tolerant plants, including a grass, in the central highlands of Mexico, accumulate trehalose. Aust J Bot 48:153–158

Kappen L, Valladares F (2007) Opportunistic growth and desiccation tolerance: the ecological success of poikilohydrous autotrophs. In: Pugnaire F, Valladares F (eds) Functional plant ecology, 2nd edn. CRC/Taylor and Francis Group, Boca Raton/London, pp 7–65

Kluge M, Brulfert J (2000) Ecophysiology of vascular plants on inselbergs. In: Porembski S, Barthlott W (eds) Inselbergs: biotic diversity of isolated rock outcrops in tropical and temperate regions, vol 146, Ecological Studies. Springer, Berlin, pp 143–174

Korall P, Kenrick P (2002) Phylogenetic relationships in Selaginellaceae based on rbcL sequences. Am J Bot 89:506–517

Kornás J (1977) Life-forms and seasonal patterns in the pteridophytes of Zambia. Acta Soc Bot Pol 46:669–690

Nitta JH (2006) Distribution, ecology and systematics of the filmy ferns (Hymenophyllaceae) of Moorea (French Polynesia). University of California, Department of Integrative Biology, Berkeley, CA

Phillips JR, Oliver MJ, Bartels D (2002) Molecular genetics of desiccation tolerant systems. In: Black M, Pritchard HW (eds) Desiccation and survival in plants: drying without dying. CABI Publishing, Wallingford, pp 319–341

Porembski S (2005) Epiphytic orchids on arborescent Velloziaceae and Cyperaceae: extremes of phorophyte specialisation. Nord J Bot 23:505–513

Porembski S (2006) Vegetative architecture of desiccation-tolerant arborescent monocotyledons. Aliso 22:129–134

Porembski S, Barthlott W (1995) On the occurrence of a velamen radicum in tree-like Cyperaceae and Velloziaceae. Nord J Bot 15:625–629

Porembski S, Barthlott W (eds) (2000) Inselbergs: biotic diversity of isolated rock outcrops in tropical and temperate regions, vol 146, Ecological studies. Springer, Berlin

Porembski S, Biedinger N (2001) Epiphytic ferns for sale: influence of commercial plant collection on the frequency of *Platycerium stemaria* (Polypodiaceae) in coconut plantations on the southeastern Ivory Coast. Plant Biol 3:72–76

Porembski S, Brown G, Barthlott W (1996) A species-poor tropical sedge community: *Afrotrilepis pilosa* mats on inselbergs in West Africa. Nord J Bot 16:239–245

Porembski S, Martinelli G, Ohlemüller R, Barthlott W (1998) Diversity and ecology of saxicolous vegetation mats on inselbergs in the Brazilian Atlantic rainforest. Divers Distrib 4:107–119

Pryer KM, Schuettpelz E, Wolf PG, Schneider H, Smith AR, Cranfill R (2004) Phylogeny and evolution of ferns (monilophytes) with a focus on the early leptosporangiate divergences. Am J Bot 91:1582–1598

Sazima M, Sazima I (1990) Humming bird pollination in two species of *Vellozia* (Liliiflorae, Velloziaceae) in southeastern Brazil. Bot Acta 103:83–86

Schiller P, Wolf R, Hartung W (1999) A scanning electron microscopical study of hydrated and desiccated submerged leaves of the aquatic resurrection plant *Chamaegigas intrepidus*. Flora 194:97–102

Szarzynski J (2000) Xeric islands. Environmental conditions on inselbergs. In: Porembski S, Barthlott W (eds) Inselbergs: biotic diversity of isolated rock outcrops in tropical and temperate regions, vol 146, Ecological Studies, Springer, Berlin, pp 37–48

Tuba Z, Proctor MCF, Csintalan Z (1998) Ecophysiological responses of homiochlorophyllous desiccation tolerant plants: a comparison and an ecological perspective. Plant Growth Regul 24:211–217

van Vuuren SF (2008) Antimicrobial activity of South African medicinal plants. J Ethnopharmacol 119:462–472

Vicré M, Farrant JM, Driouich A (2004) Insights into the cellular mechanisms of desiccation tolerance among angiosperm resurrection plant species. Plant Cell Environ 27:1329–1340

Walter H (1931) Die Hydratur der Pflanze und ihre physiologisch-ökologische Bedeutung. Gustav Fischer, Jena

Walters C, Farrant JM, Pammenter NW, Berjak P (2002) Desiccation stress and damage. In: Black M, Pritchard HW (eds) Desiccation and survival in plants: drying without dying. CABI Publishing, Wallingford, pp 263–291

Weber A (2004) Gesneriaceae. In: Kadereit JW (ed) The families and genera of vascular plants, vol 7, Flowering plants, dicotyledons: Lamiales (except Acanthaceae including Avicenniaceae). Springer, Berlin

Zotz G, Andrade J-L (1998) Water relations of two co-occurring epiphytic bromeliads. J Plant Physiol 152:545–554

Chapter 9
Ecophysiology of Homoiochlorophyllous and Poikilochlorophyllous Desiccation-Tolerant Plants and Vegetations

Zoltán Tuba and Hartmut K. Lichtenthaler

9.1 Introduction

In this chapter, we are dealing with the vascular plants, which are desiccation-tolerant in their photosynthetically active green leaves and vegetative tissues. Of course, if unavoidable, we also touch the subject of nonvascular, cryptogamic desiccation-tolerant plants (first of all bryophytes, lichens, and algae). Desiccation-tolerant (DT) plants can survive the loss of 80–95% of the cell water content in their photosynthetically active green leaf tissues, so that the plants appear completely dry and no liquid phase remains in their cells. After a shorter or longer period in the desiccated state, they revive and resume their normal metabolism when they are rehydrated, e.g., at rainfall. This is a qualitatively different phenomenon from drought tolerance as ordinarily understood in vascular-plant physiology. Indeed, desiccation tolerance could be seen as a drought-avoiding mechanism according to Levitt's concepts of water stress and hardiness (Levitt 1972).

Desiccation-tolerant (DT) plants may be subdivided into homoiochlorophyllous (HDT) and poikilochlorophyllous (PDT) types (Tuba et al. 1994). The HDT plants (which are the majority) retain their chlorophyll during desiccation and can quickly regenerate their photosynthetic function in the next rain when water is available again. In contrast, in PDT plants the desiccation results in the loss of chlorophylls and thylakoids, which must be resynthesized following remoistening (Hambler 1961; Bewley 1979; Farrant 2000; Gaff 1977, 1989; Gaff and Hallam 1974; Hetherington and Smillie 1982a, b; Tuba et al. 1994; Sherwin and Farrant 1996, 1998). During evolution, PDT plants developed a unique and special type of chloroplast, the desiccoplast that, during droughts, develops all signs of an advanced senescence (as known from gerontoplasts), but recovers upon remoistening to fully functional chloroplasts (Tuba et al. 1994, 1996b). HDT and PDT plants apparently represent contrasting strategies to solve the same ecological problem at drought conditions (Tuba et al. 1994, 1998), whereby HDT plants usually survive shorter drought periods of several days and weeks (up to a few months in some cases), whereas PDT plants

The first author of this contribution died on July 4, 2009, at the age of only 58 years after having finished his part, i.e., the major part, of this manuscript. With great strength and a strong will he managed, despite his progressing illness, to write a very competent review.

U. Lüttge et al. (eds.), *Plant Desiccation Tolerance*, Ecological Studies 215,
DOI 10.1007/978-3-642-19106-0_9, © Springer-Verlag Berlin Heidelberg 2011

can endure longer drought periods of 6–11 months. Much has been published on the photosynthetic responses of HDT plants, especially on the cryptogamic plants (see Proctor and Tuba 2002 and Chaps. 2–8), and a good deal of information is also available on the tolerance limits of HDT desiccation-tolerant cryptogams and phanerogams (Proctor and Tuba 2002; Georgieva et al. 2005). However, far less is known about the ecology, ecophysiology, distribution, and abundance of PDT plants, which appear to be restricted to the monocotyledonous plants.

9.2 Distribution and Evolutionary Aspects of Desiccation Tolerance in Plants

Desiccation tolerance (DT) occurs widely in the plant kingdom. It is commonplace among bryophytes and lichens and is found sporadically among vascular plants of diverse taxonomic affinities. Here, we overview the taxonomic occurrence of desiccation-tolerant vascular plants based on Proctor and Tuba (2002) with inclusion of more recently detected species. In vascular plants, desiccation tolerance of the vegetative tissues has unequivocally been demonstrated in some 400 species, comprising less than 0.2% of the total terrestrial flora, of which nearly 90% occur on inselbergs (rock outcrops) known for their limited water availability and drought conditions. Desiccation-tolerant vascular plants are best represented within the monocotyledons and ferns. Lists of the DT species of higher plants are given by Bewley and Krochko (1982), Porembski and Barthlott (2000), and a more detailed list by Proctor and Pence (2002). The newest list of the genera with vascular DT species, based on Proctor and Tuba (2002) but extended by recently detected species, is presented in Table 9.1 (see also Table 8.1). However, this list is continuously being extended as many new DT species are detected. This particularly applies to pteridophytes, since most of the numerous ferns, detected more recently on various inselbergs, might be DT plants (see also Chap. 8). That DT plants occur in most parts of the world such as Australia, Africa, South America, South Europe, China, India, and tropical Asia is summarized in Chap. 8 for some newly detected species (Table 9.1). In Chap. 16, it is documented that within the same genus many different DT species can be found as is indicated for Linderniaceae (Table 16.1).

From the list in Table 9.1, one recognizes that desiccation tolerance of green vegetative tissue is widely but thinly and unevenly distributed among vascular plants. Virtually all vascular plants have desiccation-tolerant spores (including pollen) or seeds; thus, the potentiality for desiccation tolerance is probably universal (Oliver et al. 2000). In bryophytes, it is a reasonable assumption that desiccation tolerance of vegetative tissue is present in an early primordial way, but this was lost during the evolution of tracheophytes. However, as the authors cited above emphasized, the expression of desiccation tolerance in vegetative tissues is rare, and it must have evolved independently in *Selaginella*, in ferns, and in angiosperms. In view of the general desiccation tolerance of bryophytes, although in a more

Table 9.1 List of genera with vascular dessication-tolerant (DT) plant species (see also Table 8.1)

Pteridophytes
Lycopsida: *Isoetes* (corms only, at least 1 terrestrial sp.), *Selaginella* (13)
Pteropsida: *Actiniopteris* (2), *Adiantum* (1), *Anemia* (1), *Arthropteris* (1), *Asplenium* (11), *Ceterach* (2), *Cheilanthes* (27), *Ctenopteris* (1), *Drynaria* (1), *Dryopteris* (3), *Hemionitis* (1), *Hymenophyllum* (3), *Mohria* (1), *Notholaena* (3), *Paraceterach* (1), *Pellaea* (13), *Phymatosorus* (1), *Platycerium* (1), *Pleopeltis* (1), *Pleurosorus* (1), *Polypodium* (4), *Schizaea* (1), *Woodsia* (1)
Angiosperms
Monocotyledons
Cyperaceae: *Afrotrilepis* (1), *Bulbostylis* (1), *Carex* (1), *Cheilanthes* (1), *Coleochloa* (2), *Cyperus* (1), *Doryopteris* (1), *Fimbristylis* (2), *Kyllingia* (1), *Mariscus* (1), *Microdracoides* (1), *Trilepis* (1)
Boryaceae: *Borya* (3)
Poaceae: *Brachyachne* (1), *Eragrostiella* (3), *Eragrostis* (4), *Micrairia* (5), *Microchloa* (3), *Oropetium* (3), *Poa* (1), *Sporobolus* (7), *Trilepis* (1), *Tripogon* (10)
Velloziaceae: *Aylthonia* (1), *Barbacenia* (4), *Barbaceniopsis* (2), *Burlemarxia* (1), *Nanuza* (1), *Pleurostima* (1), *Talbotia* (1), *Vellozia* (~124), *Xerophyta* (~28)
Dicotyledons
Myrothamnaceae: *Myrothamnus* (2)
Cactaceae: *Blossfeldia* (1)
Gesneriaceae: *Boea* (1), *Haberlea* (1), *Jancaea* (1), *Ramonda* (3), *Streptocarpus* (20)
Scrophulariaceae: *Ilysanthes* (1)
Plantaginaceae: *Limosella* (1)
Linderniaceae: *Chamaegigas* (1), *Craterostigma* (9), *Lindernia* (17)
Lamiaceae: *Micromeria* (1), *Satureja* (1)

The number of currently known DT species is indicated in parentheses behind the genus name. The list is based on Proctor and Tuba (2002) with inclusion of recently detected species

primordial form, one could also speak of an independent re-evolution of desiccation tolerance in different vascular plant groups, and this time in a more advanced form.

Oliver et al. (2000) postulated that the initial evolution of vegetative desiccation tolerance was a crucial step required for the colonization of the land by primitive plants from a freshwater origin (Mishler and Churchill 1985). During evolution, the loss of an original desiccation tolerance might have been favored in parallel with the internalization of the water transport system that developed as the vascular plants became more complex. Within the angiosperms, at least eight independent cases of evolution (or re-evolution) of desiccation tolerance occurred. In *Selaginella* and in the ferns, however, at least one independent evolution (or re-evolution) of desiccation tolerance happened. In vascular plants, desiccation tolerance is clearly a derived, characteristic condition, and DT vascular angiosperms often have evidently evolved from precursors adapted in other ways to tolerate severe drought. Thus, DT angiosperms include species showing some indication of CAM activity or with close CAM relatives. In addition, DT species and drought-tolerant C_4 species can be found within the same genus (e.g., in *Eragrostis* and *Sporobolus*), although in general DT plants belong to the C_3 category. Each time the general desiccation-tolerance phenotype was re-evolved in different plant lineages, the timescales of desiccation, rehydration, and responsiveness are different.

9.3 Habitats and Vegetation of Desiccation-Tolerant Plants

Desiccation-tolerant vascular plants occur particularly on rock outcrops in the tropics, partially in the subtropics, and to a lesser extent in temperate zones (Porembski and Barthlott 2000). DT plants are found from sea level up to 2,800 m, but most of the species occur below 1,000 m a.s.l. The diversity of desiccation-tolerant plant species is highest in the southern hemisphere, such as East Africa, Madagascar, and Brazil, where many granitic and gneissic inselbergs and outcrops exist. Inselbergs frequently occur as isolated monoliths characterized by extreme environmental conditions (i.e., extremely severe edaphic dryness and high degrees of insolation). On tropical inselbergs, desiccation-tolerant monocotyledons (i.e., Cyperaceae and Velloziaceae) dominate in mat-like plant communities, which cover even steep slopes. Mat-forming desiccation-tolerant species can be extremely old in age (hundreds of years) and size (several meters in height, for pseudostemmed species). Both homoiochlorophyllous and poikilochlorophyllous DT plant species occur under these circumstances. In their natural habitats, both DT plant groups survive dry periods of several months (5–11 months) and resume their photosynthetic activity within a few days after the beginning of rainfall. The mat-forming monocotyledons are particularly remarkable due to the possession of roots with a velamen radicum, which had been reported for the first time in the genus *Borya* (Porembski and Barthlott 2000).

The second DT plant dominated community can be found below 1,000 m, in very dry, exposed rocky places, e.g., in the Uluguru Mountains (Tanzania, East Africa), where there is a peculiar xerophytic vegetation, formed by the PDT model plant, the shrubby monocotyledon *Xerophyta scabrida* (Pócs 1976). This open bush dries up completely during the dry season, but turns into a vivid green during the rain. The species are highly adapted to very dry conditions. Below the shrubs, there is usually a dense mat formed by the poikilohydric *Selaginella dregei*, often accompanied by *S. mittenii*. In the herbaceous layer, poikilohydric ferns (*Actinopteris dimorpha*, *Pellaea schweinfurthii*, and *Piptadenia adiantoides*), geophytes (*Ophioglossum costatum*, *O. gomezianum*, and *O. lanczfolium*), or therophytes are present (*Borreria arvensis*, *Oldenlandia herbacea*, and *Heliotropium strigosum*). The fibrous and persistent leaf sheaths of *Xerophyta* absorb some humidity even during the driest periods and provide conditions for epiphytes to live in these places. Along with mosses and lichens, even epiphytic orchids, e.g., *Polystachya tayloriana*, can survive there. A similar pioneer community colonizing bare rock faces has been described by Jackson (1956) in the Imatong Mountains on the Sudan-Uganda border.

9.4 The Poikilochlorophyll Desiccation-Tolerance Strategy

The majority of nonvascular and vascular desiccation-tolerant plants (DT plants) are homoiochlorophyllous (HDT), retaining their photosynthetic pigments and keeping their thylakoids and chloroplast structure intact throughout a drying and rewetting cycle. The HDT strategy is based on the preservation of the integrity of the photosynthetic apparatus in classical mesophyll chloroplasts by protective and/ or repair mechanisms. Dicotyledons and ferns are always chlorophyll retainers, i.e., HDT plants (Hambler 1961; Gaff and Hallam 1974). In contrast, the monocotyledons include both types of the DT strategy and also contain plants with poikilochlorophyllous desiccation tolerance (PDT), which is an advanced way of surviving severe drought periods. During evolution, PDT plants seem to have evolved later than HDT plants. The PDT strategy evolved relatively recently in geological times, in response to the rigors of high irradiance and intermittently arid habitats. Its mechanism relies on specific properties of the chloroplasts, which, under drying stress, degenerate to membrane-free desiccoplasts (e.g., Tuba et al. 1994) and reverse back to functional chloroplasts after rewetting of their leaves (Tuba et al. 1996b). In fact, the PDT strategy is strictly based on the possession of these very specialized chloroplasts, known as desiccoplasts (Tuba et al. 1994). Their unique properties and ultrastructural changes are described in detail below in Sect. 9.5.

The phenomenon of chlorophyll loss of PDT plant species during desiccation was first described by Vassiljev (1931) in *Carex physodes* from central Asia. The term poikilochlorophylly was coined by Hambler (1961), and it appears to be restricted to certain DT monocotyledonous plants (Hambler 1961; Bewley 1979; Bewley and Krochko 1982; Gaff 1977, 1989; Gaff and Hallam 1974; Hetherington and Smillie 1982a). The PDT strategy evolved in plants that are anatomically complex and include the largest in size of all DT species. This can be seen as an evolutionarily different and probably youngest strategy of desiccation tolerance. It is based on the dismantling of internal chloroplast structures by an ordered deconstruction process during drying and its resynthesis upon rehydration by an ordered reconstruction process. These processes can thus be thought of as not only being superimposed on an existing cellular protection mechanism of vegetative desiccation tolerance (Oliver et al. 2000), but as a distinct new class of DT mechanism, as a new DT strategy (Tuba et al. 1994). New plants with the PDT strategy can and will certainly be detected in future ecological research of inselberg plants. Poikilochlorophylly is currently known in eight genera of four families (Cyperaceae, Liliaceae (Anthericaceae), Poaceae, and Velloziaceae). Most occupy the almost soil-less rocky outcrops known as inselbergs and in strongly seasonal subtropical climates (Porembski and Barthlott 2000). See in this respect also Chap. 8 of this book. The most thoroughly studied physiology of PDT plants is that of the African plants *X. scabrida*, *X. viscosa*, and *X. humilis* and that of the Australian plant *Borya nitida* (see Proctor and Tuba 2002; and also Gaff and Churchill 1976; Hetherington and Smillie 1982a, b; Gaff and Loveys 1984).

The PDT strategy has significant ecological consequences. Such higher plants have evolved in habitats where the duration of the desiccated state lasts for 6–10 months or even longer. Such long desiccation periods may reach the limit of survivability for most HDT plants. Under these conditions, it is evidently more advantageous to dismantle the whole photosynthetic pigment and thylakoid apparatus and to reconstitute it after rehydration. In contrast, the HDT strategy is evidently more favorable in habitats where the desiccated periods are shorter, and wet and dry periods alternate more frequently. The DT strategy also exists, although in a different form, in unicellular cyanobacteria where the desiccated state can last even up to several decades or even very much longer (see Chap. 3).

9.5 The Desiccoplast, a Very Specialized, New Type of Chloroplast

In the course of evolution, photosynthetically functioning green chloroplasts have differentiated into various forms, e.g., mesophyll and bundle sheath chloroplasts, sun- and shade-type chloroplasts, or the chloro-amyloplasts with large starch grains in starch storing leaf tissue. During autumnal or induced senescence of green leaves, the chloroplasts of practically all plants, except PDT plants, degenerate to gerontoplasts, whereby their chlorophylls and thylakoids are broken down, osmiophilic plastoglobuli (either some large or numerous small ones) show up in the stroma (Lichtenthaler 1968, 1969, 2007), and the photosynthetic function is irreversibly lost (Lichtenthaler and Babani 2004). Such gerontoplasts are regarded as premortal plastid stages that no longer have the competence or capacity to regreen and regain photosynthetic activity.

The *desiccoplasts* of PDT plants, in turn, behave quite different. Their ultrastructure in the green chloroplast stage of watered plants resembles, e.g., in *X. scabrida* and in other green PDT plants that of normal chloroplasts of the majority of plants (Tuba et al. 1994). During desiccation, however, desiccoplasts dismantle their photosynthetic apparatus, lose all of their chlorophyll and thylakoids, and accumulate osmiophilic or translucent plastoglobuli, but they possess the unique capacity to regreen and resynthesize their functional photosynthetic apparatus following rehydration (Tuba et al. 1993a, b, 1994, 1996a, b; Sherwin and Farrant 1996). The typical desiccoplast in its nongreen stage, after having lost its chlorophylls and thylakoids during desiccation, exhibits the specific ultrastructure of gerontoplasts of senescing leaves. With its unique characteristic to regreen, regain photosynthetic activity, and redevelop into a fully functional chloroplast upon rehydration (Tuba et al. 1996a, b), the desiccoplast enables the PDT plants to survive the prolonged seasonal severe drought conditions of inselbergs and other very dry rocky land areas and to return to photosynthetically active green plants in the rainy season.

The desiccoplast has evolved in plants with the most highly derived form of desiccation tolerance, i.e., the PDT plants, which repeatedly undergo prolonged (and often seasonal) desiccation of 6–10 months and rewatering of their tissue. In general, desiccation-tolerant (DT) plants can survive drying to a point where no liquid phase remains in the cells and the water content may be no more than 5–10% DW (equivalent to an equilibrium water potential of −100 MPa or less). After remoistening dried leaves of PDT plants, an essentially normal metabolism, such as mitochondrial respiration, returns within a few hours. Other activities return after one or a few more days and the desiccoplast becomes green and photochemically active within 24 h. Its thylakoid ultrastructure and photosynthetic capacity are completed after 2–3 days. The structural and functional changes of desiccoplasts have been studied in detail in desiccation/rehydration cycles of leaves of the PDT plant *X. scabrida*, a C_3 plant, as described by Tuba et al. (1993a, b, 1994, 1996a, b) and Csintalan et al. (1998). These reversible changes are briefly reviewed here.

9.5.1 Desiccation of Leaves and Desiccoplast Formation

During desiccation, the relative water content of detached *Xerophyta* leaves declines rather fast; after 12 h ca. 64% of the original leaf water (100%) is lost. The stomata close even more rapidly and the values of the stomatal conductance, g_S, decline from 260 (well-watered state) to values of 48 mmol m^{-2} s^{-1} within 12 h of desiccation (Tuba et al. 1996b). The breakdown of chlorophylls and carotenoids proceeds more slowly than the water loss; yet after 48 h of desiccation 87% of the total chlorophylls (a + b) are lost, however, only 57% of total carotenoids (Table 9.2). As a consequence, the leaves turn yellow and the pigment ratio chlorophylls/carotenoids, (a + b)/(x + c), declines from values of 5.2 (±0.2), which is the normal range for the green leaf tissue of most plants, to values of 1.55 and 1.53 after 36 and 48 h of desiccation, respectively (Table 9.2). After 2 more days of desiccation, the chlorophylls of *Xerophyta* are completely lost, whereas the carotenoids are retained at ca. 25–30% of the level of fully green leaves. Thus, this decline in photosynthetic pigments proceeds in the same way as during autumnal or induced leaf senescence (Lichtenthaler and Babani 2004).

Table 9.2 Decline in the level of total chlorophylls (a + b) and total carotenoids (x + c), and in the pigment weight ratio chlorophylls to carotenoids, (a + b)/(x + c), in green photosynthetically active leaves of *Xerophyta scabrida* during desiccation

Hours	Chlorophylls (a + b)	Carotenoids (x + c)	Weight ratio (a + b)/(x + c)
0	4.20	0.81	5.19
2	3.75	0.73	5.14
4	3.54	0.69	5.13
6	3.24	0.64	5.06
8	2.84	0.59	4.81
12	2.27	0.62	3.66
18	1.82	0.53	3.43
26	0.97	0.45	2.16
36	0.68	0.44	1.55
48	0.55	0.36	1.53

The pigment levels are expressed in mg g^{-1} dry matter. The standard deviation of pigment levels ranged from 5 to 8% and the values of the weight ratio (a + b)/(x + c) ranged from 3 to 4% [Based on Tuba et al. (1996a, b) and Csintalan et al. (1998)]

The photochemically active thylakoids are broken down, together with pigments, and the photosynthetic net CO_2 assimilation P_N drops to zero already after 12 h of desiccation (Tuba et al. 1996b).

The *ultrastructure of desiccoplasts* in the air-dried state of leaves very much resembles that of gerontoplasts of senescent leaves, but is also clearly different from the latter. After the breakdown of thylakoids, one finds few osmiophilic plastoglobuli and in many desiccoplast cross sections a smear of osmiophilic lipid material in the place of former grana and stroma thylakoids as well as large groups of translucent plastoglobuli (Fig. 9.1a) as described in detail by Tuba et al. (1993b). The breakdown of chlorophylls and thylakoids apparently proceeds so fast during desiccation that part of the osmiophilic thylakoid lipids, including the remaining carotenoids, stays at the site where the thylakoids are destroyed. This osmiophilic lipid material, usually present more so in the center parts of the desiccoplasts, is embedded and surrounded by a dense, granular stroma. In contrast to the green chloroplast stage, the desiccoplasts show an oval and often isodiametric form and they are tightly pressed against each other. Their plastid envelope (with the outer and inner membranes) seems to be preserved fully intact as one can judge by magnifications of some points. However, the usual thin cytoplasm parts separating different chloroplasts from each other in the green cells are not observed in the air-dried desiccated state. The fact is that under these conditions the cells have lost most of their cytoplasmic water.

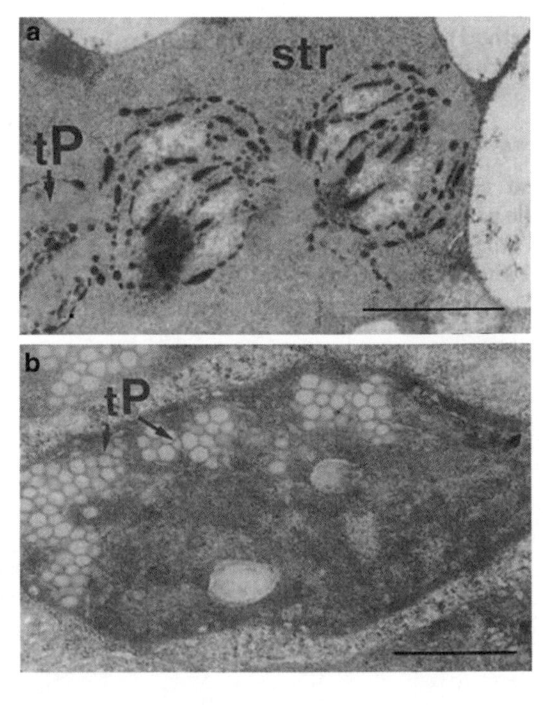

Fig. 9.1 (**a**) Cross section through a mesophyll cell of an air-dried leaf of the PDT plant *Xerophyta scabrida*. The three desiccoplasts exhibit elongated osmiophilic lipid material on the site of former thylakoids and also translucent plastoglobuli (tP) surrounded by a dense granular stroma (str). (**b**) Desiccoplast with translucent plastoglobuli (tP) 6–10 h after rewetting air-dried leaves. The *bars* equal 1 μm (based on Tuba et al. 1993b, modified)

The remaining carotenoids, mainly xanthophylls and low amounts of β-carotene, seem to be localized in the few osmiophilic or the numerous translucent plastoglobuli. The exact composition of the latter is not yet known. They apparently contain less osmiophilic lipid material, such as glycerolipids or even neutral lipids and carotenoids, which are known to be less osmiophilic than plastoquinol-9 and α-tocopherol, which are the major constituents of osmiophilic plastoglobuli in green chloroplasts (Lichtenthaler and Sprey 1966; Lichtenthaler 1969, 2007). In the air-dried state, the desiccoplast stroma and the cytosol appear dense and granular; ribosomes may be preserved intact, but, despite staining with uranylacetate, the definite structure of plastidic or cytosolic ribosomes cannot be clearly differentiated. Mitochondria have lost a major part of their tubuli; however, some persist in the desiccated state, and this later facilitates the fast restart of respiration when the desiccated leaves are rewetted.

9.5.2 Rehydration of Leaves and Resynthesis of Functional Chloroplasts

A few hours after rehydration of air-dried leaves, the desiccoplasts show again a more elongated form and they are anew separated by cytoplasmic strands. At this stage, the translucent plastoglobuli can be perceived much better (Fig. 9.1b) than in the air-dried state (Tuba et al. 1993b). Ribosomes and polyribosomes can clearly be identified. Mitochondria show again many tubuli, and respiration is operating at high rates (see below). The restitution of tubuli in mitochondria is much faster than that of the internal thylakoid membranes of chloroplasts (Tuba et al. 1993b; see also below Sect. 9.6). By the onset of full water saturation after rewetting, reconstruction of thylakoids in *Xerophyta* leaves has already started. The first thylakoids, and also incomplete ones, are usually detected ca. 10–12 h after rewetting. In this stage, the intrathylakoidal spaces are slightly swollen; some flattened vesicles attach to the thylakoid ends and help to extend the thylakoids (Tuba et al. 1993). These primary thylakoids are partially appressed to each other; they still exhibit a low stacking degree of about 18% only. A clear differentiation between stroma and grana thylakoids is not yet possible. The osmiophilic lipid material of the former desiccoplasts disappears and is apparently used for the biosynthesis of the new thylakoids. The number of translucent plastoglobuli and their size is strongly reduced, and they almost completely disappear already 24 h after rewetting of air-dried leaves (Fig. 9.2).

The whole process of the thylakoid biosynthesis phase in desiccoplasts of rewetted *Xerophyta* leaves resembles exactly the light-induced biosynthesis of thylakoids from young etioplasts as had first been described for etiolated barley seedlings by Sprey and Lichtenthaler (1966) and Lichtenthaler (1967). In this early phase of thylakoid biosynthesis of the desiccoplasts, small elongated starch grains show up whose appearance is well known to precede the biosynthesis of

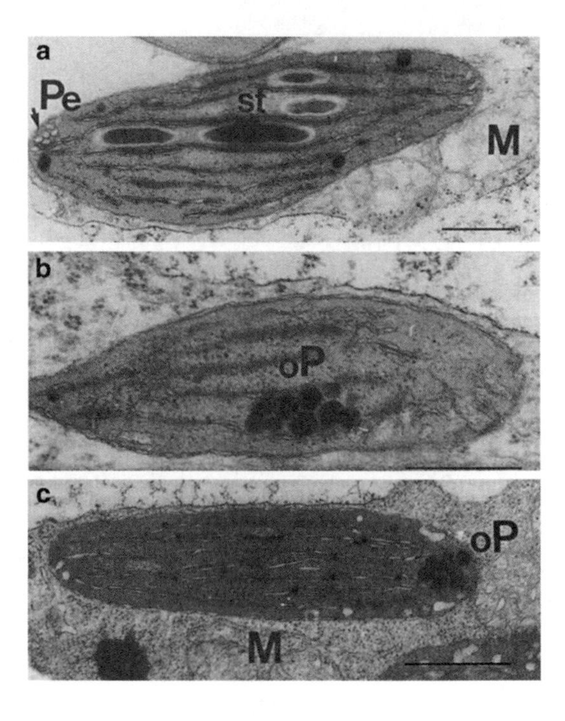

Fig. 9.2 Regreened chloroplasts after rewetting of air-dried leaves of the PDT plant *Xerophyta scabrida*. (**a** and **b**) 24 h after rewetting and (**c**) 72 h after rewetting. Pe = peristromium, oP = osmiophilic plastoglobuli, st = starch, M = mitochondrion with tubular membranes. Ribosomes (staining with uranylacetate) are visible in the chloroplast and in the cytoplasm. *Bars* equal 1 μm (based on Tuba et al. 1993b, modified)

thylakoids during the regular chloroplast development from proplastids in embryonic cells of illuminated plant seedlings. In addition, during the whole resynthesis phase of thylakoids, one sees at the chloroplast envelopes in several places peristromium sites where the envelope membranes fold in and form new vesicles (Fig. 9.2). These are sites where extraplastidic materials are imported and special biosynthetic processes take place, including the formation of membrane vesicles that may later transform to new thylakoids. Thus, also in this respect, the reconversion of desiccoplasts into functional chloroplasts shows principally the same steps and responses as the chloroplast development in young cells of developing leaves. The essential difference, however, is that desiccoplasts are not proplastids but former, fully functional chloroplasts that go through a gerontoplast stage, a senescence-type, desiccated stage, and possess the unique capacity of completely reviving and redifferentiating into regular chloroplasts after rehydration of desiccated leaves.

From 24 to 72 h after rewetting, the thylakoid system increasingly develops and completes, clear grana thylakoid regions show up, osmiophilic plastoglobuli can appear again (Fig. 9.2a–c), although a few translucent plastoglobuli may persist. Seventy-two hours after the start of rehydration, the stacking degree of thylakoids reaches the normal value of 56% corresponding to a ratio of appressed to exposed thylakoid membranes of 1.27. Also these parameters are in full agreement with the development of the thylakoid system in leaf chloroplasts of non-DT plants. Thus, the stacking degree of thylakoids in chloroplasts of beech leaves amounts to 62%

(sun leaf) and 78% (shade leaf), in radish cotyledons 55% (high light) and 64% (low light), and in mesophyll cells of maize leaves 55% (high light) and 77% (low light) (Lichtenthaler et al. 1981, 1984).

Respiration of the rewetted leaves of *Xerophyta* begins 20 min after immersing the air-dried leaves into water. The emission of respiratory CO_2 shows a maximum after 2 h of rewetting with a dark respiration rate R_D of 2 μmol CO_2 m^{-2} s^{-1} and then decreases continuously. The fast restitution of tubuli in mitochondria and the high respiration rates provide the high energy needed for the metabolic revival of the desiccated leaves and the reconstruction of their photosynthetic pigment and thylakoid apparatus and their full photosynthetic capacity. The full water content of rewetted leaves is reached after ca. 12 h. The appearance and accumulation of newly synthesized chlorophylls (a + b) start about 7–10 h after rewetting the leaves, and this is paralleled by a corresponding accumulation of carotenoids (x + c). About 40% of the chlorophylls are present 24 h after rehydration, and 3 days after rewetting the maximum level of chlorophylls and carotenoids is reached (Fig. 9.3a, b). The weight ratio of photosynthetic pigments (a + b)/(x + c) starts from low values of 1.70 at 8 h after rewetting and reaches again the regular values of green leaf tissue of 5–5.2 three days after rewetting.

Net photosynthetic CO_2 assimilation P_N begins 1 day after rehydration of air-dried *Xerophyta* leaves at a chlorophyll level of ca. 12 μg (a + b) cm^{-2} and reaches its maximum almost shortly before the third day (Fig. 9.4b, c). However,

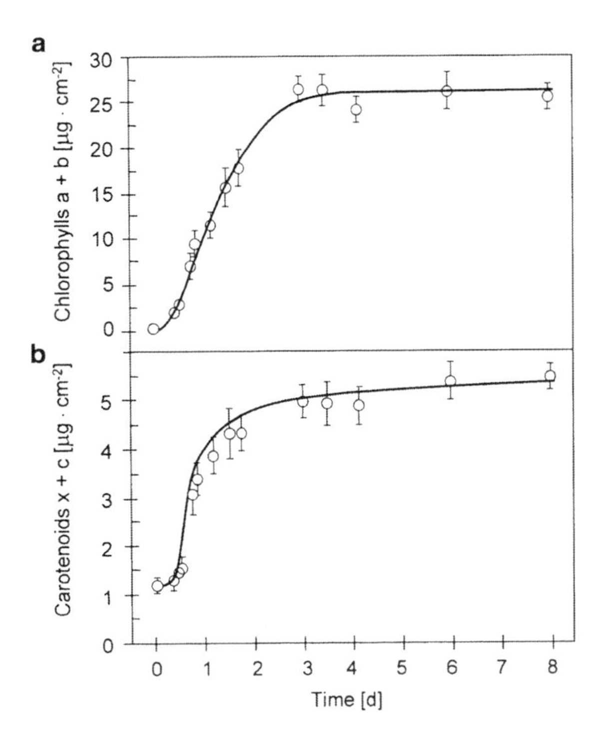

Fig. 9.3 Resynthesis and accumulation of (**a**) chlorophylls (a + b) and (**b**) total carotenoids (x + c) in air-dried, achlorophyllous leaves of *Xerophyta scabrida* during revival in aerated water. The *dark bars* represent the standard deviation (based on Tuba et al. 1993a, 1994)

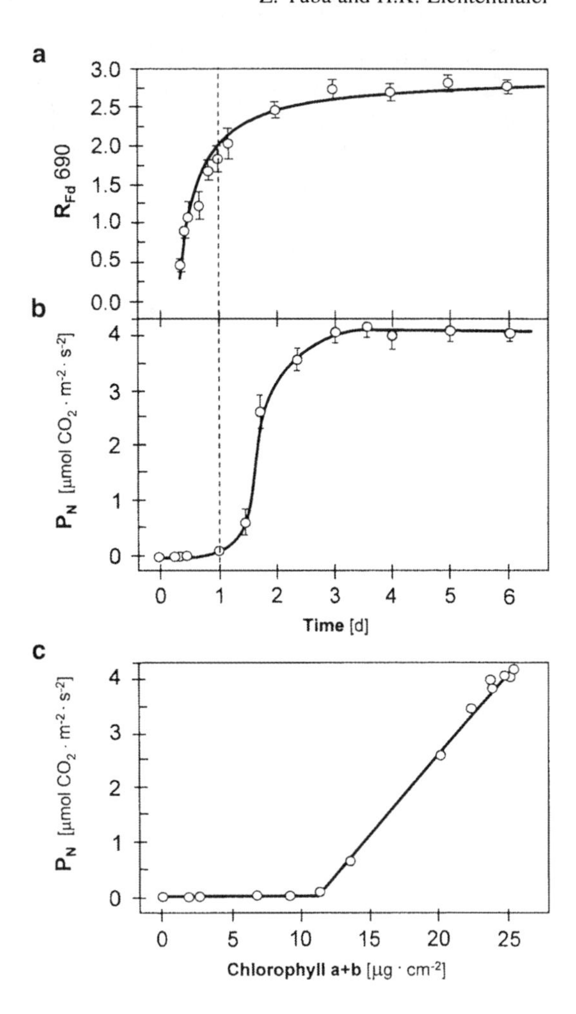

Fig. 9.4 Reappearance of photosynthetic quantum conversion and net CO_2 assimilation rates P_N in air-dried desiccated leaves of *Xerophyta scabrida* during rehydration and regreening of leaves. (**a**) Rise of the chlorophyll fluorescence decrease ratio R_{Fd} 690, (**b**) net photosynthesis rates P_N at saturating light conditions (800 μmol photons m^{-2} s^{-1}) against time after rewetting, and (**c**) P_N as a function of the rising chlorophyll (a + b) content of leaves. The *dark bars* represent the standard deviation (based on Tuba et al. 1993a, 1994)

the newly synthesized chlorophyll is photochemically active much earlier. This can be seen in the rapid rise of the values of the chlorophyll fluorescence decrease ratio R_{Fd} 690, which clearly precedes the appearance and rise of P_N as contrasted in Fig. 9.4a, b. This chlorophyll fluorescence ratio is strongly correlated to the net photosynthesis rates P_N of leaves and is a well-known indicator of the photosynthetic activity and capacity of green leaves (Lichtenthaler and Babani 2004; Lichtenthaler et al. 2005; Sarijeva et al. 2007). Already 1 day after rehydration, the *Xerophyta* leaves show R_{Fd} values of 1.8–2.0, and the final and regular values of 2.7, as also found in the green leaf tissue of other plants, are reached after 2.5–3 days after rewetting the air-dried leaves. The R_{Fd} values are known to be independent of the total chlorophyll content of a leaf, and their values indicate if and that the chlorophyll, present at a certain stage of the regreening process, is active in the photochemical quantum conversion.

In fact, the rapidly rising R_{Fd} values indicate that photosynthetic quantum conversion and CO_2 fixation occur in rewetted *Xerophyta* leaves long before a net P_N rate can be measured. As a matter of fact, due to the high respiration rates in the first 24 h after rehydration (Tuba et al. 1994) there is more than enough respiratory CO_2 available inside the leaf and this is used for photosynthetic CO_2 fixation. In the PDT *X. scabrida*, net CO_2 assimilation is first measurable after 24 h of rehydration. At this point, the chlorophyll content is just 35%, and the R_{Fd} value amounts to 68% in fully active control plants (Tuba et al. 1993b) as indicated in Fig. 9.4. From the first to the third day, the net CO_2 assimilation P_N rises to maximum rates and is paralleled by an increasing opening of stomata as seen in the rise of stomata conductance, g_S, from a value of 70 mmol m^{-2} s^{-1} on the first day to regular values of around 260–270 mmol m^{-2} s^{-1} on the third day after the start of rehydration. Thus, complete regeneration of the photosynthetic apparatus and its function, including gas exchange parameters, occurs 72 h after rewetting (Tuba et al. 1993b).

Three days after rewetting *Xerophyta* leaves, the ultrastructure of chloroplasts is back to normal and their photosynthetic apparatus is again fully active in photosynthetic quantum conversion and net CO_2 assimilation. This is also the time range in nature when the rainy season begins after a longer dry period. In the nature, desiccation as well as regreening and revival of desiccated leaves can proceed various times for the same leaf. Even with detached desiccated *Xerophyta* leaves, collected at the natural habitat, we could perform several desiccation and rewetting cycles in the laboratory. This again demonstrates that the desiccoplast is a unique chloroplast that is different in its revival capacity from the gerontoplast stage of the chloroplasts of other plants. The PDT strategy can be thought of as not only being superimposed on an existing cellular protection mechanism of vegetative desiccation tolerance (Oliver et al. 2000), but as a distinct new class of DT strategy. The selective advantage of poikilochlorophylly, in minimizing photooxidative damage and not having to maintain an intact photosynthetic system through long inactive periods of desiccation, presumably outweighs the disadvantage of slow recovery and the considerably high energy costs of reconstruction (Tuba et al. 1998).

Comparison of PDT plants with C_3, C_4, and CAM plants: PDT plants are C_3 species with a normal CO_2 fixation via Rubisco as key enzyme of the Calvin cycle, although they can occur in the same genera as species possessing the C_4 pathway of carbon assimilation. It is well known that C_4 plants exhibit a *structural compartmentation in space* (C_4 chloroplast dimorphism), whereas CAM plants possess a *functional compartmentation in time* (CAM pathway). PDT plants, in turn, form a third group of plants that are characterized by a *structural compartmentation in time* (desiccoplast/chloroplast stage). They have evolved in habitats where the duration of the desiccated state lasts for 6–8 months or even longer. Such long desiccated periods may reach the limit of survivability for most HDT plants. Under these conditions, it is evidently more advantageous to dismantle the whole photosynthetic apparatus and to reconstitute it after rehydration. On the other hand, the HDT strategy is evidently more favorable in habitats where the desiccated periods are shorter, and wet and dry periods alternate more frequently. It seems that desiccoplasts and the PDT strategy evolved distinctively later in response to an ecological problem, which was straining the capabilities of the HDT strategy and the survival of HDT plants. DT plants with desiccoplasts are both theoretically interesting and potentially of considerable practical and ecological significance.

9.6 Differential Physiological Responses of Individual Vascular HDT and PDT Plants Under Desiccation

Various aspects of physiological responses of homoiochlorophyllous (HDT) and poikilochlorophyllous (PDT) plants have been studied in detail, indicating that individual DT plants can respond in quite differential ways to desiccation and rehydration showing a rather large variability. These results suggest that there is not in all cases and physiological responses a clear distinction between PDT and HDT plants. In addition, the results also demonstrate that the physiological responses largely depend upon the general environmental conditions during the dehydration process. Thus, considerable differences can occur, even within one DT plant species, when the desiccation process proceeds in full light or in darkness or when it is a fast or slow process. Since such investigations provide an insight into the flexibility of the responses and into some of the underlying dehydration response mechanisms, some of these investigations are contrasted here.

9.6.1 Chlorophyll Content and Chloroplast Ultrastructure

There are clear-cut examples where homoiochlorophyllous (HDT) and poikilochlorophyllous plants (PDT) respond in different ways to desiccation, which is reflected in chloroplast ultrastructure. Thus, the HDT *Pellaea calomelanos* retains chlorophyll and chloroplasts with discernible grana when dry (Gaff and Hallam 1974). In contrast, *X. scabrida*, a characteristic PDT plant, typically loses all of its chlorophylls and thylakoids of the chloroplast during desiccation (Tuba et al. 1993a, b). However, dicotyledonous DT plants, although all retaining chlorophyll, vary a good deal in the details of their behavior. Thus, the HDT plant *Myrothamnus flabellifolia* preserves its thylakoids but loses half of the chlorophyll (Farrant et al. 1999). An unusual feature of *M. flabellifolia* leaves is the stacking of the thylakoid membranes of chloroplasts in desiccated leaves, described as "staircase grana" by Wellburn and Wellburn (1976), which has been proposed to minimize photooxidative stress (Sherwin and Farrant 1996; Koonjul et al. 2000; Farrant et al. 1993, 2003). On the other hand, there are no significant changes in chlorophyll content of the HDT plant *Haberlea rhodopensis* through dehydration, but part of the thylakoid system disappears (Markovska et al. 1994). Some species may be seen as exhibiting an intermediate strategy

(Farrant et al. 1999, 2008), or a range of possibilities may be envisaged between the two extremes. The HDT *Ramonda nathaliae* lost 20% of its chlorophyll when desiccated in the greenhouse and 70% in its natural habitat (Drazic et al. 1999). Therefore, the extent of chlorophyll loss is not a species-specific constant, but it may be influenced by environmental factors, particularly irradiance. The activity of chlorophyllase, the key enzyme in chlorophyll catabolism, has its peak at a water content of about 50% in the HDT *Ramonda serbica*. After further dehydration, this enzyme activity gradually ceases and chlorophyll is retained throughout the desiccated state (Drazic et al. 1999).

9.6.2 Abscisic Acid and Chlorophyll Breakdown

Despite the involvement of abscisic acid (ABA) in some PDT plants and the structural resemblance of desiccoplast formation with gerontoplast formation, the desiccation-induced breakdown of the photosynthetic apparatus in PDT plants seems to be clearly different from the processes involved in leaf senescence. Indeed, Gaff (1986) found that senescence reduced desiccation tolerance. The dismantling of the photosynthetic apparatus can be seen as a strictly organized protective mechanism rather than damage to be repaired after rehydration. If the PDT *B. nitida* is desiccated rapidly, it does not have the time to break down chlorophyll, but loses viability (Gaff and Churchill 1976). On the other hand, *X. scabrida* preserves most of its chlorophyll when desiccated in the dark, so in this case most of the chlorophyll loss seems to be due to photooxidation under natural circumstances (Tuba et al. 1997). Chlorophyll breakdown itself does not appear to be controlled by ABA (Gaff and Loveys 1984). This suggests that PDT plants in general probably do not decompose their chlorophyll enzymatically, but rather do not invest in preserving it through the dry state.

9.6.3 Photosystem II Electron Transport and Thermoluminescence

Desiccation-tolerant plants provide a very suitable model for the study of photosynthetic activity during dehydration and rehydration (Drazic et al. 1999; Deng et al. 2003). The European flowering HDT species are restricted to the two genera *Ramonda* and *Haberlea* (Alpert and Oliver 2002). Both genera could be an excellent model species of higher HDT plants. The recent paper of Peeva and Maslenkova (2004) showed some unique characteristics of *H. rhodopensis* leaves, judging by thermoluminescence (TL) investigations of the photosynthetic photosystem II (PSII). Further detailed studies by Georgieva et al. (2005) described the functional peculiarities and responses of *H. rhodopensis* and nondesiccation-tolerant spinach during desiccation and rehydration. Subsequent rehydration was investigated in order to characterize some of the mechanisms that allow such plants to survive desiccation stress. During the first period of desiccation, the decline in the net CO_2 assimilation rate was influenced largely by stomatal closure. Later it was caused by the decrease both in the stomatal conductance and in the photochemical activity of photosystem II (PSII). Desiccation clearly modified the kinetic parameters of chlorophyll fluorescence induction, thermoluminescence emission of PSII, far-red-induced P700 oxidation, and oxygen evolution in the leaves of both species. Predominantly, a charge recombination of the radical pair $S_2Q_A^-$ in the photosynthetic PS II took place. PS II was converted to a nonfunctional state in desiccated spinach in accordance with the observed changes in membrane permeability, malondialdehyde, proline, and H_2O_2 content. The values of these parameters returned nearly to the control level after a 24 h rehydration, but only in the leaves of the HDT plant *Haberlea*. In addition, the blue and green fluorescence emission, which under physiological conditions preferentially comes from ferulic acid molecules covalently bound to cell walls (Lichtenthaler and Schweiger 1998; Buschmann and Lichtenthaler 1998), was strongly increased in desiccated leaves. This suggests that at unchanged chlorophyll content and chlorophyll-protein amounts, reversible modifications in PSII electron transport, an enhanced probability for nonradiative energy dissipation, as well as increased synthesis and accumulation of phenolic substances during desiccation contribute to the desiccation tolerance of HDT *H. rhodopensis* and its fast recovery after rehydration (Georgieva et al. 2005). These adaptive characteristics of *H. rhodopensis* enable this HDT plant to rapidly take advantage of frequent changes in water availability.

9.6.4 CO_2 Assimilation

Measurements of CO_2 assimilation provide a good tool for investigating the transition of the plant from the biotic to the abiotic state and the revival speed. After the cessation of watering plants or leaves, net photosynthesis in *H. rhodopensis* even increased until a water deficit (WD) of 50% had occurred. After this point, stomatal resistance increased gradually and net CO_2 gas exchange became negative at a WD of 80%. This marked tolerance of photosynthesis toward a high water deficit is attributed to the rather minor destruction of the chloroplasts' thylakoid system (Kimenov et al. 1989). In HDT *R. serbica* drying caused an immediate decrease in CO_2 assimilation and a negative carbon exchange at a water deficit of about 25%. Stomatal resistance showed a gradual decrease in this species; thus, stomatal regulation did not seem to play a role in slowing down the desiccation (Markovska et al. 1994). The rate of CO_2 photoassimilation was greatly reduced during dehydration, but the recovery was complete with rehydration when the relative water content of the leaves reached similar values as the control leaves (Degl'Innocenti et al. 2008). The authors stated that the response of HDT *R. serbica* leaves to severe water stress involves two different mechanisms: (a) CO_2

assimilation is limited by stomata closure creating an excess proton concentration in the lumen and activating nonphotochemical quenching; and (b) when dehydration became severe, the electron transport rate (ETR) decreased markedly, whereas the q_{NP} (nonphotochemical quenching) was greatly reduced. Therefore, it appears that severely dehydrated leaves of HDT *R. serbica* are protected against the formation of reactive oxygen species (ROS) by mechanisms that quench chlorophyll triplet formation via xanthophylls.

9.6.5 CO_2 Gas Exchange and Respiration

The CO_2 gas exchange pattern of desiccating PDT *X. scabrida* leaves was studied in detail by Tuba et al. (1997). Photosynthetic activity declines dramatically to zero throughout the first 4 days of dehydration. By this time the leaves have lost about 60% of their water content (on a dry weight basis). The fall in net CO_2 assimilation is caused mainly by the rapid stomatal closure, but it is the decrease in chlorophyll content and in the chlorophyll fluorescence signal that finally brings photosynthesis to a halt (Tuba et al. 1996b). Dark respiration continued until the 14th day of dehydration, at a water content of 24% (fresh weight basis). The declines in water content and respiration were in a close-to-linear correlation. The prolonged desiccation respiration is thought to cover the energy demand of a controlled disassembly of the internal chloroplast membrane structures in PDT plants (Tuba et al. 1997). Respiratory processes are completely operational before full turgor is achieved in rehydrating *X. scabrida* leaves (Tuba et al. 1994). This corresponds well with the high preservation ratio of respiratory enzymes during desiccation in resurrection plants (Harten and Eickmeier 1986). The importance of the fast recovery of respiration is underlined by the fact that it is the only ATP source during the primary phase of rehydration (Gaff 1989). The phenomenon called "resaturation respiration" (Tuba et al. 1996a, b), involving an elevated respiratory CO_2-emission, also occurs in higher DT plants. In contrast to HDT plants, it lasts much longer in PDT species: 30 h in the case of *X. scabrida* (Tuba et al. 1994). In the PDT *Xerophyta villosa*, not only the internal chloroplast membranes but even most of the mitochondrial cristae disappear after dehydration and the remaining ones decompose within 30 min after rewetting. This dismantling is mirrored by a loss in insoluble or structural proteins (almost 50%), which is not always obvious in HDT plants (generally ~10%) (Gaff and Hallam 1974).

9.6.6 Leaf Responses

Sherwin and Farrant (1996) and Farrant et al. (2003) stated that the most conspicuous change occurring upon desiccation is a fan-type closure of the tropical HDT *M. flabellifola* leaves with their adaxial surfaces appressed against the stem. The loss of protoplasmic water results in a considerable anatomical and ultrastructural reorganization of the leaf tissue (Sherwin and Farrant 1996; Moore et al. 2006). A recent study of the leaf cell wall of *M. flabellifolia* showed that the majority of leaf mesophyll cells folded their walls in response to desiccation, whereas the sclerenchyma and vascular cells did not fold and thereby provided a rigid support. It is proposed that the high concentration of arabinose polymers, in the form of arabinans and arabinogalactan proteins associated with the pectin matrix, is responsible for the high degree of flexibility of the mesophyll cell walls of HDT plants (Moore et al. 2006).

9.7 Recovery and Reestablishment of Physiological Activity of Vascular Homoiochlorophyllous and Poikilochlorophyllous Plants

HDT bryophytes and lichens (see Chaps. 5–7) growing in exposed situations typically remoisten quickly and diffusely. The thin leaves of bryophytes equilibrate rapidly with the humidity of the surrounding air and can regain turgor from overnight dewfall even if no obvious deposition of water has taken place (Lange et al. 1990; Csintalan et al. 2000). Water from rain or cloudwater deposition is directly imbibed by the leaves as it spreads by capillarity over the surface of the shoots. In species with well-developed internal conduction and water-repellent leaf surfaces, such as many large Polytrichaceae and Mniaceae, imbibition is slower and may take an hour or two to complete.

As a result of their larger size and greater complexity, HDT and PDT vascular plants tend to dry out and rehydrate more slowly than HDT lichens and bryophytes. In most cases, DT vascular plants have a fully functional vascular system, which can support the needs of the photosynthetic tissues as long as the water supply is reasonably plentiful. Recovery entails refilling of the xylem to reestablish the water conduction system as well as the return to normal cell metabolism. Details on this topic are given in Chap. 10. Recovery is not a problem for small plants, in which (given the time) xylem embolisms can be repaired by root pressure and/or from neighboring tissues when water is freely available. However, a DT tree is hardly conceivable since the height limits recorded for DT angiosperms (monocotyledonous pseudoshrubs) are ~3–4 m (Porembski and Barthlott 2000).

Vascular desiccation-tolerant plants in general require liquid water for rehydration (Gaff 1977), and the continuity of water within the xylem must be restored as the tissues rehydrate if the leaves are to remain turgid for more than a few hours after rain. Embolism in xylem vessels has been assumed to hinder resaturation of the photosynthetic tissues in some species (Sherwin and Farrant 1996). The fact that desiccation-tolerant plants are generally low herbs or shrubs (Gaff 1977; Kappen and Valladares 1999) is consistent with the idea that restitution of xylem transport is a real challenge for these desiccation-tolerant plants and probably imposes a practical limit on their height (see Porembski and Barthlott 2000). If a normal function of the vascular system is to be reestablished, it is vital that the vascular bundles remain essentially undamaged in the desiccated state.

Rosetto and Dolder (1996) found that vascular bundles of the HDT *Nanuza plicata* showed no significant changes due to desiccation. In contrast, drought stress severely impaired vascular bundles in non-DT barley leaves (Pearce and Beckett 1987), which would preclude the reestablishment of water flow even if all the other factors necessary for rehydration were present. In *M. flabellifolia*, an HDT shrub with a height of a few decimeters at most, capillary rise is enough to achieve continuity of the water column. Narrow reticulate xylem vessels contribute to this ability (Sherwin et al. 1998). On the other hand, in *P. calomelanos* capillary rise is not able to transport water even into the lowest leaves; therefore, root pressure is assumed to contribute (Gaff and Hallam 1974). The leaves on the upper half of PDT *B. nitida* shoots had not yet rehydrated from the soil after 6 days (Gaff and Churchill 1976), so in this case rehydration through the leaf cuticle also seems to be essential. The roots of the PDT plant *X. scabrida* die back when desiccated; therefore, rehydration can only occur via the leaves. In this species, leaf turgor is regained within 6 h and maximum water content after 12 h in immersed leaves (Tuba et al. 1994). After the reestablishment of metabolism in the leaves, adventitious roots develop and ensure a continuous water supply (Tuba et al. 1993a, b). The relative importance of the water received by the roots and water absorbed through the leaves of desiccation-tolerant plants varies from species to species (Gaff 1977). Further, particular requirements and mechanisms for full rehydration and refunctioning of DT plants are found in Chap. 10.

9.8 Revival of Metabolism: Reassembly or Repair?

This part is based on the work of Proctor and Tuba (2002): "In desiccation-tolerant mosses that have been air dried for a few days, respiration recommences almost instantaneously on remoistening and net photosynthetic carbon fixation typically reaches two-thirds or more of normal levels within the first few minutes. The initial respiratory carbon loss is typically compensated within 30 min or so (Tuba et al. 1996a, b; Proctor and Pence 2002). Chlorophyll fluorescence measurements show that the initial recovery of the photosynthetic photosystems is extraordinarily rapid and independent of the protein synthesis (Csintalan et al. 1999; Proctor and Smirnoff 2000). A complete return to unstressed levels of chlorophyll fluorescence parameters may take a number of hours, but is unaffected by protein-synthesis inhibitors. There are indications that cytoplasmic protein synthesis following remoistening is needed for a return to predesiccation levels of CO_2 fixation (Proctor and Smirnoff 2000). Over all, the results suggest that a recovery of respiration and photosynthesis in these plants is largely a matter of reassembly and reactivation of components which have survived drying and re-wetting essentially intact. A substantial 'repair' element in the recovery of DT bryophytes has been postulated

(Bewley and Oliver 1992; Oliver and Bewley 1984, 1997), largely on the evidence of membrane leakage following re-wetting, and electron micrographs of freshly-rehydrated leaf cells. However, leakage of membranes for a short time after re-wetting appears to occur in all desiccation-tolerant organisms (Crowe et al. 1992). The potential difficulties of getting a good fixation of bryophyte cells in the phase of rapidly-changing concentrations and solute potentials immediately following remoistening counsel caution in interpreting electron-microscopical data without corroborative physiological evidence. Recent electron micrographs suggest that bryophytes can withstand drying and re-wetting without a visible fine-structural damage (J.G. Duckett, personal communication)."

The ATP amount conserved in the desiccated state is highly variable among the HDT and PDT desiccation-tolerant species. Besides this, there is no indication of extreme energy costs of a desiccation–rehydration cycle (Gaff and Ziegler 1989). Therefore, the speed of recovery is not a matter of available energy. It depends rather on the extent a certain species retains its internal structures during desiccation. This, in turn, is determined mostly by the strategy (i.e., HDT or PDT) adopted by the plant in order to avoid harmful excitation of chlorophyll in the drought-stressed state (Sherwin and Farrant 1996, 1998).

The conflicting data about the main saccharide constituents (see also Chap. 16) of the tropical HDT *M. flabellifolia* may be a result of the collection of plants from different locations in southern Africa, namely Zimbabwe, South Africa, and Namibia, respectively. It has recently been confirmed that South African plants have high relative concentrations of sucrose and trehalose (Moore et al. 2006, 2007). A detailed study of the desiccation response of *M. flabellifolia* supports the hypothesis that saccharides play an important role in the desiccation tolerance of this HDT species as partial mRNA transcripts for putative carbohydrate biosynthesis. Sugar transport proteins were found to be upregulated upon desiccation (Koonjul 1999). High tannin (polyphenol) contents have also been described to be present in the leaves of *M. flabellifolia* (Koonjul et al. 1999). The synthesis of the polyphenol pigment cyanidin 3-glucoside is induced upon desiccation, and it is suggested that this functions in synergy with 3,4,5-tri-*O*-galloylquinic acid to protect the leaves from photooxidative stress during desiccation (Moore et al. 2004).

9.9 Constitutive and Induced Tolerance

Oliver et al. (1998) suggested that all vascular "resurrection plants" fall into the "modified desiccation-tolerant" category. This strategy implies the extension of the active period during dehydration with the help of physiological and morphological mechanisms. This gives the plants time to build up desiccation tolerance involving inducible protective mechanisms (Oliver et al. 1998). *H. rhodopensis* and *R. serbica* were described to have inducible CAM metabolisms (Markovska et al. 1989). Xeromorphism can also hinder resaturation and therefore decrease the ability of the plant to take advantage of favorable wet periods (Kappen and Valladares 1999). Therefore, the prolongation of the active period as a characteristic of "modified desiccation-tolerant strategy" is meaningful rather in comparison to cryptogamic desiccation-tolerant species and does not entail an extreme capacity to conserve water in the context of vascular plants. In general, vascular resurrection plants dehydrate within 2–3 days after the depletion of soil moisture content (Gaff 1977).

Many of these features tend to lengthen the period that vascular DT plants spend in their partially dehydrated state, whereas DT bryophytes and lichens typically dry rapidly once extracellular water is exhausted.

A distinction has been drawn between "fully" desiccation-tolerant plants, such as the moss *Tortula ruralis*, and "modified" DT plants exemplified by many vascular "resurrection plants" (Bewley and Oliver 1992; Oliver and Bewley 1997; Oliver et al. 1998). The form of tolerance appears to be essentially constitutive and influenced to only a limited extent by the previous desiccation history or rate of drying. The latter group may tolerate severe and prolonged desiccation under appropriate conditions, but show little or no tolerance if they are dried rapidly. Tolerance is induced in the course of slow drying, and ABA seems generally to be implicated in its induction. This statement may at least in part reflect different evolutionary origins of desiccation tolerance: it appears to be more primordial in bryophytes (and lichens) and secondary and polyphyletic in vascular plants, where it has probably often evolved as a fallback in plants already more or less tolerant of drought stress. Slow drying is a natural consequence of the vascular-plant habit. Leaf cells of mosses in exposed sunny locations may switch from full turgor to air dryness within a few minutes, but many forest bryophytes dry much more slowly, and a degree of drought hardening is readily demonstrated (Abel 1956; Proctor 1972; Dilks and Proctor 1976). ABA is undetectable in *T. ruralis* (Oliver et al. 1998), but effects of ABA on desiccation tolerance have been demonstrated in mosses (Bopp and Werner 1993; Beckett et al. 2000) and liverworts (Hellwege et al. 1996). A systematic search for endogenous ABA in bryophytes is much needed.

Whereas it seems to be a well-established fact that ABA produced in roots due to water deficit induces an array of water-conservation responses in the shoots of vascular plants (Schulze 1986), there is no such general role of ABA in evoking desiccation tolerance in DT plants (but see Chap. 12). Nevertheless, some communication is evident, since the induction of desiccation tolerance fails to occur in the detached leaves of some species (e.g., the DT grass, *Sporobolus stapfianus*). Only detachment below an RWC of 61% results in desiccation-tolerant separate leaves. In this case, ABA cannot be responsible for the induction, since its peak in shoots shows up only at an RWC of 15%. In addition to this, ABA is produced in the *S. stapfianus* leaves themselves and does not originate from the roots (Gaff and Loveys 1992). In this species, desiccation tolerance is promoted only marginally by exogenous ABA, whereas other growth inhibitors, brassinolide and methyljasmonic acid, had much larger effects (Ghasempour et al. 1998). The concentration of ABA does not increase in desiccating leaves of *Polypodium virginianum*, but externally applied ABA helps in surviving rapid desiccation, which would normally be lethal (Oliver et al. 1998). Gaff and Loveys (1984) recorded a substantial rise in ABA levels within the detached leaves of both *M. flabellifolia* and *B. nitida* when equilibrated with

96% RH. However, equilibration of *B. nitida* with 98% relative humidity induces complete desiccation tolerance without a peak in ABA concentration. Exogenously applied ABA could substitute for drought stress in stimulating desiccation tolerance in *B. nitida*.

In some DT species, ABA makes no contribution and has no ability to develop desiccation tolerance. In other species, ABA does not evoke desiccation tolerance in vivo, but has the capacity to do so if added exogenously. In other vascular DT plants, ABA clearly promotes desiccation tolerance, but its necessity is questionable. The ambiguous role of ABA underlines the polyphyletic origin of desiccation tolerance. In the search for the coordinating signal, investigations should concentrate on the other two growth inhibitors used by Ghasempour et al. (1998), because their existence in DT plants is still a question. A new horizon has been opened in this field by Frank et al. (2000), who suggest that phospolipase D (a phospholipid cleaving enzyme) has a role in tolerance induction as a secondary, intracellular messenger.

9.10 Importance of Scale and Ecological Context

Tuba et al. (1998) strongly emphasized the importance of scale in the optimal adaptation for an organism. Raven (1977) described the essential role of supracellular transport (xylem and phloem) systems in the evolution of land vegetation. Raven's (1977) supracellular transport system theory is primarily a limitation for the evolution of large land plants, which require highly organized internal conducting systems to maintain the integrated functioning of a bulky plant body in the presence of steep external water-potential gradients. For small plants, such as lichens and bryophytes with a far lower ratio of mass to linear dimension and surface area, a diffuse conducting system (which may be external) may be as good or better – and for these small plants, poikilohydry and desiccation tolerance may be an easy adaptive option (Tuba et al. 1996a, 1998). For vascular plants, a larger physical scale and greater anatomical complexity probably impose limits on the possible timescale of response to wetting and drying events. According to Proctor and Tuba (2002) and Tuba et al. (1998), it is interesting that some large bryophytes, Polytrichaceae and Dawsoniaceae, have well-developed internal conducting systems approaching the vascular-plant pattern, whereas some small pteridophytes, such as *Hymenophyllum*, and some small ferns and *Selaginella* species of intermittently dry habitats (Oppenheimer and Halevy 1962; Stuart 1968; Eickmeier 1979) appear to rely little on internal conduction and approach the typical bryophyte pattern in their ecological adaptation (Proctor and Tuba 2002; Tuba et al. 1998). In this respect, see also Chap. 8.

The HDT and PDT strategies have the desiccation tolerance in common, permitting colonization of habitats inaccessible to perennial vascular plants because these habitats are hard and impenetrable for roots, or they cannot provide a year-round supply of water for other reasons (Tuba et al. 1998; Proctor and Tuba 2002).

However, both strategies contrast greatly in the timescale of rehydration and desiccation events to which they are adapted – a timescale measured in weeks or many months for the achlorophyllous PDT plants, whereas air-dried plants with an HDT strategy, i.e., plants that still possess chlorophyll, can utilize even short moist periods lasting for only an hour or two for photosynthesis. Of course, the HDT and PDT categories overlap in their ecological adaptation, and two or more may coexist in one habitat (Ibisch et al. 1995).

Both ends of this ecological spectrum have particular points of interest. In *Xerophyta* and other PDT species (representing the "high inertia" end), the selective advantage of poikilochlorophylly, minimizing photooxidative damage (Smirnoff 1993; Tuba et al. 1994), and not having to maintain an intact photosynthetic apparatus through long periods of desiccation, presumably outweighs the disadvantage of slow recovery (2–3 days), and still leaves *Xerophyta* and other PDT species in permanent possession of a habitat with a clear advantage over ephemeral plants. The majority of HDT plants in the intermediate part of the range have adapted to periods of drying and remoistening, ranging from hours to weeks.

Acknowledgments The authors are indebted to Dr. Michael C. F. Proctor for valuable discussions on the plants' desiccation-tolerance strategies during the last 2 decades. We are also grateful to Dr. Zoltán Szabó for preparation of the reference list, to Szilvia Hajdu for her technical help, and to Ms. Gabrielle Johnson for English language assistance. H.K. Lichtenthaler wants to thank his friend and scientific colleague, the late Zoltán Tuba, for 25 years of excellent, very stimulating discussions, and close scientific cooperation in the ecophysiology of plants and their photosynthetic apparatus.

References

Abel WO (1956) Die Austrocknungsresistenz der Laubmoose. Sitzungsberichte. Österreichische Akademie der Wissenschaften. Mathematisch-naturwissenschaftliche Klasse, Abt.I 165: 619–707

Alpert P, Oliver MJ (2002) Drying without dying. In: Black M, Prichard HW (eds) Desiccation and survival in plants: drying without dying. CABI, Wallingford, UK, pp 3–43

Beckett RP, Csintalan Z, Tuba Z (2000) ABA treatment increases both the desiccation tolerance of photosynthesis, and nonphotochemical quenching in the moss *Atrichum undulatum*. Plant Ecol 151:65–71

Bewley JD (1979) Physiological aspects of desiccation tolerance. Annu Rev Plant Physiol 30:195–238

Bewley JD, Krochko JE (1982) Desiccation-tolerance. In: Lange OL, Nobel PS, Osmond CB, Ziegler H (eds) Encyclopaedia of plant physiology, new series, vol 12B. Springer, Berlin, pp 325–378

Bewley JD, Oliver MJ (1992) Desiccation tolerance in vegetative plant tissues and seeds: protein synthesis in relation to desiccation and a potential role for protection and repair mechanisms. In: Somero GN, Osmond CB, Bolis CL (eds) Water and life: comparative analysis of water relationships at the organismic cellular and molecular levels. Springer, Heidelberg, pp 141–160

Bopp M, Werner O (1993) Abscisic acid and desiccation tolerance in mosses. Bot Acta 106:103–106

Buschmann C, Lichtenthaler HK (1998) Principles and characteristics of multi-colour fluorescence imaging of plants. J Plant Physiol 152:297–314

Crowe JH, Hoekstra FA, Crowe LM (1992) Anhydrobiosis. Annu Rev Physiol 54:579–599

Csintalan Z, Tuba Z, Lichtenthaler HK (1998) Changes in laser-induced chlorophyll fluorescence ratio F690/F735 in the poikilochlorophyllous desiccation tolerant plant *Xerophyta scabrida* during desiccation. J Plant Physiol 152:540–544

Csintalan Z, Proctor MCF, Tuba Z (1999) Chlorophyll fluorescence during drying and rehydration in the mosses *Rhytidiadelphus loreus* (Hedw.) Warnst., *Anomodon viticulosus* (Hedw.) Hook. & Tayl. and *Grimmia pulvinata* (Hedw.) Sm. Ann Bot 84:235–244

Csintalan Z, Takács Z, Proctor MCF, Nagy Z, Tuba Z (2000) Early morning photosynthesis of the moss *Tortula ruralis* following summer dew fall in a Hungarian temperate dry sandy grassland. Plant Ecol 151:51–54

Degl'Innocenti E, Guidi L, Stevanovic B, Navari F (2008) CO_2 fixation and chlorophyll a fluorescence in leaves of Ramonda serbica during a dehydration–rehydration cycle. J Plant Physiol 165:723–733

Deng X, Hu Z-A, Wang H-X, Wen X-G, Kuang T-Y (2003) A comparison of photosynthetic apparatus of the detached leaves of the resurrection plant Boea hygrometrica with its non-tolerant relative Chirita heterotrichia in response to dehydration and rehydration. Plant Sci 165:851–861

Dilks TJK, Proctor MCF (1976) Effects of intermittent desiccation on bryophytes. J Bryol 9:49–264

Drazic G, Mihailovic N, Stevanovic B (1999) Chlorophyll metabolism in leaves of higher poikilohydric plants *Ramonda service* Panc. and *Ramonda nathaliae* Panc. et Petrov. during dehydration and rehydration. J Plant Physiol 154:379–384

Eickmeier WG (1979) Photosynthetic recovery in the resurrection plant *Selaginella lepidophylla* after wetting. Oecologia 39:93–106

Farrant JM (2000) A comparison of mechanisms of desiccation tolerance among three angiosperm resurrection plant species. Plant Ecol 151:29–39

Farrant JM, Pammenter NW, Berjak P (1993) Seed desiccation in relation to desiccation tolerance: a comparison between desiccation-sensitive (recalcitrant) seeds of *Avicennia marina* and desiccation-tolerant types. Seed Sci Res 3:1–13

Farrant JM, Cooper K, Kruger LA, Sherwin HW (1999) The effect of drying rate on the survival of three desiccation-tolerant angiosperm species. Ann Bot 84:371–379

Farrant JM, Vander Willigen C, Loffell DA, Bartsch S, Whittaker A (2003) An investigation into the role of light during desiccation of three angiosperm resurrection plants. Plant Cell Environ 26:1275–1286

Farrant JM, Lehner A, Cooper K, Wiswedel S (2008) Desiccation tolerance in the vegetative tisues of the fern *Mohria cafforum* is seasonally regulated. Plant J 57:65–79

Frank W, Munnik T, Kerkmann K, Salamini F, Bartels D (2000) Water deficit triggers phospholipase D activity in the resurrection plant *Craterostigma plantagineum.* Plant Cell 12:111–124

Gaff DF (1977) Desiccation tolerant plants of southern Africa. Oecologia 31:95–109

Gaff DF (1986) Desiccation tolerant resurrection grasses from Kenya and West Africa. Oecologia 70:118–120

Gaff DF (1989) Responses of desiccation-tolerant "resurrection plants" to water stress. In: Kreeb KH, Richter H, Hinkley TM (eds) Adaptation of plants to water and high temperature stress. Academic Publishing, The Hague, pp 207–230

Gaff DF, Churchill DM (1976) *Borya nitida* Labill. – an Australian species in the Liliaceae with desiccation-tolerant leaves. Aust J Bot 24:209–224

Gaff DF, Hallam ND (1974) Resurrecting desiccated plants. R Soc New Zealand Bull 12:389–393

Gaff DF, Loveys BR (1984) Abscisic acid content and effects during dehydration of detached leaves of desiccation tolerant plants. J Exp Bot 35:1350–1358

Gaff DF, Loveys BR (1992) Abscisic acid levels in drying plants of a resurrection grass. Trans Malays Soc Plant Physiol 3:286–287

Gaff DF, Ziegler H (1989) ATP and ADP contents in leaves of drying and rehydrating desiccation-tolerant plants. Oecologia 78:407–410

Georgieva K, Maslenkova L, Peeva V, Markovska Yu, Stefanov D, Tuba Z (2005) Comparative study on the changes in photosynthetic activity of the homoiochlorophyllous desiccation-tolerant *Haberlea rhodopensis* and spinach leaves during desiccation and rehydration. Photosynth Res 85:191–203

Ghasempour HR, Gaff DF, Williams RPW, Gianello RD (1998) Contents of sugars in leaves of drying desiccation tolerant flowering plants, particularly grasses. Plant Growth Regul 24:185–191

Hambler DJ (1961) A poikilohydrous, poikilochlorophyllous angiosperm from Africa. Nature 191:1415–1416

Harten JB, Eickmeier WG (1986) Enzyme dynamics of the resurrection plant *Selaginella lepidophylla* (Hook and Grev.) Spring during rehydration. Plant Physiol 82:61–64

Hellwege EM, Dietz KJ, Hartung W (1996) Abscisic acid causes changes in gene expression involved in the induction of the land form of the liverwort *Riccia fluitans* L. Planta 198:423–432

Hetherington SE, Smillie RM (1982a) Humidity-sensitive degreening and regreening of leaves of *Borya nitida* Labill. as followed by changes in chlorophyll fluorescence. Aust J Plant Physiol 9:587–599

Hetherington SE, Smillie RM (1982b) Tolerance of *Borya nitida*, a poikilohydrous angiosperm, to heat, cold and highlight stress in the hydrated state. Planta 155:76–81

Ibisch PL, Rauer G, Rudolph D, Barthlott W (1995) Floristic, biogeographical, and vegetational aspects of Precambrian rock outcrops (inselbergs) in eastern Bolivia. Flora 190:299–314

Jackson JK (1956) The vegetation of the Imatong Mountains, Sudan. J Ecol 44:341–374

Kappen L, Valladares F (1999) Opportunistic growth and desiccation tolerance: the ecological success of poikilohydrous autotrophs. In: Pugnaire FI, Valladares F (eds) Handbook of functional plant ecology. Marcel Dekker, New York, pp 9–80

Kimenov GP, Markovska YK, Tsonev TD (1989) Photosynthesis and transpiration of *Haberlea rhodopensis* FRIV. in dependence on water deficit. Photosynthetica 23:368–371

Koonjul PK (1999) Investigating the mechanisms of desiccation tolerance in the resurrection plant *Myrothamnus flabellifolius* (Welw.). South Africa: University of Cape Town; 1999. PhD Thesis

Koonjul PK, Brandt WF, Farrant JM, Lindsey GG (1999) Inclusion of polyvinylpyrrolidone in the polymerase chain reaction reverses the inhibitory effects of polyphenolic contamination of RNA. Nucleic Acids Res 27:915–916

Koonjul PK, Brandt WF, Lindsey GG, Farrant JM (2000) Isolation and characterisation of chloroplasts from the resurrection plant *Myrothamnus flabellifolius* Welw. J Plant Physiol 156:584–594

Lange OL, Meyer A, Zellner H, Ullmann I, Wessels DCJ (1990) Eight days in the life of a desert lichen: water relations and photosynthesis of *Teloschistes capensis* in the coastal fog zone of the Namib desert. Madoqua 17:17–30

Levitt J (1972) Responses of plants to environmental stress. Academic, New York

Lichtenthaler HK (1967) Beziehungen zwischen Zusammensetzung und Struktur der Plastiden in grünen und etiolierten Keimlingen von *Hordeum vulgare* L. Z Pflanzenphys 56:273–281

Lichtenthaler HK (1968) Plastoglobuli and the fine structure of plastids. Endeavour XXVII:144–149

Lichtenthaler HK (1969) Die Plastoglobuli von Spinat, ihre Größe und Zusammensetzung während der Chloroplastendegeneration. Protoplasma 68:315–326

Lichtenthaler HK (2007) Biosynthesis, accumulation and emission of carotenoids, α-tocopherol, plastoquinone and isoprene in leaves under high photosynthetic irradiance. Photosynth Res 92:163–179

Lichtenthaler HK, Babani F (2004) Light adaption and senescence of the photosynthetic apparatus: changes in pigment composition, chlorophyll fluorescence parameters and photosynthetic activity during light adaptation and senescence of leaves. In: Papageorgiou G and Govindjee (eds.), Chlorophyll Fluorescence: A Signature of Photosynthesis (Chapter 30), pp. 713–736. Springer, Dordrecht, 2004

Lichtenthaler HK, Schweiger J (1998) Cell wall bound ferulic acid, the major substance of the blue-green fluorescence emission of plants. J Plant Physiol 152:272–282
Lichtenthaler HK, Sprey B (1966) Über die osmiophilen globulären Lipideinschlüsse der Chloroplasten. Z Naturforsch 21b:690–697
Lichtenthaler HK, Buschmann C, Döll M, Fietz H-J, Bach T, Kozel U, Meier D, Rahmsdorf U (1981) Photosynthetic activity, chloroplast ultrastructure, and leaf characteristics of high-light and low-light plants and of sun and shade leaves. Photosynth Res 2:115–141
Lichtenthaler HK, Meier D, Buschmann C (1984) Development of chloroplasts at high and low light quanta fluence rates. Isr J Bot 33:185–194
Lichtenthaler HK, Buschmann C, Knapp M (2005) How to correctly determine the different chlorophyll fluorescence parameters and the chlorophyll fluorescence decrease ratio RFd of leaves with the PAM fluorometer. Photosynthetica 43:379–393
Markovska YK, Kimenov GP, Tsonev TD (1989) Presence of crassulacean acid metabolism in higher poikilohydric plants *Haberlea rhodopensis* Friv. and Ramonda serbica Panč. Photosynthetica 23:364–367
Markovska YK, Tsonev TD, Kimenov GP, Tutekova AA (1994) Physiological changes in higher poikilohydric plants *Haberlea rhodopensis* Friv. and *Ramonda serbica* Panč. during drought and rewatering at different light regimes. J Plant Physiol 144:100–108
Mishler BD, Churchill SP (1985) Transition to a land flora: phylogenetic relationships of the green algae and bryophytes. Cladistics 1:305–328
Moore JP, Ravenscroft N, Lindsey GG, Farrant JM, Brandt WF (2004) Galloylquinate ester: anthocyanin complexes in the leaves of the desiccated resurrection plant Myrothamnus flabellifolius. In: Hoikkalo A, Soidinsalo O (eds) Polyphenols communications: XXII international conference on polyphenols, Helsinki, Finland, 25–28 August 2004
Moore JP, Nguema-Ona E, Chevalier L, Lindsey GG, Brandt WF, Lerouge P et al (2006) Response of the leaf cell wall to desiccation in the resurrection plant *Myrothamnus flabellifolius*. Plant Physiol 141:651–662
Moore JP, Lindsey GG, Farrant JM, Brandt WF (2007) An overview of the biology of the desiccation-tolerant resurrection plant *Myrothamnus flabellifolia*. Ann Bot 99:211–217
Oliver MJ, Bewley JD (1984) Desiccation and ultrastructure in bryophytes. Adv Bryol 2:91–131
Oliver MJ, Bewley JD (1997) Desiccation-tolerance of plant tissues: a mechanistic overview. Hortic Rev 18:171–213
Oliver MJ, Wood AJ, O'Mahony P (1998) "To dryness and beyond" – preparation for the dried state and rehydration in vegetative desiccation-tolerant plants. Plant Growth Regul 24:193–201
Oliver MJ, Tuba Z, Mischler BD (2000) The evolution of vegetative desiccation tolerance in land plants. Plant Ecol 151:85–100
Oppenheimer HR, Halevy A (1962) Anabiosis of *Ceterach officinarum* Lam. et DC. Bull Res Counc Isr D3(11):127–147
Pearce RS, Beckett A (1987) Cell-shape in leaves of drought-stressed barleyexamines by low-temperature scanning electron-microscopy. Ann Bot 59:191–195
Peeva V, Maslenkova L (2004) Thermoluminescence study of photosystem II activity in *Haberlea rhodopensis* and spinach leaves during desiccation. Plant Biol 6:1–6
Pócs T (1976) Vegetation mapping in the Uluguru Mountains (Tanzania, East Africa). Boissiera 24b:477–498, + 1 map
Porembski S, Barthlott W (2000) Granitic and gneissic outcrops (inselbergs) as centers of diversity for desiccation-tolerant vascular plants. Plant Ecol 151:19–28
Proctor MFC (1972) An experiment on intermittent desiccation with *Anomodon viticulosus* (Hedw.) Hook. & Tayl. J Bryol 7:181–186
Proctor MCF, Pence VC (2002) Vegetative tissues: bryophytes, vascular 'resurrection plants' and vegetative propagules. In: Black M, Pritchard HW (eds) Desiccation and survival in plants: drying without dying. CABI, Wallingford, UK, pp 207–237
Proctor MCF, Smirnoff N (2000) Rapid recovery of photosystems on rewetting desiccation-tolerant mosses: chlorophyll fluorescence and inhibitor experiments. J Exp Bot 51:1695–1704

Proctor MCF, Tuba Z (2002) Poikilohydry and homoihydry: antithesis or spectrum of possibilities? New Phytol 156:327–349
Raven JA (1977) The evolution of land plants in relation to supracellular transport processes. Adv Bot Res 5:153–219
Rosetto ES, Dolder H (1996) Comparison between hydrated and desiccated leaves of a Brazilian resurrection plant. A light and scanning electron microscopy approach. Rev Bras Biol 56:553–560
Sarijeva G, Knapp M, Lichtenthaler HK (2007) Differences in photosynthetic activity, chlorophyll and carotenoid levels, and in chlorophyll fluorescence parameters in green sun and shade leaves of *Ginkgo and Fagus*. J. Plant Physiology 164:950–955
Schulze E-D (1986) Whole-plant responses to drought. Aust J Plant Physiol 13:127–141
Sherwin HW, Farrant JM (1996) Rehydration of three desiccation-tolerant species. Ann Bot 78:703–710
Sherwin HW, Farrant JM (1998) Protection mechanisms against excess light in the resurrection plants *Craterostigma wilmsii* and *Xerophyta viscosa*. Plant Growth Regul 24:203–210
Sherwin HW, Pammenter NW, February E, Willigen CV, Farrant JM (1998) Xylem hydraulic characteristics, water relations and wood anatomy of the resurrection plant *Myrothamnus flabellifolius* Welw. Ann Bot 81:567–575
Smirnoff N (1993) The role of active oxygen in the response of plants to water deficit and desiccation. New Phytol 125:27–58
Sprey B, Lichtenthaler HK (1966) Zur Frage der Beziehungen zwischen Plastoglobuli und Thylakoidgenese in Gerstenkeimlingen. Z Naturforsch 21b:697–699
Stuart TS (1968) Revival of respiration and photosynthesis in dried leaves of *Polypodium polypodioides*. Planta 83:185–206
Tuba Z, Lichtenthaler HK, Csintalan Z, Pócs T (1993a) Regreening of desiccated leaves of the poikilochlorophyllous *Xerophyta scabrida* upon rehydration. J Plant Physiol 142:103–108
Tuba Z, Lichtenthaler HK, Maroti I, Csintalan Z (1993b) Resynthesis of thylakoids and functional chloroplasts in the desiccated leaves of the poikilochlorophyllous plant *Xerophyta scabrida* upon rehydration. J Plant Physiol 142:742–748
Tuba Z, Lichtenthaler HK, Csintalan Z, Nagy Z, Szente K (1994) Reconstitution of chlorophylls and photosynthetic CO_2 assimilation upon rehydration in the desiccated poikilochlorophyllous plant *Xerophyta scabrida*. Planta 192:414–420
Tuba Z, Csintalan Z, Proctor MCF (1996a) Photosynthetic responses of a moss, *Tortula ruralis* ssp. *ruralis*, and the lichens *Cladonia convoluta* and *C. furcata* to water deficit and short periods of desiccation, a baseline study at present-day CO_2 concentration. New Phytol 133:353–361
Tuba Z, Lichtenthaler HK, Csintalan Z, Nagy Z, Szente K (1996b) Loss of chlorophylls, cessation of photosynthetic CO_2 assimilation and respiration in the poikilochlorophyllous plant *Xerophyta scabrida*. Physiol Plant 96:383–388
Tuba Z, Smirnoff N, Csintalan Z, Nagy Z, Szente K (1997) Respiration during slow desiccation of the poikilochlorophyllous desiccation tolerant plant *Xerophyta scabrida* at present-day CO_2 concentrations. Plant Physiol Biochem 35:381–386
Tuba Z, Proctor MCF, Csintalan Z (1998) Ecophysiological responses of homoichlorophyllous and poikilichlorophyllous desiccation tolerant plants: a comparison and an ecological perspective. Plant Growth Regul 24:211–217
Vassiljev JM (1931) Über den Wasserhaushalt von Pflanzen der Sandwüste im sudöstliche Kara-Kum. Planta 14:225–309
Wellburn FAM, Wellburn AR (1976) Novel chloroplasts and cellular ultrastructure in the 'resurrection' plant *Myrothamnus flabellifolia* Welw. (Myrothamnaceae). Bot J Linn Soc 72:51–54

Chapter 10
Hydraulic Architecture of Vascular Plants

Ernst Steudle

10.1 Introduction

The water balance of a plant is given by the input of water across the root and the losses spent by transpiration. When there is a surplus on the uptake side, plants may grow and develop. When losses prevail, plants will suffer from desiccation that results in damage of cellular organs and eventually die. In the past, there have been many efforts to understand the mechanisms by which plants regulate transpiration in terms of both water balance and carbon gain. In higher plants, these efforts have centered on the regulation of stomatal conductance and how it would be affected by water status, light intensity, nutrition, and how gas exchange would change under stress conditions such as water shortage or drought (Cowan 1977; Schulze 1986; Kramer and Boyer 1995). Much less is known about the regulation of carbon gain in poikilohydric plants, where the hydration of the cytoplasm is matched with the humidity (water potential) of the surroundings. At a certain degree of desiccation, photosynthesis stops and recovers during episodes of sufficient hydration (Lange et al. 1970; Larcher 1995). In higher plants, the output function (transpiration) has been studied in detail. Fewer efforts have been made to study the input function. The acquisition of water by plant roots can be quite variable depending on the conditions in the soil (water shortage and salinity) and also the water demands from transpiring shoots (Brewig 1937; Brouwer 1954; Weatherley 1982; Fiscus 1975; Peterson et al. 1993; Steudle and Peterson 1998; Hose et al. 2001; Steudle 2000a, b, 2001). Other important factors that adjust or even regulate the hydraulics of roots are nutrient deficiency, anoxia, temperature, and heavy metals (Munns and Passioura 1984; Azaizeh et al. 1992; Birner and Steudle 1993; Maggio and Joly 1995; Carvajal et al. 1996; Peyrano et al. 1997; Schreiber et al. 1999; Henzler et al. 1999; Ye and Steudle 2006). Although there has been some progress, we are still lacking important details in the hydraulic architecture of roots to provide quantitative models for its functioning based on root anatomy, and how this would change under conditions of water shortage or other stresses.

In leaf hydraulics, a quantitative understanding is required of how the different xylar component resistances (petioles, major, and minor veins) affect overall leaf hydraulics trying to unravel the situation (Cochard et al. 2002). Just recently, the

U. Lüttge et al. (eds.), *Plant Desiccation Tolerance*, Ecological Studies 215,
DOI 10.1007/978-3-642-19106-0_10, © Springer-Verlag Berlin Heidelberg 2011

role of extraxylar pathways (living tissue) got into the focus of studies (Sack and Holbrook 2006). Although root work is handicapped by the fact that roots are less accessible than shoots, research on roots seems to be ahead in that latter point of unraveling. Leaf and root hydraulics do have in common that they are quite variable depending on many factors and that responses differ depending on circadian and developmental time scales.

The schematic Fig. 10.1 shows how the water flow across a transpiring plant would be regulated in terms of a catenary system of variable resistors (soil–plant–air continuum, SPAC). Usually, the vapor pathway in the air spaces of leaf mesophyll, across stomata, and the boundary layer outside of leaves represents by far the highest resistance within the SPAC. Most of the water potential difference will drop here (Nobel 1999). Transpiration dominates the scene. Therefore, the top shoot resistor is the biggest in Fig. 10.1 (but not to scale). Besides the diffusive transport of water vapor in the leaf, it incorporates the hydraulic resistance of the leaf, i.e., the resistance of the liquid phase, which then continues down the stem and root into the soil. Within the liquid phase, the root usually contributes to the most of

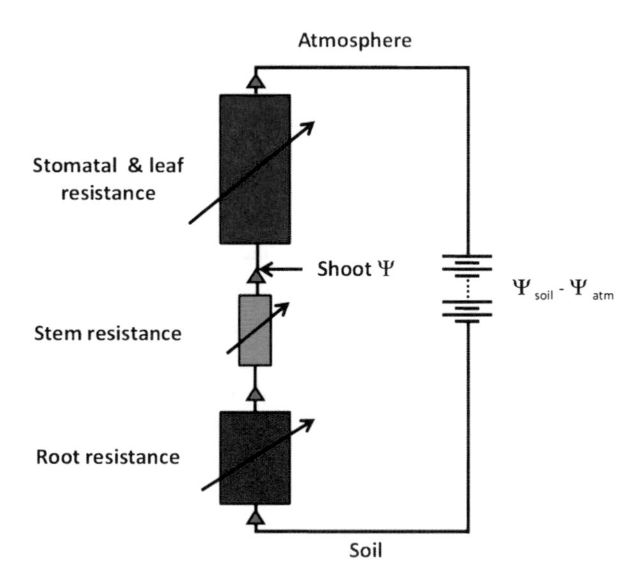

Fig. 10.1 Simple model of steady water flow across the soil–plant–air continuum (SPAC). Water uptake by roots, long-distance flow in the stem xylem, and flow in the shoot (leaves) are denoted by three variable resistances. The top resistance includes both the stomatal resistance (gas phase including boundary layers) and the leaf hydraulic resistance (liquid phase). Usually, the stomatal resistance dominates in the SPAC. The resistance along the stem is rather small. At a given flow across the SPAC, the water potential in the shoot (Ψ_{shoot}) depends on the overall resistance of root and stem in relation to the stomatal resistance. Ψ_{shoot} will affect the open/closed state of stomata and transpiration. Hence, the resistances in the liquid phase play an important role in the overall water balance. In order to improve the water balance, the plant can either reduce ($R_{stem} + R_{root}$) at the water input or increase R_{shoot} by closing stomata at the output. When the soil gets dry, the hydraulic resistance of the soil will be significant as well and may then even dominate (for further explanation, see text)

the resistance, but the axial (stem) resistance may become significant under conditions of water stress, when vessels cavitate and SPAC is interrupted.

Although gas exchange is usually dominating the SPAC, water relations within the liquid phase is nevertheless an important issue. This is so, because the water status of the shoot (leaves) affects the open/closed state of stomata. At a given rate of transpiration, water relations within the liquid phase determine the supply of water to the shoot and the water status of the shoot, which, in turn, regulates the opening of stomata, carbon gain, and water loss in a feedback loop.

This chapter on "hydraulic architecture" summarizes some aspects of water flow across higher plants, focusing on the variability of flows and the mechanisms by which this is achieved. By definition, "hydraulics" take place in the liquid phases of tissues and vessels. There is no "hydraulics" in desiccated plants, when the forces that drive water movement are just matrix in nature and vessels – when existing – do not function. To the best of the author's knowledge, there are no detailed quantitative studies about force/flow relations of water in the organs of desiccated plants as they exist in detail for many higher plants. A chapter on "hydraulic architecture" may be, thus, considered to be not too important for people dealing with desiccation tolerance. However, some of the mechanisms by which plants maintain their water balance under conditions of "normal" water supply may be of interest, when dealing with extreme water shortage such as during early phases of desiccation. On the other hand, when desiccation-tolerant plants restore their hydraulic system (cell, tissue, organ level, and xylem) following conditions of extreme water shortage, the recovery of roots, stems, and leaves is of great importance besides the xylem, which is usually dealt with. By dealing with the hydraulics of vascular plants, the focus of the chapter is on the hydraulics of roots, which acquire water which is then largely spent by transpiration. Another focus is on the cavitation and the composite transport (CT) of water flow in plants. This will also include how scientists think or better guess on refilling of cavitated vessels in plants, a topic which may be of interest during the rewatering of resurrection plants. Eventually, there will be some remarks on leaf hydraulics.

10.2 Water Uptake at Water Shortage: Role of Apoplast and of Composite Transport

Root anatomy underlies transport processes. Root hydraulics can be only interpreted with sufficient knowledge about the anatomy of roots. Root anatomy varies a lot depending on the way plants are grown and in response to the developmental state of roots (Steudle and Peterson 1998). For example, in maize, rates of uptake of water and that of nutrient ions were significantly reduced during drought stress (Stasovsky and Peterson 1993). To minimize water losses, roots developed a suberized interface between living tissue and soil, which reduced both the ability of roots to take up water and water losses, both measured as the hydraulic conductivity of roots (root Lp_r in $m^3\ m^{-2}\ MPa^{-1}\ s^{-1}$). Taleisnik et al. (1999) grew plants under conditions of water shortage, where their ability to retain water was increased in the presence of an exodermis, which was not present in the controls. Older roots being more suberized released less water to the dry environment than younger ones. It was concluded that the changes in root hydraulics were brought about by the presence of apoplastic barriers rather than by changes in membrane properties of cells. Similarly, earlier results from Kramer and coworkers (summarized by Kramer and Boyer 1995, on pp. 130, and 184/5) indicated that suberization substantially reduced the ability of roots of woody species to take up water. However, it was stressed that older suberized roots did still contribute to the overall uptake of water and nutrients.

Zimmermann and Steudle (1998) compared the hydraulics of young roots of corn with and without an exodermis. When the seedlings were grown in hydroponics, they did not develop a continuous exodermis. But when grown in mist culture they did, which reduced the hydraulic conductivity by a factor of 3.6 at constant water permeability of root cells. Recently, the thicker roots of *Iris germanica* have been studied by Meyer et al. (2010). When exposed to air, these roots developed a continuous multiseriate exodermis, which was uniseriate, when roots were brought up in hydroponics. As for the young corn roots, the treatment caused substantial changes in the root hydraulics measured by combining the root pressure probe and the pressure chamber. When present, the multiseriate exodermis was limiting water flow (as for corn grown in mist culture).

The above examples suggest that the simple textbook notion that the endodermis would usually represent the dominating hydraulic resistance in roots (and the osmotic barrier) may be questioned, at least for roots that develop an exodermis. For roots grown under harsh conditions of water shortage, the effects of suberization may be dramatic. This means that, during restoration from the desiccated state, roots may have a problem to acquire water from the rewetted soil in view of a suberized exo- and endodermis. In addition, plants grown at low water potentials of the soil develop less extended root systems. Unfortunately, there are, for technical reasons, no quantitative data for these roots to show whether or not their hydraulics may encounter severe problems or may even become limiting during water uptake following periods of desiccation. This knowledge would be of great importance when judging about role of root pressure during the recovery of the function of xylem during these time periods.

The discovery of aquaporins (water channels; AQPs) by Agre and coworkers around 1990 (Denker et al. 1988; Zeidel et al. 1992; Preston et al. 1992) caused a focus on the cell-to-cell rather than on the apoplastic passage of water across roots as emphasized in the previous paragraph (Maurel 1997; Maurel et al. 2009). AQPs are proteins of a molecular weight of about 30 kDa, which span the membrane six times to form the pore by a folding back of two loops of the protein into the bilayer. The actual pore of a length of about 2 nm (20 Å) has a diameter, which is adapted to the size of the water molecule (about 4 Å). It has a mouth part and a constriction part in the center, which determines its selectivity according to size and polarity. Molecules which are similar in size and polarity as water may also pass through the AQPs. The break in the water structure at the constriction and electrostatic barriers provide that no charge can pass (Burykin and Warshel 2004). The latter point is of major importance with respect to a potential passage of protons or hydroxyl ions (Grotthus mechanism), which would otherwise short circuit pH gradients across membranes and substantially affect membrane potentials, proton motive forces, and the like.

The open/closed state or activity of AQPs may be affected by different external and internal factors that cause a conformational transition from open to closed state. This gating of AQPs happens under conditions of stress such as in the presence of heavy metals, high concentrations of osmolytes or salinity, low temperature, oxidative stress and hypoxia, or mechanical stress (Azaizeh et al. 1992; Tyerman et al. 1999; Ye and Steudle 2006; Henzler et al. 2004; Ye et al. 2004; Lee et al. 2005a, b; Birner and Steudle 1993; Wan et al. 2004). The water stress hormone abscisic acid has been shown to have an ameliorative effect on AQPs tending to reopen closed AQPs (Freundl et al. 1998; Hose et al. 2000). ABA is also of key importance in physiology of desiccation as described in Chaps. 9, 12, and 16. For technical reasons, it has not been possible yet to measure effects of just low water potential as it occurs during drought or desiccation. However, the results obtained in the presence of high concentrations of osmolytes ("physiological drought") for roots or individual algal cells used as model systems may mimic the situation at low water potential, at least to some extent. The results obtained so far indicate that low water potential induces a "shrinkage" of AQPs. The osmolytes present on both sides of the membrane are excluded from the pores and, therefore, extract water from the pore, which would eventually and reversibly collapse under a certain stress (Ye et al. 2004). The experimental results obtained from both corn root cells and *Chara* internodes were in favor of the dehydration or "cohesion/tension" model, which also allowed to estimate pore volumes (Ye et al. 2005). When one extrapolates to the effects of low water potential and high concentration to conditions as they occur, and when plant tissue is subjected to desiccation (or just recovers from it), AQP activity should then soon become zero. The remaining water permeability of the bilayer in the absence of AQP activity may be around 10% of that in its presence. However, this should also disappear, when protoplasts collapse.

It is evident that in tissues such as the root, the radial flow of water may be either apoplastic or cell-to-cell, and the contribution of each pathway would depend on the hydraulic resistance along the two parallel pathways. In other words, due to the composite structure of roots (or of other tissue) transport should be composite in nature. This is dealt with in detail in the next section.

10.3 The Nature of Water Movement in Roots

The radial transport of water in the root cylinder is often described as an osmotic process, whereby the endodermis is functioning as a "root membrane," i.e., as an osmotic barrier with selective properties. As a "cell membrane," the root is often considered as a fairly perfect osmometer with a reflection coefficient of close to unity, as is true for a cell membrane. The root hydraulic resistance is thought to be constant in this simple picture. However, evidence collected during the past two decades indicated that this view has to be modified. The modification is necessary because we have a composite structure with two parallel pathways of quite different osmotic and hydraulic properties. Along the apoplast, no membranes have to be crossed and, therefore, osmotic gradients (such as between root xylem and medium) cannot operate. In other words, the reflection coefficient along this path (σ_s^{apo}) is close to zero. Osmotic gradients should not operate along this pipe, but hydrostatic gradients will cause a flow, which would be, therefore, purely hydraulic in nature.

Along the cell-to-cell passage, on the other hand, osmotic gradients cause an osmotic water flow with a reflection coefficient, σ_s^{cc}, of close to unity, and pressure gradients work as well. For the entire passage, we end up with a mixed passive selectivity, as expressed by the root's overall reflection coefficient, σ_{sr}. This should be substantially smaller than unity (as found experimentally; for references, see Steudle and Peterson 1998). Depending on the species used, the range differs from 0.1 to 0.9, and this may reflect difference in root structure and, namely, in the "tightness" of Casparian bands. For a cell membrane, a $\sigma_s = 0.9$ would already denote structure highly permeable to the given solute. The osmotic barrier in the root is, therefore, qualitatively different from that of a plasma membrane.

This picture of the root suggests that the water permeability (hydraulic conductivity) of the apoplast including that of apoplastic barriers in the endo- and exodermis is not completely zero, but contributes to the overall root Lp_r. It should be noted that the term "composite transport" has been used for plant tissue in analogy to the terminology used in irreversible thermodynamics to describe transport across membranes, which are composed of arrays with different transport properties (Kedem and Katchalsky 1963; Steudle and Henzler 1995). It should be also noted that a $\sigma_{sr} < 1$ does not mean that the osmometer model of the root does not hold any more. It just says that the root is no ideal osmometer with a passive selectivity, which is sufficiently low to allow, for example, the active accumulation of nutrient ions from the soil solution with no substantial leakage. On the other hand, composite transport has important advantages with respect to the water balance of plants, namely, under conditions of water shortage, which is dealt with below.

There is a caveat that the reflection coefficient cannot be taken as a quantitative measure of solute permeability, namely, in composite systems. For example, in tree roots suberization causes a reduction of root Lp_r by an order of magnitude (as compared with roots of herbaceous plants). At the same time, root σ_{sr} is much smaller ($\sigma_{sr} = 0.1$–0.4), but the permeability of roots for solutes was not measurable, because solute permeability was more affected by suberization than water permeability (for references, see Steudle and Peterson 1998).

10.4 Pathways for Water and Solutes and Composite Transport

According to the previous section on flows of water and solutes (nutrients), there should be parallel cellular (cell-to-cell) and apoplastic components. However, the passive cell-to-cell component of solute flow should be small compared with that of water, because of the low permeability of membranes to solutes such as nutrient ions. Water, on the other hand, moves more rapidly by several orders of magnitude across membranes (in part, due to the existence of AQPs). In principle, the cell-to-cell or "protoplastic" component has to be split up into a transcellular (across membranes) and a symplastic (across plasmodesmata) component. However, to date, the latter two components cannot be separated experimentally and are, therefore, summarized as a cell-to-cell component. The cell-to-cell component can be determined from measurements with the cell pressure probe. These results have been compared with measurements at the root level, which allowed to judge about pathways, at least semiquantitatively. There is, to date, no technique to measure the contribution of the apoplastic water flow across the root cylinder or in other tissue. However, Zhu and Steudle (1991) compared hydraulic conductivities at the level of cortical cells of young corn seedlings (cell pressure probe) with those measured at the root level (root pressure probe). It turned out that values at the cell level were not much bigger than those measured at the level of entire roots. This indicated that, at least for hydroponically grown corn, there was a substantial "apoplastic bypass" of water.

When root pressure is established under conditions of zero transpiration or when a manometer is placed on top of an excised root (such as a root pressure probe), there will be an uptake of water along the cell-to-cell passage causing an increase of root pressure. This, in turn, will result in a backflow of water along the apoplastic passage tending to cause a root pressure smaller than expected from the osmotic pressure gradient between xylem and root medium, i.e., the overall reflection coefficient will be smaller than unity (Fig. 10.2). Eventually, a steady state will be attained where the opposing flows are equal. This results in a circulation flow of water across the root. To create additional extracellular bypasses, the system may be manipulated by puncturing the endodermis with needles as done with young corn roots (Steudle et al. 1993). As a response, the steady-state root pressure and the reflection coefficient were reduced, as one would expect in the presence of an additional bypass. On the other hand, solute permeability increased. In other experiments, the apoplastic path could be occluded using nanoparticles or causing salt precipitations in cell walls. These experiments were done with rice roots, where the outer part of the root with its well-developed exodermis provided a model system (Ranathunge et al. 2004, 2005a, b). As expected, blockage resulted in an increase of reflection coefficients and in a decrease of water permeability.

10.5 Roles of the Exo- and Endodermis

On its radial passage across the root cylinder water has to cross several layers of cells, which are arranged in series and form serial hydraulic resistances. In the tissues (rhizodermis, cortex, and stele), water may travel in the parallel apoplastic, symplasmic, and transcellular pathways, but there should be, of course, combinations of pathways for the flow of water. For example, water may travel within the

a

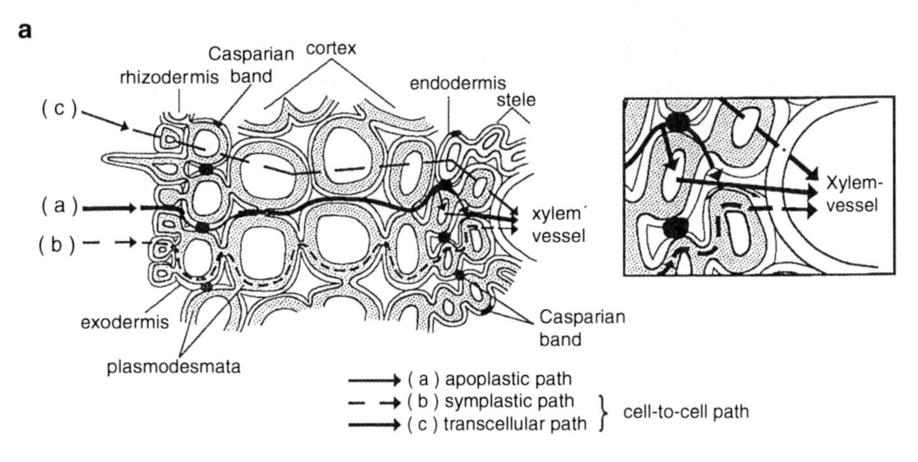

b

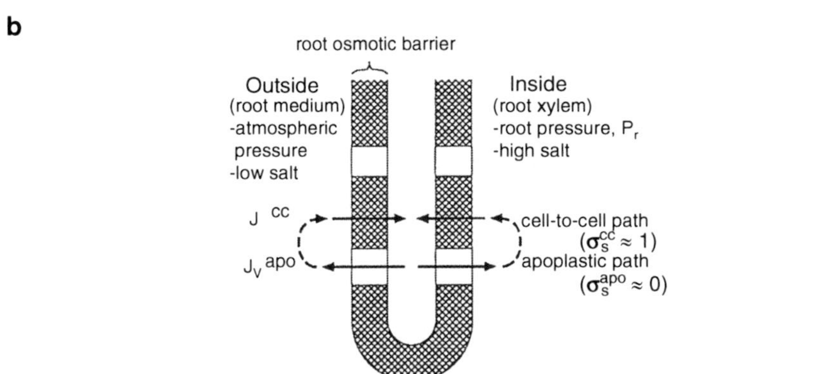

Fig. 10.2 (**a**) Radial flow of water across the root cylinder. There are three different pathways. The apoplastic pathway denotes water flow in the cell wall space and the transcellular pathway the water movement across cells, whereby the cell membrane has to be crossed twice per cell layer. The symplastic route is across plasmodesmata, i.e., within the symplasmic continuum. The transcellular and symplasmic components cannot be separated experimentally and are, therefore, summarized as the cell-to-cell component. In the endodermis and exodermis, Casparian bands cause a more or less complete interruption of the apoplastic path. Suberin lamellae may also affect the cell-to-cell component. (**b**) As shown schematically, the complex composite structure of the root causes composite water transport. Along the apoplast, there is virtually no selectivity, and the reflection coefficient is $\sigma_s^{apo} \approx 0$. The water flow in the apoplast is hydraulic in nature. Along the cell-to-cell path, water flow is both hydraulic and osmotic, and $\sigma_s^{cc} \approx 1$. In the absence of transpiration, the root pressure built up in the system (because of the active pumping of nutrient ions) causes a backflow of water along the apoplast. Water balance is achieved at a root pressure, which is smaller than expected from the osmotic pressure difference between xylem and root medium. This results in an overall root reflection coefficient of smaller than unity. The apoplastic bypass flow of water is hindered by the existence of Casparian bands, which, however, do not completely interrupt it. In other words, roots do behave like osmometers, but not like ideal ones

symplastic pathway for some time and then cross the plasma membrane and move within the cell wall, etc. These combinations due to lateral exchange have been included in a model of composite transport as long as allowance is made for "local equilibrium," i.e., rapid exchange of water between pathways, which is a reasonable or even excellent assumption (Molz and Ikenberry 1974; Molz and Ferrier 1982; Westgate and Steudle 1985; Steudle 1989, 1992, 1994; Steudle and Frensch 1996). Hence, the passage across tissues is usually considered to have a relatively low hydraulic resistance, namely, due to high AQP activity. Similarly, the resistance offered to water uptake by vessel walls (early and late metaxylem) is usually small, though not negligible (Peterson and Steudle 1993).

In the root, the exo- and endodermis with its walls modified by Casparian bands and suberin lamellae represent a substantial hydraulic resistance, which is due to the deposit of suberin and lignin (Casparian bands) or just suberin (Schreiber et al. 1994; Zeier and Schreiber 1998). By comparing roots with and without an exodermis, perfusion and puncturing experiments with different plant species have shown that the contribution of the exodermis and endodermis can be substantial, depending on the species and the developmental state of roots (Zimmermann and Steudle 1998; Peterson et al. 1993; Steudle et al. 1993; Ranathunge et al. 2004, 2005a, b; Meyer et al., unpublished). The development of Casparian bands should affect the apoplastic movement of water and ions (actively taken up and retained in the stele). In contrast, the presence of suberin lamellae would affect the cell-to-cell passage of water. In some roots, both apoplastic barriers can develop thickened roots, which are suberized and/or lignified; this may result in substantial increases of hydraulic resistance (see e.g., North and Nobel 1991; Melchior and Steudle 1993; Zimmermann and Steudle 1998). However, there are also exceptions, where the presence of suberin lamellae did not decrease the water permeability of the exodermis (Sanderson 1983; Clarkson et al. 1987).

In roots of corn and rice, Ranathunge et al. (2005a, b) found that the exo- and endodermis did show some permeability for ions, although Casparian bands were well developed. They used a precipitation technique, and the aerenchyma of rice roots was perfused with a solution that contained either copper cations or ferrocyanide anions ($[Fe(CN)_6]^{4-}$). The two ions form a brown precipitation when they "meet" (as in Pfeffer cells). In the solution, the opposite ion was offered. Precipitations were found, which indicated that the ions could permeate the well-developed Casparian bands in the outer part of the root. Most interestingly, the occurrence of precipitations was asymmetric, indicating that the ferrocyanide ion with its four negative charges was less permeant in the apoplast, most likely, because it was repelled by negative fixed charges in the walls. Similar experiments with young corn roots showed that the precipitation technique could also be used to demonstrate some permeability of Casparian bands. These results are in line with comparisons of the cellular and overall hydraulic conductivity of corn roots, which suggested some apoplastic transport of water, even in the endodermis. However, the situation may vary between species. In some species such as barley, the comparison between cell Lp and root Lp_r was in line with a dominating cell-to-cell passage, due to a relatively high cell Lp (Steudle and Jeschke 1983).

10.6 Physiological Consequences of Composite Transport

The most obvious consequence of composite transport is that it provides a variable root Lp_r, which can be increased according to an increasing demand from the shoot. On the other hand, it can be decreased at low rates of transpiration and when water is tending to be lost to a dry soil. In a transpiring plant, the high demand from the shoot will cause a high gradient in hydrostatic pressure across the root cylinder (tensions in root xylem). As a consequence, root Lp_r will be high (both passages used). However, in the absence of transpiration, at low demands from the shoot, root Lp_r should be low (only the cell-to-cell passage used). When the soil is dry, the mechanism should prevent big losses of water to the soil, which can otherwise not be prevented. The idea that plants use suberization to prevent losses of water to a dry soil may be easily accepted. However, composite transport would allow plants to still acquire some water under adverse conditions, when tensions in the root xylem are favorable for water uptake, by switching to a higher root Lp_r. It would tend to optimize the water balance by adjusting root hydraulics according to the demand from the shoot and the availability of water in the soil. The mechanism provides a short-term adjustment within a few minutes, which plants may use besides others such as suberization as water stress occurs (time scale of a few days) or reduced root growth (time scale of weeks). Eventually, the situation of a heavily suberized root resembles that of a cutinized shoot in air, usually a rather dry environment, where the existence of a cuticle minimizes nonstomatal water losses. Stomatal water loss is a highly regulated one, but plants have to open stomata to acquire the nutrient CO_2 at the expense of some unavoidable water loss. In the root, composite transport allows for adjustable recruitment of both water and nutrients, even under harsh conditions, when plants may tend to desiccate.

10.7 Consequences of Composite Transport for Growth Under Conditions of Severe Water Stress

There are, to date, no detailed data (cell, tissue, and whole organ level) of the hydraulics of roots or shoots measured under conditions of water stress, which would allow us to compare data with well-watered or under conditions of mild water stress. This is so, because it is difficult to measure water flows and resistances at the cell and organ level without rewatering, which may change resistances such as AQP activity. It is usually anticipated, though, that, as soils dry and plants suffer from severe water stress, root growth is reduced and there are changes in root anatomy tending to increase suberization. Under these conditions, apoplastic barriers should dominate the resistance, which should reduce extracellular water flow. The cell-to-cell passage may then dominate, but nothing is known about the gating of AQPs at severe water stress. What happens under conditions of severe drought is that the water transport in the soil rather than that in the root dominates water uptake. As soil dries, its hydraulic conductance is reduced by several orders of magnitude (Nobel 1999; Kramer and Boyer 1995). A rapid uptake of water by roots may cause a decrease of ψ_{soil} in the vicinity of roots. This may result in substantial decreases of the hydraulic conductivity of the soil. In addition, roots shrink in dry soil as they dehydrate, which may result in an additional resistance as the hydraulic contact is lost between soil and root surface (Herkelrath et al. 1977; Faiz and Weatherley 1978). However, this resistance may be less than expected, provided that gaps are not too big (Ranathunge et al. 2003). Overall, dry soil imposes conditions, at which composite transport extends into the soil, and the root and soil form composite hydraulic resistances arranged in series. In the future, the model would have to be extended by additional serial components to include the soil with a variable Lp_{soil} and a component referring to the gap resistance. Different from the others, the latter is not hydraulic, but diffusional in nature.

When roots grow under adverse conditions, this slows down root extension. It will cause suberization and, usually, the formation of an exodermis with additional Casparian bands and suberin lamellae. As a result, root Lp_r should be reduced. The ratio of osmotic to hydraulic water flow should change as well (Frensch et al. 1996). At the first sight, one would expect that the formation of apoplastic barriers should reduce the hydraulic component in favor of the osmotic. However, things are more complex, because the formation of suberin lamellae would also reduce the cell-to-cell component. The definite phenomenon is significant reduction of the overall root Lp_r. For roots grown under adverse conditions such as water shortage, root Lp_r may be reduced severalfold (Zimmermann and Steudle 1998; Rieger and Litvin 1999; Stasovsky and Peterson 1993; Meyer et al. 2011).

10.8 Variability of Axial Hydraulic Resistance

In Fig. 10.1, another candidate for an overall variable hydraulic resistance is the resistance offered to water flow in the conduits of xylem in the stem (vessels or tracheids), which connect roots with shoots. As indicated in the figure, this resistance is usually the smallest. However, it can increase substantially, when the continuity of vessels breaks under tensions, where water columns in vessels are replaced by air/gas (Steudle 2001). Cavitation then causes dysfunction (Tyree and Sperry 1989). Dysfunction may result in a runaway cavitation. Runaway cavitation occurs when, at a given rate of transpiration, some vessel members cavitate. This, in turn, causes a decrease of the axial hydraulic conductivity (K_h) and steeper gradients of pressure, which induce more cavitations and a loss of functional xylem cross section in a vicious cycle. However, trees rarely suffer runaway embolism.

The reason for this is that stomata tend to reduce water flow as K_h decreases and the water status of shoot gets worse (Tyree and Zimmermann 2002).

According to the capillary equation, cavitation occurs, when gas seeds in vessels (usually at vessel walls) exceed a critical size, which is usually quite small (Steudle 2001). More likely, they occur when air is forced into functioning vessels across pores in pit membranes due to pressure differences between cavitated and functioning vessels or between vessels and surroundings. As this process is also governed by the capillary equation, air-seeding across pores depends on the diameter of pores, and there is a correlation that plants adapted to dry habitats have smaller diameters of pit pores than others (Tyree 1997). In roots, K_h is usually larger than that in stems, because of the larger diameter of vessels and of vessels length (Tyree and Zimmermann 2002). In root hydraulics, the hydraulic conductivity decreases as the endodermis develops (Melchior and Steudle 1993; Frensch et al. 1996). In this case, the major resistance is in the living cells around the vessels rather than in the axial resistance of developed xylem, which has been shown to be much smaller (Frensch and Steudle 1989). It may be stated that, during long-distance transport, the points of water entry through tissue into vessels offer a resistance bigger than that for the axial transport in vessels, which is a movement of water in capillaries ranging in diameter from about 10 to several 100 μm. According Poiseuille's fourth power law (valid with some limitations), this covers a range of fairly large values of K_h.

Limits imposed at high rates of transpiration are usually worked out in vulnerability curves (Tyree et al. 1994). Other reasons for a variability of axial K_h, although not such drastic as during substantial water stresses, would be changes in temperature and in the concentration of solutes (minerals) dissolved in xylem sap. Since the flow along the xylem is convective in nature, changes of the viscosity of water may change flow rates and K_h. Compared to other factors (existence of cross walls or variations in vessel diameter), this effect is relatively small and accounts for about 30% at a 10° change in temperature, i.e., the Q_{10} is 1.3. Xylem sap at higher temperature is more fluid and less viscose.

The effect of changes in concentration of xylem has been first studied by Zimmermann (1978) and later by van Ieperen et al. (2000) and Zwieniecki et al. (2001). Apparently, the effect was not always present. It was shown that, when xylem sap concentration decreased by dilution of the sap at high rates of transpiration (as the concentration decreases at constant uptake of mineral salts), the axial K_h increases. Most authors think that this is caused by a shrinking of hydrogels (pectins) in pit membranes between xylem vessels as the concentration of xylem sap increases. The passage of water across pit membrane and cross walls is thought to be a significant fraction of overall vessel resistance tending to cause most of the deviation from Poiseuille's law.

10.9 Embolism and Refilling of Xylem Vessels

For plants being subjected to severe drought, embolism is an important issue, when the input *via* the root lags behind transpiration (Fig. 10.1). Cavitated vessels offer a substantially high resistance to water flow (having a low K_h). The cohesion–tension (CT) theory of the ascent of sap in xylem vessels is now widely accepted after some questioning and subsequent discussions in the past two decades (Balling and Zimmermann 1990; Melcher et al. 1998). There is a lot of indirect evidence in favor of the CT theory, which has been collected during the period of its existence (for references, see Tyree and Zimmermann 2002). Using the cell pressure probe, xylem pressures (tensions) have been measured directly for the first time in transpiring plants (Wei et al. 1999a, b; Steudle 2001). It has been shown that tensions of at least 1 MPa (10 bar) can exist in the xylem, which is less than what was predicted by indirect measurements (tensions of up to 100 bar), but nevertheless verified the CT theory, at least for the range of low or moderate tensions. It was shown that the tension in the xylem was affected by changes in transpiration, as predicted by the CT theory and the results obtained indirectly for xylem pressure using the pressure chamber. Technical limitations in the range of measuring tensions were due to the fact that the pressure probe itself had an upper limit to withstand tensions where it cavitated (Wei et al. 1999a). The results indicated that plant xylem is less vulnerable to cavitation than pressure probes, because the walls of xylem consist of polymers (cellulose, lignin, etc.) that imbibe water without leaving gas seeds at the surface of walls, which "catalyze" cavitation depending on the diameter of voids (Steudle 2001). On the other hand, the internal surface of probes will be hardly free of such seeds. It should be realized that the critical seed diameter equivalent to a tension of 10 bar would be 0.3 μm, which is already smaller than the wavelength of visible light. The highest tensions of 100 bar, as measured in plant xylem, would be equivalent to only 30 nm. Hence, the walls of operating xylem have to be quite "clean." It is unlikely that events of seeding occur in bulk solution, even when gases such as O_2 or N_2 are dissolved at high pressure (Kenrick et al. 1924). The upper limit of the measured tensile strength of water is at 140 MPa, at which the density of water is only half of the usual (Zheng et al. 1991).

When it was demonstrated that xylem could operate under tension, the question was left how cavitated vessels are able to refill even under conditions, where the rest of the tissue is still under tension with water potentials that are significantly smaller than zero. However, the water potential in cavitated vessels would be around zero, i.e., the xylem sap at atmospheric pressure or close to. In other words, can embolism dissolve in plant without using root pressure in the absence of transpiration? There is experimental evidence that they can (Canny 2001; Canny et al. 2001; Salleo et al. 1996; Holbrook et al. 2001). During refilling of cavitated vessels under these conditions, water would have to move uphill from surrounding tissue (at negative water potential) to cavitated vessels (water potentials of close to zero). Despite the experimental evidence that water does fill cavitated vessels under the conditions outlined, there has been, to date, no convincing mechanism, by which water could be persuaded to move in the prospected direction without violating basic thermodynamics. Perhaps, the only way to do so would be to propose a solute movement into cavitated vessels first and that then the water follows downhill. However, the experimental proof for such a mechanism is lacking. To understand the recovery of dehydrated or even desiccated plants, it would be most important to also understand the mechanism(s) of refilling of cavitated vessels. An osmotic filling would require to move huge amounts of solutes (e.g., mobilized sugars) into the lumen of vessels, which is unlikely (see discussion in Tyree and Zimmermann 2002). Although there has been a debate about possible mechanism, existing paradigms seem to be unlikely. Some are improbable, because they violate basic laws of physics. We still wait for the clue.

In the past, mechanisms of refilling of vessels have been an issue with resurrection plants such as the woody South African shrub *Myrothamnus*, which is one of the most spectacular examples of a quick and effective restoration of the hydraulic system following long periods of desiccation. In the literature, different mechanisms have been discussed for the refilling of the rather narrow vessels such as capillarity, root pressure, and the existence of large amounts of lipids either lining vessel walls with relatively thick layers or being present as lipid globules (Schneider et al. 1999, 2003). An indication that capillary pressure may be sufficient to refill the vessels came from experiments of Vieweg and Ziegler (1969) showing that cut twigs of *Myrothamnus flabellifolia*, when put in water, quickly rehydrate. In hydraulic experiments, the plants developed root pressures of 24 kPa, which would be sufficient to drive water up to the top of the 0.80 m tall plants (Sherwin et al. 1998). However, the Zimmermann group demonstrated that the existence of lipids in the xylem affected its wettability and, hence, the effect of capillary pressure. As a result of the lining of xylem vessels by lipids, water was initially rising fastest within the narrow strands of xylem (where low wettability was compensated for by low diameters). As the lipids were swept away in other parts within the cross section, this resulted in a continuous increase in the strands of xylem available for the ascent of sap. Schneider et al. (2003) claim that transients in the distribution of lipids may provide a useful mechanism during the rapid resurrection in that it transiently reduced water losses by transpiration/evaporation of the shoot. The example shows that mechanisms of resurrection consider the same factors such as during ordinary refilling. However, additional factors may play a role as well.

It is curious that the mechanisms discussed so far for the restoring of the hydraulic system of resurrection plants focus just on the resistance of the axial (long distance) flow during refilling. However, the resistance for the radial water uptake by the roots may be dominant as well, namely, when roots were exposed for some time to a dry soil (see Sect. 10.7). This is evident from the fact that, when roots of resurrection plants had died off during periods of desiccation, rehydration took place *via* the leaves (Chap. 9). In this case, the time required for rehydration was much prolonged. Another important issue during the restoration of the hydraulic system (xylem) of resurrection plants is the fact that capillarity and root pressure should usually be sufficient. There are no tricky mechanisms required such as a refilling of vessels in the presence of low water potentials in its surroundings, which still puzzles people working in the field.

As discussed in Chap. 9 of this book, mechanisms of desiccation and refilling of xylem vessels are an important issue during the drying and rehydration of homoiochlorophyllous (HDT) and poikilochlorophyllous (PDT) desiccation-tolerant vascular plants as compared with lichens and bryophytes. In general, HDT and PDT vascular plants seem to have a greater capacity to withstand desiccation due to the water stored in vessel lumina. However, when water supply from vessels is interrupted by cavitation, the functioning of photosynthetic tissue ceases. Refilling of interrupted vessels may then occur by root pressure provided that there is enough water available around the roots and plants are not too tall (see above). Capillarity may also help during refilling. According to the capillary equation, a vessel diameter of 10 μm is equivalent to a capillary rise of 3 m water column. At a vessel diameter of 100 μm, however, the rise of the water meniscus would be only 0.3 m, etc. This requires completely wettable surfaces of vessels, which may not be the case during prolonged periods of desiccation (see above). Namely in taller species, root pressure may be more efficient. Again, both mechanisms require that enough water is radially supplied across the root cylinder (see Discussion in Chap. 9.7 and in Canny 2000).

10.10 Leaf Hydraulics and Overall Leaf Resistance

Leaf hydraulics is usually characterized by the "leaf hydraulic conductance" (K_{leaf} in units of a conductivity of mmol m^{-2} MPa^{-1} s^{-1}), which is referred to unit driving force and leaf lamina area. Its inverse is the leaf hydraulic resistance (R_{leaf}). The hydraulic conductance (resistance) of the leaf, K_{leaf} (R_{leaf}), is variable. According to a recent review of Sack and Holbrook (2006), it varies by more than 65-fold between species from different habitats depending on factors such as leaf anatomy and venation architecture, which determine the pathways for water within leaves and along petioles. As in roots, leaf hydraulics is fairly dynamic over different time scales in response to leaf development, water supply from the root, irradiance, temperature, etc. The important issue is that K_{leaf} refers to just the liquid phase of water. The contribution of water flow across stomata is excluded when measuring K_{leaf}, but the stomatal resistance would usually dominate being larger by two orders of magnitude than R_{leaf}. K_{leaf} is usually measured using different techniques (Sack et al. 2002), but the preferred technique is the high pressure flow meter (HPFM; Tyree et al. 1995). During HPFM measurements, water is injected at a known pressure (applied in a ramp to the xylem of an excised shoot or leaf), and the resulting flow rates are recorded until water drips out of stomata. By referring the measured rate of water flow to unit pressure and leaf surface area, K_{leaf} can be determined. One may debate whether or not the situation during the measurement really denotes the situation in the transpiring leaf, namely, when it is suffering from some water shortage.

Despite the fact that K_{leaf} refers to just the liquid phase, it is nevertheless an important parameter. At a given flow rate through the system (transpiration rate, E), it determines the water status (water potential) of the leaf in the linear system of resistances of Fig. 10.1 (analogous to a voltage divider in electronics). In other words, when the drop in water potential across the root (including the soil resistance) is large at a given transpiration, this lowers ψ_{leaf} accordingly. Hence, the overall resistance for the transfer of water from soil to root xylem and up to the shoot ("Nachleitwiderstand") determines the water potential of the shoot. This, in turn, affects the open/closed state of stomata and acts on E. Similarly, the catenary system of Fig. 10.1 predicts that ψ_{leaf} would be reduced as the transpiration rate increases (Tyree and Zimmermann 2002):

$$\psi_{leaf} = \psi_{soil} - \rho g h - \frac{E}{K_{plant}}, \tag{10.1}$$

where E, Ψ_{soil}, ρ, g, h, and K_{plant} represent transpiration rate, soil water potential, the density of water, acceleration due to gravity, plant height, and the whole-plant conductance in the liquid phase, respectively. This means that, at a given transfer resistance (water supply from the root), Ψ_{leaf} declines linearly with increasing transpiration. To sustain Ψ_{leaf} at a high level to keep stomata open, K_{plant} must be sufficiently high. As K_{leaf} is a major component of K_{plant}, this means that K_{leaf} should also be high.

$K_{leaf}(R_{leaf})$ is usually a function of the arrangement, density, and dimensions of vessel conduits in leaves (Sack and Holbrook 2006). In the diverging amount of studies on the topic, the conclusion has been made that the smaller conduits are a major constraint on the conduction of the venation, although the distribution of water by the larger vessels affects the overall hydraulic conductance as well.

Little is known about the effects of desiccation and related damages on K_{leaf}. There are, however, many reports that indicate that K_{leaf} declines during drought (e.g., Brodribb and Holbrook 2003a, b, 2006; Cochard 2002; Hacke and Sauter 1996; Logullo et al. 2003; Nardini et al. 2003; Salleo et al. 2001; Trifilo et al. 2003; Zwieniecki et al. 2000). Declines are largely due to xylem cavitation. However, during recovery of cavitated vessels, the usual mechanisms of refilling have been discussed such as root pressure for small plants or highly speculative mechanisms (see above).

In the past few years, there has been a tendency to suggest that the water movement outside the xylem, i.e., within the leaf parenchyma, bundle sheath, and mesophyll plays an important role in leaf hydraulics, which may be also important during the refilling of cavitated vessels following severe drought. In the older literature, living tissue surrounding vessels was hardly thought to contribute. It was thought that, when water was leaving vessels crossing their walls, it was flowing in walls around protoplasts, as the latter were thought to have a relatively high hydraulic resistance. Things changed with the discovery of AQPs tending to make a focus on the significance of the hydraulic conductivity of cell membranes in tissue transport of water. In part, this may just be trendy. People also realized that the cross-sectional areas available for a cell-to-cell transport of water are usually larger than those for the apoplastic pathway. This has stimulated work in this area as well, although, different from roots, leaf cells are usually less densely packed than those of roots (e.g., when comparing the stele with the spongy mesophyll), which reduces the contact area available for cell-to-cell transport.

Evidence for a substantial contribution of the membrane path on K_{leaf} came from the responses of this parameter to changes in temperature and irradiance, which were difficult to explain just in terms of hydraulic processes such as a viscous flow of water in veins. Increase in temperature increased leaf hydraulics by more than what was expected due to increases of the viscosity of water (Q_{10} of 1.3). However, concluding that the absence of an effect would point to an absence of transport across living cells would be premature. It has been shown that there are species in which the flow of water across AQPs may have a Q_{10} similar to that of viscosity of water (Lee et al. 2005a, b). In others, however, effects could be much more pronounced. For example, for the chilling-sensitive cucumber a $Q_{10} = 4.8$ was found, which was attributed to T-dependent collapse of AQPs rather than an increased Q_{10} for the passage of water across the molecular pores.

The hydraulic conductance of leaves strongly increases with increasing irradiance, for example, when comparing irradiances of ca. 10 μmol m^{-2} s^{-1} with those of around 1000 μmol m^{-2} s^{-1} (see Fig. 3b of Sack and Holbrook 2006; Cochard et al. 2007). (The irradiance on a sunny summer day would be around 2000 μmol m^{-2} s^{-1}.) Effects of up to severalfold have been reported in the literature. The data suggested an activation of the passage of water outside xylem vessels and most likely involves an action of AQPs, which have been shown elsewhere to follow a circadian rhythm (as stomata; Henzler et al. 1999). The light effect can be reversibly removed by adding $HgCl_2$, an inhibitor of AQP activity. It seems to be related to the pretreatment of plants, i.e., when plants are kept at darkness for some time; effects are stronger than when plants are already light-acclimated.

Recently, Kim and Steudle (2007, 2009) have investigated the light effects in detail studying the responses at the cell level (pressure probe) in midrib tissue of corn leaves close to xylem vessels. Cells were free of chloroplasts but were located in the vicinity of green cells. As light intensity increased, the cell hydraulic conductivity increased, but then decreased as turgor fell due to increased transpiration. The effect of cell turgor (water status) on AQP activity was counteracting the light effect. When leaves were perfused with 0.5 mM $CaCl_2$ solution, to maintain their water status high, the effect of light could be measured at constant turgor. This showed an increase for light intensities of up to 650 μmol m^{-2} s^{-1}, but then decreased again. This indicated an inhibition of AQP activity at high irradiance, which was observed for the first time. The authors thought that the maximum in AQP response to light may indicate a regulatory response of the plant to the environment. At low light, the increase in AQP activity would provide an enhanced water supply to the mesophyll tending to keep the water status of the leaf high and stomata open. On the other hand, the closure of AQPs at high irradiance would mean that the plant anticipates too high water losses, which are avoided in some kind of a feedforward reaction. The results indicated a light-dependent regulation of the water status based on an interaction between the turgor and light dependence of AQPs, whereby the latter could be inhibited by high light. With respect to a gating of AQPs, Kim and Steudle (2009) showed that, most likely, reactive oxygen species (as they would form at high irradiance in a leaf) triggered the open/closed state of AQPs. In the leaf perfusate, H_2O_2 could be detected at high irradiance. On the other hand, when leaves were perfused with glutathione, a scavenger for reactive oxygen species, effects could be suppressed.

Perhaps, the light effect may be integrated in the future into diurnal rhythms of AQP activity as they were demonstrated for roots of *Lotus japonicus* some time ago (Henzler et al. 1999). The findings agree with observations of large midday depressions of transpiration despite high irradiance (Brodribb and Holbrook 2003a; Nardini et al. 2003), most likely as a response to leaf dehydration, which may cause a dehydration of membranes and a reduction of the activity of AQPs.

10.11 Overall Consequences of Whole-Plant Hydraulics for Desiccation Tolerance

For a long time, the issue of desiccation tolerance has been largely related to long-distance transport and to the ability of plant xylem to withstand cavitation. Safety *vs.* efficiency was related to the fact that in bigger volumes, events of homogenous cavitation were more likely than in smaller. Recent results indicate that the resistance to withstand cavitation relates to the ability to avoid cavitation at the inner xylem surface and pit membranes. The refilling of embolized vessels is far from being understood, but is an issue which is as important as the cavitation. However, in desiccation-tolerant resurrection plants the case may be easier, as pointed out. There is a tendency, that in leaves, the extraxylar water transport across living cells plays a dominant role in the regulation or adjustment of long-distance transport, for example, by light and temperature. In other words, there is a focus on the loading and deloading of xylem across living tissue, which may, eventually, turn out to be some kind of a composite transport, where AQP activity is involved. New techniques are required to study the phenomenon. At the level of leaves, more information is required on the detailed anatomy (venation architecture and pathways of extraxylary water) and the role of AQP activity in the latter, which requires measurement at the cell level. This would be a prerequisite to understand mechanisms of diurnal and developmental dynamics. The latter are much better understood in roots on a good anatomical basis. Root hydraulics show that the composite structure of roots is the basis of composite root transport. The model is quite flexible and provides a way by which roots can adapt on a short time scale to changes in the environment by an interaction between cell-to-cell and apoplastic transport components, whereby the former may involve some interaction with metabolism. Composite transport appears to integrate into root growth and development, which are on longer time scales.

With respect to a tolerance to desiccation, it is difficult to judge from experimental results largely obtained under conditions in which plants were well-watered.

However, for the input (root) side the composite transport model may offer some interesting issues relating to the flexible role of apoplastic *vs.* cell-to-cell passages, which is largely dictated by root anatomy. In leaf hydraulics, the diversity is even bigger, but the role of living cells seems to be important as well. The role of hydraulics awaits to be quantitatively worked out under conditions of water shortage, which is difficult for technical reasons.

Acknowledgment I thank Drs. Yangmin Kim, Department of Soil Physics, Helmholtz Centre for Environmental Research, Halle, Germany, and Kosala Ranathunge, Institute of Cellular and Molecular Botany, University of Bonn, Germany, for reading and discussing the manuscript.

References

Azaizeh H, Gunse B, Steudle E (1992) Effects of NaCl and $CaCl_2$ on water transport across root cells of maize (*Zea mays* L.) seedlings. Plant Physiol 99:886–894

Balling A, Zimmermann U (1990) Comparative measurements of the xylem pressure of *Nicotiana* plants by means of the pressure bomb and pressure probe. Planta 182:325–338

Birner T, Steudle E (1993) Effects of anaerobic conditions on water and solute relations, and on active transport in roots of maize (*Zea mays* L.). Planta 190:474–483

Brewig A (1937) Permeabilitätsänderungen der Wurzelgewebe, die vom Spross beeinflusst werden. Z Bot 31:481–540

Brodribb TJ, Holbrook NM (2003a) Diurnal depression of leaf hydraulic conductance in a tropical tree species. Plant Cell Environ 27:820–827

Brodribb TJ, Holbrook NM (2003b) Stomatal closure during leaf dehydration, correlation with other leaf physiological traits. Plant Physiol 132:2166–2177

Brodribb TJ, Holbrook NM (2006) Diurnal depression of leaf hydraulic conductance in a tropical tree species. Plant Cell Environ 27:820–827

Brouwer R (1954) The regulating influence of transpiration and suction tension on the water and salt uptake by roots of intact *Vicia faba* plants. Acta Bot Neerl 3:264–312

Burykin A, Warshel A (2004) On the origin of the electrostatic barrier for proton transport in aquaporin. FEBS Lett 570:41–46

Canny MJ (2000) Water transport at the extreme – restoring the hydraulic system in a resurrection plant. New Phytol 148:187–189

Canny MJ (2001) Contributions to the debate on water transport. Am J Bot 88:43–46

Canny MJ, McCully ME, Huang CX (2001) Cryo-scanning electron microscopy observations of vessel content during transpiration in walnut petioles. Facts or artefacts? Plant Physiol Biochem 39:7–8

Carvajal M, Cooke DT, Clarkson DT (1996) Responses of wheat plants to nutrition deprivation may involve the regulation of water-channel function. Planta 199:372–381

Clarkson DT, Robards AW, Stephens JE, Stark M (1987) Suberinlamellae in the hypodermis of maize (*Zea mays*) roots: development and factors affecting the permeability of hypodermal layers. Plant Cell Environ 10:83–93

Cochard H (2002) Xylem embolism and drought-induced stomatal closure in maize. Planta 215:466–471

Cochard H, Coll L, Le Roux X, Ameglio T (2002) Unraveling the effects of plant hydraulics on stomatal closure during water stress in walnut. Plant Physiol 128:282–290

Cochard H, Venisse JS, Barigah TS, Brunel N, Herbette S, Guilliot A, Tyree MT, Sakr S (2007) Putative role of aquaporins in variable conductance of leaves in response to light. Plant Physiol 143:122–133

Cowan IR (1977) Stomatal behaviour and environment. Adv Bot Res 4:117–228
Denker BM, Smith BL, Kuhadja FP, Agre P (1988) Identification, purification, and partial characterization of a novel MW 28, 000 integral membrane-protein from erythrocytes and renal tubules. J Biol Chem 263:15634–15642
Faiz SMA, Weatherley PE (1978) Further investigation into the location and magnitude of the hydraulic resistances in the soil–plant system. New Phytol 81:19–28
Fiscus EL (1975) The interaction between osmotic- and pressure-induced water flow in plant roots. Plant Physiol 55:917–922
Frensch J, Steudle E (1989) Axial and radial hydraulic resistance to roots of maize (*Zea mays* L.). Plant Physiol 91:719–726
Frensch J, Hsiao TC, Steudle E (1996) Water and solute transport along developing maize roots. Planta 198:348–355
Freundl E, Steudle E, Hartung W (1998) Water uptake by roots of maize and sunflower affects the radial transport of abscisic acid and the ABA concentration in the xylem. Planta 207:8–19
Hacke U, Sauter JJ (1996) Drought-inuced xylem dysfunction in petioles, branches, and roots of *Populus balsamifera* L. and *Alnus glutinosa* (L.) Gaertn. Plant Physiol 111:413–417
Henzler T, Waterhouse RN, Smyth AJ, Carvajal M, Cooke DT, Schäffner AR, Steudle E, Clarkson DT (1999) Diurnal variations in hydraulic conductivity and root pressure can be correlated with the expression of putative aquaporins in the root of *Lotus japonicus*. Planta 210:50–60
Henzler T, Ye Q, Steudle E (2004) Oxidative gating of water channels (aquaporins) in *Chara* by hydroxyl radicals. Plant Cell Environ 27:1184–1195
Herkelrath WN, Miller EE, Gardner WR (1977) Water uptake by plants. II. The root contact model. Soil Sci Soc Am J 41:1039–1043
Holbrook NM, Ahrens ET, Burns MJ, CB ZMA (2001) *In vivo* observation of cavitation and embolism repair using magnetic resonance imaging. Plant Physiol 126:27–31
Hose E, Steudle E, Hartung W (2000) Abscisic acid and the hydraulic conductivity of roots: a cell- and root-pressure probe study. Planta 211:874–882
Hose E, Clarkson DT, Steudle E, Schreiber L, Hartung W (2001) The exodermis: a variable apoplastic barrier. J Exp Bot 52:2245–2264
Kedem O, Katchalsky A (1963) Permeability of composite membranes. Part 2. Parallel elements. Trans Faraday Soc 59:1931–1940
Kenrick FB, Wismer KL, Wyatt KS (1924) Supersaturation of gases in liquids. J Phys Chem 28:1308–1315
Kim YX, Steudle E (2007) Light and turgor affect the water permeability (aquaporins) of parenchyma cells in the midrib of leaves of *Zea mays*. J Exp Bot 58:4119–4129
Kim YX, Steudle E (2009) Gating of aquaporins by light and reactive oxygen species in leaf parenchyma cells of the midrib of *Zea mays*. J Exp Bot 60:547–556
Kramer PJ, Boyer JS (1995) Water relations of plants and soils. Academic, Orlando
Lange OL, Schulze ED, Koch W (1970) Experimentell-ökologische Untersuchungen an Flechten der Negev-Wüste. II. CO_2-Gaswechsel und Wasserhaushalt von *Ramalina maciformis* (Del.) Bory am natürlichen Standort während der sommerlichen Trockenperiode. Flora 159:38–62
Larcher W (1995) Physiological plant ecology. Springer, Berlin
Lee SH, Chung GC, Steudle E (2005a) Gating of aquaporins by low temperature in roots of chilling-sensitive cucumber and chilling-tolerant figleaf gourd. J Exp Bot 56:985–995
Lee SH, Chung GC, Steudle E (2005b) Low temperature and mechanical stresses differently gate aquaporins of root cortical cells of chilling-sensitive cucumber and -resistant figleaf gourd. Plant Cell Environ 28:1191–1202
Logullo MA, Nardini A, Trifilo P, Salleo S (2003) Changes in leaf hydraulics and stomatal conductance following drought stress and irrigation in *Ceratonia siliqua* (caob tree). Physiol Plant 117:186–194
Maggio A, Joly RJ (1995) Effects of mercuric chloride on the hydraulic conductivity of tomato root systems: evidence for a channel-mediated pathway. Plant Physiol 109:332–335

Maurel C (1997) Aquaporins and water permeability of plant membranes. Annu Rev Plant Physiol Plant Mol Biol 48:399–429
Maurel C, Santoni V, Luu DT, Wudick MM, Verdoucq L (2009) The cellular dynamics of plant aquaporin expression and functions. Curr Opin Plant Biol 12:690–698
Melcher PJ, Meinzer FC, Yount DE, Goldstein G, Zimmermann U (1998) Comparative measurements of xylem pressure in transpiring and non-transpiring leaves by means of the pressure chamber and the xylem pressure probe. J Exp Bot 49:1757–1760
Melchior W, Steudle E (1993) Water transport in onion (*Allium cepa* L.) roots. Changes of axial and radial hydraulic conductivity during root development. Plant Physiol 101:1305–1315
Meyer CJ, Peterson CA, Steudle E (2011) Permeability of *Iris germanica*'s multiseriate exodermis to water, NaCl and ethanol. J Exp Bot 62, Publ. online Dec. 3, 2010. doi: 10.1093/jxb/erg380
Molz FJ, Ferrier JM (1982) Mathematical treatment of water movement in plant cells and tissues: a review. Plant Cell Environ 5:191–206
Molz FJ, Ikenberry E (1974) Water transport through plant cells and walls: theoretical development. Soil Sci Am Proc 38:699–704
Munns R, Passioura JB (1984) Hydraulic resistance of plants. III. Effects of NaCl in barley and lupin. Aust J Plant Physiol 11:351–359
Nardini A, Salleo S, Raimondo F (2003) Changes in leaf hydraulic conductance correlate with leaf vein embolism in *Cercis siliquatrum* L. Trees 17:529–534
Nobel PS (1999) Physiochemical and environmental plant physiology. Academic, San Diego
North GB, Nobel PS (1991) Hydraulic conductivity of concentric root tissue of *Agave desertii* Engelm. Under wet and drying conditions. New Phytol 130:47–57
Peterson CA, Steudle E (1993) Lateral hydraulic conductivity of early metaxylem vessels in *Zea mays* L. roots. Planta 189:288–297
Peterson CA, Murrmann M, Steudle E (1993) Location of major barriers to water and ion movement in young roots of *Zea mays* L. Planta 190:127–136
Peyrano G, Taleisnik E, Quiroga M, de Forchetti SM, Tigker H (1997) Salinity effects on hydraulic conductance, lignin content and peroxydase activity in tomato roots. Plant Physiol Biochem 35:387–393
Preston GM, Carroll TP, Guggino WB, Agre P (1992) Appearance of water channels in Xenopus oocytes expressing red-cell CHIP28 protein. Science 256:385–387
Ranathunge K, Steudle E, Lafitte R (2003) Control of water uptake by rice (*Oryza sativa* L.): role of the older part of the root. Planta 217:193–205
Ranathunge K, Steudle E, Lafitte R (2004) Water permeability and reflection coefficient of the outer part of young rice roots are differently affected by closure of water channels (aquaporins) or blockage of apoplastic pores. J Exp Bot 55:433–447
Ranathunge K, Steudle E, Lafitte R (2005a) Blockage of apoplastic bypass-flow of water in rice roots by insoluble salt precipitates analogous to a Pfeffer cell. Plant Cell Environ 28:121–133
Ranathunge K, Steudle E, Lafitte R (2005b) A new precipitation technique provides evidence for the permeability of Casparian bands to ions in young roots of corn (*Zea mays* L.) and rice (*Oryza sativa* L.). Plant Cell Environ 28:1450–1462
Rieger M, Litvin P (1999) Root system hydraulic conductivity in species with contrasting root anatomy. J Exp Bot 50:201–209
Sack L, Holbrook NM (2006) Leaf hydraulics. Annu Rev Plant Physiol Plant Mol Biol 57:361–381
Sack L, Melcher PJ, Zwieniecki MA, Holbrook NM (2002) The hydraulic conductance of the angiosperm leaf lamina: a comparison of three measurement methods. J Exp Bot 53:2177–2184
Salleo S, LoGullo MA, DePaoli D, Zippo M (1996) Xylem recovery from cavitation-induced embolism in young plants of *Laurus nobilis*: a possible mechanism. New Phytol 132:47–56
Salleo S, LoGullo MA, Raimondo F, Nardini A (2001) Vulnerability to cavitation of leaf minor veins: any impact on leaf gas exchange? Plant Cell Environ 24:851–859
Sanderson J (1983) Water uptake by different regions of the barley root. Pathways for radial flow in relation to development of the endodermis. J Exp Bot 34:240–253

Schneider H, Thürmer F, Zhu JJ, Wistuba N, Gessner P, Lindner K, Herrmann B, Zimmermann G, Hartung W, Bentrup FW, Zimmermann U (1999) Diurnal changes in xylem pressure off the hydrated resurrection plant *Myrothamnus flabellifolia*: evidence for lipid bodies in conducting xylem vessels. New Phytol 143:471–484

Schneider H, Manz B, Westhoff M, Mimietz S, Szimtenings M, Neuberger T, Faber C, Krohne G, Haase A, Volke F, Zimmermann U (2003) The impact of lipid distribution, composition and mobility on xylem water refilling of the resurrection plant *Myrothamnus flabellifolia*. New Phytol 159:487–505

Schreiber L, Breiner HW, Riederer M, Düggelin M, Guggenheim R (1994) The Casparian strip of *Clivia miniata* Reg. roos: fine structure and chemical nature. Bot Acta 107:353–361

Schreiber L, Hartmann K, Skrabs M, Zeier J (1999) Apoplastic barriers in roots: chemical composition of endodermal and hypodermal cell walls. J Exp Bot 50:1267–1280

Schulze ED (1986) Carbon dioxide and water vapor exchange in response to drought in the atmosphere and soil. Annu Rev Plant Physiol 37:247–274

Sherwin HW, Pammenter NW, February E, van der Willigen C, Farrant JM (1998) Xylem hydraulic characteristics, water relations and wood anatomy of the resurrection plant Myrothamnus flabellifolia Welw. Ann Bot 81:567–575

Stasovsky E, Peterson CA (1993) Effects of drought and subsequent rehydration on the structure, vitality and permeability of *Allium cepa* adventitious roots. Can J Bot 71:700–707

Steudle E (1989) Water flows in plants and its coupling with other processes: an overview. Meth Enzymol 174:183–225

Steudle E (1992) The biophysics of plant water: compartmentation: coupling with metabolic processes, and flow of water in plant roots. In: Somero GN, Osmond CB, Bolis CL (eds) Water and life: comparative analysis of water relationships at the organismic, cellular, and molecular levels. Springer, Heidelberg, pp 173–204

Steudle E (1994) Water transport across roots. Plant Soil 167:79–90

Steudle E (2000a) Water uptake by roots: effects of water deficit. J Exp Bot 51:1532–1542

Steudle E (2000b) Water uptake by plant roots: an integration of views. Plant Soil 226:45–56

Steudle E (2001) The cohesion-tension mechanism and the acquisistion of water by plant roots. Annu Rev Plant Physiol Plant Mol Biol 52:847–875

Steudle E, Frensch J (1996) Water transport in plants: role of the apoplast. Plant Soil 187:67–79

Steudle E, Henzler T (1995) Water channels in plants: do basic concepts of water transport change? J Exp Bot 46:1067–1076

Steudle E, Jeschke WD (1983) Water transport in barley roots. Planta 158:237–248

Steudle E, Peterson CA (1998) How does water get through roots? J Exp Bot 49:775–788

Steudle E, Murrmann M, Peterson CA (1993) Transport of water and solutes across maize roots modified by puncturing the endodermis. Further evidence for the composite transport model of the root. Plant Physiol 103:335–349

Taleisnik E, Peyrano G, Cordoba A, Arias C (1999) Water retention capacity in root segments differing in the degree of exodermis development. Ann Bot 83:19–27

Trifilo P, Gasco A, Raimondo F, Nardini A, Salleo S (2003) Kinetics of recovery of leaf hydraulic conductance and vein functionality from cavitation-induced embolism in sunflower. J Exp Bot 119:4009–4417

Tyerman SD, Bohnert HJ, Maurel C, Steudle E, Smith JAC (1999) Plant aquaporins: their molecular biology, biophysics and significance for plant water relations. J Exp Bot 50:1055–1071

Tyree MT (1997) The cohesion-tension theory of sap ascent. Current controversies. J Exp Bot 48:1753–1765

Tyree MT, Sperry JS (1989) Vulnerability of xylem to cavitation and embolism. Annu Rev Plant Physiol Plant Mol Biol 14:19–38

Tyree MT, Zimmermann MH (2002) Xylem structure and the ascent of sap. Springer, Berlin

Tyree MT, Davis SD, Cochard H (1994) Biophysical perspectives of xylem evolution: is there a tradeoff of hydraulic efficiency for vulnerability dysfunction? IAWA J 15:335–360

Tyree MT, Patino S, Bennink J, Alexander J (1995) Dynamic measurements of root hydraulic conductance using a high-pressure flow meter in the laboratory and field. J Exp Bot 46:83–94
van Ieperen W, van Meeren U, van Gelder H (2000) Fluid ionic composition influences hydraulic conductance of xylem conduits. J Exp Bot 51:769–776
Vieweg GH, Ziegler H (1969) Zur Physiologie von *Myrothamnus flabelliflora*. Ber. Dtsch. Bot. Ges. 82:29–36
Wan X, Steudle E, Hartung W (2004) Gating of water channels (aquaporins) in cortical cells of young corn roots by mechanical stimuli (pressure pulses): effects of ABA and of $HgCl_2$. J Exp Bot 55:411–422
Weatherley PE (1982) Water uptake and flow into roots. In: Lange OL, Nobel PS, Osmond CB, Ziegler H (eds) Encyclopedia of plant physiology, vol 12B. Springer, Berlin, pp 79–109
Wei C, Steudle E, Tyree MT (1999a) Water ascent in plants: do ongoing controversies have a sound basis? Trends Plant Sci 4:372–375
Wei C, Tyree MT, Steudle E (1999b) Direct measurement of xylem pressure in leaves of intact maize plants: a test of cohesion-tension theory taking into account hydraulic architecture. Plant Physiol 121:1191–1205
Westgate ME, Steudle E (1985) Water transport in the midrib tissue of maize leaves. Direct Maesurement of the propagation of changes in cell turgor across a plant tissue. Plant Physiol 78:183–191
Ye Q, Steudle E (2006) Oxidative gating of water channels (aquaporins) in corn roots. Plant Cell Environ 29:459–470
Ye Q, Wiera B, Steudle E (2004) A cohesion/tension mechanism explains the gating of water channels (aquaporins) in *Chara* internodes by high concentration. J Exp Bot 55:449–461
Ye Q, Muhr J, Steudle E (2005) A cohesion/tension model for the gating of aquaporins allows estimation of water channel pore volumes in *Chara*. Plant Cell Environ 28:525–535
Zeidel ML, Ambudakar SV, Bl S, Agre P (1992) Reconstitution of functional water channels in liposomes containing purified red-cell CHIP28 protein. Biochemistry 31:7436–7440
Zeier J, Schreiber L (1998) Comparative investigation of primary and tertiary endodermal cell walls isolated from the roots of five monocotyledoneous species: chemical composition in relation to root fine structure. Planta 206:349–361
Zheng Q, Durben DJ, Wolf GH, Angell CA (1991) Liquids at large negative pressures: water at the homogenous nucleation limit. Science 254:829–832
Zhu GL, Steudle E (1991) Water transport across maize roots: simultaneous measurement of flows at the cell and root level by double pressure probe technique. Plant Physiol 95:305–315
Zimmermann MH (1978) Hydraulic architecture of some diffuse-porous trees. Can J Bot 56:2286–2295
Zimmermann HM, Steudle E (1998) Apoplastic transport across young maize roots: effects of the exodermis. Planta 206:7–19
Zwieniecki MA, Hutyra L, Thompson MV, Holbrook NM (2000) Dynamic changes in petiole specific conductivity in red maple (*Acer rubrum* L.), tulip tree (*Liriodendon tulipifera* L.) and northern fox grape (*Vitis labrusca* L.). Plant Cell Environ 23:407–414
Zwieniecki MA, Melcher PJ, Holbrook NM (2001) Hydrogel control of xylem hydraulic resistance in plants. Science 291:1059–1062

Chapter 11
Drought, Desiccation, and Oxidative Stress

Renate Scheibe and Erwin Beck

11.1 Introduction

Apart from the differentiation between homoiohydrous and poikilohydrous plant constitutional types, a clear demarcation of drought and desiccation is not possible, since dehydration of a plant tissue is a gradual process (Chen et al. 2004). Only the degree to which a plant can tolerate loss of tissue water distinguishes between both functional groups. Most of the studies on dehydration and oxidative stress have used homoiohydrous plants, in particular *Arabidopsis*, tobacco and a few crops, and one resurrection plant (see Chap. 16), and therefore this chapter focuses mainly on plants that can survive drought situations, but cannot completely desiccate. In many publications, the degree of dehydration has even not been mentioned, but nevertheless from the treatment it is clear that these plants have been subjected to experimental drought stress. A special case that differs from the environmentally imposed drought stress is the intrinsic, closely controlled dehydration of ripening seeds that leads to desiccation (Bailly 2004) and merits extra consideration (see also Chap. 14).

Physiological water shortage is not only a result of insufficient water supply to the plant, but can also emerge upon osmotic stress from high salt concentrations or from the freezing of tissue water at subzero temperatures (Verslues et al. 2006). Irrespective of stressor-specific responses, a bunch of similar or even identical reactions to these stresses indicate the common problem of shortage of cellular water. The best-known example of such a general response is the involvement of abscisic acid (ABA) in the transduction of the osmotic stress signal (see also Chap. 15). Another case is the enhancement of the oxidative stress syndrome under drought stress. The production of *r*eactive *o*xygen *s*pecies (ROS) is inevitably connected with a plant's aerobic metabolism. Due to a multitude of scavenging mechanisms, ROS and the oxidative potential of a cell can be maintained at a low level, even under stress. Upon drought-induced closure of the stomata, light absorption by the photosynthetic apparatus produces an excess of reducing power (and energy equivalents) that can be dissipated through reduction of molecular oxygen to the reactive superoxide anion or by a spin conversion of the normal triplet oxygen to the extremely reactive singlet oxygen (Bray 1997; Smirnoff 1993).

U. Lüttge et al. (eds.), *Plant Desiccation Tolerance*, Ecological Studies 215,
DOI 10.1007/978-3-642-19106-0_11, © Springer-Verlag Berlin Heidelberg 2011

Especially organelles with high rates of electron flow, such as chloroplasts, peroxi- and glyoxisomes, and mitochondria are major sites of enhanced reactive oxygen species generation. In addition, other ROS-producing systems, such as the NADPH oxidases or the apoplastic amine and oxalate oxidases and peroxidases, are likewise candidates for the increase of ROS under abiotic, especially drought stress (Mittler 2002).

Due to considerable improvement of methods for ROS detection in vivo (e.g., Fryer et al. 2002) in combination with advanced molecular genetics techniques, and because of the immense importance of ROS in animal and human diseases, active oxygens have attracted enormous attention by biochemists, physicists, medical doctors, and plant biologists. Although the toxicity of ROS is uncontested, their potential of damage depends on the actual species, their concentrations and lifetimes, their ability to diffuse through membranes, and the activity of ROS scavengers in the various compartments of the cell. Mobility of ROS carrying an unpaired electron ($O_2^{\bullet-}$ and the $OH^{\bullet}$ radical) is limited by biomembranes, while the uncharged singlet oxygen and H_2O_2 can permeate through membranes and aquaporins, respectively (Henzler and Steudle 2000). Interconversion of ROS is another fact that merits consideration: Detoxification of the superoxide anion requires two metabolic steps: disproportionation by *s*uper*o*xide *d*ismutases (SOD) to oxygen and hydrogen peroxide, and reduction of H_2O_2 to water by catalase or peroxidase. All cellular compartments maintain ample amounts of SOD and one or more membrane-bound or soluble H_2O_2-decomposing enzymes. Thylakoid membranes contain α-tocopherol as a scavenger of $O_2^{\bullet-}$ and zeaxanthin (perhaps also lutein) for discharging singlet oxygen to its triplet form. Among the dangerous reactions of the superoxide anion is that with H_2O_2 (Haber–Weiss reaction) to yield the most reactive radical $OH^{\bullet}$, a reaction that is catalyzed by Fenton reagents such as Fe^{2+}, Cu^{+}, or Mn^{2+}.

Upon dehydration stress, increase of the steady-state levels of ROS, in particular of the relatively stable H_2O_2, may indeed trigger uncontrolled or *p*rogrammed *c*ell *d*eath (PCD). In particular, the hypersensitive reaction to pathogen attack requires an oxidative burst, which usually results from the activation of an apoplastic or plasma-membrane-bound NADPH oxidase. Membrane-lipid peroxidation, protein oxidation (e.g., formation of disulfides or higher oxidation states of cysteine), enzyme inhibition, and damage of nucleic acids finally cause drying-up and death of the cells (Mitsuhara et al. 1999).

Concomitantly with the drought-induced increase of ROS, the armada of scavengers is often boosted, raising the question of the signals that cause this adaptive reaction. One possible candidate is the general signal of osmotic stress, ABA; another candidate or set of candidates may be ROS itself (Van Breusegem et al. 2008; Jaspers and Kangasjärvi 2010). As discussed in detail in this chapter, ROS indeed act as a signal, but in addition to this appear as modulator of the ABA-signal transduction (Miller et al. 2008). This mode of feedback regulation results in, and requires, a tightly controlled redox status of the cell and its compartments. This fact attracts nowadays much attention (Foyer and Noctor 2005a, b; Cruz de Carvalho 2008; Potters et al. 2010). As a result, the dual action of ROS in plant dehydration stress, namely, as a toxic set of compounds and as signals involved in the regulation of the stress response is now well accepted. ROS as well as the detoxification systems represent complex networks of interacting reactions and metabolites that render disentangling of the individual players and antagonists very difficult (see Chap. 15). In *Arabidopsis*, these interacting networks comprise at least 152 genes whose regulation warrants its flexibility and dynamics (Mittler et al. 2004). With regard to the differentiation of homoiohydrous and poikilohydrous plants, it is worth mentioning that desiccation and revival of resurrection plants require a comprehensive and closely controlled antioxidant system of membrane-bound and soluble enzymes and compounds (Kranner et al. 2002; see also Chap. 16).

11.2 Avoiding ROS Production Under Drought Stress

Avoidance of ROS production by avoiding overreduction of the photosynthetic electron transport chain might be as important as the scavenging of these products (Robinson and Bunce 2000; Mittler 2002; Kranner and Birtić 2005; see also Chaps. 7 and 9). This can be achieved by export of reducing equivalents via the malate valve (Scheibe 2004; Scheibe et al. 2005) in cooperation with the cytosolic 2-oxo-acid dehydrogenases and the mitochondrial electron flow either to complex IV or to the *a*lternative *ox*idase AOX (Atkin and Macharel 2009). Export of excess reduction equivalents from the chloroplast and reoxidation of the NADH in the mitochondria is a powerful way of stress avoidance, and upregulation of the respiratory AOX pathway has been demonstrated under drought (Bartoli et al. 2005; Yoshida et al. 2007). *Arabidopsis* mutants lacking AOX showed acute sensitivity to drought stress (Giraud et al. 2008). In addition to AOX, mitochondria possess two more energy-dissipating systems, an ATP-sensitive potassium channel, and the plant uncoupling protein. In wheat, only the latter two were activated by drought-triggered ROS (Pastore et al. 2007). ROS production, in particular of $O_2^{\bullet -}$, by the mitochondrial electron transport chain is biochemically likely (Møller 2001), but it is a minor component in comparison with chloroplastic reactions (Apel and Hirt 2004). Nevertheless can the expression of ROS-scavenging enzymes, e.g. SOD or *a*scorbate *p*ero*x*idase (APX), be enhanced by ROS, especially by H_2O_2 (Møller 2001).

Another way is the endogenous production of CO_2 in combination with consumption of oxygen in the photorespiratory pathway or, as Tolbert (1994) termed it, the oxidative photosynthetic carbon cycle. This is an ambivalent pathway with respect to ROS, as it helps to avoid overreduction and ROS formation in the chloroplast, but produces stoichiometric amounts of H_2O_2 in the peroxisome, and in addition to this has the potential of further ROS generation in the peroxisome as well as in the mitochondria. Peroxisomal decomposition of H_2O_2 is unique in a twofold respect: (1) It does not require another substrate besides H_2O_2 itself, (2) it is catalyzed by catalase, an enzyme which in plants appears to be confined to the microbodies (Mittler 2002; Mittler et al. 2004). NADH produced upon the oxidation of glycine in the mitochondria can be reoxidized, e.g. by AOX, already at the site of its origin, but then reduction of hydoxypyruvate to glycerate requires reducing equivalents that have to be imported from the chloroplast, e.g., by the malate valve, thus relieving photosynthetic electron pressure. Some upregulation of catalase gene expression upon drought stress has been reported (Mittler et al. 2004), which, however, might also be due to a proliferation of peroxisomes under such conditions (Lopez-Huertas et al. 2000). From gas exchange measurements, Noctor et al. (2002) calculated the relative contributions of photorespiration and the Mehler reaction (see Sect. 11.3) to the oxidative load on a photosynthesizing leaf cell and came up with a contribution of the former of 70% to the total H_2O_2 formed. Downregulation of catalase resulted in considerable damage in high light, in spite of the upregulation of the ascorbate peroxidase and *g*lutathione *p*ero*x*idase (GPX) levels (Willekens et al. 1997). At least one APX isoenzyme is known from the peroxisomes (Mittler 2002).

11.3 Cell Biology and Biochemistry of ROS-Producing and ROS-Detoxifying Systems and Their Relation to Water Deficit

ROS-detoxifying mechanisms are a prerequisite of organisms to survive in an oxygen-containing environment as ROS production is inevitably a component of an aerobic metabolism. A large number not only of original papers but also of reviews have been published addressing generation and detoxification of ROS (e.g., Smirnoff 1993; Buchanan and Balmer 2005; Apel and Hirt 2004; Mittler 2002; Mittler et al. 2004; Zimmermann and Zentgraf 2005; Van Breusegem et al. 2008; Dietz 2005; Foyer and Noctor 2005a, b; Vieira Dos Santos and Rey 2006), and many of them emphasize the connection between ROS and abiotic stress like drought (Smirnoff 1993; Mittler 2006; Chen et al. 2004; Jaspers and Kangasjärvi 2010; Dat et al. 2000; Cruz de Carvalho 2008; Reddy et al. 2004; Miller et al. 2008). In view of this widespread knowledge, we do not expand on the basics of ROS biochemistry but rather address some special aspects of cell biology related to drought and desiccation. ROS-generating and ROS-scavenging reactions are known from almost all membrane-surrounded organelles of a plant cell and the apoplast, except the vacuole (Asada and Takahashi 1987; Mittler 2002; Mittler et al. 2004). Upon drought and stomatal closure, the major cause of oxidative stress is the low internal CO_2 concentration that finally results in the overreduction of the components of the photosynthetic electron transport chain (see also Chap. 7). At several stations of that electron pathway, the elevated electron pressure can result in the formation of ROS, either at PS II [singlet oxygen; perhaps also H_2O_2 (Mubarakshina et al. 2006)] or more commonly by the Mehler reaction at the acceptor side of photosystem I where electrons can jump over from reduced iron–sulfur proteins to molecular oxygen, generating the oxygen radical $O_2^{\bullet -}$ (Asada 1999; Apel and Hirt 2004). The rate at which these reactions take place depends on the extent of overreduction of the electron carriers and on the availability of electron acceptors such as 3-phosphoglycerate in the Calvin cycle, nitrite, sulfite, oxidized thioredoxins, or oxygen.

11.4 ROS, Antioxidative Systems, and Drought

11.4.1 The Oxygen Radical $O_2^{\bullet-}$

In general, plants are equipped with a multiplicity of reactions to avoid overreduction of the photosynthetic electron transport chain and hence one may wonder why it is oxygen which appears as a prioritized Hill oxidant. One reason is the high concentration of oxygen in the photosynthesizing chloroplast (Steiger et al. 1977); another may be the fact that O_2 is uncharged and therefore can easily permeate through biomembranes and react with many kinds of molecules, and last but not least the autoxidability of the iron–sulfur proteins. At saturating CO_2 concentration, the leakage rate to oxygen is very low [3.5 μmol/mg chlorophyll/h, (Steiger and Beck 1981; Hosein and Palmer 1983)], but under shortage of carbon dioxide up to 50% of the entire photosynthetic electron flow may end up as $O_2^{\bullet-}$ (Biehler and Fock 1996). Also, damage of the photosynthetic membranes and addition of methylviologen (Paraquat) as catalyst greatly enhance the production of the oxygen radical anion $O_2^{\bullet-}$ (Iturbe-Ormaetxe et al. 1998). Measurement of the rates of ROS production is difficult because of the powerful battery of antioxidants as well as ROS-scavenging enzymes of chloroplast membranes and stroma that maintain photosynthetic ROS at low levels. Toxicity of $^1O_2$, $O_2^{\bullet-}$, H_2O_2, and $OH^{\bullet}$ depends, apart from their reactivity, on their half-lives and concentrations. Half-life of the extremely toxic species, $OH^{\bullet}$, is ~1 μs (Dat et al. 2000), while that of the other species is longer and determined by the high activity of specific scavenging systems. It is only the hydroxyl radical for which the chloroplast does not possess a specific antagonist. Thus, avoidance of its formation by fast enzymatic decomposition of the other reactant, $O_2^{\bullet-}$, by superoxide dismutase is the only way out of the dilemma. The oxygen radical anion $O_2^{\bullet-}$ spontaneously disproportionates to oxygen and peroxide at a rate of $\sim 10^5\ M^{-1}\ s^{-1}$ at pH 7.0, which is enhanced by the chloroplastic SOD by four orders of magnitude [$2 \times 10^9\ M^{-1}\ s^{-1}$ (Ogawa et al. 1996)]. Investigating the thylakoid-bound CuZn-SOD of spinach, Ogawa et al. (1995) have compared the decomposition potential of the chloroplast SOD with the maximum photosynthetic electron flow and have concluded that each $O_2^{\bullet-}$ molecule is enzymatically disproportionated faster than the next one is formed. Therefore, in the chloroplast, which is the major organelle of $O_2^{\bullet-}$ formation, this radical does not accumulate and hence also the Haber–Weiss reaction is prevented. $O_2^{\bullet-}$ can originate not only in the chloroplast but also in the respiratory electron transport in the mitochondria and from several oxidases (e.g., NADPH oxidases and xanthine oxidase) in the plasma membrane and the peroxisomes, respectively, and by peroxidases in the cell wall. Accordingly, SOD isozymes have also been demonstrated in these organelles and in the cytosol (Mittler 2002). While, e.g., the cytosolic CuZn-SOD was strongly induced by drought, the chloroplastic isoenzyme remained unchanged, indicating the large excess of the latter with respect to $O_2^{\bullet-}$ production by the Mehler reaction (Bowler et al. 1992).

11.4.2 Hydrogen Peroxide (H_2O_2)

The least reactive of a plant's ROS is hydrogen peroxide with a half-life of 1 ms (Levine et al. 1994) and a redox potential of +0.85 at pH 7 (Asada 1999). It originates from direct reduction of oxygen (two-electron transfer to O_2, e.g., in the glycolate oxidase reaction in the peroxisome or by the plasma-membrane-bound NADPH oxidases) and upon disproportionation of $O_2^{\bullet-}$ catalyzed by SOD. Thus, it can arise in chloroplasts, mitochondria, glyoxi- and peroxisomes, in the cytosol, the cell wall, and the apoplast. However, because of its high ability to permeate membranes, occurrence of H_2O_2 in an organelle does not necessarily indicate its formation at the particular site. It is now well established that H_2O_2 serves a dual function in cell biology, as a toxic compound, on the one hand, mediating programmed cell death (Apel and Hirt 2004), e.g., by oxidizing the unsaturated lipids of the biomembranes, and as a signal, on the other hand, triggering gene expression or modulating signals, e.g., from ABA (Miller et al. 2008; see Sect. 11.5). In order to function in this way, its concentration must be closely controlled by a balance between formation and scavenging of H_2O_2. Plants constantly sense the level of ROS and reprogramme their gene expression as to optimally respond to changes in their environment (Miller et al. 2008). Plant cells contain a plethora of H_2O_2-detoxifying compounds and more or less organelle-specific enzyme systems (Asada 1999). The most effective ones of these are the ascorbate peroxidases, represented by compartment-specific membrane-bound and soluble isoenzymes that have been demonstrated in all cellular compartments (Mittler et al. 2004). For regeneration of reduced ascorbate, the APX system is coupled to monodehydro- and dehydroascorbate reductases, which in turn can be connected to further redox systems such as the glutathione-peroxidase cycle or, as in the thylakoid-bound APX, can receive the electrons directly from ferredoxin. As mentioned above, H_2O_2 in the microbodies is decomposed by catalase. Activities of APX and glutathione reductase have been shown to increase under drought stress in many plant species (for a compilation, see Cruz de Carvalho 2008), presumably to counteract the increased formation of H_2O_2. Maintaining hydrogen peroxide at a low level instead of its complete decomposition means that some ROS-mediated damage will occur continuously (Halliwell 2007). Increase of that level results in an increase of the damage or inactivation, e.g., of the thioredoxin-modulated enzymes of the Calvin cycle. A concentration of 10 μM H_2O_2, e.g., inhibits photosynthetic CO_2 fixation (Kaiser 1976, 1979), and in the absence of ascorbate 2 μM H_2O_2 inactivates APX (Miyake and Asada 1996). Although such a case is very unlikely, it suggests a basic level of H_2O_2 in the submicromolar range.

11.4.3 Singlet Oxygen

The only component of ROS confined to the chloroplast is singlet oxygen. It results mainly from quenching of $^3P_{680}$ by molecular oxygen. The triplet P_{680} originates from charge recombination ($P_{680}{}^+Q_A{}^-$) if stabilization of charge separation by electron flow to Q_B is not possible because of the lack of (oxidized) plastoquinone and an empty Q_B-binding site. The lifetime of $^1O_2$ in cells has been estimated between 0.2 μs (Gorman and Rodgers 1992) and 3 μs (Skovsen et al. 2005; Hatz et al. 2007). However, due to its uncharged nature, it could easily diffuse out of the thylakoids (Fischer et al. 2007), and oxidation products were even suggested to trigger the expression of a glutathione-peroxidase gene in *Chlamydomonas* (Fischer et al. 2006, 2007). However, the majority of the $^1O_2$ molecules are trapped already in the thylakoids and their energy is dissipated through reaction with zeaxanthin and α-tocopherol (Halliwell and Gutteridge 1999). In addition to the membrane-bound scavengers, *w*ater-*s*oluble *c*hlorophyll-binding *p*roteins (WSCPs) have been described, which can also quench the chlorophyll-triggered $^1O_2$ production (Schmidt et al. 2003) and whose formation is induced by abiotic stress such as drought (Downing et al. 1992) and heat (Annamalai and Yanagihara 1999).

11.4.4 The Cellular Thiol/Disulfide Redox State as a Regulator of a Cell's Response to Oxidative Stress and Drought

A correlation between water deficit and an altered thiol/disulfide redox state of the protein and nonprotein thiols has been demonstrated for wheat leaves (Zagdańska and Wiśniewski 1996). These leaves reached a drought-acclimated state by increasing glutathione content and glutathione reductase activity. By analyzing stomatal closure activity in GSH-deficient mutants, the involvement of thiol/disulfide changes downstream of ROS production in the ABA-signaling cascade was demonstrated (Jahan et al. 2008). Redox proteomics also revealed the link between drought and oxidative stress (Hajheidari et al. 2007). Thioredoxins were shown to be responsible for stress protection in cells of all organisms, where the redox state of transcription factors is linked to apoptosis (Tanaka et al. 2000). In potato plants subjected to water deficit, a new stromal protein of 32 kDa (*c*hloroplastic *d*rought-induced *s*tress *p*rotein: CDSP32) was identified that carried a thioredoxin active-site motif and was found to accumulate under cold, drought, and salt stress (Rey et al. 1998; Broin et al. 2000). Analysis of the targets of CDSP32 showed that this stress-induced member of the thioredoxin family interacts with stromal proteins and components of antioxidant systems such as peroxiredoxins (Rey et al. 2005). In general, plants possess multiple members of the various "redoxin" families (Buchanan and Balmer 2005). Glutathione peroxidases that belong to the peroxiredoxin family (Rodriguez Milla et al. 2003; Navrot et al. 2006) are expressed upon stress and were shown to be involved in many different signaling pathways and also in ABA signaling upon water stress. In order to maintain a balanced H_2O_2 concentration, scavengers such as glutathione peroxidase (GPX3) are required, which contribute to the ABA-signal transduction chain by oxidative inactivation of a protein phosphatase C2 (ABI2, "*a*bscisic *a*cid *i*nsensitive") under drought (Miao et al. 2006). A 1-Cys peroxiredoxin was discovered in the resurrection plant *Xerophyta viscosa* as a stress-induced antioxidant enzyme (Mowla et al. 2002). The large cyclophilin family in plants might also be involved in their dynamic responses under changing environmental conditions (Romano et al. 2004). During the last decades, it became apparent that members of the thioredoxin superfamily are present in all compartments and in all organisms, contributing to the redox regulation upon oxidative stress (Tanaka et al. 2000; Buchanan and Balmer 2005; Ahsan et al. 2009; Lukosz et al. 2010). The redox signal is translated into a changed gene expression by redox-responsive components of the signal transduction chain and DNA-binding factors.

In general, the synthesis of compatible solutes is increased upon osmotic stress: The amino acid proline is the classical osmolyte whose levels frequently increase upon abiotic stress. However, there is some evidence that this amino acid not only is a protectant for membrane and protein surfaces under water stress, but also acts as redox buffer (Bellinger and Larher 1987; Szabados and Savouré 2010; Yang et al. 2009). For *Myrothamnus flabellifolia*, a higher plant that tolerates complete desiccation, a polyphenol was shown to act as protectant of membranes in the presence of free radicals (Moore et al. 2005). Furthermore, citrulline (Akashi et al. 2001) and metallothionein (Akashi et al. 2004) have been discussed as $OH^{\bullet}$ scavengers. In general, many compatible solutes are also considered as potential ROS scavengers (Smirnoff and Cumbes 1989; Couée et al. 2006).

11.5 Involvement of ROS in Dehydration-Signal Transduction

11.5.1 Interactions of ROS and ABA

From many studies using gene- knockout mutants, it becomes evident that there is a complex regulatory network to control stress-related gene expression (Nakashima and Yamaguchi-Shinozaki 2006). Upon water stress and desiccation, ABA-dependent and ABA-independent signaling pathways have been identified, which are further modified by ROS levels and also by the sugar status of the cells (Couée et al. 2006; Yu et al. 2003; see also Chap. 15).

Drought and desiccation cause enhanced ROS production, in particular in high light. Plants can cope with this situation, if sufficient antioxidant systems are present or are induced during an adaptation phase (Jiang and Zhang 2002b; Khanna-Chopra and Selote 2007). Induction of ROS-scavenging enzymes upon osmotic stress has been demonstrated (Morabito and Guerrier 2000; Munné-Bosch and Peñuelas 2004). The content of antioxidants is particularly important during rewatering after drought (Baroli et al. 1999). If ROS increases above a certain threshold, it can serve as a signal to induce or enhance acclimation to water stress. The level at which ROS becomes toxic, causing damage and finally cell death, exceeds that of ROS acting as signal. The role of ROS in signal transduction upon stress is highly interlinked with ROS scavenging, because nondeleterious levels are required for signal transduction from chloroplasts or mitochondria to the nucleus for the induction of antioxidant enzymes (Halliwell 2006; Fujita et al. 2006; Van Breusegem et al. 2008; Miller et al. 2008; Jaspers and Kangasjärvi 2010).

The common second messenger in all oxidative stress responses is H_2O_2, because it is produced and consumed at high rates allowing for a fast attainment of a steady-state level, it is sufficiently stable, and it can diffuse between compartments. In signaling, its action might be accomplished by oxidatively modifying cysteine residues in proteins (Xiong and Zhu 2002), thus changing their conformation and activity as DNA-binding proteins or channels (Pei et al. 2000).

Drought-induced signal generation and transduction represent a complex network of interactions between the downstream signal transduction components of ABA and ROS. Impact of ABA on the level of ROS – by inducing elevated levels of ROS-detoxifying systems, e.g., cytosolic APX (Fryer et al. 2003; Yoshimura et al. 2000) – is as well known, since ROS effect ABA biosynthesis (Zhao et al. 2001) and signaling (Jiang and Zhang 2002a).

11.5.2 Involvement of ROS in Drought Sensing and Signal Transduction

Drought and other abiotic and biotic impacts on the plants, all lead to deviations from redox homeostasis maintained by a balance between ROS/RNS ("*r*eactive *n*itrogen *s*pecies")-generating reactions and the adaptation of the antioxidative systems. Any imbalance will also serve as a signal inducing the various stress responses leading to tolerance or even resistance (Fig. 11.1). In the following, examples of the participation of ROS/RNS in the ABA-dependent and ABA-independent signal transduction networks that are triggered by drought are presented.

Osmotic changes occurring upon water loss during dehydration are sensed by a transmembrane hybrid-type histidine kinase (*Arabidosis thaliana* HK1) transmitting the signal to a downstream MAPK cascade (Urao et al. 1999). ROS may play a major role in the regulation of MAPK cascades downstream of the primary signal transmitter ABA, but also of a signal transduction pathway that acts ABA-independently (Yamaguchi-Shinozaki and Shinozaki 2004; Saibo et al. 2009; Jaspers and Kangasjärvi 2010).

In addition, some MAP-kinase kinases have been shown to be stabilized by H_2O_2 (Dóczi et al. 2007). One interesting target gene of the H_2O_2-activated *Arabidopsis* MAP kinases MPK3 and MPK6 is the *n*ucleoside-*d*iphosphate *k*inase 2 (NDPK2), a housekeeping enzyme that maintains the intracellular levels of all desoxy-nucleoside-triphosphates except ATP. In addition, there is evidence that it is also involved in the H_2O_2-mediated MAPK phosphorylation (Moon et al. 2003).

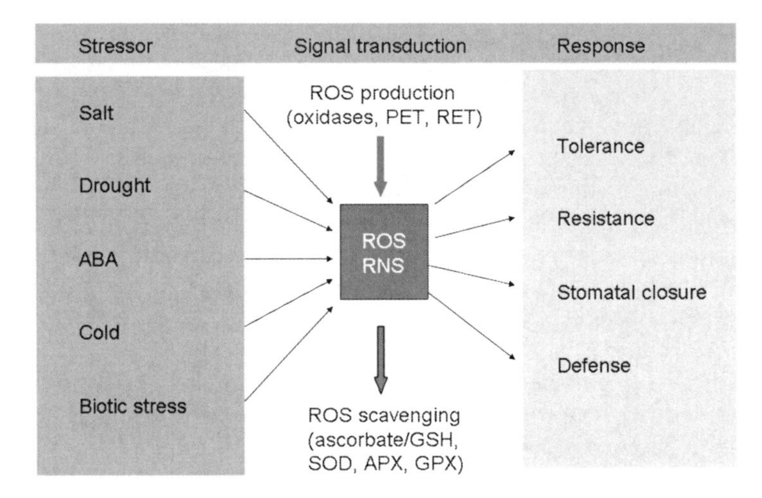

Fig. 11.1 Temporal and spatial deviations from ROS/RNS homeostasis act as signals for the induction of adaptation to the changed environment. *PET, RET* photosynthetic and respiratory electron transfer chains, respectively. *ROS, RNS* reactive oxygen and reactive nitrogen species, respectively

Several class *A* *h*eat *s*hock *f*actors (HsfA2, A4a, and others), which are responsive to oxidative stress and are capable of gene activation, might act as H_2O_2 sensors (Miller and Mittler 2006). APX1-knockout mutants that maintain higher levels of ROS have substantially elevated levels of the transcripts of these Hsf genes (Davletova et al. 2005; Miller and Mittler 2006), while the abundance of HsfA5 transcript (an inhibitor of the activity of HsfA2 and A4a) remains low under abiotic stress conditions (Miller and Mittler 2006). Downstream transcription factors that could integrate ROS signals belong to the plant-specific Cys2/His2-type zinc finger protein Zat family and the WRKY-domain transcription factor family (Rodriguez Milla et al. 2006), or are coactivators like the *m*ultiprotein *b*ridging *f*actor MBF1c (Miller et al. 2008). Target genes of these signal transduction pathways are among many others APX, FeSOD, and AOX genes. Other types of transcription factors are the DREB proteins (*d*rought-*r*esponsive *e*lement *b*inding proteins; see also Chap. 15), which contain a redox-responsive AP2 domain (Stockinger et al. 1997; Nakashima et al. 2000; Gadjev et al. 2006; Sakuma et al. 2006). Members of the DREB and related families can bind to the *cis*-elements such as ABRE (*ab*scisic acid *r*esponse *e*lement), DRE/CRT (*d*ehydration-*r*esponsive *e*lement/*C*-*r*epeat element), MYBR, MYCR, and NACR (MYB-, MYC-, and NAC-*r*ecognition sites) in promoters of the stress-responsive genes (Guan et al. 2000; Nakashima and Yamaguchi-Shinozaki 2006; Saibo et al. 2009). Interestingly, a gene mutation in a subunit of the polyadenylation specificity factor (*c*leavage and *p*olyadenylation *s*pecificity *f*actor, CPSF30) resulted in a stress-sensitive phenotype. The mutant was characterized by an increased expression of a number of proteins containing thioredoxin and glutaredoxin domains (Zhang et al. 2008), pointing to their role in signaling and/or stress protection.

There is crosstalk between cold, salt, and osmotic stress caused by dehydration, and by making use of various combinations of *cis*-regulatory elements and transcription factors. Responses may vary depending on the developmental state. The *cis*-elements ABRE and DRE were shown to interact in response to dehydration and high salinity stress (Narusaka et al. 2003). In a differential screening approach, various genes have been identified that are induced upon osmotic stress and show differential patterns of expression (Takahashi et al. 2000). As one of the responsive gene codes for a Ca^{2+}-binding protein, Ca^{2+} as second messenger is likely to be involved in both ABA-dependent and ABA-independent signaling. The interconnections between Ca^{2+} homeostasis and ROS have been reviewed as part of a crosstolerance phenomenon (Bowler and Fluhr 2000). Crossprotection and cross-talk between different signal transduction pathways, even with biotic stress signaling, can be assumed as ROS are involved in various kinds of osmotic stress as well as in the *h*ypersensitive *r*esponse (HR) upon biotic stress (Chinnusamy et al. 2004; Fujita et al. 2006).

11.5.3 NO as a Component of the ROS-Signaling Network

The NO radical and peroxynitrite belong to the *r*eactive *n*itrogen *s*pecies (RNS) found to be involved in many cellular processes, both as toxic compounds formed under oxidative stress conditions, and in signaling. In parallel to H_2O_2, NO generation by nitrate reductase seems to contribute to ABA-induced stomatal closure (Desikan et al. 2004; Neill et al. 2008), although the exact sequence of molecular events has still to be unraveled. Both, ROS and NO have been suggested as the primary signals that sense water deficiency in root tips and trigger increased ABA synthesis (Zhao et al. 2001). Suppressing the generation of ROS and NO in wheat root tips prior to water stress prevents ABA accumulation and signal transduction. NO production under osmotic stress involves nitrate reductase (Rockel et al. 2002), however, possibly only in a later stage of stress response when lateral roots are initiated, and not as a signal to reduce primary root length (Kolbert et al. 2010). The role of NO in signaling might also be in posttranslational modification of cysteine and tyrosine residues (Wilson et al. 2008; Moreau et al. 2010). Both, ROS and NO production and removal, occurring in plant mitochondria, are tightly linked and controlled by multiple redundant systems, resulting in a fine-tuning of redox homeostasis and signaling (Blokhina and Fagerstedt 2010).

11.6 ROS, ABA, and the Regulation of the Stomates

Stomata closure and inhibition of stomata opening is controlled by ABA and the cytosolic calcium concentration. ABA signaling in guard cells has been extensively studied and is one of the best understood signaling networks in plants. For a comprehensive overview, the reader is referred to "The clickable Guard Cell, Version II" by Kwak et al. (2010). This chapter focuses only on the contribution of ROS to stomata closure and maintenance of that state (Fig. 11.2). Hydrogen

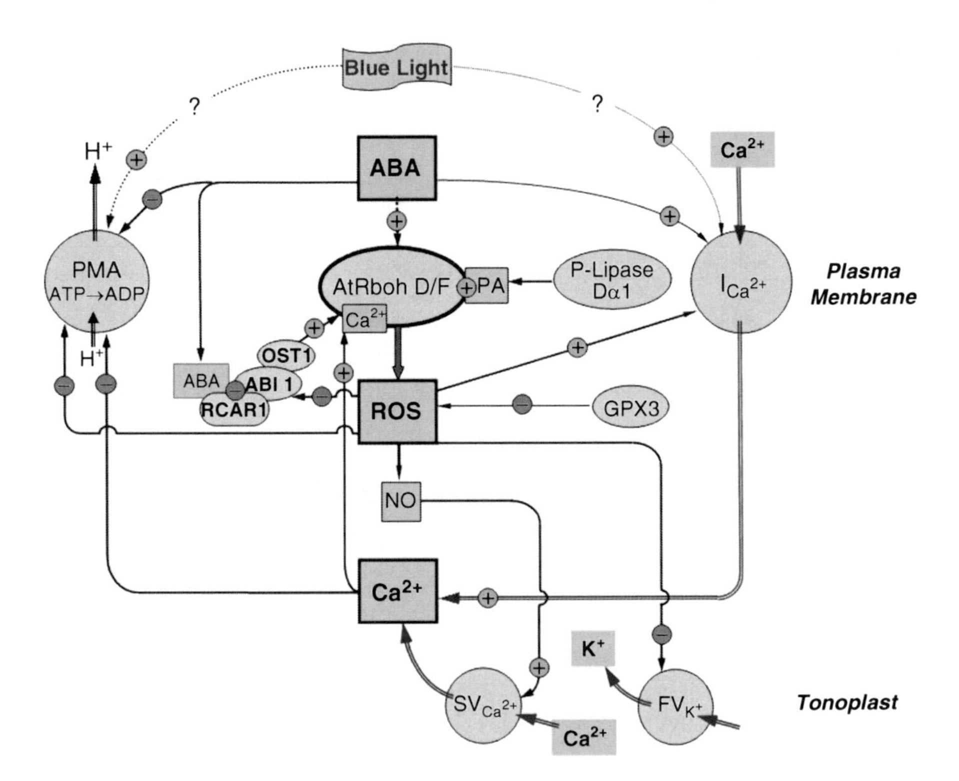

Fig. 11.2 ROS as an element of the network mediating ABA signaling of stomata closure. Only those interactions are shown where at least one functional target element has been identified. There are more proteins that are directly or indirectly affected by ROS, but the final outcomes of these interactions have still to be unraveled. In the clickable Guard Cell Version II, Kwak et al. (2010) presented 64 elements of guard cell signal transduction for stomata closure, and there are several more in the literature. The scheme shown here is based on the publication by Kwak et al. (2010), thematically focused on ROS, including results from Yoshida et al. (2006), Ma et al. (2009), Park et al. (2009), and Zhang et al. (2009). Note that RCAR1 ("Regulatory Components of ABA Receptor"), which is synonymous with PYR1 ("Pyrabactin Resistance 1"), is a recently detected ABA receptor protein (Ma et al. 2009 and Park et al. 2009, respectively). Abbreviations: AtRbohD/F plasma membrane-bound oxidases that are activated by Ca^{2+}, phosphatidic acid, and by phosphorylation. OST1 ("open stomates 1") is a protein kinase that phosphorylates membrane-bound oxidases but also autophosphorylates. ABI1 is a protein phosphatase ("ABA-insensitive") that can bind to OST1 and must be inhibited for a positive ABA signal. It is inactivated by ROS but also by ABA bound to RCAR1. RCAR also binds to ABI2 protein (not shown), which is likewise inactivated by ROS; however, the final target element of this phosphatase is not yet identified. PMA is a plasma membrane H^+-ATPase, which is potentially activated by blue light, but inhibited by ABA, Ca^{2+}, and ROS. $I_{Ca^{2+}}$ is a calcium influx channel that is activated by ABA and by ROS, resulting in an elevated intracellular calcium level. In the same direction operates NO, generated by nitrate reductase that activates the *s*low *v*acuolar calcium channel ($SV_{Ca^{2+}}$). The level of ROS is controlled by glutathione peroxidase (GPX3)

peroxide and ABA were shown to activate hyperpolarization-activated Ca^{2+} channels (I_{Ca}) in the guard cell plasma membrane, thus mediating the ABA signal by increasing the cytosolic free Ca^{2+} concentration (Pei et al. 2000), which finally leads to stomata closure. Similarly, NO enhances cytosolic calcium concentrations by activating slow tonoplast calcium channels (Garcia-Mata et al. 2003). ROS generation, on the other hand, appears to be initiated by plasma membrane NAD (P)H oxidases (Rboh: *r*espiratory *b*urst *o*xidase *h*omologues), which are induced upon ABA exposure (Murata et al. 2001) and activated by binding Ca^{2+} (Ogasawara et al. 2008) and phosphatidic acid (Zhang et al. 2009) at the N-terminus. Further activation appears to be by phosphorylation, e.g., by the SnRK2-type[1] protein kinase OST1 (*o*pen *st*omates 1; Kobayashi et al. 2007; Nuhse et al. 2007). This protein kinase is inactivated by PP2C *p*rotein *p*hosphatases, e.g., ABI1 or HAB1 (*h*omology to *AB*I1), which act as negative regulators of ABA signaling (Santiago et al. 2009). A recent study indicates that ABI1 may bind OST1 protein, suggesting that it could directly modulate its phosphorylation status (Yoshida et al. 2006) and thus negatively regulate ABA signal transduction upstream of OST1. ABI1 is inactivated by ROS and by proteins of a family termed START (PYR/PYLs; Park et al. 2009) or RCARs (*r*egulatory *c*omponents of *A*BA *r*eceptor; Ma et al. 2009). These proteins have been identified as ABA sensors that can bind to the PP2C protein phosphatases. By increasing Ca^{2+}-influx, ROS activates the plasma membrane oxidases (Jiang and Zhang 2002b), thus boosting an autocatalytic cycle. On the other hand, the level of H_2O_2 is controlled by glutathione peroxidase(s). The role of the NADPH oxidase genes *AtRbohD* and *F* for ABA-induced ROS generation and stomata closure as well as for other ABA-dependent processes has been demonstrated with double mutants lacking both genes (Kwak et al. 2003). ROS also inhibits blue-light-induced proton pumping by the plasma membrane-bound H^+-ATPase, thus supporting the effects of ABA, triggering the dephosphorylation of blue-light-induced phosphorylation of that enzyme (Zhang et al. 2004). It also prevents a rise of the cytosolic K^+ level by inhibiting the vacuolar fast potassium channel. ABA signaling in guard cells has also recently been shown to function by phosphorylation/dephosphorylation of the *sl*ow *a*nion *c*hannel SLAC1 (Geiger et al. 2009). The protein kinase/phosphatase pair interacting with the channel was identified as the above-mentioned protein kinase OST1 and the protein phosphatase PP2CA (Geiger et al. 2009; Lee et al. 2009). Another 2C-type protein phosphatase ABI2 was described as redox-regulated target of the H_2O_2 signal, generated as a consequence of ABA-mediated processes such as stomata closure or seed dormancy (Meinhard et al. 2002).

[1]*S*ucrose *n*onfermenting-*r*elated protein *k*inase 2.

11.7 Dehydration of Seeds: A Special Case

Desiccation of ripening seeds, germination, and aging may lead to oxidative stress (Bailly 2004). Orthodox seeds are able to cope with this stress; recalcitrant seeds can tolerate dehydration only to that extent as repair systems are present or induced, similar to the situation of water stress in vegetative parts of a plant (Pammenter and Berjak 1999). The importance of ascorbate and glutathione metabolism during desiccation of recalcitrant *Acer saccharinum* seeds has been demonstrated (Pukacka and Ratajczak 2006). With respect to glutathione, the significance of the formation of mixed disulfides with proteins was shown in dormant seeds (Kranner and Grill 1996). Thioredoxins are of special importance during germination when disulfide bridges (as protection from oxidative stress in the dehydrated state) need to be reduced to allow for renaturation and activation of the proteins (Buchanan and Balmer 2005).

11.8 Improvement of Stress Tolerance by GeneTransfer: The Role of ROS

In order to better understand the patterns of responses to water stress and desiccation, various "omics" approaches and genetic screens have been undertaken (Tabaei-Aghdaei et al. 2000; Eckardt 2001; Watkinson et al. 2003; Oono et al. 2003; Rabbani et al. 2003; Hazen et al. 2003; Shinozaki et al. 2003; Zheng et al. 2004; Takahashi et al. 2004; Seki et al. 2004; Devi et al. 2006; Hajheidari et al. 2007). Since many drought-induced genes in wheat are thioredoxin targets, it has been even suggested to use them as genetic markers for breeding of more resistant lines (Hajheidari et al. 2007). This study underlines again the importance of redox-dependent processes in achieving drought tolerance. Also, during the rehydration process, expression profiling was performed to detect new genes that are induced in this critical phase (Oono et al. 2003). These studies are aimed at an understanding of the complex mechanisms of stress acclimation and tolerance. The final goal is to identify a set of genes or even a master gene that would transfer tolerance to water stress and related abiotic stress factors in genetically modified crops (Holmberg and Bülow 1998; Kasuga et al. 1999; Moore et al. 2009). Recently, expression of a chloroplast glutathione S-transferase from *Prosopis* in tobacco resulted in an increase of drought tolerance (George et al. 2010). As one of the stress-activated genes, an aldose aldehyde reductase had the potential to confer tolerance to oxidative stress and dehydration by reducing the ROS levels (Bartels 2001; see also Chap. 16). However, since ROS have a dual function in plant stress, as toxic products and as signaling compounds, their levels have to be tightly controlled (Dat et al. 2000). Keeping the levels below the threshold required to be active to transmit the signal of an increased stress load might even impede an adequate response. Systems-based approaches are used to find the relevant genes for obtaining drought-resistant plants without impact on growth and yield. But it always has to be kept in mind that drought tolerance is a complex trait interlinked with multiple aspects of plant metabolism and development (Moore et al. 2009).

References

Ahsan MK, Lekli I, Ray D, Yodoi J, Das DK (2009) Redox regulation of cell survival by the thioredoxin superfamily: an implication of redox gene therapy in the heart. Antioxid Redox Signal 11:2741–2758

Akashi K, Miyake C, Yokota A (2001) Citrulline, a novel compatible solute in drought-tolerant wild watermelon leaves, is an efficient hydroxyl radical scavenger. FEBS Lett 508:438–442

Akashi K, Nishimura N, Ishida Y, Yokota A (2004) Potent hydroxyl radical-scavenging activity of drought-induced type-2 metallothionein in wild watermelon. Biochem Biophys Res Commun 323:72–78

Annamalai P, Yanagihara S (1999) Identification and characterization of a heat stress induced gene in cabbage encodes a Kunitz type protease inhibitor. J Plant Physiol 155:226–233

Apel K, Hirt H (2004) Reactive oxygen species: metabolism, oxidative stress, and signal transduction. Annu Rev Plant Biol 55:373–399

Asada K (1999) The water-water cycle in chloroplasts: scavenging of active oxygens and dissipation of excess photons. Annu Rev Plant Physiol Plant Mol Biol 50:601–639

Asada K, Takahashi M (1987) Production and scavenging of active oxygen in photosynthesis. In: Kyle DJ, Osmond CB, Arntzen CJ (eds) Photoinhibition. Elsevier, Amsterdam, pp 227–287

Atkin OK, Macharel D (2009) The crucial role of plant mitochondria in orchestrating drought tolerance. Ann Bot 103:581–597

Bailly C (2004) Acitve oxygen species and antioxidants in seed biology. Seed Sci Res 14:93–107

Baroli CG, Simontacci M, Tambussi E, Beltrano J, Montaldi E, Puntarulo S (1999) Drought and watering-dependent oxidative stress: effect on antioxidant content in *Triticum aestivum* L. leaves. J Exp Bot 50:375–383

Bartels D (2001) Targeting detoxification pathways: an efficient approach to obtain plants with multiple stress tolerance? Trends Plant Sci 6:284–286

Bartoli CG, Gomez F, Gergoff G, Guiamét JJ, Puntarulo S (2005) Up-regulation of the mitochondrial alternative oxidase pathway enhances photosynthetic electron transport under drought conditions. J Exp Bot 56:1269–1276

Bellinger Y, Larher F (1987) Proline accumulation in higher plants: a redox buffer? Plant Physiol 6:23–27

Biehler K, Fock H (1996) Evidence for the contribution of the Mehler-peroxidase reaction in dissipating excess electrons in drought-stressed wheat. Plant Physiol 112:265–272

Blokhina O, Fagerstedt KV (2010) Reactive oxygen species and nitric oxide in plant mitochondria: origin and redundant regulatory systems. Physiol Plant 138:447–462. doi:10.1111/j.1399-3054.2009.01340.x

Bowler C, Fluhr R (2000) The role of calcium and activated oxygens as signals for controlling cross-tolerance. Trends Plant Sci 5:241–246

Bowler C, van Montagu M, Inzé D (1992) Superoxide dismutase and stress tolerance. Annu Rev Plant Physiol Plant Mol Biol 43:83–116

Bray EA (1997) Plant responses to water deficit. Trends Plant Sci 2:48–54

Broin M, Cuiné S, Peltier G, Rey P (2000) Involvement of CDSP 32, a drought-induced thioredoxin, in the response to oxidative stress in potato plants. FEBS Lett 467:245–248

Buchanan BB, Balmer Y (2005) Redox regulation: a broadening horizon. Annu Rev Plant Biol 56:187–220

Chen KM, Gong HJ, Chen GC, Wang SM, Zhang CL (2004) Gradual drought under field conditions influences the glutathione metabolism, redox balance and energy supply in spring wheat. J Plant Growth Regul 23:20–28
Chinnusamy V, Schumaker K, Zhu J-K (2004) Molecular genetic perspectives on cross-talk and specificity in abiotic stress signalling in plants. J Exp Bot 55:225–236
Couée I, Sulmon C, Gouesbet G, El Amrani A (2006) Involvement of soluble sugars in reactive oxygen species balance and response to oxidative stress in plants. J Exp Bot 57:449–459
Cruz de Carvalho MH (2008) Drought stress and reactive oxygen species. Plant Signal Behav 3:156–165
Davletova S, Rizhsky L, Liang H, Shengqiang Z, Oliver DJ, Coutu J, Shulaev V, Schlauch K, Mittler R (2005) Cytosolic ascorbate peroxidase 1 is a central component of the reactive oxygen network of *Arabidopsis*. Plant Cell 17:268–281
Dat J, Vandenabeele S, Vranová E, van Montagu M, Inzé D, van Breusegem F (2000) Dual action of the active oxygen species during plant stress responses. Cell Mol Life Sci 57:779–795
Desikan R, Cheung M-K, Bright J, Henson D, Hancock JT, Neill SJ (2004) ABA, hydrogen peroxide and nitric oxide signalling in stomatal guard cells. J Exp Bot 55:205–212
Devi SR, Chen X, Oliver DJ, Xiang C (2006) A novel high-throughput genetic screen for stress-responsive mutants of *Arabidopsis thaliana* reveals new loci involving stress responses. Plant J 47:652–663
Dietz K-J (2005) Plant thiol enzymes and thiol homeostasis in relation to thiol-dependent redox regulation and oxidative stress. In: Smirnoff N (ed) Antioxidants and reactive oxygen species in plants. Blackwell, NJ, USA, pp 25–52
Dóczi R, Brader G, Pettko-Szandtner A, Rajh I, Djamei A, Pitzschke A, Teige M, Hirt H (2007) The *Arabidopsis* mitogen-activated protein kinase kinase MKK3 is upstream of group C mitogen-activated protein kinases and participates in pathogen signaling. Plant Cell 19:3266–3279
Downing WL, Mauxion F, Fauvarque MO, Reviron MP, de Vienne D, Vartanian N, Giraudat J (1992) A *Brassica napus* transcript encoding a protein related to the Kunitz protease inhibitor family accumulates upon water stress in leaves, not in seeds. Plant J 2:685–693
Eckardt N (2001) *Luc* genetic screen illuminates stress-responsive gene regulation. Plant Cell 13:1969–1972
Fischer BB, Eggen RIL, Trebst A, Krieger-Liszkay A (2006) The glutathione peroxidase homologous gene Gpxh in *Chlamydomonas reinhardtii* is upregulated by singlet oxygen produced in photosystem II. Planta 223:583–590
Fischer BB, Krieger-Liszkay A, Hideg E, Snyrychová I, Wiesendanger M, Eggen RIL (2007) Role of singlet oxygen in chloroplast to nucleus retrograde signaling in *Chlamydomonas reinhardtii*. FEBS Lett 581:5555–5560
Foyer CH, Noctor G (2005a) Oxidant and antioxidant signaling in plants: a re-evaluation of the concept of oxidative stress in physiological context. Plant Cell Environ 28:1056–1071
Foyer CH, Noctor G (2005b) Redox homeostasis and antioxidant signaling: a metabolic interface between stress perception and physiological responses. Plant Cell 17:1866–1875
Fryer MJ, Oxborough K, Mullineaux PM, Baker NR (2002) Imaging of photo-oxidative stress responses in leaves. J Exp Bot 53:1249–1254
Fryer MJ, Ball L, Oxborough K, Karpinski S, Mullineaux PM, Baker NR (2003) Control of *Ascorbate Peroxidase 2* expression by hydrogen peroxide and leaf water status during excess light stress reveals a functional organization of *Arabidopsis* leaves. Plant J 33:691–705
Fujita M, Fujita Y, Noutoshi Y, Takahashi F, Naruasaka Y, Yamaguchi-Shinozaki K, Shinozaki K (2006) Crosstalk between abiotic and biotic stress responses: a current view from the points of convergence in the stress signaling networks. Curr Opin Plant Biol 9:436–442
Gadjev I, Vanderauwera S, Gechev TS, Laloi C, Minkov IM, Shulaev V, Apel K, Inzé D, Mittler R, Van Breusegem F (2006) Transcriptomic footprints disclose specificity of reactive oxygen species signaling in Arabidopsis. Plant Physiol 141:436–445

Garcia-Mata C, Gay R, Sokolovski S, Hills A, Lamattina L, Blatt MR (2003) Nitric oxide regulates K^+ and Cl^- channels in guard cells through a subset of abscisic acid-evoked signaling pathways. Proc Natl Acad Sci USA 100:11116–11121

Geiger D, Scherzer S, Mumm P, Stange A, Marten I, Bauer H, Ache P, Matschi S, Liese A, Al-Rasheid KAS, Romeis T, Hedrich R (2009) Activity of guard cell anion channel SLAC1 is controlled by drought-stress signaling kinase-phosphatase pair. Proc Natl Acad Sci USA 106:21425–21430

George S, Venkataraman G, Parida A (2010) A chloroplast-localized and auxin-induced glutathione S-transferase from phreatophyte *Prosopis juliflora* confers drought tolerance on tobacco. J Plant Physiol 167:311–318

Giraud E, Ho LHM, Clifton R, Carroll A, Estavillo G, Tan Y-F, Howell KA, Ivanova A, Pogson BJ, Millar AH, Whelan J (2008) The absence of ALTERNATIVE OXIDASE1A in Arabidopsis results in acute sensitivity to combined light and drought stress. Plant Physiol 147:595–610

Gorman AA, Rodgers MA (1992) Current perspectives of singlet oxygen detection in biological environments. J Photochem Photobiol 14:159–176

Guan LM, Zhao J, Scandalios JG (2000) *Cis*-elements and *trans*-factors that regulate expression of the maize *Cat1* antioxidant gene in response to ABA and osmotic stress: H_2O_2 is the likely intermediary signaling molecule for the response. Plant J 22:87–95

Hajheidari M, Eivazi A, Buchanan BB, Wong JH, Majidi I, Salekdeh GH (2007) Proteomics uncovers a role for redox in drought tolerance in wheat. J Prot Res 6:1451–1460

Halliwell B (2006) Reactive species and antioxidants. Redox biology is a fundamental theme of aerobic life. Plant Physiol 141:312–322

Halliwell B (2007) Biochemistry of oxidative stress. Biochem Soc Trans 35:1147–1150

Halliwell B, Gutteridge JMC (1999) Free Radicals in Biology and Medicine. Oxford University Press, Oxford

Hatz S, Lambert JDC, Ogilby PR (2007) Measuring the lifetime of singlet oxygen in a single cell: addressing the issue of cell viability. Photochem Photobiol Sci 6:1106–1116

Hazen SP, Wu Y, Kreps JA (2003) Gene expression profiling of plant responses to abiotic stress. Funct Integr Genom 3:105–111

Henzler T, Steudle E (2000) Transport and metabolic degradation of hydrogen peroxide in *Chara corallina*: model calculations and measurements with the pressure probe suggest transport of H_2O_2 across water channels. J Exp Bot 51:2053–2066

Holmberg N, Bülow L (1998) Improving stress tolerance in plants by gene transfer. Trends Plant Sci 3:61–66

Hosein B, Palmer G (1983) The kinetics and mechanism of reaction of reduced ferredoxin by molecular oxygen and its reduced products. Biochem Biophys Acta 723:383–390

Iturbe-Ormaetxe I, Escuredo PR, Arrese-Igor C, Becana M (1998) Oxidative damage in pea plants exposed to water deficit or Paraquat. Plant Physiol 116:173–181

Jahan MS, Ogawa K, Nakamura Y, Shimoishi Y, Mori IC, Murata Y (2008) Deficient glutathione in guard cells facilitates abscisic acid-induced stomatal closure but does not affect light-induced stomatal opening. Biosci Biotechnol Biochem 72:2795–2798

Jaspers P, Kangasjärvi J (2010) Reactive oxygen species in abiotic stress signaling. Physiol Plant 138:405–413. doi:10.111/j1399-3054.2009.01321.x

Jiang M, Zhang J (2002a) Involvement of plasma-membrane NADPH oxidase in abscisic acid- and water stress-induced antioxidant defense in leaves of maize seedlings. Planta 215:1022–1030

Jiang M, Zhang J (2002b) Water stress-induced abscisic acid accumulation triggers the increased generation of reactive oxygen species and up-regulates the activities of antioxidant enzymes in maize leaves. J Exp Bot 53:2401–2410

Kaiser WM (1976) The effect of hydrogen peroxide on CO_2-fixation of isolated chloroplasts. Biochim Biophys Acta 440:476–482

Kaiser WM (1979) Reversible inhibition of the Calvin Cycle and activation of oxidative pentose phosphate cycle in isolated intact chloroplasts by hydrogen peroxide. Planta 145:377–382

Kasuga M, Liu Q, Miura S, Yamaguchi-Shinozaki K, Shinozaki K (1999) Improving plant drought, salt, and freezing tolerance by gene transfer of a single stress-inducible transcription factor. Nat Biotechnol 17:287–291

Khanna-Chopra R, Selote DS (2007) Acclimation to drought stress generates oxidative stress tolerance in drought-resistant than – susceptible wheat cultivar under field conditions. Environ Exp Bot 60:276–283

Kobayashi M, Ohura I, Kawakita K, Yokota N, Fujiwara M, Shimamoto K, Doke N, Yoshida H (2007) Calcium-dependent protein kinases regulate the production of reactive oxygen species by potato NADPH oxidase. Plant Cell 19:1065–1080

Kolbert Z, Ortega L, Erdei L (2010) Involvement of nitrate reductase (NR) in osmotic stress-induced NO generation of *Arabidopsis thaliana* L. roots. J Plant Physiol 167:77–80

Kranner I, Beckett RP, Wornik S, Zorn M, Pfeifhofer HW (2002) Revivval of a resurrection plant correlates with its antioxidant status. Plant J 31:13–24

Kranner I, Birtić S (2005) A modulating role for antioxidants in desiccation tolerance. Integr Comp Biol 45:734–740

Kranner I, Grill D (1996) Significance of thiol-disulfide exchange in resting stages of plant development. Bot Acta 109:8–14

Kwak JM, Mori IC, Pei Z-M, Leonhardt N, Torres MA, Dangl JL, Bloom RE, Bodde S, Jones JDG, Schroeder JI (2003) NADPH oxidase *AtrbohD* genes function in ROS-dependent ABA signaling in *Arabidopsis*. EMBO J 22:2623–2633

Kwak JM, Mäser P, Schroeder JI (2010) The clickable guard cell, Version II: Interactive model of guard cell signal transduction mechanisms and pathways. http://www-biology.ucsd.edu/labs/schroeder/index.html

Lee SC, Lan W, Buchanan BB, Luan S (2009) A protein kinase-phosphatase pair interacts with an ion channel to regulate ABA signaling in plant guard cells. Proc Natl Acad Sci USA 106:21419–21424

Levine A, Tenhaken R, Dixon R, Lamb C (1994) H_2O_2 from the oxidative burst orchestrates the plant hypersensitive disease resistance response. Cell 79:583–593

Lopez-Huertas E, Charlton WL, Johnson B, Graham IA, Baker A (2000) Stress induces peroxisome biogenesis genes. EMBO J 19:6770–6777

Lukosz M, Jakob S, Büchner N, Zschauer T-C, Altschmied J, Haendeler J (2010) Nuclear redox signaling. Antioxid Redox Signal 12:713–742

Ma Y, Szostkiewicz I, Korte A, Moes D, Yang Y, Christmann A, Grill E (2009) Regulators of PP2C phosphatase activity function as abscsic acid sensors. Science 324:1064–1068

Meinhard M, Rodriguez PL, Grill E (2002) The sensitivity of ABI2 to hydrogen peroxide links the abscisic acid-response regulator to redox signalling. Planta 214:775–782

Miao Y, Lv D, Wang P, Wang X-C, Chen J, Miao C, Song C-P (2006) An *Arabidopsis* glutathione peroxidase functions as both a redox transducer and a scavenger in abscisic acid and drought stress responses. Plant Cell 18:2749–2766

Miller G, Mittler R (2006) Could heat shock transcription factors function as hydrogen peroxide sensors in plants? Ann Bot 98:279–288

Miller G, Shulaev V, Mittler R (2008) Reactive oxygen signaling and abiotic stress. Physiol Plant 133:481–489

Mitsuhara I, Malik KA, Miura M, Ohashi Y (1999) Animal cell-death suppressors Bcl-x_L and Ced-9 inhibit cell death in tobacco plants. Curr Biol 9:775–778

Mittler R (2002) Oxidative stress, antioxidants and stress tolerance. Trends Plant Sci 7:405–410

Mittler R, Vanderauwera S, Gollery M, Van Breusegem F (2004) Reactive oxygen gene network of plants. Trends Plant Sci 9:490–498

Mittler R (2006) Abiotic stress, the field environment and stress combination. Trends Plant Sci 11:15–19

Miyake C, Asada K (1996) Inactivation mechanism of ascorbate peroxidase at low concentrations of ascorbate; hydrogen peroxide decomposes Compound I of ascorbate peroxidase. Plant Cell Physiol 37:423–430

Moon H, Lee B, Choi G, Shin D, Prasad DT, Lee O, Kwak S-S, Kim DH, Nam J, Bahk J, Hong JC, Lee SY, Cho MJ, Lim CO, Yun D-J (2003) NDP kinase 2 interacts with two oxidative stress-activated MAPKs to regulate cellular redox state and enhances multiple stress tolerance in transgenic plants. Proc Natl Acad Sci USA 100:358–363

Moore JP, Westall KL, Ravenscroft N, Farrant JM, Lindsey GG, Brandt WF (2005) The predominant polyphenol in the leaves of the resurrection plant *Myrothamnus flabellifolius*, 3, 4, 5 tri-O-galloylquinic acid, protects membranes against desiccation and free radical-induced oxidation. Biochem J 385:301–308

Moore JP, Le NT, Brandt WF, Driouich A, Farrant JM (2009) Towards a systems-based understanding of plant desiccation tolerance. Trends Plant Sci 14:110–117

Morabito D, Guerrier G (2000) The free oxygen radical scavenging enzymes and redox status in roots and leaves of *Populus* x *euramericana* in response to osmotic stress, desiccation and rehydration. J Plant Physiol 157:74–80

Moreau M, Lindermayr C, Durner J, Klessig DF (2010) NO synthesis and signaling in plants – where do we stand? Physiol Plant 138:372–383

Mowla SB, Thomson JA, Farrant JM, Mundree SG (2002) A novel stress-inducible antioxidant enzyme identified from the resurrection plant *Xerophyta viscosa* Baker. Planta 215:716–726

Møller IM (2001) Plant mitochondria and oxidative stress: electron transport, NADPH turnover, and metabolism of reactive oxygen species. Annu Rev Plant Physiol Plant Mol Biol 52:561–591

Mubarakshina M, Khorobrykh S, Ivanov B (2006) Oxygen reduction in chloroplast thylakoids results in production of hydrogen peroxide inside the membrane. Biochim Biophys Acta 1757:1496–1503

Munné-Bosch S, Peñuelas J (2004) Drought-induced oxidative stress in strawberry tree (*Arbutus unedo* L.) growing in Mediterranean field conditions. Plant Sci 166:1105–1110

Murata Y, Pei Z-M, Mori IC, Schroeder J (2001) Abscisic acid activation of plasma membrane Ca^{2+} channels in guard cells requires cytosolic NAD(P)H and is differentially disrupted upstream and downstream of reactive oxygen species production in *abi1-1* and *abi2-1* protein phosphatase 2C mutants. Plant Cell 13:2513–2523

Nakashima K, Shinwari ZK, Sakuma Y, Seki M, Miura S, Shinozaki K, Yamaguchi-Shinozaki K (2000) Organization and expression of two *Arabidopsis DREB2* genes encoding DRE-binding proteins involved in dehydration- and high-salinity-responsive gene expression. Plant Mol Biol 42:657–665

Nakashima K, Yamaguchi-Shinozaki K (2006) Regulons involved in osmotic stress-responsive and cold stress-responsive gene expression in plants. Physiol Plant 126:62–71

Narusaka Y, Nakashima K, Shinwari ZK, Sakuma Y, Furihata T, Abe H, Narusaka M, Shinozaki K, Yamaguchi-Shinozaki K (2003) Interaction between two *cis*-acting elements, ABRE and DRE, in ABA-dependent expression of *Arabidopsis rd29A* gene in response to dehydration and high-salinity stresses. Plant J 34:137–148

Navrot N, Collin V, Gualberto J, Gelhaye E, Hirasawa M, Rey P, Knaff DB, Issakidis E, Jacquot J-P, Rouhier N (2006) Plant glutathione peroxidases are functional peroxiredoxins distributed in several subcellular compartments and regulated during biotic and abiotic stresses. Plant Physiol 142:1364–1379

Neill S, Barros R, Bright J, Desikan R, Hancock J, Harrison J, Morris P, Ribeiro D, Wilson I (2008) Nitric oxide, stomatal closure, and abiotic stress. J Exp Bot 59:165–176

Noctor G, Veljovic-Jovanovic S, Driscoll S, Novitskaya L, Foyer CH (2002) Drought and oxidative load in the leaves of C_3 plants: a predominant role for photorespiration? Ann Bot 89:841–850

Nuhse TS, Bottrill AR, Jones AM, Peck SC (2007) Quantitative phosphoproteomic analysis of plasma membrane proteins reveals regulatory mechanisms of plant innate immune responses. Plant J 51:931–940

Ogasawara Y, Kaya H, Hiraoka G, Yumoto F, Kimura S, Kadota Y, Hishinuma H, Senzaki E, Yamagoe S, Nagata K, Nara M, Suzuki K, Tanokura M, Kuchitsu K (2008) Synergistic activation of the *Arabidopsis* NADPH oxidase AtrbohD by Ca^{2+} and phosphorylation. J Biol Chem 283:8885–8892

Ogawa K, Kanematsu S, Takabe K, Asada K (1995) Attachment of CuZn-superoxide dismutase to thylakoid membranes at the site of superoxide generation (PS I) in spinach chloroplasts: detection by immunogold labeling after rapid freezing and substitution method. Plant Cell Physiol 36:565–573

Ogawa K, Kanematsu S, Asada K (1996) Intra- and extra-cellular localization of "cytosolic" CuZn-superoxide dismutase in spinach leaf and hypocotyls. Plant Cell Physiol 37:790–799

Oono Y, Seki M, Nanjo T, Narusaka M, Fujita M, Satoh R, Satou M, Sakurai T, Ishida J, Akiyama K, Iida K, Maruyama K, Satoh S, Yamaguchi-Shinozaki K, Shinozaki K (2003) Monitoring expression profiles of *Arabidopsis* gene expression during rehydration process after dehydration using *ca.* 7000 full-length cDNA microarray. Plant J 34:868–887

Pammenter NW, Berjak P (1999) A review of recalcitrant seed physiology in relation to desiccation-tolerance mechanisms. Seed Sci Res 9:13–37

Park S-Y, Fung P, Nishimura N, Jensen DR, Fujii H, Zhao Y, Lumba S, Santiago J, Rodrigues A, Chow T-f F, Alfred SE, Bonetta D, Finkelstein R, Provart NJ, Desveaux D, Rodriguez D, McCourt R, Zhu J-K, Schroeder JI, Volkman BF, Cutler SR (2009) Abscisic acid inhibits Type 2C protein phosphatases via the PYR/PYL family of START proteins. Science 324:1068–1071

Pastore D, Trono D, Laus MN, Di Fonzo N, Flagella Z (2007) Possible plant mitochondria involvement in cell adaptation to drought stress. A case study: Durum wheat mitochondria. J Exp Bot 58:195–210

Pei ZM, Murata Y, Benning G, Thomine S, Klusener B, Allen GJ, Grill E, Schroeder JI (2000) Calcium channels activated by hydrogen peroxide mediate abscisic acid signaling in guard cells. Nature 406:731–734

Potters G, Horemans N, Jansen MAK (2010) The cellular redox state in plant stress biology – a charging concept. Plant Physiol Biochem

Pukacka S, Ratajczak E (2006) Antioxidative response of ascorbate-glutathione pathway enzymes and metabolites to desiccation of recalcitrant *Acer saccharinum* seeds. J Plant Physiol 163:1259–1266

Rabbani MA, Maruyama K, Abe H, Khan A, Katsura K, Ito Y, Yoshiwara K, Seki M, Shinozaki K, Yamaguchi-Shinozaki K (2003) Monitoring expression profiles of rice genes under cold, drought, and high-salinity stresses and abscisic acid application using cDNA microarray and RNA gel-blot analyses. Plant Physiol 133:1755–1767

Reddy AR, Chaitanya KV, Vivekanandan M (2004) Drought-induced responses of photosynthesis and antioxidant metabolism in higher plants. J Plant Physiol 161:1189–1202

Rey P, Pruvot G, Besuwe N, Eymery F, Rumeau D, Peltier G (1998) A novel thioredoxin-like protein located in the chloroplast is induced by water deficit in *Solanum tuberosum* L. plants. Plant J 13:97–107

Rey P, Cuiné S, Eymery F, Garin J, Court M, Jacquot J-P, Rouhier N, Broin M (2005) Analysis of the proteins targeted by CDSP32, a plastidic thioredoxin participating in oxidative stress responses. Plant J 41:31–42

Robinson JM, Bunce JA (2000) Influence of drought-induced water stress on soybean an spinach leaf ascorbate-dehydroascorbate level and redox status. Int J Plant Sci 161:271–279

Rockel P, Strube F, Rockel A, Wildt J, Kaiser WM (2002) Regulation of nitric oxide (NO) production by plant nitrate reductase *in vivo* and *in vitro*. J Exp Bot 53:103–110

Rodriguez Milla MA, Maurer A, Rodriguez Huete A, Gustafson JP (2003) Glutathione peroxidase genes in *Arabidopsis* are ubiquitous and regulated by abiotic stresses through diverse signaling pathways. Plant J 36:602–615

Rodriguez Milla MA, Townsend J, Chang I-F, Cushman JC (2006) The Arabidopsis *AtDi19* gene family encodes a novel type of Cys2/His2 zinc-finger protein implicated in ABA-independent dehydration, high-salinity stress and light signaling pathways. Plant Mol Biol 61:13–30

Romano PGN, Horton P, Gray JE (2004) The Arabidopsis cyclophilin gene family. Plant Physiol 134:1268–1282

Saibo NJM, Lourenço T, Oliveira MM (2009) Transcription factors and regulation of photosynthetic and related metabolism under environmental stresses. Ann Bot 103:609–623
Sakuma Y, Maruyama K, Osakabe Y, Qin F, Seki M, Shinozaki K, Yamaguchi-Shinozaki K (2006) Functional analysis of an *Arabidopsis* transcription factor, DREB2A, involved in drought-responsive gene expression. Plant Cell 18:1292–1309
Santiago J, Dupeux F, Round A, Antoni R, Park S-Y, Jamin M, Cutler SR, Rodriguez PL, Márquez JA (2009) The abscisic acid receptor PYR1 in complex with abscisic acid. Nature 462:665–668
Scheibe R (2004) Malate valves to balance cellular energy supply. Physiol Plant 120:21–26
Scheibe R, Backhausen JE, Emmerlich V, Holtgrefe S (2005) Strategies to maintain redox homeostsis during photosynthesis under changing conditions. J Exp Bot 56:1481–1489
Schmidt K, Fufezan C, Kieger-Liszkay A, Satoh H, Paulsen H (2003) Recombinant water-soluble chlorophyll protein from *Brassica oleracea* var *botrys* binds various chlorophyll derivatives. Biochemistry 42:7427–7433
Seki M, Satou M, Sakurai T, Akiyama K, Iida K, Ishida J, Nakajima M, Enju A, Narusaka M, Fujita M, Oono Y, Kamei A, Yamaguchi-Shinozaki K, Shinozaki K (2004) RIKEN *Arabidopsis* full-length (RAFL) cDNA and its applications for expression profiling under abiotic stress condition. J Exp Bot 55:213–223
Shinozaki K, Yamaguchi-Shinozaki K, Seki M (2003) Gene networks involved in drought stress response and tolerance. J Exp Bot 58:221–227
Skovsen E, Snyder JW, Lambert JDC, Ogilby PR (2005) Lifetime and diffusion of singlet oxygen in a cell. J Phys Chem B 109:8570–8573
Smirnoff N (1993) The role of active oxygen in the response of plants to water deficit and desiccation. New Phytol 125:27–58
Smirnoff N, Cumbes Q (1989) Hydroxyl radical scavenging activity of compatible solutes. Phytochemistry 28:1957–1960
Steiger HM, Beck E, Beck R (1977) Oxygen concentration in isolated chloroplasts during photosynthesis. Plant Physiol 60:903–906
Steiger HM, Beck E (1981) Formation of hydrogen peroxide and oxygen dependence of photosynthetic CO_2 assimilation by intact chloroplasts. Plant Cell Physiol 22:561–576
Stockinger EJ, Gilmour SJ, Thomashow MF (1997) *Arabidopsis thaliana CBF1* encodes an AP2 domain-containing transcriptional activator that binds to the C-repeat/DRE, a cis-acting DNA regulatory element that stimulates transcription in response to low temperature and water deficit. Proc Natl Acad Sci USA 94:1035–1040
Szabados L, Savouré A (2010) Proline: a multifunctional amino acid. Trends Plant Sci 15:89–97
Tabaei-Aghdaei SR, Harrison P, Pearce RS (2000) Expression of dehydration-stress-related genes in the crowns of wheatgrass species [*Lophopyrum elongatum* (Host) A. Love and *Agropyron desertorum* (Fisch. ex Link.) Schult.] having constrating acclimation to salt, cold and drought. Plant Cell Environ 23:561–571
Takahashi S, Katagiri T, Yamaguchi-Shinozaki K, Shinozaki K (2000) An *Arabidopsis* gene encoding a Ca^{2+}-binding protein is induced by abscisic acid during dehydration. Plant Cell Physiol 41:898–903
Takahashi S, Seki M, Ishida J, Satou M, Sakurai T, Narusaka M, Kamiya A, Nakajima M, Enju A, Akiyama K, Yamaguchi-Shinozaki K, Shinozaki K (2004) Monitoring the expression profiles of genes induced by hyperosmotic, high salinity, and oxidative stress and abscisic acid treatment in Arabidopsis cell culture using a full-length cDNA microarray. Plant Mol Biol 56:29–55
Tanaka T, Nakamura H, Nishiyama A, Hosoi F, Masutani H, Wada H, Yodoi J (2000) Redox regulation by thioredoxin superfamily; protection against oxidative stress and aging. Free Radical Res 33:851–855
Tolbert NE (1994) Role of photosynthesis and photorespiration in regulating atmospheric CO_2 and O_2. In: Tolbert NE, Preiss J (eds) Regulation of atmospheric CO_2 and O_2 by photosynthetic carbon metabolism. Oxford University Press, New York, Oxford, pp 8–33

Urao T, Yakubov B, Satoh R, Yamaguchi-Shinozaki K, Seki M, Hirayama T, Shinozaki K (1999) A transmembrane hybrid-type histidine kinase in Arabidopsis functions as an osmosensor. Plant Cell 11:1743–1754
Van Breusegem F, Bailey-Serres J, Mittler R (2008) Unraveling the tapestry of networks involving reactive oxygen species in plants. Plant Physiol 147:978–984
Vasquez-Robinet C, Mane SP, Ulanov AV, Watkinson JI, Stromberg VK, De Koeyer D, Schafleitner R, Willmot DB, Bonierbale M, Bohnert HJ, Grene R (2008) Physiological and molecular adaptation to drought in Andean potato genotypes. J Exp Bot 59:2109–2123
Verslues PE, Agarwal M, Katiyar-Agarwal S, Zhu J, Zhu J-K (2006) Methods and concepts in quantifying resistance to drought, salt and freezing, abiotic stresses that affect plant water status. Plant J 45:523–539
Vieira Dos Santos C, Rey P (2006) Plant thioredoxins are key actors in the oxidative stress response. Trends Plant Sci 11:329–334
Watkinson JI, Sioson AA, Vasquez-Robinet C, Shukla M, Kumar D, Ellis M, Heath LS, Ramakrishnan N, Chevone B, Watson LT, van Zyl L, Egertsdotter U, Sederoff RR, Grene R (2003) Photosynthetic acclimation is reflected in specific patterns of gene expression in drought-stressed loblolly pine. Plant Physiol 133:1702–1716
Willekens H, Chamnongpol S, Davey M, Schraudner M, Langebartels C, Van Montagu M, Inzé D, Van Camp W (1997) Catalase is a sink for H_2O_2 and is indispensable for stress defence in C_3 plants. EMBO J 16:4806–4816
Wilson ID, Neill SJ, Hancock JT (2008) Nitric oxide synthesis and signaling in plants. Plant Cell Environ 31:622–631
Xiong L, Zhu JK (2002) Molecular and genetic aspects of plant responses to osmotic stress. Plant Cell Environ 25:131–139
Yamaguchi-Shinozaki K, Shinozaki K (2004) Organization of *cis*-acting regulatory elements in osmotic- and cold-stress-responsive promoters. Trends Plant Sci 10:88–94
Yang S-L, Lan S-S, Gong M (2009) Hydrogen peroxide-induced proline and metabolic pathway of its accumulation in maize seedlings. J Plant Physiol 166:1694–1699
Yoshida R, Umezawa T, Mizoguchi T, Takahashi S, Takahashi F, Shinozaki K (2006) The regulatory domain of SRK2E/OST1/SnRK2.6 interacts with ABI1 and integrates abscisic acid (ABA) and osmotic stress signals controlling stomatal closure in Arabidopsis. J Biol Chem 281:5310–5318
Yoshida K, Terashima I, Noguchi K (2007) Up-regulation of mitochondrial alternative oxidase concomitant with chloroplast over-reduction by excess light. Plant Cell Physiol 48:606–614
Yoshimura K, Yabuta Y, Ishikawa T, Shigeoka S (2000) Expression of spinach ascorbate peroxidase isoenzymes in responses to oxidative stresses. Plant Physiol 123:223–233
Yu C-W, Murphy TM, Lin C-H (2003) Hydrogen peroxide-induced chilling tolerance in mung beans mediated through ABA-independent glutathione accumulation. Funct Plant Biol 30:955–963
Zagdańska B, Wiśniewski K (1996) Change in the thiol/disulfide redox potential in wheat leaves upon water deficit. J Plant Physiol 149:462–465
Zhang X, Wang H, Takemiya A, Song CP, Kinoshita T, Shimazaki K (2004) Inhibition of blue light-dependent H+ pumping by abscisic acid through hydrogen peroxide-induced dephosphorylation of the plasma membrane H+-ATPase in guard cell protoplasts. Plant Physiol 136:4150–4158
Zhang J, Addepalli B, Yun K-Y, Hunt AG, Xu R, Rao S, Li QQ, Falcone DL (2008) A polyadenylation factor subunit implicated in regulating oxidative signaling in *Arabidopsis thaliana*. PLoS ONE 3:e2410
Zhang Y, Zhu H, Zhang Q, Li M, Yan M, Wang R, Wang L, Welti R, Zhang W, Wang X (2009) Phospholipase Dα1 and phosphatidic acid regulate NADPH oxidase activity and production of reactive oxygen species in ABA-mediated stomatal closure in *Arabidopsis*. Plant Cell 21:2357–2377

Zhao Z, Chen G, Zhang C (2001) Interaction between reactive oxygen species and nitric oxide in drought-induced abscisic acid synthesis in root tips of wheat seedlings. Aust J Plant Physiol 28:1055–1061

Zheng J, Zhao J, Tao Y, Wang J, Liu Y, Fu J, Jin Y, Gao P, Zhang J, Bai Y, Wang G (2004) Isolation and analysis of water stress induced genes in maize seedlings by subtractive PCR and cDNA macroarray. Plant Mol Biol 55:807–823

Zimmermann P, Zentgraf U (2005) The correlation between oxidative stress and leaf senescence during plant development. Mol Cell Biol Lett 10:515–534

Chapter 12
Chamaegigas intrepidus DINTER: An Aquatic Poikilohydric Angiosperm that Is Perfectly Adapted to Its Complex and Extreme Environmental Conditions

Hermann Heilmeier and Wolfram Hartung

12.1 Introduction

The majority of poikilohydric vascular plants ("resurrection plants") occur in arid and semi-arid regions of southern and southwestern Africa, southern America and Western Australia (Gaff 1977, 1987). They predominantly colonise shallow rocky soils in their sub-tropical habitats, often on inselbergs (Porembski and Barthlott 2000). Desiccation-tolerant angiosperms are found within both monocotyledons (Cyperaceae, Poaceae and Velloziaceae) and dicotyledons (predominantly the ex-Scrophulariaceae). These poikilohydric cormophytes are exposed to severe drought during long dry seasons (5–10 months per year). Hartung et al. (1998) pointed out that the mechanisms for drought tolerance from the molecular to the physiological and anatomical level are rather costly and with the consequence of a selective disadvantage under most less extreme growing conditions.

Among the Scrophulariaceae *Chamaegigas intrepidus* Dinter, formerly *Linderniaintrepidus* (Dinter) Oberm. (Fig. 12.1), is rather unique. This aquatic resurrection plant occurs in shallow, only temporarily water-filled rock pools on granite outcrops in Namibia (Giess 1969). Hundred years ago this plant was discovered in 1909 by the German botanist Kurt Dinter 12 km east of Okahandja in Central Namibia. He had mentioned this plant already in 1909 (Dinter 1909). He gave the plant the scientific name, which means in German "Unerschrockener Zwergriese" ("undaunted dwarf giant"), since he was very much impressed by the plant surviving the extreme conditions at its natural habitat (Dinter 1918). High temperature and high evaporative demand caused the bottom of the pool in which the plant grew to remain permanently dry for at least half a year. The plants survived as tiny rhizomes (diameter about 1 mm) and shrivelled leaves densely covering the bottom of the pool in a 1 cm thick layer of sand grains, dehydrated algae, dead daphnias, animal faeces and leaf litter. In spite of the high solar irradiation, extreme temperatures and nearly completely dry air, a dense mat of small green *Chamaegigas* leaves covered the bottom of the pools within minutes after the first summer rainfall had filled the small pools. Two days after that Dinter could see pink-coloured flowers in the midst of small rosette leaves floating on the water on top of a thin stem.

U. Lüttge et al. (eds.), *Plant Desiccation Tolerance*, Ecological Studies 215,
DOI 10.1007/978-3-642-19106-0_12, © Springer-Verlag Berlin Heidelberg 2011

Fig. 12.1 Flowering rosette of *Chamaegigas intrepidus* with lanceolate submerged leaves (8–15 mm long; *open arrow*), inserting on short rhizomes, and two pairs of decussate floating leaves (*filled arrow*) on a 10 cm long stem with two leaves. For a more detailed description of the plant, especially the flowers, refer to Fischer (1992) and Woitke et al. (2006)

Chamaegigas is adapted to the fluctuations of wet and dry conditions at its habitat through its ability for fast de- and rehydration. When water from the pools has been lost through evaporation, plants dry within less than 2 h (Gaff and Giess 1986). After rewatering, the vegetative organs regain full metabolic activity within 2 h (Hickel 1967). Rosettes kept for 4 years desiccated could be revived successfully including the formation of flowers (O.H. Volk, personal communication); after 6 years, however, rosettes only became green. Floating leaves and flower buds were not formed (own observation). The extremely fast recovery after rehydration and the long survival in the desiccated state makes *Chamaegigas* by far the most impressive of all poikilohydric angiosperms. In the following, habitat conditions and the plant's mechanisms of adaptation to its complex stressful environmental conditions at the anatomical, biochemical and physiological level are described as well as implications of its isolated geographical distribution for generative reproduction and gene flow.

12.2 Distribution and Habitat

Chamaegigas grows endemically in Namibia (Fischer 1992), exclusively in areas with granite outcrops (inselbergs) in the semi-desert and savanna transition zone (Giess 1969, 1997). The habitats of the species are in the arid to semi-arid region with 160–570 mm precipitation per year, with rainfall on only 20–70 days during summer (November to April), and a high variability from year to year. During the wet season, a few (5–12) rainy days alternate with a number (up to 60) of dry days (Hickel 1967). The shallow rock pools (maximum water level ca. 15 cm) usually dry out completely during dry days. Over the whole wet season, these ephemeral pools may be filled with water for some 40–85 days in total. Thus, the plant may experience 15–20 rehydration–dehydration cycles during a single rainy season (Gaff and Giess 1986). Annual average temperature is 20°C. During the dry season, air temperatures may rise up to 42°C, and sun-exposed rocks heat up to 50°C at least (Dinter 1918). Average air humidity is 40% and at the end of the dry season (September) 22% only (Hickel 1967).

Due to the thin layer of debris at the bottom of the rock pools, the water is very poor in nutrients. Furthermore, after extensive rainfall the shallow pools may overflow, thus leaching mineral nutrients from the sediment (Gaff and Giess 1986). This can cause more severe nutrient deficiencies, especially when there is plenty of water for sustained growth. On the other hand, wild and domestic animals deposit urine and dung on the rock surface, which are subsequently washed into the pools by rain water (Heil 1924). Thus, especially during the early part of the wet season, there may be high amounts of urea and perhaps other dissolved organic nitrogen (DON) compounds present in the water, which are steadily diluted by leaching and plant uptake during the growth period.

Chamaegigas is physiologically active during the warm rainy season, whereas it survives the long (up to 11 months) dry season in the dehydrated state (Hickel 1967). Apart from the intense solar irradiation, high air temperature and low air humidity during the dry season, the plants are exposed to recurrent flooding and drought, low nutrient contents and drastic diurnal fluctuations of pH of the water during the wet season. Thus, this species suffers from a complex set of extremely harsh and interacting environmental conditions.

12.3 Site Description

Most of the data described here originate predominantly from field experiments performed at one of the species' natural growing sites on the farm Otjua (Omaruru District, Namibia, 21°10′S, 16°E). The farm lies in the thornbush savanna at an elevation of about 1,400 m above sea level. Vegetation is dominated by *Acacia* spp. In between the vegetated area, granite outcrops rise to a height of about 20 m with a length of several hundred metres. Due to high temperature fluctuations between day and night, thin sections of granite burst from the rock surface. Subsequent erosion by wind deepens the shallow holes to a maximum depth of about 0.3 m. During the wet season, these small depressions become filled with rain and run-off water from the outcrops and partly from overflow from adjacent pools.

In pools with a water level of more than 15 cm (with the homoiohydric aquatic plant *Limosella grandiflora* and/or *C. intrepidus*), there was a high abundance of tadpoles.

Pools with a high nutrient input from dung were heavily populated by algae of the genus *Spirogyra*. At the foot and in cracks of the outcrops, the terrestrial resurrection plants *Craterostigma plantagineum*[1], *Xerophyta humilis* and *Eragrostis nindensis*, the poikilohydric liverwort *Exormotheca bulbigena* (Marchantiales) and the *aufi Riccia* species *R. angolesis, atropurpurea, crinita, nigrella, okahandiana, rosea* and *runssoriensis* occurred (Bornefeld and Volk 2002).

12.4 Environmental Stress Conditions

12.4.1 Air Temperature and Humidity at the Rock Surface

When pools are not filled with water, the dry *Chamaegigas* plants are exposed to temperatures up to 60°C, which is approximately 20–30°C higher than water temperature when the plants are physiologically active. Diurnal amplitudes of air temperature between day and night may exceed 30°C, due to large losses of energy during cloudless nights.

During days without rainfall, minimum air humidity decreases to 5–10%. Maximum air humidity at night (ca. 40%) does not indicate dewfall. As a consequence of the high evaporative demand of the air, actual evaporation exceeds 10 mm day^{-1}. On more cloudy days, average minimum humidity is 10–20%, with 70 to nearly 100% at night.

12.4.2 Water Level and Conductivity

Water level in water-filled pools with *Chamaegigas* ranged from 4 to 13 cm (median 7.5 cm). In pools drying out loss of water caused the dissolved compounds to be concentrated as measured by the electrical conductivity. A decrease in water level by a factor of two increased conductivity on average up to twofold (Fig. 12.2a) and in some cases up to fourfold. After refilling by rain, conductivity decreased to rather low values (<50 μS cm^{-1}). This indicates that towards the end of the wet season, only minute amounts of nutrients are washed into the pools from the surrounding rock surface.

[1]Plants and cultures of *C. plantagineum* used in European laboratories for ecophysiological and molecular studies originate from a plant brought from this site to Germany by Prof. Dr. O.H. Volk.

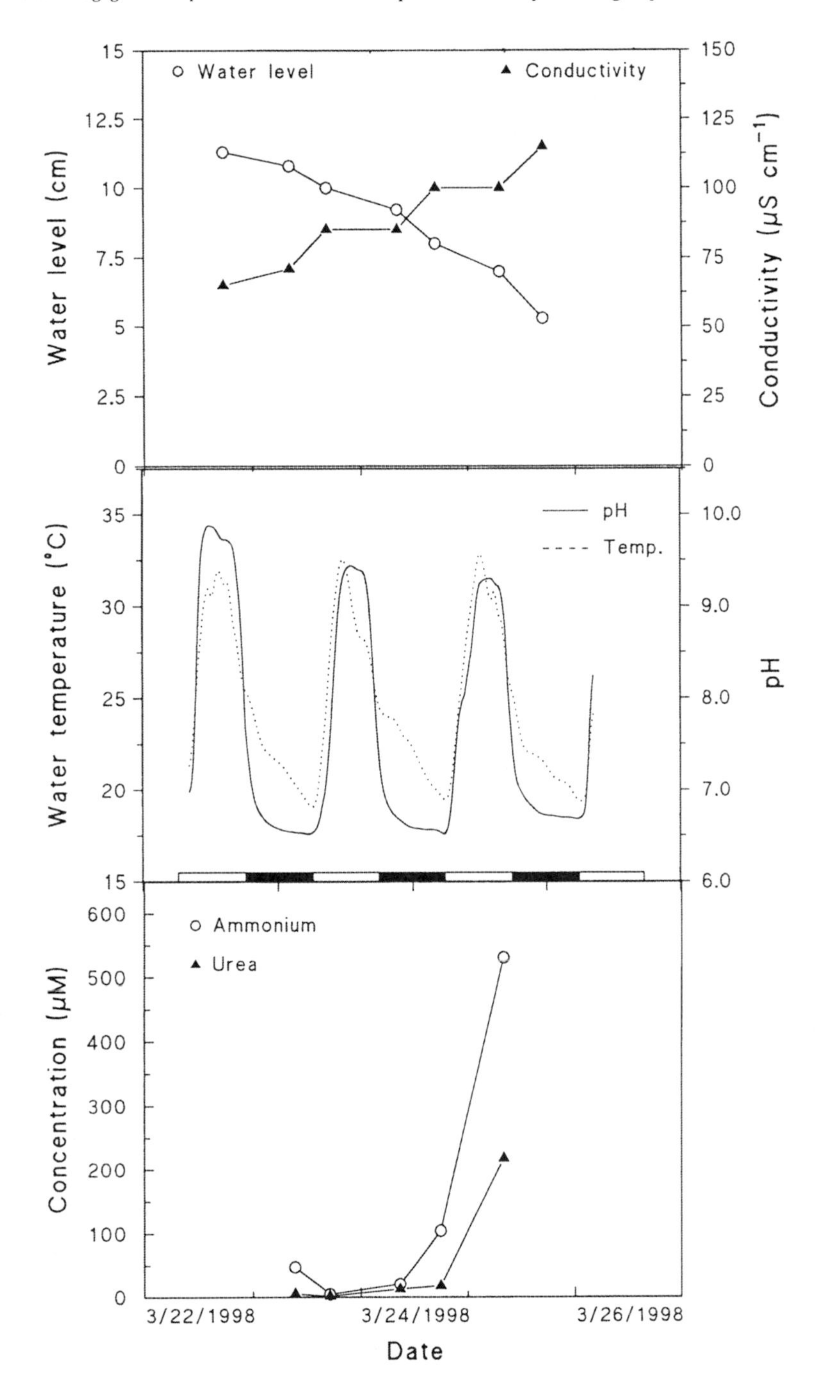

Fig. 12.2 Temporal course of (**a**) water level and electric conductivity, (**b**) water temperature and pH, (**c**) ammonium and urea concentration during desiccation in a temporarily water-filled rock pool on a granite outcrop on the farm Otjua (Omaruru District, Namibia). Floating leaves of *Chamaegigas intrepidus* covered the pool surface by ca. 25%. *Filled bars* indicate nighttime hours, *open bars* daytime hours (modified after Schiller et al. 1997; Heilmeier and Hartung 2001)

12.4.3 Temperature and pH of the Pool Water

Water temperature showed similar diurnal oscillations as air temperature, however, with a smaller amplitude. Maximum values reached ca. 35°C during the day and minimum temperatures at night were comparable to nocturnal air temperature (Fig. 12.2b). Parallel to diurnal fluctuations of water temperature were oscillations in pH of the pool water, with maximum values at afternoon and minimum values during night. While minimum pH, similarly to minimum water temperature, stayed constant for several nights, maximum pH values varied in accordance with maximum water temperature Table 12.1. Thus, during hotter days, diurnal amplitudes in pH were larger than during cooler days. Maximum pH values recorded in flooded pools in 1998 were near 10 in the afternoon and minimum near 5 in the morning. Still higher pH values in the afternoon (pH 12) were measured in 1995 (Schiller et al. 1997).

12.4.4 CO_2 and HCO_3^- Concentration of the Pool Water

Diurnal oscillations of pH were mainly a consequence of changing solubility of CO_2 due to fluctuations of water temperature. During the daytime CO_2 was completely lost from the pools by diffusion when water temperature was above 30°C. This raised the pH to strong alkaline values up to pH 10. Most of the inorganic carbon was present in the form of HCO_3^-, with peak concentrations about 100 times higher than maximum CO_2 concentrations during the night. After sunset, when water temperature steadily decreased, CO_2 concentration increased, initially with a high rate, throughout the whole night due to respiration. This caused a rapid decrease in pH down to slightly acidic values. However, as soon as water

Table 12.1 Amplitudes of temperature and pH in different temporarily water-filled rock pools on granite outcrops on the farm Otjua (Omaruru District, Namibia), inhabiting the aquatic resurrection plant *Chamaegigas intrepidus* (T_{min}, minimum temperature; T_{max}, maximum temperature; pH_{min}, minimum pH; pH_{max}, maximum pH; n.d., no data)

Rock pool	Date	T_{min} (°C)	T_{max} (°C)	pH_{min}	pH_{max}
1	3/16/1996	15.2	34.4	7.1	9.3
1	3/17/1996	16.9	34.6	7.2	9.9
1	3/18/1996	18.8	n.d.	8.0	n.d.
2	3/11/1997	n.d.	32.9	n.d.	9.0
2	3/12/1997	18.6	30.6	6.2	8.2
2	3/13/1997	19.8	27.0	6.0	7.5
2	3/14/1997	18.9	n.d.	6.0	n.d.
3	3/22/1998	n.d.	32.3	n.d.	9.9
3	3/23/1998	19.0	33.3	6.5	9.5
3	3/24/1998	19.5	32.9	6.6	9.2
3	3/25/1998	19.3	n.d.	6.7	n.d.

temperature increased in the morning, CO_2 concentration rapidly dropped to zero. Thus, for submerged *Chamaegigas* plants CO_2 is not available for photosynthesis during most part of the daytime.

12.4.5 Concentration of Mineral Nutrients in the Pool Water and the Sediment

The most abundant metallic cation in the water of pools with *Chamaegigas* was sodium (average concentration 190 mmol m^{-3}), followed by calcium (70 mmol m^{-3}), potassium and magnesium (30 mmol m^{-3} each). Median concentrations of chloride and sulphate were about half that of Na^+ and Ca^{2+}, respectively. Phosphate concentrations were below detection limits in most of the cases.

Mineral nutrient contents in the sediment of the rock pools at the end of the dry season are in most cases higher than those given by Gaff and Giess (1986) for *Chamaegigas* rock pools. This is especially true for potassium, magnesium (ca. 5 g kg^{-1} dw each) and phosphorus (ca. 0.6 g kg^{-1} dw), whereas the content of total nitrogen (3.5 g kg^{-1} dw) is somewhat lower than the value from Gaff and Giess (1986).

Among nitrogenous compounds, both inorganic and organic N species were considered as possible nitrogen sources for *Chamaegigas*. In contrast to temporary rock pools in the Namib-Naukluft Park (Kok and Grobbelaar 1985), nitrate was below detection limit in most of the cases. Ammonium, however, was much more abundant, with maximum concentrations in drying pools of 0.6 mM (Fig. 12.2c). Amino acids were also present in the low micromolar concentration range (Schiller et al. 1998b). In flooded pools, the most abundant amino acids were glycine, serine and asparagine. Less common, although also with high concentrations when present, were glutamine and alanine. During drying, the relative proportions of amino acids changed, with asparagine, serine and glycine found in nearly all samples. Four amino acids could be detected in almost dry pools only, namely glutamine, threonine, arginine and tyrosine.

The most abundant N compound in both flooded and drying pools was urea. The general time course of urea concentration during drying out of the pools was similar to NH_4^+ (Fig. 12.2c). Evaporation of the water caused a dramatic increase for both compounds, which was, however, more pronounced in NH_4^+.

12.5 Anatomical Features of *C. intrepidus*

Fully hydrated leaves of *Chamaegigas* do not exhibit xeromorphic anatomical features. Their submerged leaves show, however, a remarkable shrinkage during desiccation. The length of the desiccated leaves is 10–20% of the hydrated leaves

(Heil 1924; Hickel 1967; Schiller et al. 1999). Thereby, they withdraw to the sediment where they are protected against the extremely high light intensities. This drastic shrinkage has been attributed to the unique "contractive tracheids". These xylem elements that exist in the submerged leaves only [although Hickel (1967) reports that a few may exist in floating leaves also] have thin longitudinal cell walls with broad spiral strengthenings and terminal plates with a small lumen. The shrinking process is mainly due to these specialised xylem elements that contract like an accordion. Maximum shrinkage is limited to living leaves. The floating mesophytic, homoiohydric leaves that exhibit a normal bifacial anatomy shrink by only 15–20% during desiccation, as it is usually the case for mesophytic plants.

Only little information has been published about the root anatomy of *Chamaegigas*. Heil (1924) and Hickel (1967) described the rhizodermis, which is absolutely free of root hairs, and the dimorphic exodermis with short cells. Heil (1924) observed a distinct aerenchyma with radial rows of tiny cells in the cortex. Different from submerged leaves, a minute longitudinal shrinkage of *Chamaegigas* roots by 3% only was observed with our plant material (Heilmeier et al. 2002). In contrast to Hickel (1967), however, lateral shrinkage between 30 and 35% was recorded, which is the result of an exclusive shrinkage of the rhizodermis and the exodermis. This conclusion is supported by scanning electron microscopic investigations of desiccated *Chamaegigas* roots (Heilmeier et al. 2002). Contrarily to Heil (1924), no aerenchyma could be detected. The root cortex consists of a single layer of extremely large cortical cells with extremely thin cell walls and an endodermis (Fig. 12.3). The central stele exhibits a small and simple diarchic xylem. Despite the

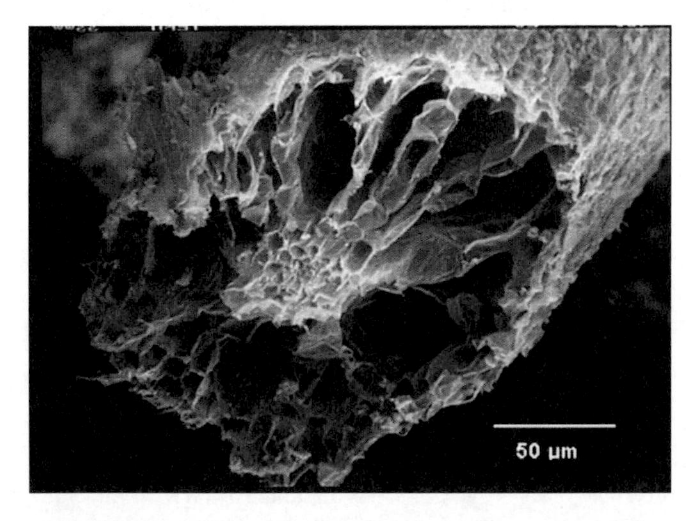

Fig. 12.3 Scanning electron micrograph of a cross section from a desiccated root with large cortex cells. The shrunken rhizodermis and hypodermis form a velamen-type layer

extreme fragile appearance of the large cortical cells, no shrinkage or collapse could be observed after complete desiccation. The thin walls of the large cortex cells must be extremely stiff to resist shrinkage. No remarkable staining that could point to an extraordinarily chemical composition of cell walls could be seen as it seems to be the case with the cell walls of leaves of *Craterostigma wilmsii* that contain both xyloglucans and unesterified pectins particularly in dehydrated cells (Vicrè et al. 1999). As both these compounds are known to strengthen cell walls, it would be interesting to examine whether cortical cell walls of *Chamaegigas* roots also exhibit a similar chemical composition. The unique stability of the large cortical cell walls seems to be an additional part of the adaptation of the plant to its extreme environment as it minimises shrinking of the roots during desiccation and thus maintains their physical structure within the sediment. Intracellular spaces are small and vacuum infiltration of hydrated roots with water resulted in a fresh weight increase of approximately 10–15% only (Heilmeier et al. 2002). This also supports the conclusion that the cortex of *Chamaegigas* is not formed by an aerenchyma.

As demonstrated already by Heil (1924) and Hickel (1967), the root exodermis contains passage cells with outer cell wall thickenings which are known from aerial roots of orchids and of some roots of plants within the family of Asclepiadaceae (Guttenberg 1968). They are believed to be an adaptation of roots to drought and are not known for roots of homoiohydric aquatic plants. Hickel (1967) concludes that in desiccated roots, when the collapsed rhizodermis and exodermis form a layer that resembles a velamen radicum, the outer pad, which would close the passage cells, acts like a valve and slows down the water loss.

12.6 Physiological, Biochemical and Molecular Adaptations to Stress in *C. intrepidus*

12.6.1 Intracellular pH Stability

Chamaegigas has to cope with substantial diurnal fluctuations in the pool water pH. ^{31}P NMR spectroscopy has been used to investigate the effect of external pH and dehydration on intracellular pH of *Chamaegigas* roots and leaves (Schiller et al. 1998a). High external pH (10.0) caused only a negligible alkalisation of the root cytoplasm, and drastic dehydration caused a small alkalisation of leaf vacuoles at pH 10. These results imply an unusually effective pH regulation consistent with the adaptation of *Chamaegigas* to a large number of adverse environmental factors. The NMR analysis also showed that dehydration had no effect on the pools of inorganic phosphate and phosphocholine. This indicated that membranes are most effectively protected from damages due to low water potentials, because membrane damage is usually accompanied by an increase of inorganic phosphate and phosphocholines.

12.6.2 Photosynthesis

Heilmeier and Hartung (2001) determined oxygen and bicarbonate concentration, pH and temperature of a pool covered by 25% with *Chamaegigas* and observed O_2 production even under conditions when the pool water was nearly free of CO_2. They concluded that HCO_3^- is a carbon source for submerged *Chamaegigas* leaves. Floating leaves can take up CO_2 via stomata on the upper side of the leaves, which seem to be locked open because most of the stomata stay open all the time, even in darkness. This has been observed for many other floating photosynthesising systems like fronds of *Lemna* (Landolt and Kandeler 1987).

Woitke et al. (2004) concluded from measurement of chlorophyll fluorescence that 75% of total plant photosynthesis takes place in floating leaves. Using this technique, they also detected some desiccation tolerance of floating leaves within 1–2 days. Whether such tolerance exists also over longer periods, as it is the case for submerged leaves, remains very doubtful.

12.6.3 Nitrogen Nutrition

As shown above, *Chamaegigas* has to cope with extreme conditions of nitrogen deficiency. Nitrate is virtually absent in the rock pools. Urea, ammonium ions and the amino acids glycine and serine occur in low concentrations, in the micromolar range. Urea cannot be utilised by *Chamaegigas* directly. Heilmeier et al. (2000) performed uptake experiments with ^{14}C- and ^{15}N-labelled urea. No incorporation and utilisation of urea could be observed, even after long incubation periods (up to 5 days), when the root systems were cleaned carefully. When the natural sediment or jack bean urease was present in the medium, urea-N was accumulated in tissues of *Chamaegigas*. Furthermore, ^{15}N NMR spectra performed with *Chamaegigas* roots after incubation with ^{15}N urea did not show any ^{15}N signals such as ammonium, glutamine and glutamate in the absence of urease. However, when ammonium was released by the action of urease ^{15}N could be utilised by *Chamaegigas*. Urease, therefore, plays an essential role for the acquisition of urea-N. It is important to note in this context that urease can survive the harsh conditions in the desiccated rock pools during the Namibian winter at temperature up to 60°C, complete dryness and high UV irradiation. This remarkable resistant enzyme seems to be a key factor for survival of *Chamaegigas* (and also of other plants that depend on urea-N deposited from animals).

Ammonium, which originates from urea as shown above especially in drying pools (cf. Fig. 12.2c), may become extremely deficient, especially in highly alkaline pool water where it can be lost as ammonia to the atmosphere. Then *Chamaegigas* switches to utilisation of glycine, which is most abundant among amino acids in the pool water, as shown after incubation with ^{15}N glycine, both under laboratory and field conditions (Schiller et al. 1998b). ^{14}C glycine can be taken up by *Chamaegigas* roots with a high-affinity glycine uptake system (K_M = 16 μM). Since uptake of glycine is strongly reduced under alkaline conditions (Schiller et al. 1998b), nitrogen utilisation must be expected to occur mainly during the morning hours only.

When *Chamaegigas* rosettes or isolated root systems were incubated with ^{15}N glycine, ^{15}N NMR spectra showed a glycine, serine, glutamine and glutamate signal (Hartung and Ratcliffe 2002). This agrees with the action of glycine decarboxylase (GDC) and mitochondrial serine hydroxy methyl transferase (SHMT; C1-metabolism, Mouillon et al. 1999) in the *Chamaegigas* roots. In this case, GDC would release CO_2 and ammonium from glycine, and NH_4^+ would be incorporated into glutamine and glutamate. The second product of glycine metabolism, methylenetetrahydrofolate (M-THF) would then transfer one carbon to another glycine molecule to form serine. Roots are believed to have extremely low (Walton and Woodhouse 1986) or even to lack GDC (Bourguignon et al. 1993), although ^{15}N NMR spectra of ^{15}N glycine supplied maize root tips showed signals similar to those of *Chamaegigas* roots. The latter, however, seem to metabolise glycine much more rapidly than maize roots (Hartung and Ratcliffe 2002). This could be a part of an adaptation to the natural habitat with glycine as one of the dominating N sources.

Serine, the other dominating amino acid of the rock pools, is taken up by *Chamaegigas* roots with similar rates as glycine (Schiller et al. 1998b). Assuming that roots of *Chamaegigas* exhibit activity of SHMT as shown for sycamore suspension cells by Mouillon et al. (1999), glycine again would be a product of serine metabolism and it could be metabolised as shown above. Indeed, different from *Zea mays*, ^{15}N NMR spectra of *Chamaegigas* roots that have been preincubated with ^{15}N serine exhibited a glutamine and a glutamate signal (Hartung and Ratcliffe 2002).

Glycine has been shown earlier to be the dominant N source for arctic and alpine plants (Chapin et al. 1993; Raab et al. 1996) and a number of sedges from different ecosystems (Raab et al. 1999). Schmidt and Stewart (1999) have demonstrated glycine as the main soil-derived N source for a wide range of wild plants in Australian communities. *Chamaegigas* is the first example of an aquatic vascular plant showing that amino acids such as glycine and serine can be an important N source.

12.6.4 Abscisic Acid

The role of abscisic acid (ABA) as a stress hormone seems to be well established. Its biosynthesis is increased especially under drought stress conditions. After transport to the target cells ABA mediates responses that are major components of plant survival mechanisms under stress conditions. Of particular importance is the ABA-controlled stomatal movement, which, however, should be of minor importance in *Chamaegigas*. The leaves are either submerged or floating on the water surface. Stomatal movement cannot be expected under those conditions. Additionally stomata of water plants very often are non-functional. The anatomy of *Lemna* guard cells does not allow movement; they are "locked open" (Landolt and Kandeler 1987). Both Heil (1924) and Hickel (1967) investigated stomata of *Chamaegigas*, however, without any conclusions concerning their function. Epidermal preparations of *Chamaegigas* floating leaves, published by Heil (1924), only show open stomata. Own unpublished scanning electron microscopical studies of the surface of floating leaves also show open stomata only.

12.6.4.1 ABA Content Under Control Conditions and Drought Stress

The ABA content of leaves of the poikilohydric angiosperms *C. plantagineum*, *Myrothamnus flabellifolia* and *Borya nitida* is very high under both hydrated and desiccated conditions (0.3–3 nmol g^{-1} dw; Schiller 1998). Drought-dependent fluctuations, however, are very small. Desiccation causes not more than a two- to threefold increase. In contrast, the ABA concentration in hydrated organs of *Chamaegigas* is much lower (0.05–0.25 nmol g^{-1} dw), similar to what is found in a wide range of mesophytic plants. During desiccation, however, a 20- to 30-fold increase in submerged leaves and roots can be observed. The desiccation-dependent ABA increase in the homoiohydric floating leaves, however, is never more than fivefold.

The ABA content of both roots and submerged leaves is well related to the tissue osmotic potential (Fig. 12.4). Different from many mesophytic species no threshold of osmotic potential has to be reached in *Chamaegigas* to trigger ABA accumulation. It seems that ABA biosynthesis responds more sensitively to water shortage in *Chamaegigas* than in other plants (Schiller et al. 1997).

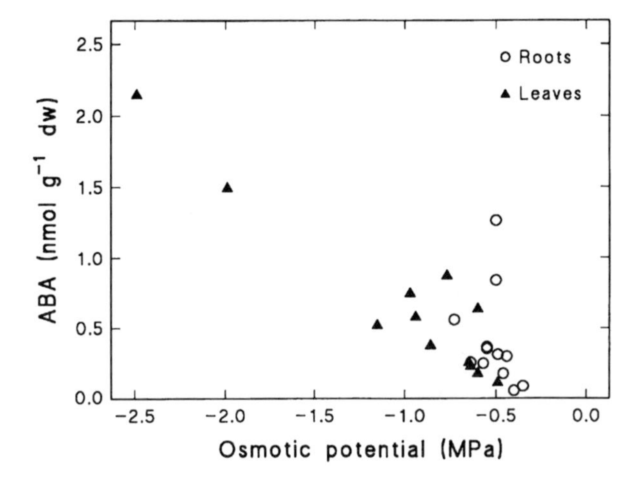

Fig. 12.4 The relationship between osmotic potential and tissue ABA concentration in roots and submerged leaves of *Chamaegigas intrepidus* (modified after Schiller et al. 1997)

12.6.4.2 Oxidative Degradation of ABA

When *Chamaegigas* tissues were pre-incubated with ^{14}C-ABA, phaseic acid (PA), dihydrophaseic acid (DPA) and conjugates (predominantly glucose esters) of ABA, PA and DPA could be extracted – the same ABA metabolites as in any other plant (Schiller 1998). The response of *Chamaegigas* to the inhibitor of oxidative ABA degradation, tetcyclacis, is also similar to reports from the literature. Degradation, however, is much slower in *Chamaegigas* tissues than in those of mesophytic plants such as *Valerianella locusta*. After 24 h pre-incubation with labelled ABA 3% of the label were recovered in PA in *Chamaegigas*, but 31% in *V. locusta*. Leaves of many other plants metabolise ABA even more rapidly (90–100% within 1 day). The slow degradation rates of ABA in *Chamaegigas* contribute to maintain high ABA concentrations, similarly as it happens when PA formation is inhibited by substances such as tetcyclacis or paclobutrazol.

12.6.4.3 ABA Conjugates

In *Chamaegigas*, the content of ABA conjugates, predominantly the glucose ester of ABA (ABA-GE), is low without distinct fluctuations of ABA-GE. Roots contained less ABA-GE than submerged leaves (Schiller et al. 1997). Conjugated ABA compounds have been regarded earlier as an indicator of stress history of plant tissues. They are deposited in the vacuoles where no further degradation is possible. This is not the case in *Chamaegigas*, which can release excess ABA to the surrounding medium. Deposition into the vacuoles is not necessary.

12.6.4.4 Distribution of ABA Between the Plants and the Pool Water

Apart from cycles of serious desiccation *Chamaegigas* is also exposed to drastic pH fluctuations, reaching alkaline pH values up to 12. We must, therefore, expect serious losses of ABA to the pool water considering that ABA distributes within *Chamaegigas* according to the anion trap concept (Hartung and Slovik 1991; Slovik et al. 1995). A significant loss of ABA during the afternoon hours might be a serious problem, especially because during this time pools usually dry up. *Chamaegigas*, however, seems to be well adapted to these conditions. Rosettes preloaded with labelled ABA lost significantly less ABA to the medium than a mesophytic terrestrial rosette (*V. locusta*) when transferred to an alkaline medium (pH 10).

With an increase of the external pool water pH by one pH unit, ABA in the pool water increased less than twofold (Schiller et al. 1997). According to the Henderson–Hasselbalch equation a tenfold increase should be expected. Uptake of external ABA was strongly pH dependent with high uptake rates under acid conditions, indicating that external ABA can be taken up during the night. Membranes of *Chamaegigas* roots must have a remarkably low permeability coefficient for undissociated ABA. Additionally one cannot exclude the existence of an effective hypodermal apoplastic barrier, which also could slow down ABA loss to the medium drastically (Freundl et al. 2000). This, together with a high cytosolic pH of nearly 7.6 (Schiller et al. 1998a), could explain the low external ABA concentration, which is below those predicted by computer simulations (Slovik et al. 1995) and found in soil solutions under different crops (Hartung et al. 1996).

In conclusion, the role of ABA in *Chamaegigas* is not in regulating stomatal aperture as in most other plants. Rather, the most sensitive ABA biosynthesis and the maintenance of high tissue ABA concentrations point to its essential function, e.g. in triggering metabolic processes (see below).

12.6.5 Dehydrins

Fully hydrated submerged leaves of *Chamaegigas* synthesise proteins that show similarities with the ABA-inducible desiccation-related proteins of *C. plantagineum* (Bartels et al. 1990), the dehydrins. Even after 4 weeks of optimal hydration, the dehydrin amount is high. In fully hydrated roots, however, dehydrins are not detectable. When hydrated roots were treated with ABA dehydrins were formed, as it was the case when roots desiccated. (The function of dehydrins is described in Chaps. 14 and 16.)

12.6.6 Carbohydrates

Poikilohydric plants exhibit high amounts of carbohydrates, predominantly sucrose, fructose and glucose, but also glucopyranosyl-β-glycerol, trehalose and arbutin (*M. flabellifolia*, Bianchi et al. 1991) and octulose in *C. plantagineum* where it is converted during desiccation to sucrose (Bianchi et al. 1991, 1992; Norwood et al. 1999, 2000; Scott 2000). In *Chamaegigas*, besides sucrose (140–180 μmol g^{-1} dw) stachyose (approx. 180–250 μmol g^{-1} dw) also becomes very dominant. Raffinose is present, however, at a relatively low level (approx. 20 μmol g^{-1} dw). Glucose and fructose are substantially higher in leaves (180–200 and 60–120 μmol g^{-1} dw, respectively) than in roots (20–30 μmol g^{-1} dw). During dehydration, the monosaccharides of the leaves decrease together with increasing stachyose and sucrose content (Fig. 12.5). Heilmeier and Hartung (2001) suggested that glucose and fructose are incorporated into sucrose and stachyose. In roots, stachyose content is similar to leaves, and the changes of the amounts of sugars mentioned above during desiccation are far less distinct.

In air dry tissues, the total content of soluble carbohydrates of roots and submerged leaves ranged from 14 to 26% with stachyose (roots) and sucrose (leaves) as the dominant sugars. In roots, stachyose can account for 14% of dry weight. ABA treatment caused soluble carbohydrates to rise in roots, especially as far as sucrose is involved (Schiller 1998).

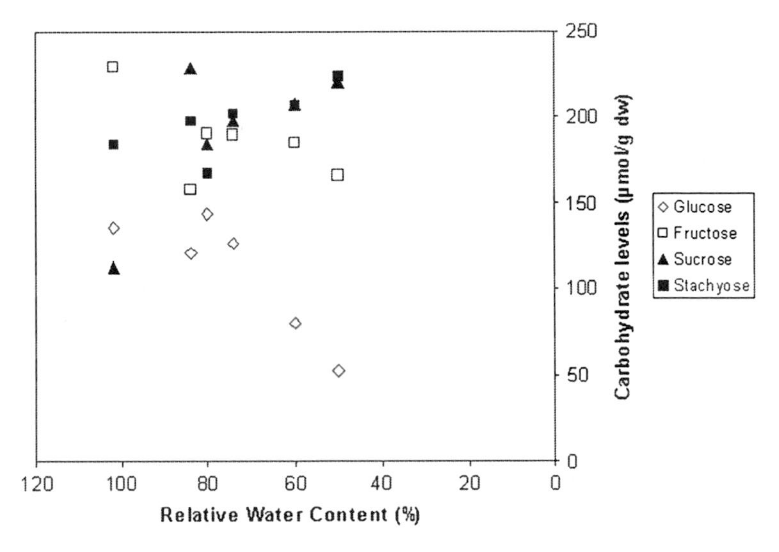

Fig. 12.5 Levels of fructose, glucose, sucrose and stachyose in submerged leaves of *Chamaegigas inrepidus* as a function of decreasing leaf water content (modified after Heilmeier and Hartung 2001)

The function of sugars such as raffinose or stachyose or other galactosyl-sucrose oligosaccharides is the suppression of crystallisation of protoplastic constituents and the promotion of glass formation at low water content (Bruni and Leopold 1991; seed embryos). At a glassy state, a liquid has a high viscosity. Chemical reactions are slowed down, and interactions between cell components are prevented. Therefore, a glassy state is highly stable and ideal to survive desiccation. Furthermore, these raffinose family oligosaccharides (RFOs) may serve as storage carbohydrates for immediate regrowth of floating leaves and flowers after rewatering (*C. plantagineum*; Norwood et al. 2000).

12.7 Breeding System and Genetic Diversity in *Chamaegigas* Populations

As an endemic plant *Chamaegigas* grows in an extremely small area of central Namibia. Compared to plants of a wide distribution, such populations suffer from genetic impoverishment, i.e., a small number of polymorphic gene loci and a loss of alleles. Inbreeding within small populations and genetic drift may cause a reduced degree of heterozygosity and a fixation of harmful alleles. Additionally the isolated occurrence of *Chamaegigas* on inselbergs of approximately 25 km distance may reduce genetic exchange via seed dispersal and pollen transfer drastically. Thus, restricted genetic diversity should severely limit the potential for adaptation in *Chamaegigas* to its stressful environmental conditions.

However, molecular genetic investigation by means of AFLP-marker (amplified fragment length polymorphism) has shown a surprisingly high genetic diversity within the plants of one inselberg, even within one pool (Durka et al. 2004). This high genetic diversity is very likely due to pollination by four insects, two bee species (*Liotrigona bottegoi* and *Apis mellifera*) and mainly two beetle species of the genus *Condylops* (*C. erongoensis* and a newly identified species) (Woitke et al. 2006). The zygomorphous and slightly protandric flowers show a typical insect pollination syndrome: they are distinctly coloured, intensively scenting and provide a rich floral reward (abundant pollen grains), but they do not produce any nectar. Rather, dense layers of trichomes (400–1,600 per mm^2) can be found on the lower lip, similar to well-known oil flowers. However, there are no indications that *Chamaegigas* lives in an oil flower/oil-collecting bee symbiosis (Woitke et al. 2006). Pollination experiments indicated that *Chamaegigas* is a predominantly outcrossing species, and a large number of pollen could be found both on the wild bee and beetle species (Durka et al. 2004).

In conclusion, this endemic species does not show any genetic impoverishment. The high genetic variability may be a result of high UV radiation at high altitude, which should result in many mutations and a high gene flow both within inselbergs by pollinating insects and among inselbergs due to seed dispersal by wind or animals (Heilmeier et al. 2005). This may provide the genetic basis for a successful adaptation of *Chamaegigas* to its extreme habitat conditions.

12.8 Concluding Remarks

Chamaegigas is often considered to represent the most spectacular resurrection plant (Gaff and Giess 1986; Hartung et al. 1998) as it (1) shows the most rapid recovery after rewetting among all poikilohydric angiosperms, and (2) possesses two types of mature leaves with contrasting drought tolerance on the same plant. Therefore, this species has been considered to be well adapted to its natural habitat, shallow, only temporarily water-filled rock pools. However, as shown here, desiccation is only one aspect of a multi-dimensional complex of environmental stress conditions, which also include low partial pressures of carbon dioxide during the day when plants are flooded, drastic diurnal oscillations in pH and low nutrient contents of the pool water, high temperatures and solar radiation, especially in the ultraviolet range. Thus, both acquisition and conservation of resources such as carbon and nitrogen when they are scarce and protection of cellular and molecular structures from damage due to dehydration, alkaline, heat and UV stress are essential for the survival of this tiny plant at its extreme habitat.

Similar complex environmental stress conditions also influence other resurrection plants such as *C. plantagineum* and *X. humilis*, which grow in the direct vicinity of *Chamaegigas* bearing outcrops. Except pH fluctuations and CO_2 availability, very similar environmental stresses (temperature, solar radiation, unfavourable pH of the soil solution and extreme nutrient deficiency) act on those plants as well. Salt stress may be even an additional problem. Growth conditions of poikilohydric study plants, grown in glasshouses or growth cabinets, may necessarily have to be adapted accordingly. More extremely tissues cultures, as they are often used in the case of *C. plantagineum*, grow under conditions that are extremely far from any reality, even if they may have been treated with osmotica.

In a recent opinion article, Marris (2008) has shown convincingly that molecular biology, the use of transgenic plants, and gene technology have not brought significant gain and improvement for increase of yield and performance of crops that are cultivated in arid regions. One reason may be that in many of these studies, the complexity of drought tolerance has not been considered sufficiently. To engineer successful transgenic crops not only drought-responsive genes should be considered, but also those of ABA metabolism, of transporters of nitrogenous compounds, of enzymes involved in cell wall biochemistry and membrane integrity, of enzymes involved in the acquisition of DON, of enzymes involved in salt and alkaline pH tolerance and pH homeostasis and of enzymes and metabolites involved in detoxifying reactive oxygen species. Research with plants such as *Chamaegigas* that have developed a complex system of strategies to survive the extremely harsh conditions can teach us the necessary direction of future research.

Acknowledgements This work was supported by Schimper-Stiftung (H.H.) and DFG-SFB 251 (W.H.). A. and W. Wartinger and B. Dierich excellently assisted in the field work and performed laboratory experiments. We thank D. Morsbach (Ministry of Wildlife, Conservation and Tourism) and Dr. B. Strohbach (National Botanical Research Institute, Windhoek) for their support. We are indebted to Mrs. Arnold and Mrs. and Mr. Gaerdes for their great hospitality on Otjua farm. E. Brinckmann was a great help in any respect. We appreciate the great interest of Prof. Dr. O.L. Lange in all aspects of this study.

References

Bartels D, Schneider K, Terstappen G, Piatkowski D, Salamini F (1990) Molecular cloning of abscisic acid modulated genes which are induced during desiccation of the resurrection plant *Craterostigma plantagineum*. Planta 181:27–34

Bianchi G, Gamba A, Murelli C, Salamini F, Bartels D (1991) Novel carbohydrate metabolism in the resurrection plant *Craterostigma plantagineum*. Plant J 1:355–359

Bianchi G, Gamba A, Murelli C, Salamini F, Bartels D (1992) Low molecular weight solutes in desiccated and ABA treated calli and leaves of *Craterostigma plantagineum*. Phytochem 31:1917–1922

Bornefeld T, Volk OH (2002) Annotations to a collection of liverworts (Hepaticae, Marchantiales) from Omaruru District, Namibia, during summer 1997. Dinteria 27:13–17

Bourguignon J, Vauclare P, Merand V, Forest E, Neuburger M, Douce R (1993) Glycine decarboxylase complex from higher plants. Molecular cloning, tissue distribution and mass spectrometry analysis of the T-protein. Eur J Biochem 217:377–386

Bruni F, Leopold AC (1991) Glass transitions in soybean seeds. Relevance to anhydrous biology Plant Physiol 96:660–663

Chapin FS III, Mollanen L, Kielland K (1993) Preferential use of organic nitrogen for growth by a non-mycorrhizal arctic sedge. Nature 361:150–153

Dinter K (1909) Deutsch-Südwest-Afrika. Flora, Forst- und landwirtschaftliche Fragmente. Theodor Oswalt Weigel, Leipzig

Dinter K (1918) Botanische Reisen in Deutsch-Südwest-Afrika. Feddes Repert. Beiheft 3

Durka W, Woitke M, Hartung W, Hartung S, Heilmeier H (2004) Genetic diversity in *Chamaegigas intrepidus* (Scrophulariaceae). In: Breckle SW, Schweizer B, Fangmeier A (eds) Results of worldwide ecological studies Proc 2nd Symp Schimper Foundation. Günter Heimbach, Stuttgart, pp 257–265

Fischer E (1992) Systematik der afrikanischen Lindernieae (Scrophulariaceae). Trop Subtrop Pflanzenwelt 81. Fritz Steiner, Stuttgart

Freundl E, Steudle E, Hartung W (2000) Apoplastic transport of abscisic acid through roots of maize: effect of the exodermis. Planta 210:222–231

Gaff DF (1977) Desiccation tolerant vascular plants of Southern Africa. Oecologia 31:95–109

Gaff DF (1987) Desiccation tolerant plants in South America. Oecologia 74:133–136

Gaff DF, Giess W (1986) Drought resistance in water plants in rock pools of Southern Africa. Dinteria 18:17–36

Giess W (1969) Die Verbreitung von *Lindernia intrepidus* (Dinter) Oberm. (*Chamaegigas intrepidus* Dinter) in Südwestafrika. Dinteria 2:23–27

Giess W (1997) A preliminary vegetation map of Namibia. 3 rd rev edn. Dinteria 4:1–112

Guttenberg H von (1968) Der primäre Bau der Angiospermenwurzel. Handbuch der Pflanzenanatomie 8, Teil 5. Borntraeger, Berlin-Stuttgart

Hartung W, Ratcliffe RG (2002) The utilization of glycine and serine as nitrogen sources in roots of *Zea mays* and *Chamaegigas intrepidus*. J exp Bot 53:2305–2314

Hartung W, Slovik S (1991) Physico-chemical properties of plant growth regulators and plant tissues determine their distribution and redistribution. New Phytol 119:361–382

Hartung W, Sauter A, Turner NC, Fillery I, Heilmeier H (1996) Abscisic acid in soils: what is its function and which factors and mechanisms influence its concentration? Plant Soil 184:105–110

Hartung W, Schiller P, Dietz KJ (1998) Physiology of poikilohydric plants. Prog Bot 59:299–327

Heil H (1924) *Chamaegigas intrepidus* Dtr., eine neue Auferstehungspflanze. Beih bot Zbl 41:41–50

Heilmeier H, Hartung W (2001) Survival strategies under extreme and complex environmental conditions: the aquatic resurrection plant *Chamaegigas intrepidus*. Flora 196:245–260

Heilmeier H, Ratcliffe RG, Hartung W (2000) Urea: a nitrogen source for the aquatic resurrection plant *Chamaegigas intrepidus* Dinter. Oecologia 123:9–14

Heilmeier H, Wolf R, Wacker R, Hartung W (2002) Observations on the anatomy of hydrated and desiccated roots of *Chamaegigas intrepidus* Dinter. Dinteria 27:1–12

Heilmeier H, Durka W, Woitke M, Hartung W (2005) Ephemeral pools as stressful and isolated habitats for the endemic aquatic resurrection plant *Chamaegigas intrepidus*. Phytocoenologia 35:449–468

Hickel B (1967) Zur Kenntnis einer xerophilen Wasserpflanze: *Chamaegigas intrepidus* DTR. aus Südwestafrika. Int Revue Ges Hydrobio 52:361–400

Kok OB, Grobbelaar JU (1985) Notes on the availability and chemical composition of water from the gravel plains of the Namib-Naukluft Park. J Limnol Soc South Afr 11:66–70

Landolt E, Kandeler R (1987) The family of *Lemnaceae* – a monographic study. II Phytochemistry, physiology, application bibliography. Veröff Geobot Inst ETH. Stiftung Rübel 95:270–272

Marris E (2008) More crop per drop. Nature 452:273–277

Mouillon JM, Aubert S, Bourguignon J, Gout E, Douce R, Rébeillé F (1999) Glycine and serine catabolism in non-photosynthetic higher plant cells: their role in C1 metabolism. Plant J 20:197–205

Norwood M, Truesdale MR, Richter AM, Scott P (1999) Metabolic changes in leaves during dehydration of the resurrection plant *Craterostigma plantagineum* (Hochst). South Afr J Bot 65:1–7

Norwood M, Truesdale MR, Richter AM, Scott P (2000) Photosynthetic carbohydrate metabolism in the resurrection plant *Craterostigma plantagineum*. J exp Bot 51:159–165

Porembski S, Barthlott W (2000) Granitic and gneissic outcrops (inselbergs) as centers of diversity for desiccation-tolerant vascular plants. Plant Ecol 151:19–28

Raab TK, Lipson DA, Monson RK (1996) Non-mycorrhizal uptake of amino acids by roots of the alpine sedge *Kobresia myosuroides*: implications for the alpine nitrogen cycle. Oecologia 108:488–494

Raab TK, Lipson DA, Monson RK (1999) Soil amino acid utilization among species of the Cyperaceae: plant and soil processes. Ecology 80:2408–2419

Schiller P (1998) Anatomische, physiologische und biochemische Anpassungen der aquatischen Auferstehungspflanze *Chamaegigas intrepidus* an ihren extremen Standort. PhD Dissertation, Julius-Maximilians-Universität Würzburg

Schiller P, Heilmeier H, Hartung W (1997) Abscisic acid (ABA) relations in the aquatic resurrection plant *Chamaegigas intrepidus* under naturally fluctuating environmental conditions. New Phytol 136:603–611

Schiller P, Hartung W, Ratcliffe RG (1998a) Intracellular pH stability in the aquatic resurrection plant *Chamaegigas intrepidus* in the extreme environmental conditions that characterize its natural habitat. New Phytol 140:1–7

Schiller P, Heilmeier H, Hartung W (1998b) Uptake of amino acids by the aquatic resurrection plant *Chamaegigas intrepidus* and its implication for N nutrition. Oecologia 117:63–69

Schiller P, Wolf R, Hartung W (1999) A scanning electromicroscopical study of hydrated and desiccated submerged leaves of the aquatic resurrection plant *Chamaegigas intrepidus*. Flora 194:97–102

Schmidt S, Stewart GR (1999) Glycine metabolism in plant roots and its occurrence in Australian plant communities. Austr J Plant Physiol 26:253–264

Scott P (2000) Resurrection plants and the secrets of eternal leaf. Ann Bot 85:159–166

Slovik S, Daeter W, Hartung W (1995) Compartmental redistribution and long distance transport of abscisic acid (ABA) in plants as influenced by environmental changes in the rhizosphere. A biomathematical model J exp Bot 46:881–894

Vicrè M, Sherwin HW, Driouich A, Jaffer MA, Farrant JM (1999) Cell wall characteristics and structure of hydrated and dry leaves of the resurrection plant *Craterostigma wilmsii*, a microscopical study. J Plant Physiol 155:719–726

Walton NJ, Woodhouse HW (1986) Enzymes of serine and glycine metabolism in leaves and non photosynthetic tissues of *Pisum sativum* L. Planta 167:119–128

Woitke M, Hartung W, Gimmler H, Heilmeier H (2004) Chlorophyll fluorescence of the submerged and floating leaves of the aquatic resurrection plant *Chamaegigas intrepidus*. Funct Plant Biol 31:53–62

Woitke M, Wolf R, Hartung W, Heilmeier H (2006) Flower morphology of the resurrection plant *Chamaegigas intrepidus* Dinter and some of its potential pollinators. Flora 201:281–286

Part III
The Cell Biological Level

Chapter 13
Molecular Biology and Physiological Genomics of Dehydration Stress

Ruth Grene, Cecilia Vasquez-Robinet, and Hans J. Bohnert

Abbreviations

ABA	Abscisic acid
ABI	ABA Insensitive
AREB	ABA Response Element Binding Factor
ABF	ABA-Responsive Binding Factors
ABI5 3	*ABI-INSENSITIVE 5, 3*
APETALA	*Transcription factor gene of the AP2 family*
APX1	Cytosolic ascorbate peroxidase 1
ASK1	Arabidopsis skp1-like1-1
AtCPK	*Arabidopsis thaliana* calcium-dependent protein kinase
AtCyp	*Arabidopsis thaliana* cyclophilin encoding gene
AtHB	*Arabidopsis thaliana* Homeobox Factor
AtMYC	*Arabidopsis thaliana* transcription factor with a helix-loop-helix and a bZip domain
AtSUC	*Arabidopsis thaliana* Sucrose Transporter
AtTLP	*Arabidopsis thaliana* Tubby-Like Protein
bZIP TF	Basic Leucine Zipper Domain transcription factor
CCA1	*Circadian Clock Associated1 gene*
CNV	Copy number variation
COL1	*Constans-Like 1 gene*
DEAD RNA helicase	ATP-Dependent RNA Helicase
DREB2A	Drought-Responsive Element Binding Protein 2A
DRIP1	DREB2A-Interacting Protein1
ESTs	Expressed sequence tags
FAR1	Far-Red-Impaired-Response
FHY3	Far-Red-Elongated-Hypocotyl
FUS	*Arabidopsis gene encoding a FUSCA protein involved in signalling networks*
GA	Giberellic Acid
GFP	Green Fluorescent Protein
HAB1	*Hypersensitive to ABA1 gene*
LHY	*Late Elongated Hypocotyl gene*
LEAs	Late Embryogenesis Active Proteins

U. Lüttge et al. (eds.), *Plant Desiccation Tolerance*, Ecological Studies 215,
DOI 10.1007/978-3-642-19106-0_13, © Springer-Verlag Berlin Heidelberg 2011

LEC1	*Leafy Cotyleydon1gene*
MAPK	Mitogen-activated Protein Kinase
MSTR	Multiple Stress Regulatory Genes
NCED3	9-***cis***-epoxycarotenoid dioxygenase
NF-Y	Plant Nuclear Factor Y
NILs	Near Isogenic Lines
OST1	Open Stomata1
PICKLE	Encodes a chromatin remodelling protein (CHD3)
PLD	Phospholipase D
PP2C	Protein phosphatase2C
RD29	*Responsive To Drought 29gene*
RILs	Recombinant Inbred Lines
RING	*Really Interesting New Gene*
ROS	Reactive Oxygen Species
SDIR1	Salt and Drought-Induced RING FINGER 1
SFN1	Regulatory Subunit of SnRK1
SnRKs	Sucrose non-fermenting protein (SNF-1)-related kinases
SUMO	Small Ubiquitin-like Modifier
TF	Transcription factor
WUE	Water Use Efficiency
XERICO	RING-H2 zinc finger factor promoting ABA synthesis
YUCC	Arabidopsis HYPERTALL
ZAT1	Putative Zinc Transporter1

13.1 Introduction

An organism's response to changes in the environment is characterized by tissue and time-specific components that represent developmental windows of relative sensitivity to the condition. What this response is in detail can be gauged most easily from alterations in gene expression profiles and metabolome adjustments. More recently, epigenomics, copy number variation (CNV, both on the genome structure and the transcript levels), allele structure, protein assemblies, protein: DNA interactions, and post-translational protein modifications have provided additional information that begin to explain physiological responses and phenotypic characters. Dehydration stress, in particular, affects every aspect of plant growth and function, modifying anatomy, morphology, physiology, and biochemistry. Such responses have often been studied and described, while presently the focus has shifted towards explaining the various phenomena in molecular and genetic terms. During seed development and, especially, seed maturation, desiccation constitutes an active, induced programme that can establish extreme drought tolerance potentially lasting millennia. To a degree, a deepened understanding of the events that confer desiccation tolerance in seeds can be used to illuminate events in other plant tissues and developmental stages. Studies that monitor plant responses to water-deficit conditions of widely variable strength have recently increased in number. The term "drought tolerance" has become a rallying topic because biotechnology companies and farmers alike recognize lack or scarcity of water as a serious yield-limiting factor, the late recognition of insightful warning by John Boyer more than 25 years ago (Boyer 1982). Meanwhile, water has become a valuable commodity that, according to some, will become a contentious political topic the world over (for a relevant reference, see the "Intergovernmental Panel on Climate Change Report"; http://www.ipcc.ch/). The reasons for the prevalence of water deficits are manyfold, with the societal and agronomic aspects having been discussed by others (Easterling et al. 2007; Parry et al. 2004).

One major reason for biologists and plant breeders to be concerned may be that the engineering or breeding of single trait protection schemes has progressed, along with improved agricultural practices, but this strategy is now seen as not sufficient. The plant drought stress response in a natural setting – like all abiotic stresses that plants constantly experience – is significantly more complex than what studies conducted under controlled conditions can reveal, because it is multigenic and varied over developmental time, while differently affecting organs and tissues. In nature, as well, responses must be integrative because signals come from many different inputs. Contrasting "lifestyles" shown by different species, or as factors that separate lines or ecotypes of one species, may be related to differences in "input sensitivity" or "response capacity" in ecophysiological terms. When phrased in genomics terms, drought-tolerant species/landraces are separated from sensitive ones by gene number (more genes for the same function in a family), gene nature (more "effective" genes in xerophytic species or landraces), gene expression strength (transcript copy number variation – CNV; e.g., Stranger et al. 2007), coding region diversity and genome-wide allele structure (combinatorial pyramiding of genes), or the structure of regulatory circuits (promoter sequences, miRNAs, siRNA, or intergenic DNA sequence complexity).

Plant physiologists – and indeed farmers throughout the millennia – have always been aware of the stupendous importance of plant water relations. Up to recently, however, integrating the complexity of drought responses and understanding what determines the degrees of tolerance to water deficits is best exemplified by the parable of the blind men touching an elephant.

The contours of the enigmatic elephant are now emerging. Decades of physiological observations have provided a framework. Water deficit, drought, and desiccation affect in particular the chloroplast machinery in multiple ways, which then lead to organismal responses. Rather than continuing such work with more sophisticated experiments in controlled environments, now may be the right time, and a more rewarding undertaking, to begin integrating this information into a genomics-based conceptual framework, especially also by paying attention to plant behaviour in natural environments. Combining physiological, biochemical, and molecular tools has, during the last decade, resulted in models of drought-responsive signalling and metabolic pathways (Bray 1997; Zhu 2001; Chaves et al. 2003; Yamaguchi-Shinozaki and Shinozaki 2006; Ribaut 2006). It is now apparent, also, that translational control is an important component of drought responses (Kawaguchi and Bailey-Serres 2002; Kawaguchi et al. 2003, 2004). The complexity of drought responses has been underscored by the detection of a myriad of genes that affect how stressed plants tolerate or avoid water stress. Parallel to these studies, the generation of marker-assisted genetic maps and QTLs analyses, the development of hybrids, NILS, and RILs resulted in the identification of novel candidate drought-responsive genes, alleles, and loci (Tuberosa and Salvi 2006; Reynolds and Tuberosa 2008; Harris et al. 2007; Szalma et al. 2007).

Our discussion focuses on mechanistic models that, we think, can be applied to the various conditions of a relative scarcity of water – developmentally programmed "dehydration" and re-hydration tolerance, water deficit with its inherent severity scale, and outright drought responses. We incorporate concepts that emerged following the introduction of "genomics" approaches, attendant physical and mathematical and statistical results, and attempts to integrate functional genomics, epigenomics, genetics, molecular physiology, and systems biology.

13.2 Physiology, Biochemistry, and Phenology of Drought and Desiccation

Each of these three terms, resistance, tolerance, and avoidance, has been and is being applied to define plant responses to water deficit. An initial distinction is made between desiccation, in which the plant is deprived of water entirely with survival depending on the evolution of mechanisms that achieve quiescence, and water deficit, in which the plant utilizes adaptation mechanisms that allow it to protect the cellular machinery, while still functioning albeit in an altered or highly diminished mode. It can be argued that the term "drought-resistant" does not, in reality, apply to any higher plant, apart from resurrection plants. Higher plants that are not resurrection plants essentially "resist" because they have evolved avoidance reactions – entering a period of dormancy, sacrificing limbs or leaves, subterranean storage organs, and accelerated flowering – to fit into a time window during which sufficient moisture is available. Avoidance is therefore the rule rather than the exception. However, it appears that tolerance exists, at least for short periods of water deficit, and that this character may be possible to engineer, based on molecular and genetic knowledge (see Sect. 13.7).

13.2.1 A Brief Summary of Drought-Response Physiology

Physiological responses to drought have been extensively studied, debated, and reviewed (e.g., Jenks 2007). A most salient point is the water-deficit-dependent closing of stomata that, while it increases water use efficiency (WUE), leads to a down-regulation of many chloroplast functions that are ultimately based on changes in the internal [CO_2] concentration, c_i (Brodribb 1996). Second, water transport and its maintenance or disruption as water deficit increases has received much attention over decades. The physical functioning of the plant "piping" system, connected to stomatal aperture changes, and the morphological/anatomical requisites that ensure continued water transport are well known (Ranathunge et al. 2004; Ye and Steudle 2006; Shao et al. 2008).

By reflecting the complexity of drought as a stress condition, physiological work has uncovered phenotypes pertaining to root architecture, xylem development, and cohesion capacity in water conducting tissues (Chap. 10), to plant habitus, to epidermis structure and composition, and finally to biochemical adaptations (Chap. 16), such as those commonly labelled as osmotic adjustments, and altered general biochemistry and pathway engagement in energy generating or, late during stress, energy-dissipating mechanisms. These details need not be recounted; a number of excellent reviews are available (Ingram and Bartels 1996; Chaves et al. 2003; Bartels and Sunkar 2005) and results of physiological research have entered textbooks and monographs (e.g., Ribaut 2006; Taiz and Zeiger 2006).

Briefly, in countless variations on a common theme, physiological reactions of plants are geared at obtaining water, facilitating water flux to distribute solutes through the stream of water, and to store and/or conserve water. Many of these variations affect and modify development through altered morphology: from epidermal wax depositions, to trichome density, vascular structures, orientation of leaves and branches, sacrificing parts of the plant, and also outright hibernation. Also, several biochemical mechanisms have been identified. Much attention has been devoted to the concept of "osmotic adjustment", the drought or osmotically induced appearance of low molecular weight metabolites, whose accumulation lowers the osmotic potential of cells, thus favouring the retention of water (Bohnert and Shen 1999; Jiang and Huang 2001; Hoekstra et al. 2001). Accumulating metabolites are many, ranging from sucrose and some non-reducing hexoses, di- and oligo-saccharides that are not used in starch biosynthesis, to complex sugars and low-complexity glycans and also polyols (Buitink et al. 2006).

These carbon-only compounds may be considered a safety valve for dealing with photosynthate that is not used for growth under stress. A similar role in channelling carbon into end-products can be assigned to accumulating metabolites of phenylpropanoid pathways that can also assume a role in defence reactions (Arfaoui et al. 2007).

Similarly, accumulating proline and other amino group carrying substances have been interpreted as N-storage metabolites in the absence of growth with an additional function as compatible osmolytes (Verslues and Bray 2006; Fan et al. 2008; Izanloo et al. 2008).

One of these metabolites, glycine betaine, may also be viewed as storage compounds for methyl groups (McNeil et al. 2000). Yet, other mechanisms include the synthesis and moderate accumulation of LEA proteins, typically heat-soluble, hydrophilic, largely unstructured proteins that are present in many organisms, prokaryotes, and eukaryotes alike (see also below). Finally, small heat shock proteins have been identified as promoters of drought and desiccation tolerance responses (Kotak et al. 2007).

When viewed in relation to time dependence of accelerating drying, initial reactions beyond the sensing that leads to reduced stomatal aperture and the shut-down, or shut-off, of the photosystems are geared towards two measures that are interconnected. These reactions generate processes that prevent protein unfolding and it is this reaction that is evolutionarily deeply connected (Chakrabortee et al. 2007), and – connected to photosynthetic performance and protein damage – measures that restrict the generation of ROS, other than the oxidative processes on which signalling events depend.

Reactive oxygen species (ROS) are both a by-product of injury responses and they also function as cellular messengers, triggering individual defence pathways at transcriptional and metabolic levels (Bailey-Serres and Mittler 2006; see also Chap. 11). The redox status of plant cells is a pivotal feature in regulatory mechanisms controlling every aspect of their function, from abiotic stress signalling to development, making clear, once again, the essential connection between growth processes and responses to the environment (Mittler et al. 2004; Foyer and Noctor 2005). ROS arise at all times from the functioning of each of the cell's electron transport chains, although primarily from the chloroplast in the light (Khandelwal et al. 2008). Individual ROS sensing mechanisms, and parts of ensuing organelle to nucleus signalling pathways, have been described (Koussevitzky et al. 2007). The apoplast is a source of ROS, generated through the action of NADPH oxidases and/or stressors such as atmospheric ozone, setting in motion a signalling pathway involving G proteins, and MAPKs (reviewed by van Breusegem et al. 2008). NADPH oxidases are now known to also play a more general role in ROS-mediated signalling pathways in plant cells (Mittler et al. 2004). Cellular responses to water deficits are no exception. ABA action on guard cell function is a clear example. ROS, generated at the plasma membrane by NADPH oxidases, is part of the ABA signalling pathway in guard cells (Kwak et al. 2003).

A ROS-responsive gene network has been identified, comprising 152 genes to date (Mittler et al. 2004). Miller et al. (2008) summarize the current state of understanding of this network by classifying ROS sensing mechanisms into (1) processes mediated by (unknown) receptor proteins, (2) those involving redox-sensitive transcription factors (TFs), such as the heat shock TFs, and (3) ROS inhibition of phosphatases. Epigenomic-related phenomena are also likely involved in ROS sensing mechanisms (see below). Heat shock TFs respond to developmental as well as to stress signals, providing a specific link between the two categories of signals (von Koskull-Doring et al. 2007).

Redox-sensitive TFs have been shown to participate in signalling pathways involving the Zn finger family, such as Zat 10, 12, and 7, multiprotein bridging factor 1e, LEA proteins, trehalose, and antioxidant defence proteins such as cytosolic ascorbate peroxidase 1 (APX1) (reviewed by Miller et al. 2008, and Fig. 13.1 therein). Over-expression of Zat 10 and 12 each results in increased stress tolerance, although the two TFs appear to control the expression of different subsets of downstream defence genes (Mittler et al. 2006). Paradoxically, Zat10 mutant plants

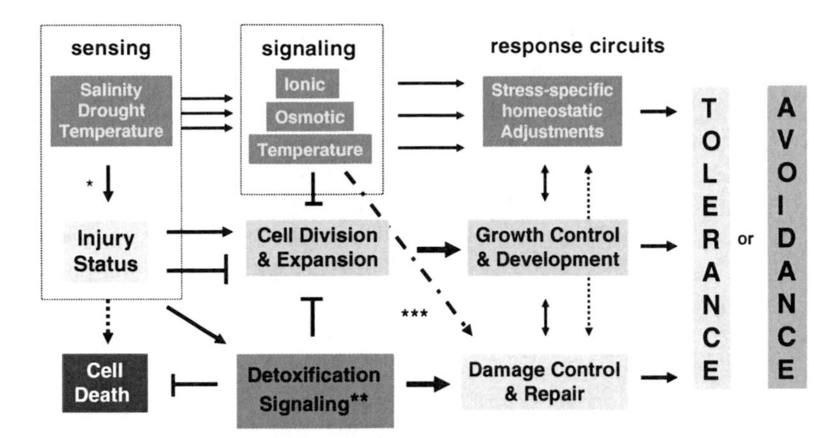

Fig. 13.1 The complexity of abiotic stress effects and responses. Integrated sensing, signalling, and response pathways. Stresses elicit specific and general, injury-based, responses leading to altered regulation of sets of genes whose functions often overlap, although different orthologs may be at the basis of the response. *protein unfolding, membrane leakage, water/ion imbalance; **distinct pathway, not specific for a particular stress; ***overlapping pathways [modified after Zhu (2001) and Bohnert and Bressan (2001)]

also showed increased tolerance to osmotic but not to heat stress, betokening, perhaps, compensation from other, as yet ill-understood, distinct stress signalling pathways (Gadjev et al. 2006).

13.2.2 Stress Response Circuits in Context

Generally, drought can be integrated into a global stress recognition and integration response network that can provide a view of underlying complexities (Fig. 13.1). Sensing of a particular stress condition includes specific reactions that identify the stress, but owing to the sensing of injury that accompanies any stress, different responses are elicited by additional signalling pathway or pathways that measure injury. While the abiotic stress-specific pathways lead to responses that are to some degree known, signal integration begins to affect cell division/expansion, growth, and development. Initiated by injury, for example, the components of endoplasmatic reticulum and denatured protein signals, detoxification signals begin to control damage and affect repair reactions. Depending on the severity of a stress, repair and stress-specific countermeasures can lead to tolerance, but may also lead to (programmed) cell and plant death (Fig. 13.1). Many components linking tolerance to signalling remain "black boxes". We do not know in detail how the initial stress sensing is transmitted into cellular signals, nor have we knowledge of intercellular signals, although some components of that pathway are already established. Other complete unknowns include the signalling connection(s) between the environment and the structure of chromatin/chromosomes.

13.2.3 What Lies at the Basis of Stress Signalling?

The signalling events that are at the basis of and set in motion molecular and then physiological stress responses are not known precisely. Conceivable, however, are changes in the (plasma) membrane potential, disturbances of lipid membranes, and/or transmembrane receptor conformational changes are among the very early events. An altered membrane potential will alter the activity of various ATPases that generate this potential and also alter activities of ion transporters (Fig. 13.2). Well documented is the stress-induced activation of Ca^{2+} channels and ATPases that can generate calcium spikes that are defined by amplitude, frequency, and location. Changes in hormone distribution likely occur early on, but most certainly involved is also the activation of protein modification in the form of phosphorylation cascades and of signals that are generated from lipids in membranes (Albrecht et al. 2003; Pandey et al. 2004; Mishra et al. 2006a, b; Qudeimat et al. 2008; Hong et al. 2008; Kant et al. 2008).

This sensing and then the initial signalling set in motion the complex machinery of transcriptional activation and repression, lead to altered protein synthesis and protein turnover, and the activation of downstream biochemical and developmental changes that are the end-points of physiological and phenological studies (Zhu 2001). While this view is supported by a number of studies in plants, and while the view is analogous to initial perception and signalling of altered homeostasis in animal, yeast, and bacterial models, it has been difficult to take hold in plants because the scenario just outlined describes rapid processes. The scenario can be generated under controlled conditions of experimentation by shock treatment of plants, and this is indeed how many of the genes and signalling pathways involved

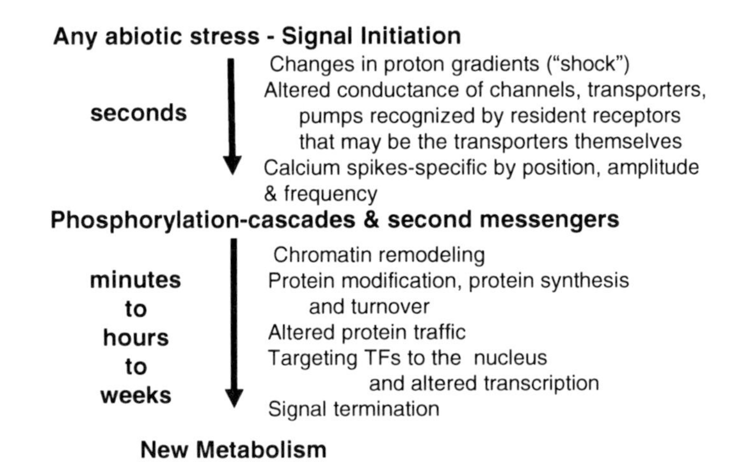

Fig. 13.2 A unifying hypothesis for stress response initiation. Different "lifestyles" distinguishing species, or lines or ecotypes of one species, seem related to different input sensitivity, response capacity, and/or the presence of paralogous genes. Long-term gradual stress measures relative acclimation success

Table 13.1 Elements of water-deficit stresses

Timing	Developmental windows during which scarcity of water affects growth, development, fruiting or yield, during:	
	Germination	Lopez-Molina et al. (2002)
	Vegetative growth	Shinozaki and Yamaguchi-Shinozaki (2007)
	Flowering	Wasilewska et al. (2008)
	Seed filling	Makela et al. (2005)
Duration		
	Development × environment	Barnabas et al. (2008)
Severity and aggravating circumstances		
	Temperature, irradiation, salinity	
	Air humidity, soil conditions, wind	Mittler (2006)

Table 13.2 Potentially distinguishing characters giving rise to relative (drought) stress tolerance

Gene number	Amplification of genes in families, orthologues, or paralogues increases with possibly altered cell-, tissue-, or condition-specificity, Sakurai et al. (2007)
Gene nature	Novel or significantly mutated genes in xerophytes, or halophytes, Hong et al. (2008)
Gene expression strength	Increased copy numbers of transcripts for evolutionarily "selected" pathways, termed CNV (copy number variation), Stranger et al. (2007)
Coding region diversity	Protein domain structure and combinations of domains (allelic difference selections)
Structure of regulatory circuits	Promoter sequence evolution, Xiao et al. (2008)
	TF dynamics in type, amount, and stability, Zhou et al. (2007)
	miRNA evolution and complexity, intergenic sequence complexity and chromatin domain structure, Chinnusamy et al. (2008)

in plant stress responses, biotic and abiotic, have been traced (Matsui et al. 2008; Hirt and Shinozaki 2004; Kreps et al. 2002; Kawasaki et al. 2001). Under natural conditions, it is often argued that plants rarely experience such shocks; rather, stress conditions – and drought stress in particular – develop gradually, over days at least (Tables 13.1 and 13.2).

13.3 Genomics

Processes outlined in Fig. 13.2 have been documented in plants in experiments that identified individual genes and proteins involved and pathways of sensing and signalling. Many of these genes are also transcriptionally induced in other organisms, animal, and fungal species alike. A recent analysis organized *Arabidopsis* genes into more than 100 clusters according to their transcription induced by a large number of different conditions and treatments (Ma and Bohnert 2007). One cluster stood out: approximately 200 of the genes included on the Affymetrix ATH1 array platform showed induction by a large number of different biotic and abiotic stresses, and many of these genes include well-known *Arabidopsis* "stress genes" that are similar stress response genes in other kingdoms. The cluster includes many transcription factors, kinases in mitogen-activated protein kinase (MAPK) pathways, and calcium- and calmodulin-dependent signalling processes. By considering such a deep evolutionary root for stress-induced pathways, it appears quite clear that all higher plants share the basal machinery of stress-responsive pathways; yet not all plants show tolerance, but rather relative tolerance spans orders of magnitude. In essence, it is not gene nature that distinguishes plants – only a limited number of truly novel genes have been described. Uncovering the basis of stress sensitivity or tolerance had to employ the new concepts that arrived with the technological breakthroughs associated with the term genomics, in which we include the terms transcriptomics, proteomics, and metabolomics.

If the nature of genes with functions in stress perception, signalling, and responses is based on an endowment of all species during the evolution of land plants, differences in tolerance must be based on other characters. Table 13.3 lists several scenarios and possibilities. Sakurai et al. (2007) compared expressed sequence tags (ESTs) from cassava, a highly drought-tolerant species, with those of *Arabidopsis* and observed lineage-specific expansions in gene families of cassava related to stress response. Taji et al. (2004) and Gong et al. (2005) each observed differences in transcript abundance between *Arabidopsis* and its more stress-tolerant relative *Thellungiella halophila*. Carjuzaa et al. (2008) report differences in dehydrin populations between quinoa with differing drought tolerances.

Genomics tools, and in particular genome sequences, transcript profiles, and metabolites, allow us to decide which of the scenarios are at the basis of differential stress tolerance phenotypes. Several plant genome sequences, foremost *Arabidopsis*, and also rice, *Physcomitrella*, sorghum, and poplar, have been completed. We have extensive knowledge about large portions of the tomato, maize, soybean, or wheat genomes, and several other plant species are in sequencing pipelines. The completed genome sequences have conclusively shown that the majority, and possibly all, of the genes guiding sensing/signalling are all encoded in the chromosomes of all plant species. With these genomics and bioinformatics resources, it is increasingly possible to investigate the genes and the regulatory mechanisms that underlie relative dehydration tolerances among different species and, importantly, among closely related genotypes that show different stress behaviour. The significance of genomics-enabling tools cannot be overestimated because of the ever-increasing speed of analysis. Data acquisition in unheard of number, complexity, and accuracy will make complete genome sequences and dynamic transcript profile images affordable to most laboratories. Once a genome sequence template or transcript profile has been established in a species, detecting allelic differences, mutations, or transgenic alterations are easily mapped and placed in the context of

Table 13.3 List of genes with regulation similar to that of ABI5

ABI3	Abscisic acid-insensitive protein 3 (ABI3)
ABI5	bZIP transcription factor family protein; similar to ABA-responsive element binding protein 1 (AREB1)
AT1G03790	Zinc finger (CCCH-type) family protein
AT1G15330	CBS domain-containing protein, low similarity to SP:Q9MYP4 5′-AMP-activated protein kinase, gamma-3 subunit (AMPK gamma-3 chain) (AMPK gamma3)
AT1G24735	Caffeoyl-CoA 3-O-methyltransferase, putative
AT1G30860	Expressed protein
AT1G65090	Similar to hypothetical protein [*Arabidopsis thaliana*]
AT1G69800	CBS domain-containing protein, low similarity to SP:Q9UGI9 5′-MP-activated protein kinase, gamma-3 subunit (AMPK gamma-3 chain) (AMPK gamma3)
AT1G72100	Late embryogenesis abundant domain-containing protein
AT1G74370	Zinc finger (C3HC4-type RING finger) family protein
AT2G03520	Expressed protein, similar to AtUPS1, an allantoin transporter
AT2G42000	Plant EC metallothionein-like family 15 protein
AT3G22490	Late embryogenesis abundant protein, putative
AT4G04870	CDP-alcohol phosphatidyltransferase family protein
AT4G16210	Enoyl-CoA hydratase/isomerase family protein
AT4G25580	Stress-responsive protein-related, contains weak similarity to low-temperature-induced 65 kDa protein (desiccation-responsive protein 29B)
AT5G04000	Expressed protein
AT5G24130	Expressed protein
AT5G39720	Avirulence-responsive protein-related/avirulence-induced gene (AIG) protein-related
AT5G43770	Proline-rich family protein, AT5G52300, low-temperature-responsive 65 kDa protein (LTI65)/desiccation-responsive protein 29B (RD29B)
AT5G55750	Hydroxyproline-rich glycoprotein family proteinAT5G56100 glycine-rich protein/oleosin
AT5G60760	2-Phosphoglycerate kinase-related
ATNRT2.7	Transporter, putative, similar to transmembrane nitrate transporter protein AtNRT2:1
CYP71B14	Cytochrome P450 71B14
GASA2	Gibberellin-regulated protein 2 (GASA2)

the organismal response to the environment. Genomics-type studies have initially been anchored on model plants, with *Arabidopsis thaliana* the most well developed, while, for reasons that are not easily understood, the rice community has been less successful. Several (crop) species have joined the genome sequence "club", which now includes the moss *Physcomitrella patens*, poplar, and sorghum (http://www.phytozome.net/index.php), and the numbers of sequenced plant genomes will rapidly increase. For example, a major addition to the maize genome database (MaizeSequence.org Release 3a.50) has just become available (12 December 2008). The release includes DNA sequence and annotations of 16,587 B73 BAC clones for close to 3 billion bases of the maize genome sequence (http://www.maizesequence.org/version.html). The challenge for the future will be the development and application of more advanced and self-learning integrative bioinformatics tools that can make sense out of the data flood.

There will be multiple species for which genome sequences, transcript profiles, and metabolite catalogues are available. Ideally, these would cover every plant family and divergently adapted (multiple) species within each family. For example, within *Arabidopsis thaliana,* hundreds of ecotypes exist that have been separated during the waxing and waning of glaciers in several cycles into fragmented habitats, in which different evolutionary adaptations were required for survival (Mckay et al. 2003; Sharbel et al. 2000; Clark et al. 2007). Taxa closely related to *Arabidopsis*, such as *Thlaspi* sp. and *Thellungiella* sp., include species adapted to extreme habitats – tundra or high mountains for example – although we begin to realize that their genetic make-up, at least in terms of protein coding regions, is only marginally more diverse than the difference between hominids of the species *Pantroglodytes* and *Homo sapiens*.

Up to now, we have discussed knowledge that followed the classical molecular biology dogma that correlated organismal response to the environment with the programmed transcription of downstream genes, their translation, and integration of newly made proteins into existing biochemical pathways that then altered physiology and phenotype. The dogma has to be revised. The revision will most probably require a more fundamental reorientation of our view of how plants work (Fig. 13.3).

Transposon-driven changes in gene number, expression, and function have been a major mechanism during evolution (Fedoroff 2000). Indications for stress-dependent transposition events and chromatin re-arrangements have been provided (Cullis 1973, 2005; Schneeberger and Cullis 1991). More recent is the recognition of mechanisms that let (all) organisms exploit challenges by the environment to promote phenotypic variation and evolutionary adaptation (Lopez-Maury et al. 2008). Apart from transposon action, mechanisms include pathways that lead to alterations in chromatin structures that initiate altered gene expression programs that may persist for short times but, in other examples, may become imprinted and thus transmitted to progeny. These mechanisms are also represented by "micro" RNAs (miRNAs) and "small interfering" RNAs (siRNAs), genome encode sequences that are processed into short – ~20–24 nucleotide long – oligonucleotides that can bind to complementary sequences and thus lead to transcriptional and translational gene silencing (Sunkar et al. 2007a, b). Relevant to the abiotic stress discussion is the detection of a large, evolutionarily partly conserved, number of these small RNA genes and the recognition that their miRNA and siRNA products target an equally large number of protein coding transcripts (Chinnusamy et al. 2008; Jones-Rhoades et al. 2006; Sunkar et al. 2006, 2007a, b; Vazquez et al. 2004). For example, Sunkar et al. (2006) showed that specific miRNAs modulate the expression of two Cu/Zn superoxide dismutases. Although representatives of mi/siRNAs were detected in *Caenorhabditis elegans* more than 15 years ago, their formidable effect on organisms has only recently been appreciated as more genome sequences became available. It seems quite clear that our vision of plant development and (abiotic) stress responses will require major revisions based on the action of miRNAs.

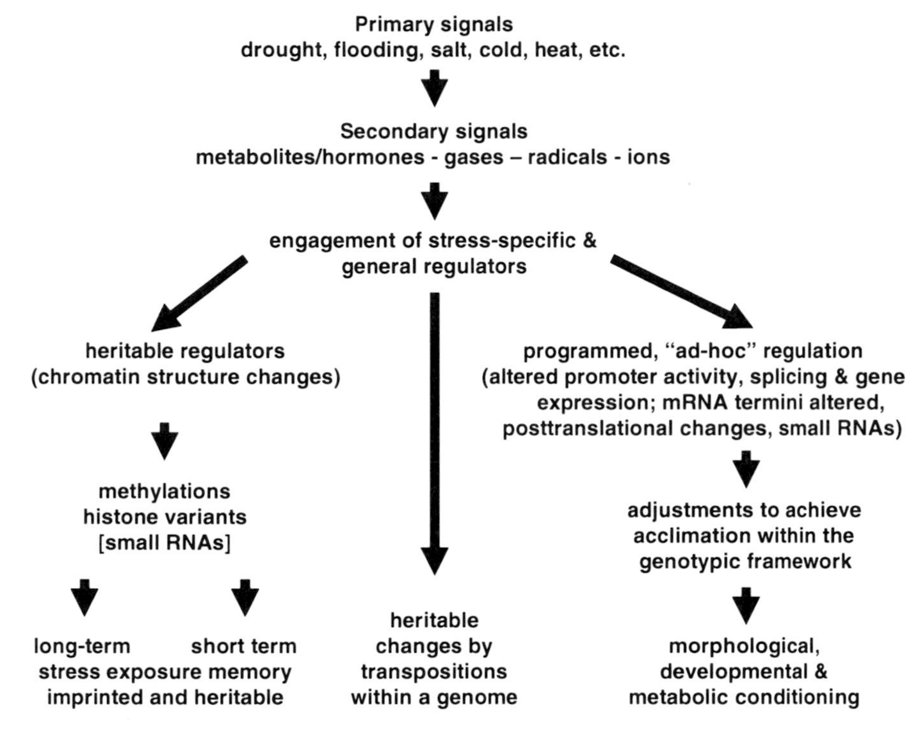

Fig. 13.3 From signals to effectors at the genome level. Programmes leading to stress tolerance or avoidance. In addition to well-studied mechanisms of altered gene expression as a response to various stresses, recent insights have revealed the extraordinary importance of (non-coding) RNA-based regulatory mechanisms. Further, stress-induced changes in genome structure and altered regulation of chromatin domains are topics that deserve attention as the tools to reliably determine such changes have matured

13.4 Drought-Responsive Molecular Mechanisms

13.4.1 Drought Signalling

From existing data, a partial picture emerges for ABA-dependent and independent drought signalling pathways, where some molecular events are understood, with a "black box" often remaining between those events and phenotypic outcomes. Drought signalling pathways involve calcium sensors calcium-dependent and independent protein kinases the AtCPK and SnRK2 families, ultimately interacting with bZIP TFs in the nucleus (Yamaguchi-Shinozaki and Shinozaki 2006; Shinozaki and Yamaguchi-Shinozaki 2007; Wasilewska et al. 2008). ABA integrates cellular responses to multiple abiotic stresses (Yamaguchi-Shinozaki and Shinozaki 2006), through what may be parallel signalling pathways (Saez et al. 2006) and hence, perhaps, the difficulty in distinguishing separate, ABA-triggered, events.

It is clear, however, that the imposition of drought stress results in increases in ABA biosynthesis (Nambara and Marion-Poll 2005), through activation of members of the NCED (9-cis epoxycarotenoid dioxygenase) gene family, and the subsequent activation of ABA-responsive transcription (Wasilewska et al. 2008), with consequent massive changes in the expression patterns of downstream genes [1,400 genes identified by Hoth et al. (2002) and 2,936 by Nemhauser et al. (2006)]. Several bZIPs may be co-ordinately controlled by AtCPK32, which appears to interact with multiple ABA-responsive bZIPs (ABFs 2, 3, and 4) all containing the core ABA-Responsive Element (ABRE) (Choi et al. 2005). All of the ABFs are expressed in vegetative tissue (Finkelstein et al. 2005). ABA-independent drought-responsive genes contain the DRE *cis*-element, which is bound by members of the *DREB* family.

Guard cell closure is mediated by ABA through SnRK family members, one of which is known as open stomata 1 (OST1) in *Arabidopsis*, in a Ca^{2+}-mediated signalling pathway, also involving ROS (Israelsson et al. 2006). Interestingly, SnRK2.8, which is similar to OST1, also responds to drought independently of ABA. The OST1-related SnRKs appear to phosphorylate bZIP TFs, among them AREB1 and ABI5 (Furihata et al. 2006). Moes et al. (2008) suggest that ABA-induced stomatal closure and ABA-mediated changes in gene expression may, in fact, constitute different modes of ABA action. Evidence for this hypothesis may be afforded by the finding that ABF3 functions in drought responses, but to a much lesser extent in stomatal regulation (Finkelstein et al. 2005).

13.4.2 Mechanisms for Modulating Sensitivity to ABA

Since constitutive expression of drought-responsive genes such *as DREB1A* lead to pleiotropic effects, such as stunted growth, attenuation mechanisms for drought stress signalling pathways are clearly of great functional significance for the plant as whole protein phosphatases (PP2Cs, Moes et al. 2008; PP2As, Pernas et al. 2007) act to modulate ABA sensitivity, among other factors. Some members of the PP2C family (ABI1, ABI2, AtPP2C, and HAB1) are known to be negative regulators of ABA action, which, when over-expressed in *Arabidopsis*, showed an ABA-hypersensitive phenotype (Schweighofer et al. 2004). Sensitivity to ABA, in the context of gene activation, is mediated by, among other factors, ABI1, which is targeted to the nucleus where it interacts with ATHB6, a TF (Himmelbach et al. 2002; Moes et al. 2008). The sites of action of protein phosphatases on ABA signalling are not restricted to the nucleus, however. In the cytosol, ABI1 has a role in inhibiting stomatal closure, perhaps through dephosphorylation of OST1 (Wasilewska et al. 2008). ABA promotes the release of phosphatidic acid (PA), itself a regulator of many stress responses (Zhang et al. 2005a, b; Wang et al. 2006), from phosphatidylcholine in the plasma membrane by the action of phospholipase D alpha 1 (PLD alpha1). PA binds to, and inactivates, the ABI1 protein resulting in an increase in ABA-mediated stomatal closure by titration of the levels of free ABI1 protein

(Zhang et al. 2004), providing a functional link between two families of important signalling enzymes, PLDs and PP2Cs. Moes et al. (2008), however, report an essential requirement of ABI1 for localization to the nucleus for any effect on ABA sensitivity. The source of this apparent contradiction is not yet understood.

Mishra et al. (2006a, b) report the existence of a bifurcating pathway for ABA effects on stomatal closing and opening. The action of a PP2A, together with SnRK3, is linked to ABA effects on plasma membrane transport (Fuglsang et al. 2006). ABA sensitivity has also been related, through mechanisms that are yet unknown, to such post-transcriptional events as mRNA capping and poly(A) tail processing (Papp et al. 2004; Kariola et al. 2006). Kant et al. (2007) report yet another mechanism for the attenuation of ABA-dependent and ABA-independent stress signalling sub-networks through the action of two DEAD RNA helicases, acting to repress the expression of upstream transcriptional activators. Mutants in either of these two genes (*STRESS RESPONSE SUPPRESSOR 1 AND 2, STR1 and STR2*) exhibit increased expression of stress-responsive genes such as *RD29A*, *DREB1A*, and *ATMYC*. Kant et al. (2007) suggest that *STR1* and *STR2* may act to degrade stress-induced mRNAs. This mechanism may be part of the various phenomena involving the regulatory action of RNA molecules on gene expression known collectively as epigenomics, although the term is most often used exclusively for the regulatory activity of small RNAs.

13.4.3 The Role of Ubiquitination in Modulation of ABA Action

Since stress responses encompass inhibition of growth, as cellular resources are diverted to defence, it is not surprising that a common mechanism underlies these two aspects of plant function. ABA modulates both stress responses and developmental transitions, e.g., the transition from vegetative to reproductive phases and from embryonic identity to germinative process (Finkelstein et al. 2002). Ubiquitination mechanisms play a central role in these regulatory processes, through effects on ABA levels, and ABA signalling (Zhang and Xie 2007). Of the 1,354 identified ABA-responsive genes, 25 are associated with proteolysis (Hoth et al. 2002). These include E3 ligases, which are thought to confer substrate specificity on the proteloysis process (e.g., RING fingers), F-, and U-box. In some cases, details of the mechanism are understood. The role of the transcription factors ABI5 and ABI3 has been well studied, using ABA-mediated inhibition of seedling growth by water deficit as an experimental system (Lopez-Molina et al. 2001, 2002, 2003). In the absence of ABA, under well-watered conditions, ABI5 and ABI3 are targeted for proteolysis by the proteasome. In the presence of ABA, ABI5 acts to bring about induction of ABA-responsive genes. ABI3 acts upstream of ABI5 in ABA signalling and exerts its gene activation function when bound to ABI5. ABA modulates ABI3 levels through regulation of the expression of its corresponding E3 ligase, AIP2 (Zhang et al. 2005a, b). Yeast two-hybrid assays revealed that ABI5 interacts with AFP1, which leads to the breakdown of ABI5. The ubiquitinating pathway is itself modulated by the action of small ubiquitin-like molecules (SUMO) that bind to ubiquitin substrates and block attachment of E3 ligases (Lois et al. 2003). Keep on Going (KEG), another RING E3 ligase, also appears to negatively regulate ABA signalling through modulation of ABI5 levels (Stone et al. 2006).

Other RING E3 ligases that are active in the modulation of ABA action have also been discovered. DREB2A specifically interacts with two such ligases (Qin et al. 2008), named DRIP 1 and DRIP2 to negatively regulate drought responses. Double mutants in DRIP1 and DRIP2 were relatively drought-tolerant, compared to wild-type plants, and showed delayed vegetative development and delayed time to flowering, illustrating, again, the link between developmental control and stress responses (Qin et al. 2008).

Two other RING proteins that have been linked to ABA action appear to exert regulatory control in the opposite, positive, direction. When SDIR1 (SALT AND DROUGHT-INDUCED RING FINGER 1), an E3 ligase, was over-expressed, plants showed increased drought tolerance. ABA treatment of SDIR1 over-expressors resulted in increased stomatal responsiveness and increased inhibition of root growth (Zhang et al. 2007). The opposite effects were observed in sdir1 mutants that could be rescued by ABI5, ABI3 and ABF4, leading Zhang et al. (2007) to propose that SDIR1 acts upstream of these TFs. No effects on PLD alpha 1 were observed by Zhang et al. (2007) that could be associated with SDIR1 action, suggesting, again, parallel pathways of ABA action. Ko et al. (2006) have described another RING protein, XERICO, which confers drought tolerance, through increasing ABA levels. Plants over-expressing XERICO, which were hypersensitive to ABA, showed increased drought-mediated expression of NCED3, the key ABA biosynthesis gene, as well as increased expression of the known drought-responsive gene, RD29A. Interestingly, a gene encoding an ABA degradation enzyme, AtCYP707A2, was also induced in XERICO over-expressors, suggesting that XERICO affects ABA homeostasis. Yeast two-hybrid assays showed an interaction between XERICO and AtTLP9, an ASK-I interacting F protein, which participates in ABA signalling (Lai et al. 2004). ASK1 is involved in both vegetative and reproductive growth (Ko et al. 2006), highlighting once again the central role of ABA in both stress responses and growth.

Figure 13.4 and Table 13.3 depict a gene co-expression network generated by the method of Ma and Bohnert (2007) for *ABI5*-associated genes. Several genes in this network graph have already been determined as important, or at least affected, in ABA response pathways. *ABI3* appears in the network, as do Responsive to dehydration (RD29B), a marker for drought-responsive gene expression, and two LEA genes, At3g22490 and At1g72100. At1g74370 and At1g30860, encoding RING proteins, are also present. At1g74370 has not, of yet, been implicated in drought signalling pathways, and At1g30860 is associated with embryo sac arrest, plausibly a drought-related phenomenon.

Three other genes that appear in the network are of potential interest for their links with other regulatory mechanisms. At1g24735 is a target gene of a phytochrome A-signalling pathway, which includes redox-associated genes. The pathway is controlled by transposase-related proteins FHY3 and FAR1

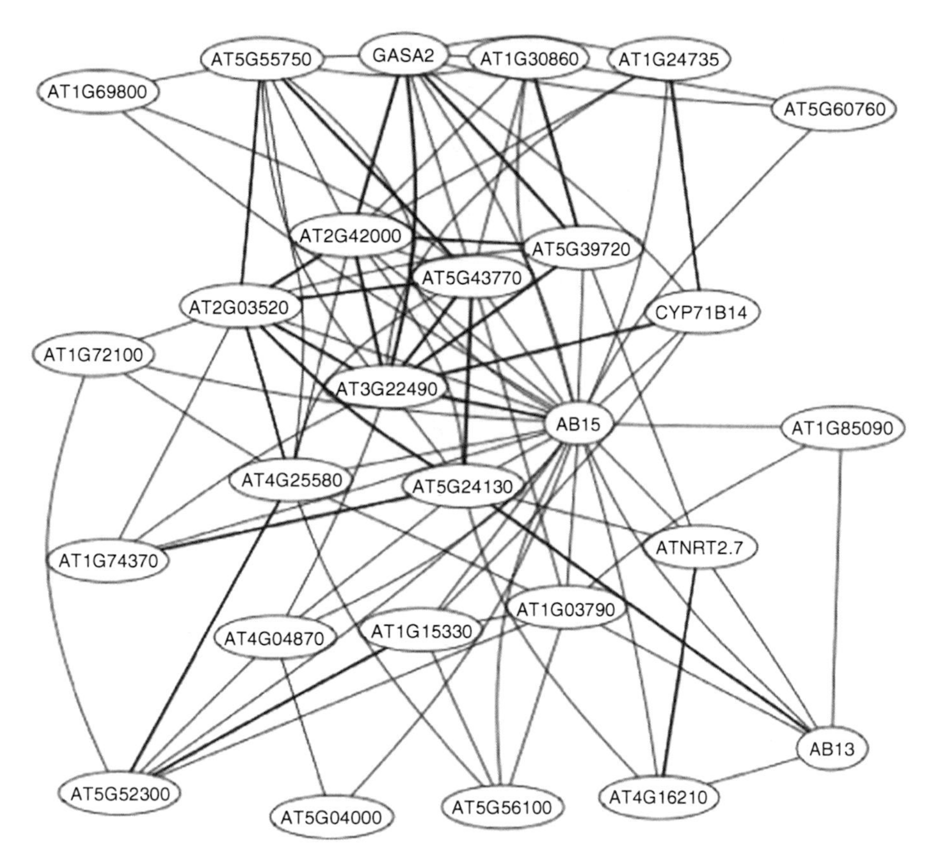

Fig. 13.4 Graphical Display of Genes Co-expressed with ABI5. The model used for generating clusters and gene networks (Ma and Bohnert 2007, 2008a, b; Ma et al. 2007; Li et al. 2008) using publicly available data for *Arabidopsis thaliana* presents the co-regulation of 22,000 genes under more than 100 experimental conditions, including abiotic stress conditions. The model is based on the Graphical Gaussian Model (GGM; Schäfer and Strimmer 2005). In one of three versions (Ma and Bohnert 2008a, b), the model includes approximately 15,000 of the genes in the network, i.e., the model is not completely scale-free

(Hudson et al. 2003), providing a possible link between stress responses, the action of transposons, and development. Adai et al. (2005) developed an algorithm to predict miRNA targets. At1g69800, which appears in the network, is one of the predicted target genes. At1g24735/*SOMNUS* encoding an enzyme of the phenylpropanoid pathway, which negatively regulates light-dependent seed germination, is reported to be responsive to drought stress by Yazaki et al. (2004). The regulatory action of *ABI5* may be a pivotal control point in the nexus of links between the perturbation exerted by stress responses and related redox processes on development. These results are testimony to the power of genomics data and bioinformatics approaches as hypothesis-generating tools.

13.4.4 An Unexpected Role for Circadian-Associated Genes in the Regulation of Stress Responses

Kant et al. (2008) mined existing data from eight different studies to identify *Arabidopsis* genes with early responses to multiple abiotic stresses. To their surprise, *CCA1*, *CIRCADIAN CLOCK ASSOCIATED 1*, a myb-related TF, scored the highest of all genes in their test for multiple stress regulatory genes (MSTR). *LHY*, *LATE ELONGATED HYPOCOTYL*, functionally a closely related gene, also scored high as an MSTR. The group then tested effects of osmotic, salt, and heat stress on *Arabidopsis* single and double mutants for genes CCA1 and LHY. The double, but not the single, mutants proved to be more stress sensitive in each case, providing good evidence for a central role of these clock-associated genes in the regulation of responses to abiotic stresses. Other clock-associated genes such as *CONSTANS-LIKE 1 (COL1)* also scored high in the MSTR test. Covington et al. (2008) report that a number of stress-responsive processes, such as the mevalonate biosynthetic pathway leading to ABA, tocopherol and carotenoid production, are regulated by the circadian clock. For ABA-responsive genes, in particular, Covington et al. (2008) observed enrichment for genes that are clock-regulated, suggesting an intimate connection between responses to the two factors (Table 13.4).

A co-expression network for CCA1, derived using the Graphical Gaussian Model, described by Ma et al. (2007), is shown in Fig. 13.5. *COL1* appears in the network, as does *COL2*, a closely related gene, *LHY*, and At3g09600, another myb-like TF, and a LHY-CCA1-like gene. At3g09600 has been reported by Yanhui et al. (2006) to be responsive to ABA, among many other plant hormones, and by Schmid et al. (2005) to be expressed in leaves, flowers, and seeds. It is possible that this TF provides one of the links between clock functions and stress regulatory processes. Another candidate for joint regulation by stress and the clock is At3g51920, encoding calmodulin-9 (*AtCML9*), a member of the plant-specific group of calcium-binding, calmoduin-like proteins (Magnan et al. 2008). Magnan et al. (2008) report that plants with mutations in AtCML9 show hypersensitivity to ABA during germination and cotyledon greening and are more drought-tolerant than the wild-type counterparts. It would be very interesting to study this gene in depth to understand its role in drought responses and the mechanism by which it is controlled through the circadian clock.

13.5 Genetically Programmed Desiccation Tolerance in Seeds

Considering the seed desiccation life maintenance programme, we feel its incorporation necessary because of its relevance to networks that govern vegetative tolerance. Events underlying the desiccation/re-hydration mechanisms that have been revealed in resurrection plants are also relevant (see Chap. 16).

Table 13.4 List of genes with regulation similar to that of the myb-related transcription factor CCA1

AT1G01520	myb family transcription factor
AT1G12580	Protein kinase family protein
AT1G26790	Dof-type zinc finger domain-containing protein
AT1G32900	Starch synthase, putative
AT1G69570	Dof-type zinc finger domain-containing protein
AT2G21320	Zinc finger (B-box type) family protein
AT2G41250	Haloacid dehalogenase-like hydrolase family protein
AT2G47490	Mitochondrial substrate carrier family protein
AT3G09600	myb family transcription factor
AT3G09600.b	Similar to myb family transcription factor
AT3G12320	Expressed protein
AT3G15310	Expressed protein
AT3G47500	Dof-type zinc finger domain-containing protein
AT3G54500	Expressed protein
AT3G55580	Regulator of chromosome condensation (RCC1) family protein
AT4G15430	ERD protein-related, similar to ERD4 protein (early responsive to dehydration stress)
AT4G38960	Zinc finger (B-box type) family protein
AT5G06980	Expressed protein
AT5G14760	L-aspartate oxidase family protein
AT5G15950	Adenosylmethionine decarboxylase family protein
AT5G17300	myb family transcription factor
AT5G22390	Expressed protein
AT5G39660	Dof-type zinc finger domain-containing protein
AT5G54130	Calcium-binding EF hand family protein
AT5G62430	Dof-type zinc finger domain-containing protein
AT5G64170	Dentin sialophosphoprotein-related
ATATH13	ABC1 family protein
ATGCN5	ABC transporter family protein
CAM9	Calmodulin-9 (CAM9)
CCA1	myb-related transcription factor (CCA1)
CIPK7	CBL-interacting protein kinase 7 (CIPK7)
COL1	Zinc finger protein CONSTANS-LIKE 1 (COL1)
COL2	Zinc finger protein CONSTANS-LIKE 2 (COL2)
LHCB2.2	Chlorophyll A-B binding protein/LHCII type II (LHCB2.2)
LHY	myb family transcription factor
SIGE	RNA polymerase sigma subunit SigE (sigE)
STH	Zinc finger (B-box type) family protein/salt tolerance-like protein

With seed maturation, a shift occurs from maternal to filial control, and, eventually, "orthodox" seeds acquire desiccation tolerance (DT). Within hours or days of germination, DT is lost (Prieto Dapena et al. 2008). It appears that many of the control mechanisms that govern the late stages of seed maturation are of the same type, often involving the same, or closely related, members of the same multi-gene families, as those occurring in vegetative tissue. Because of this similarity, some of the mechanisms that control the acquisition of DT in seeds may correspond to those occurring during drought responses in other tissues.

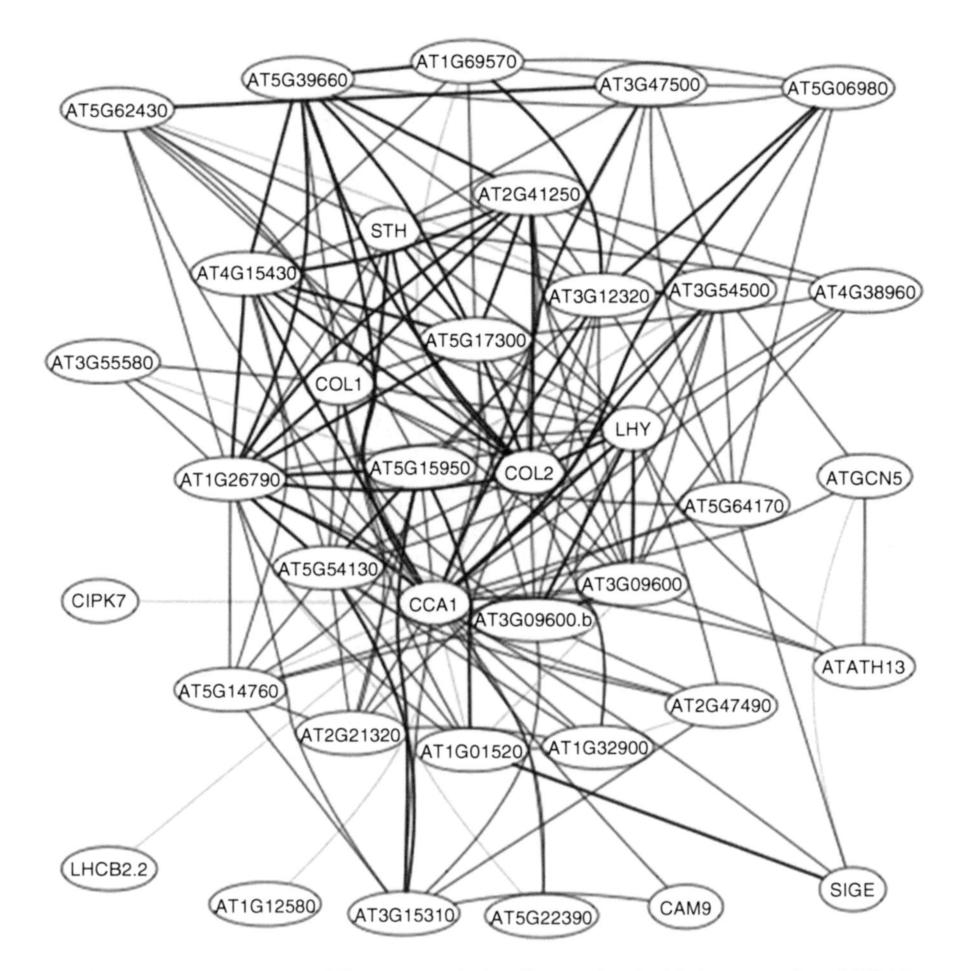

Fig. 13.5 Graphical Display of Genes transcriptionally correlated with the expression of CCA1, a myb-related transcription factor involved in circadian rhythmicity and stress responses. Methods as in Fig. 13.4

13.5.1 The Role of Hormones

The acquisition of desiccation tolerance by the seeds of some higher plants represents an evolutionary advantage since it enables plants to survive harsh environments for indefinite periods of time. In a recent report, a seed that was 2,000 years old proved to be viable (Sallon et al. 2008). Towards the end of seed maturation, cell cycle activities and embryo growth cease, cell expansion rates increase, and the seed acquires desiccation tolerance (17–20 days after flowering in *Arabidopsis* Santos-Mendoza et al. 2008). The ratio of ABA to giberellic acid (GA) in seeds regulates the maturation process and, later, germination. ABA suppresses GA biosynthesis during seed development through the induction of ATGA2ox6, a GA oxidase (Gutierrez et al. 2007). AGAMOUS-like (AGL15), a MADS box protein, is expressed in seeds and affects the expression of GA oxidation genes, thus influencing seed maturation.

ABA biosynthesis is essential for seed maturation and the onset of dormancy. ABA levels peak during maturation, later dropping when germination is initiated (Holdsworth et al. 2008). Two ABA biosynthetic genes, NCED6 and NCED9, are key to the accumulation of ABA during maturation in *Arabidopsis*, and in barley, another member of the homologous gene family, HvNCED2, is responsible for increases in ABA levels during grain development (Gutierrez et al. 2007). ABI3, one of the master regulators of seed maturation, which is also active in vegetative tissue, is expressed in seed embryos, with expression patterns characteristic of its target genes in the former tissue. As in the case of vegetative tissue, ubiquitination is a key factor in modulating ABA-mediated pathways in seeds, since AIP2, an E3 ligase, degrades ABI3 under conditions favourable for germination. PICKLE, a chromatin remodelling gene, represses ABI3 and ABI5 expression during germination (Susuki et al. 2007). ABI5, a bZip TF, belongs to a group of embryo-abundant regulatory genes (ABF family) (Kim et al. 2004), other members of which are active in vegetative tissue.

ABI5 affects ABA sensitivity and binds specifically to ABI3 in yeast two-hybrid assays. Other members of the ABI5 gene family are also involved in seed maturation and act with ABI5 to regulate the expression of seed-specific proteins. The *Arabidopsis* regulatory genes LEC1, LEC2, and FUS3, all members of the B3 domain of transcription factors, are involved in the acquisition of desiccation tolerance, since mutants in any of these genes lead to a decrease in the extent of desiccation tolerance and also in the transcription of oil and seed-specific mRNAs (Santos-Mendoza et al. 2005, 2008)

13.5.2 Sugar Signalling

A metabolic switch appears to control the acquisition of desiccation tolerance in developing seeds. In *Medicago truncatula*, it has been shown that the time frame for storage reserve accumulation and acquisition of DT overlap, and it is likely that the two processes have common regulatory mechanisms (Buitink et al. 2006). Sucrose signals have roles in many plant tissues, where they appear to control differentiation and responses to abiotic stress (Rolland et al. 2006; Gibson 2005). It would appear that the same type of signalling mechanisms that occur under other circumstances and tissues are also active in the case of desiccation tolerance. The challenge is to identify those genes/pathways/ processes that are unique to the acquisition of DT, the most extreme form of drought tolerance.

The TF APETALA2 may regulate transition into the maturation and storage phases through an effect on sugar signalling (Gutierrez et al. 2007). LEC1, LEC2, and FUS3 are all sugar-inducible. WRI1 activates a subset of sugar-responsive genes and is, itself, sugar-inducible. Specific sugars such as trehalose-6-phosphate (T6P) play a signalling function during the transition to maturation (Gomez et al. 2006).

This finding is reminiscent of the role of T6P in redox-mediated drought stress signalling in vegetative tissue (Paul et al. 2008) among its other functions, although the mechanisms operating in seeds are not yet understood. The ratio of hexose sugars to sucrose decreases in maturing seeds (Baud et al. 2002), with the ATSUC5 sucrose transporter, which is expressed in the endosperm, playing a major role in the process (Baud et al. 2005). APETALA2 appears to contribute to control of this process. Raffinose family oligosaccharides accumulate during the final stages of seed maturation, along with sucrose, with stachyose constituting the majority of total soluble sugars in dry seeds in *Medicago* (Rosnoblet et al. 2007). SFN4b, a regulatory subunit of SnRK1, acts to affect stachyose accumulation during seed maturation (Rosnablet et al. 2007).

13.5.3 Chaperones and other Protective Proteins

LEAs, a group of proteins defined originally overall by their expression during the later stages of seed maturation, rather than by their structure, are expressed throughout the plant's life cycle in response to abiotic stresses, all of which may be related to water deficit, and also during the acquisition of DT in orthodox seeds (Bray 1997; Ingram and Bartels 1996). They have now been classified into groups based on conservation of domains (Wise 2003). Their precise function in protection is not yet understood (Wise and Tunnacliffe 2004), although in vitro data suggest that LEAs can prevent protein aggregation during water stress (Goyal et al. 2005). Through data mining of publicly accessible microarray and genomic resources for *Arabidopsis*, Illing et al. (2005) identified LEAs in superfamilies 1, 6, and 9 that are only expressed in siliques and maturing seeds. Illing et al. (2005) also determined that 74% of all known LEAs are expressed in seeds, while 46% are expressed in vegetative tissue. Interestingly, seed-specific LEAs could be identified, but no LEAs were identified that were expressed only under abiotic stress in vegetative tissue, suggesting that the acquisition of desiccation tolerance in seeds has unique aspects, in addition to those processes common to water deficit responses such as occur in roots and leaves.

13.6 Roots as Sensors and Conduits of Changes in the Water Potential

There is one aspect of plant reactions to water deficit that has not been studied in sufficient detail – the response of the root system (but see Sharp et al. 2004; Spollen et al. 2008; Poroyko et al. 2004, 2007; Dinneny et al. 2008). Most drought studies conducted to date have focused on responses by the above ground parts of plants. Several studies have compared root and shoot responses (Ozturk et al. 2002; Talamè et al. 2007).

By disregarding the importance of roots in sensing, signalling, and drought defences, our views have been compromised because root/shoot comparisons indicated large, if not fundamental, differences between these tissues. Furthermore, most dehydration stress experiments have employed fast drying schemes (Ozturk et al. 2002; Umezawa et al. 2006; Taji et al. 2002), disregarding the typically slow drying experienced by plants in nature (Mane et al. 2008; Bogeat-Triboulot et al. 2007; Vasquez-Robinet et al. 2008, Watkinson et al. 2008). Irrespective of the bias that may be introduced by the differences in how stress is imposed, fast stress imposition can pinpoint processes related to perception and signalling, whereas the genes coordinating metabolic processes related to acclimation are better identified on a different time scale. Long-term gradual stress measures relative acclimation success.

The variability of root systems has consequences for drought tolerance, and root architectural differences between species are recognized as important components contributing to tolerance, or, rather, drought avoidance. Root architecture (see also Chap. 10) as an important aspect in breeding or molecular adaptation of crops might best be targeting growth regulator biosynthesis and transport, rather than focusing on biochemical pathways that have been identified in drought-tolerant species. For example, the *YUCCA6* gene affects indoleacetic acid/auxin amount and ratios and leads to altered root growth (Kim et al. 2007). ABA transport from roots to shoots is a central organizing principle in the response of the whole plant to drought stress (Schachtman and Goodger 2008; Davies et al. 2005; Dodd 2005). ABA, which is produced in the root tip, accelerates root growth, but when arriving in the shoot leads to a very rapid decline in shoot growth.

A different and important new view has been introduced by Benfey and his collaborators (Dinneny et al. 2008). Using green fluorescent (GFP) painting of different root cell lineages, or the division, elongation, and maturation (root hair development) zones and cell sorting, genes expressed at different levels in the epidermis, cortex, endodermis, and stele could be identified in *Arabidopsis* (Birnbaum et al. 2003). Gene identification could then be correlated with stress-dependent transcript expression providing a distinction between genes induced or repressed under drought, osmotic, or salt stress in different cell lineages and in the different developmental regions of the *Arabidopsis* root (Ma and Bohnert 2007). Dinneny et al. (2008) expanded on this approach and provided a catalogue of cell-specific, stress-identifying genes restricted to particular lineages of cells in the *Arabidopsis* root. It is likely that similar analyses in the future will supersede the older, physiology-based, studies, especially if analyses can be expanded to protein identification and metabolite tags.

13.7 The Potential for Engineering/Breeding Based on Knowledge

Transformation technologies, albeit still irrationally controversial in the public arena, have provided some avenues for improvements, but progress has been slow, which may indicate that single-gene approaches may not provide significant benefits in the field. The focus of many of these transformation strategies originated from extrapolation of results from biochemical and molecular physiology studies. Based on the detection of accumulation of metabolites under drought conditions, which lead to the term "osmotic adjustment" (Taiz and Zeiger 2006), schemes that attempted the accumulation of mannitol/sorbitol, sucrose, inositols, proline, or trehalose were initiated (Tarczynski et al. 1993; Sheveleva et al. 1997; Karim et al. 2007; Pasquali et al. 2008). Many of the strategies have been inspired by observations of high levels of proteins and/or metabolites in naturally drought- or salinity-adapted species, typically non-crops (Sheveleva et al. 1997, and references therein). However, osmotic adjustment has not always correlated with drought resistance (Basu et al. 2007). Although the literature now includes several hundred examples detailing how single genes can contribute to stress tolerance under strictly controlled growth conditions, success in natural settings is either lacking or included in apocryphal communications.

In part, the lack of success seems to have been due to a lack of knowledge about stress-relevant genes and gene control circuits and also about developmental and metabolic processes. In this respect, it may be that future attempts will be honoured with more success. For example, the over-expression of a transcription factor of the NF-Y family in maize has resulted in lines that show significantly improved drought tolerance when compared to lines that were severely damaged by the stress (Nelson et al. 2007). Similarly, increased hormone biosynthesis, promoted by a drought stress-inducible promoter, in tobacco generated a "stay-green" phenotype that allowed the plants to recover from a severe drought stress – administered in a greenhouse – significantly better than wild type (Rivero et al. 2007).

13.8 Where Does This Lead?

The time has come to put the information accumulated over many decades in context, especially since our knowledge on plant molecular genetics and genomics has expanded by orders of magnitude over the last 20 years. Data must be integrated over multiple scales or organizational levels, from the genomic through the whole plant level, including yield, in the case of crop plants. At least in *Arabidopsis*, the number of experiments that have analysed transcription programs during development (e.g., see Birnbaum et al. 2003; Dinneny et al. 2008 for cell-specific gene expression studies in developing roots) and environmental challenges (Ma and Bohnert 2007) is abundant enough to contemplate bioinformatics-type studies [e.g., see VirtualPlant, http://virtualplant.bio.nyu.edu/cgi-bin/vpweb2/ GGM by Ma et al. (2007, 2008) and MapMan by Sreenivasulu et al. (2007)] that begin to reveal transcriptional coupling among pathways and can be employed to posit trajectories from stress signalling to responses. A sufficiently large dataset may become useful in predicting functions of unknown genes, adding genes to pathways, or analysing in silico the consequences of gene engineering on the entire organism.

Application of the tools of the emerging field, for which the term "Systems Biology" has now been coined, are essential to a beginning understanding of the "elephant", through obtaining working mechanistic models of, and generating new hypotheses concerning, dehydration resistance in higher plants.

Acknowledgments The work has been supported by NSF DBI 0223905 and IBN0219322 and by CIP, UIUC, and VT institutional grants.

References

Adai A, Johnson C, Mlotshwa S, Archer-Evans S, Manocha V, Vance V, Sundaresan V (2005) Computational prediction of miRNAs in *Arabidopsis thaliana*. Genome Res 15:78–91

Albrecht V, Weinl S, Blazevic D, D'Angelo C, Batistic O, Kolukisaoglu U, Bock R, Schulz B, Harter K, Kudla J (2003) The calcium sensor CBL1 integrates plant responses to abiotic stresses. Plant J 36:457–470

Arfaoui A, El Hadrami A, Mabrouk Y, Sifi B, Boudabous A, El Hadrami I, Daayf F, Chérif M (2007) Treatment of chickpea with Rhizobium isolates enhances the expression of phenylpropanoid defense-related genes in response to infection by *Fusarium oxysporum* f.sp. *ciceris*. Plant Physiol Biochem 45:470–479

Bailey-Serres J, Mittler R (2006) The roles of reactive oxygen species in plant cells. Plant Physiol 141:311

Bartels D, Sunkar R (2005) Drought and Salt Tolerance in Plants Critical Reviews in Plant Sciences 24:23–58

Barnabas B, Jager K, Feher A (2008) The effect of drought and heat stress on reproductive processes in cereals. Plant Cell Environ 31:11–38

Basu PS, Ali M, Chaturvedi SK (2007) Osmotic adjustment increases water uptake, remobilization of assimilates and maintains photosynthesis in chickpea under drought. Indian J Exp Biol 45:261–267

Baud S, Boutin JP, Miquel M, Lepiniec L, Rochat C (2002) An integrated overview of seed development in *Arabidopsis thaliana* ecotype WS. Plant Physiol Biochem 40:151–160

Baud S, Wuilleme S, Lemoine R, Kronenberger J, Caboche M, Lepiniec L, Rochat C (2005) The AtSUC5 sucrose transporter specifically expressed in the endosperm is involved in early seed development in *Arabidopsis*. Plant J 43:824–836

Birnbaum K, Shasha DE, Wang JY, Jung JW, Lambert GM, Galbraith DW, Benfey PN (2003) A gene expression map of the *Arabidopsis* root. Science 302:1956–1960

Bogeat-Triboulot MB, Brosché M, Renaut J, Jouve L, Le Thiec D, Fayyaz P, Vinocur B, Witters E, Laukens K, Teichmann T, Altman A, Hausman JF, Polle A, Kangasjärvi J, Dreyer E (2007) Gradual soil water depletion results in reversible changes of gene expression, protein profiles, ecophysiology, and growth performance in *Populus euphratica*, a poplar growing in arid regions. Plant Physiol 143:876–892

Bohnert HJ, Bressan RA (2001) Abiotic stresses, plant reactions, and approaches towards improving stress tolerance. In: Nössberger J et al (eds) Crop science: progress and prospects. CABI International, Wallingford, pp 81–100

Bohnert HJ, Shen B (1999) Transformation and compatible solutes. Scientia Horticult 78:237–260

Boyer JS (1982) Plant productivity and environment. Science 218:443–448

Bray EA (1997) Plant responses to water deficit. Trends Plant Sci 2:48–54

Brodribb T (1996) Dynamics of changing intercellular CO_2 COncentration (ci) during drought and determination of minimum functional c_i. Plant Physiol 111:179–185

Buitink J, Leger JJ, Guisle I, Vu BL, Wuillème S, Lamirault G, Le Bars A, Le Meur N, Becker A, Küster H, Leprince O (2006) Transcriptome profiling uncovers metabolic and regulatory processes occurring during the transition from desiccation-sensitive to desiccation-tolerant stages in Medicago truncatula seeds. Plant J 47(5):735–750

Carjuzaa P, Castellión M, Distéfano AJ, del Vas M, Maldonado S (2008) Detection and subcellular localization of dehydrin-like proteins in quinoa (*Chenopodium quinoa* Willd.) embryos. Protoplasma 233:149–156

Chakrabortee S, Boschetti C, Walton LJ, Sarkar S, Rubinsztein DC, Tunnacliffe A (2007) Hydrophilic protein associated with desiccation tolerance exhibits broad protein stabilization function. Proc Natl Acad Sci USA 104:18073–18078

Chaves MM, Maroco JP, Pereira JS (2003) Understanding plant responses to drought – from genes to the whole plant. Funct Plant Biol 30:239–264

Chinnusamy V, Gong Z, Zhu JK (2008) Abscisic acid-mediated epigenetic processes in plant development and stress responses. J Integr Plant Biol 50:1187–1195

Choi H-I, Park H-J, Park JI, Kim S, Im M-Y, Seo H-H, Kim Y-W, Hwang I, Kim SY (2005) *Arabidopsis* calcium-dependent protein kinase AtCPK32 interacts with ABF4, a transcriptional regulator of abscisic acid-responsive gene expression, and modulates its activity. Plant Physiol 139:1750–1761

Clark RM, Schweikert G, Toomajia C, Ossowski S, Zeller G, Shinn P, Warthman N, Hu TT, Fu G, Hinds DA, Chen H, Frazer KA, Huson DH, Schölkopf B, Nordborg M, Rätsch G, Ecker JR, Weigel D (2007) Common sequence polymorphisms shaping genetic diversity in *Arabidopsis thaliana*. Science 317:338–342

Covington MF, Maloof JN, Straume M, Kay SA, Harmer SL (2008) Global transcriptome analysis reveals circadian regulation of key pathways in plant growth and development. Genome Biol 9: R130

Cullis CA (1973) DNA differences between flax genotrophs. Nature 243:515–516

Cullis CA (2005) Mechanisms and control of rapid genomic changes in flax. Ann Bot (Lond) 95:201–206

Davies WJ, Kudoyarova G, Hartung W (2005) Long-distance ABA signaling and its relation to other signaling pathways in the detection of soil drying and the mediation of the plant's response to drought. J Plant Growth Regul 24:285–295

Dinneny JR, Long TA, Wang JY, Jung JW, Mace D, Pointer S, Barron C, Brady SM, Schiefelbein J, Benfey PN (2008) Cell identity mediates the response of *Arabidopsis* roots to abiotic stress. Science 320:942–945

Dodd IC (2005) Root-to-shoot signalling: assessing the roles of up in the up and down world of long-distance signalling in planta. Plant Soil 274:251–270

Easterling WP, Aggarwal P, Batima P, Brander K, Erda L, Howden M, Kirilenko A, Morton J, Soussana J-F, Schmidhuber S, Tubiello F (2007) Food, fibre and forest products. In: Parry ML, Canziani OF, Palutikof JP, van der Linden PJ, Hanson CE (eds) Climate change 2007: impacts, adaptation and vulnerability. Contribution of working group II to the fourth assessment report of the intergovernmental panel on climate change. Cambridge, Cambridge University Press, pp 273–313

Fan XW, Li FM, Xiong YC, An LZ, Long RJ (2008) The cooperative relation between non-hydraulic root signals and osmotic adjustment under water stress improves grain formation for spring wheat varieties. Physiol Plant 132:283–292

Fedoroff NV (2000) Transposons and genome evolution in plants. Proc Natl Acad Sci USA 97:7002–7007

Finkelstein RR, Gampala SSL, Rock CD (2002) Abscisic acid signaling in seeds and seedlings. Plant Cell Suppl 14:S15–S45

Finkelstein R, Gampala S, Lynch TJ, Thomas TL, Rock C (2005) Redundant and Distinct Functions of the ABA Response Loci ABA-INSENSITIVE(ABI)5 and ABRE-BINDING FACTOR (ABF)3. Plant Mol Biol 59:253–267

Foyer and Noctor (2005) Redox homeostasis and antioxidant signaling: a metabolic interface between stress perception and physiological responses. Plant Cell 17:1866–75

Fuglsang AT, Tulinius G, Cui N, Palmgren MG (2006) Protein phosphatase 2A scaffolding subunit A interacts with plasma membrane H+-ATPase C-terminus in the same region as 14-3-3 protein. Physiol Plant 128:334–340

Furihata T, Maruyama K, Fujita Y, Umezawa T, Yoshida R, Shinozaki K, Yamaguchi-Shinozaki K (2006) Abscisic acid-dependent multisite phosphorylation regulates the activity of a transcription activator AREB1. Proc Natl Acad Sci USA 103:1988–1993

Gadjev I, Vanderauwera S, Gechev TS, Laloi C, Minkov IN, Shulaev V, Apel K, Inze D, Mittler R, Van Breusegem F (2006) Transcriptomic footprints disclose specificity of reactive oxygen species signaling in *Arabidopsis*. Plant Physiol 141:436–445

Gibson SI (2005) Control of plant development and gene expression by sugar signaling. Curr Opin Plant Biol 8:93–102

Gong Q, Li P, Ma S, Indu Rupassara S, Bohnert HJ (2005) Salinity stress adaptation competence in the extremophile *Thellungiella halophila* in comparison with its relative *Arabidopsis thaliana*. Plant J 44:826–839

Gutierrez L, Van Wuytswinkel O, Castelain M, Bellini C (2007) Combined networks regulating seed maturation. Trends Plant Sci 12:294–300

Harris K, Subudhi PK, Borrell A, Jordan D, Rosenow D, Nguyen H, Klein P, Klein R, Mullet J (2007) Sorghum stay-green QTL individually reduce post-flowering drought-induced leaf senescence. J Exp Bot 58:327–338

Himmelbach A, Hoffmann T, Leube M, Höhener B, Grill E (2002) Homeodomain protein ATHB6 is a target of the protein phosphatase ABI1 and regulates hormone responses in *Arabidopsis*. EMBO J 21:3029–3038

Hirt H, Shinozaki K (2004) Plant responses to abiotic stress topics in current genetics, vol 4. Springer, Berlin, p 300

Hoekstra FA, Golovina EA, Buitink J (2001) Mechanisms of plant desiccation tolerance. Trends Plant Sci 6:431–438

Holdsworth MJ, Finch-Savage WE, Grappin P, Job D (2008) Post-genomics dissection of seed dormancy and germination. Trends Plant Sci 13:7–13

Hong JK, Choi HW, Hwang IS, Kim DS, Kim NH, du Choi S, Kim YJ, Hwang BK (2008) Function of a novel GDSL-type pepper lipase gene, CaGLIP1, in disease susceptibility and abiotic stress tolerance. Planta 227:539–558

Hoth S, Morgante M, Sanchez JP, Hanafey MK, Tingey SV, Chua NH (2002) Genome-wide gene expression profiling in *Arabidopsis thaliana* reveals new targets of abscisic acid and largely impaired gene regulation in the abi1-1 mutant. J Cell Sci 115:4891–4900

Hudson ME, Lisch DR, Quail PH (2003) The FHY3 and FAR1 genes encode transposase-related proteins involved in regulation of gene expression by the phytochrome A-signaling pathway. Plant J 34:453–471

Illing N, Denby KJ, Collett H, Shen J, Farrant J (2005) The signature of seeds in resurrection plants: a molecular and physiological comparison of desiccation tolerance in seeds and vegetative tissues. Integr Comp Biol 45:771–787

Ingram J, Bartels D (1996) The molecular basis of dehydration tolerance in plants. Annu Rev Plant Physiol Plant Mol Biol 47:377–403

Israelsson M, Siegel RS, Young J, Hashimoto M, Iba K, Schroeder JI (2006) Guard cell ABA and CO_2 signaling network updates and Ca^{2+} sensor priming hypothesis. Curr Opin Plant Biol 9:654–663

Izanloo A, Condon AG, Langridge P, Tester M, Schnurbusch T (2008) Different mechanisms of adaptation to cyclic water stress in two South Australian bread wheat cultivars. J Exp Bot 59:3327–3346

Jenks M (2007) Advances in molecular breeding towards drought and salt tolerant crops. Lavoisier, France, p 806

Jiang Y, Huang B (2001) Osmotic adjustment and root growth associated with drought preconditioning-enhanced heat tolerance in Kentucky bluegrass crop. Science 41:1168–1173

Jones-Rhoades MW, Bartel DP, Bartel B (2006) MicroRNAS and their regulatory roles in plants. Annu Rev Plant Biol 57:19–53

Kant P, Kant S, Gordon M, Shaked R, Barak S (2007) Stress response suppressor1 and stress response suppressor2, two DEAD-box RNA helicases that attenuate *Arabidopsis* responses to multiple abiotic stresses. Plant Physiol 145:814–830

Kant P, Gordon M, Kant S, Zolla G, Davydov O, Heimer YM, Chalifa-Caspi V, Shaked R, Barak S (2008) Functional-genomics-based identification of genes that regulate *Arabidopsis* responses to multiple abiotic stresses. Plant Cell Environ 31:697–714

Karim S, Aronsson H, Ericson H, Pirhonen M, Leyman B, Welin B, Mäntylä E, Palva ET, Van Dijck P, Holmström KO (2007) Improved drought tolerance without undesired side effects in transgenic plants producing trehalose. Plant Mol Biol 64:371–378

Kariola T, Brader G, Helenius E, Li J, Heino P, Palva ET (2006) EARLY RESPONSIVE TO DEHYDRATION 15, a negative regulator of abscisic acid responses in *Arabidopsis*. Plant Physiol 142:1559–1573

Kawaguchi R, Bailey-Serres J (2002) Regulation of translational initiation in plants. Curr Opin Plant Biol 5:460–465

Kawaguchi R, Williams AJ, Bray EA, Bailey-Serres J (2003) Water-deficit-induced translational control in *Nicotiana tabacum*. Plant Cell Environ 26:221–229

Kawaguchi R, Girke T, Bray EA, Bailey-Serres J (2004) Differential mRNA translation contributes to gene regulation under non-stress and dehydration stress conditions in *Arabidopsis thaliana*. Plant J 38:823–839

Kawasaki S, Borchert C, Deyholos M, Wang H, Brazille S, Kawai K, Galbraith DW, Bohnert HJ (2001) Gene expression profiles during the initial phase of salt stress in rice. Plant Cell 13:889–906

Khandelwal A, Elvitigala T, Ghosh B, Quatrano RS (2008) *Arabidopsis* transcriptome reveals control circuits regulating redox homeostasis and the role of an AP2 transcription factor. Plant Physiol 148:2050–2058

Kim S, Kang JY, Cho DI, Park JH, Kim SY (2004) ABF2, an ABRE-binding bZIP factor, is an essential component of glucose signaling and its overexpression affects multiple stress tolerance. Plant J 40:75–87

Kim J-I, Shakhun A, Li P, Jeong J-C, Baek D, Lee S-Y, Blakeslee JJ, Murphy AS, Bohnert HJ, Hasegawa PM, Yun D-J, Bressan RA (2007) Activation of the *Arabidopsis* HYPERTALL (HYT1/YUCCA6) locus affects several auxin-mediated responses to a staygreen phenotype. Plant Physiol 145:722–735

Ko JH, Yang SH, Han KH (2006) Upregulation of an *Arabidopsis* RING-H2 gene, XERICO, confers drought tolerance through increased abscisic acid biosynthesis. Plant J 47(3):343–355

Kotak S, Larkindale J, Lee U, von Koskull-Döring P, Vierling E, Scharf K-D (2007) Complexity of the heat stress response in plants. Curr Opin Plant Biol 10:310–316

Koussevitzky S, Nott A, Mockler TC, Hong F, Sachetto-Martins G, Surpin M, Lim J, Mittler R, Chory J (2007) Signals from chloroplasts converge to regulate nuclear gene expression. Science 316:715–719

Kreps JA, Wu Y, Chang HS, Zhu T, Wang X, Harper JF (2002) Transcriptome changes for *Arabidopsis* in response to salt, osmotic, and cold stress. Plant Physiol 130:2129–4

Kwak JM, Mori IC, Pei ZM, Leonhardt N, Torres MA, Dangl JL, Bloom RE, Bodde S, Jones JD, Schroeder JI (2003) NADPH oxidase AtrbohD and AtbohF genes function in ROS-dependent ABA signaling in *Arabidopsis*. EMBO J 22:2623–2633

Lai CP, Lee CL, Chen PH, Wu SH, Yang CC, Shaw JF (2004) Molecular analyses of the *Arabidopsis* TUBBY-like protein gene family. Plant Physiol 134:1586–1597

Li P, Ma S, Bohnert HJ (2008) Co-expression characteristics of trehalose 6-phosphate phosphatase sub-family genes reveal different functions in a network context. Physiol Plant 133:544–556

Lois LM, Lima CD, Chua NH (2003) Small ubiquitin-like modifier modulates abscisic acid signaling in *Arabidopsis*. Plant Cell 15:1347–1359

Lopez-Maury L, Marguerat S, Bähler J (2008) Tuning gene expression to changing environments: from rapid responses to evolutionary adaptation. Nat Rev Genet 9:583–593
Lopez-Molina L, Mongrand S, Chua NH (2001) A postgermination developmental arrest checkpoint is mediated by abscisic acid and requires the ABI5 transcription factor in *Arabidopsis*. Proc Natl Acad Sci USA 98:4782–4787
Lopez-Molina L, Mongrand S, McLachlin DT, Chait BT, Chua NH (2002) ABI5 acts downstream of ABI3 to execute an ABA-dependent growth arrest during germination. Plant J 32:317–328
Lopez-Molina L, Mongrand S, Kinoshita N, Chua NH (2003) AFP is a novel negative regulator of ABA signaling that promotes ABI5 protein degradation. Genes Dev 17:410–418
Ma S, Bohnert HJ (2007) Integration of *Arabidopsis thaliana* stress-related transcript profiles, promoter structures, and cell-specific expression. Genome Biol 8:R49
Ma S, Bohnert HJ (2008a) Gene networks for the integration and better understanding of gene expression characteristics. Weed Sci J 56:314–321
Ma S, Bohnert HJ (2008b) Genomics data, integration, networks and systems. Mol Biosyst 4:199–204
Ma S, Gong Q, Bohnert HJ (2007) An *Arabidopsis* gene network based on the graphical Gaussian model. Genome Res 17:1614–1625
Magnan F, Ranty B, Charpenteau M, Sotta B, Galaud J-P, Aldon D (2008) Mutations in AtCML9, a calmodulin-like protein from *Arabidopsis thaliana*, alter plant responses to abiotic stress and abscisic acid. Plant J 56:575–589
Makela P, McLaughlin JE, Boyer JS (2005) Imaging and quantifying carbohydrate transport to the developing ovaries of maize. Ann Bot 96:939–949
Mane SP, Vasquez Robinet C, Ulanov A, Schafleitner R, Tincopa L, Gaudin A, Nomberto G, Alvarado C, Solis C, Avila Bolivar L, Blas R, Ortega J, Solis J, Panta A, Rivera C, Samolski I, Carbajulca DH, Bonierbale M, Pati A, Heath LS, Bohnert HJ, Grene R (2008) Molecular and physiological adaptation to prolonged drought stress in the leaves of two Andean potato genotypes Functional Plant Biology 35:669–688
Matsui A, Ishida J, Morosawa T, Mochizuki Y, Kaminuma E, Endo TA, Okamoto M, Nambara E, Nakajima M, Kawashima M, Satou M, Kim JM, Kobayashi N, Toyoda T, Shinozaki K, Seki M (2008) *Arabidopsis* transcriptome analysis under drought, cold, high-salinity and ABA treatment conditions using a tiling array. Plant Cell Physiol 49:1135–1149
Mckay JK, Richards JH, Mitchell-Olds T (2003) Genetics of drought adaptation in *Arabidopsis thaliana*: I. Pleiotropy contributes to genetic correlations among ecological traits. Mol Ecol 12:1137–1151
McNeil SD, Rhodes D, Russell BL, Nuccio ML, Shachar-Hill Y, Hanson AD (2000) Metabolic modeling identifies key constraints on an engineered glycine betaine synthesis pathway in tobacco. Plant Physiol 124:153–162
Miller G, Shulaev V, Mittler R (2008) Reactive oxygen signaling and abiotic stress. Physiol Plant 133:481–485
Mishra G, Zhang W, Deng F, Zhao J, Wang X (2006a) A bifurcating pathway directs abscisic acid effects on stomatal closure and opening in *Arabidopsis*. Science 312:264–266
Mishra NS, Tuteja R, Tuteja N (2006b) Signaling through MAP kinase networks in plants. Arch Biochem Biophys 452:55–68
Mittler R (2006) Abiotic stress, the field environment and stress combination. Trends Plant Sci 11:15–19
Mittler R, Vanderauwera S, Gollery M, Van Breusegem F (2004) Reactive oxygen gene network of plants. Trends Plant Sci 9:490–498
Mittler R, Kim Y, Song L, Coutu J, Coutu A, Ciftci-Yilmaz S, Lee H, Stevenson B, Zhu JK (2006) Gain- and loss-of-function mutations in Zat10 enhance the tolerance of plants to abiotic stress. FEBS Lett 580:6537–6542
Moes D, Himmelbach A, Korte A, Haberer G, Grill E (2008) Nuclear localization of the mutant protein phosphatase abi1 is required for insensitivity towards ABA responses in *Arabidopsis*. Plant J 54:785–964
Nambara E, Marion-Poll A (2005) ABA biosynthesis and catabolism. Ann Rev Plant Biol 56:165–185

Nelson DE, Repetti PP, Adams TR, Creelman RA, Wu J, Warner DC, Anstrom DC, Bensen RJ, Castiglioni PP, Donnarummo MG, Hinchey BS, Kumimoto RW, Maszle DR, Canales RD, Krolikowski KA, Dotson SB, Gutterson N, Ratcliffe OJ, Heard JE (2007) Plant nuclear factor Y (NF-Y) B subunits confer drought tolerance and lead to improved corn yields on water-limited acres. Proc Natl Acad Sci USA 104:16450–16455

Nemhauser JL, Hong F, Chory J (2006) Different plant hormones regulate similar processes through largely nonoverlapping transcriptional responses. Cell 126:467–475, 2936 genes

Ozturk ZN, Talamé V, Deyholos M, Michalowski CB, Galbraith DW, Gozukirmizi N, Tuberosa R, Bohnert HJ (2002) Monitoring large-scale changes in transcript abundance in drought- and salt-stressed barley. Plant Mol Biol 48:551–573

Pandey GK, Cheong YH, Kim KN, Grant JJ, Li L, Hung W, D'Angelo C, Weinl S, Kudla J, Luan S (2004) The calcium sensor calcineurin B-like 9 modulates abscisic acid sensitivity and biosynthesis in *Arabidopsis*. Plant Cell 16:1912–1924

Papp I, Mur LA, Dalmadi A, Dulai S, Koncz C (2004) A mutation in the cap binding protein 20 gene confers drought tolerance to *Arabidopsis*. Plant Mol Biol 55:679–686

Parry ML et al (2004) Effects of climate change on global food production under SRES emissions and socio-economic scenarios. Glob Environ Change Hum Policy Dimens 14:53–67

Pasquali G, Biricolti S, Locatelli F, Baldoni E, Mattana M (2008) Osmyb4 expression improves adaptive responses to drought and cold stress in transgenic apples. Plant Cell Rep 27:1677–1686

Paul MJ, Primavesi LF, Jhurreea D, Zhang Y (2008) Trehalose metabolism and signaling. Annu Rev Plant Biol 59:417–41

Pernas M, Garcia-Casado G, Rojo E, Solano R, Sánchez-Serrano JJ (2007) A protein phosphatase 2A catalytic subunit is a negative regulator of abscisic acid signaling. Plant J 51:763–778

Poroyko V, Calugaru V, Fredricksen M, Bohnert HJ (2004) Virtual-SAGE: a new approach to EST data analysis. DNA Res 4:11145–52

Poroyko V, Spollen WG, Hejlek LG, Hernandez AG, LeNoble ME, Davis G, Nguyen HT Springer GK, Sharp RE, Bohnert HJ (2007) Comparing regional transcript profiles from maize primary roots under well-watered and low water potential conditions. J Exp Bot 58:279–8

Prieto-Dapena P, Castaño R, Almoguera C, Jordano J (2008) The ectopic overexpression of a seed-specific transcription factor, HaHSFA9, confers tolerance to severe dehydration in vegetative organs. Plant J 54:1004–1014

Qin F, Sakuma Y, Tran LS, Maruyama K, Kidokoro S, Fujita Y, Fujita M, Umezawa T, Sawano Y, Miyazono K, Tanokura M, Shinozaki K, Yamaguchi-Shinozaki K (2008) *Arabidopsis* DREB2A-interacting proteins function as RING E3 ligases and negatively regulate plant drought stress-responsive gene expression. Plant Cell 20:1693–1707

Qudeimat E, Faltusz AM, Wheeler G, Lang D, Brownlee C, Reski R, Frank W (2008) A PIIB-type Ca^{2+}-ATPase is essential for stress adaptation in *Physcomitrella patens*. Proc Natl Acad Sci USA 105:19555–19560

Ranathunge K, Kotula L, Steudle E, Lafitte R (2004) Water permeability and reflection coefficient of the outer part of young rice roots are differently affected by closure of water channels (aquaporins) or blockage of apoplastic pores. J Exp Bot 55:433–447

Reynolds M, Tuberosa R (2008) Translational research impacting on crop productivity in drought-prone environments. Curr Opin Plant Biol 11:171–179

Ribaut JM (2006) Drought adaptation in cereals. Taylor and Francis, Boca Raton, FL, p 642

Rivero RM, Kojima M, Gepstein A, Sakakibara H, Mittler R, Gepstein S, Blumwald E (2007) Delayed leaf senescence induces extreme drought tolerance in a flowering plant. Proc Natl Acad Sci USA 104:19631–19636

Rolland F, Baena-Gonzalez E, Sheen J (2006) Sugar sensing and signaling in plants: conserved and novel mechanisms. Ann Rev Plant Biol 57:675–709

Rosnoblet C, Aubry C, Leprince O, Vu BL, Rogniaux H, Buitink J (2007) The regulatory gamma subunit SNF4b of the sucrose non-fermenting-related kinase complex is involved in longevity and stachyose accumulation during maturation of *Medicago truncatula* seeds. Plant J 51:47–59

Saez A, Robert N, Maktabi MH, Schroeder JI, Serrano R, Rodriguez PL (2006) Enhancement of abscisic acid sensitivity and reduction of water consumption in *Arabidopsis* by combined inactivation of the protein phosphatases type 2C ABI1 and HAB1. Plant Physiol 141:1389–1399

Sakurai T, Plata G, Rodríguez-Zapata F, Seki M, Salcedo A, Toyoda A, Ishiwata A, Tohme J, Sakaki Y, Shinozaki K, Ishitani M (2007) Sequencing analysis of 20,000 full-length cDNA clones from cassava reveals lineage specific expansions in gene families related to stress response. BMC Plant Biol 7:66

Sallon S, Solowey E, Cohen Y, Korchinsky R, Egli M, Woodhatch I, Simchoni O, Kislev M (2008) Germination, genetics, and growth of an ancient date seed. Science 320:1464

Santos Mendoza M, Dubreucq B, Miquel M, Caboche M, Lepiniec L (2005) \ LEAFY COTYLEDON 2 activation is sufficient to trigger the accumulation of oil and seed specific mRNAs in *Arabidopsis* leaves. FEBS Lett 579:4666–70

Santos-Mendoza M, Dubreucq B, Baud S, Parcy F, Caboche M, Lepiniec L (2008) Deciphering gene regulatory networks that control seed development and maturation in *Arabidopsis*. Plant J 54:608–620

Schachtman DP, Goodger JQ (2008) Chemical root to shoot signaling under drought. Trends Plant Sci 13:281–287

Schäfer J, Strimmer K (2005) An empirical Bayes approach to inferring large-scale gene association networks. Bioinformatics 21:754–64

Schmid M, Davison TS, Henz SR, Pape UJ, Demar M, Vingron M, Schölkopf B, Weigel D, Lohmann JU (2005) A gene expression map of *Arabidopsis thaliana* development. Nat Genet 37:501–506

Schneeberger RG, Cullis CA (1991) Specific DNA alterations associated with the environmental induction of heritable changes in flax. Genetics 128:619–630

Schweighofer A, Hirt H, Meskienne I (2004) Plant PP2C phosphatases: emerging functions in stress signaling. Trends Plant Sci 9:236–243

Shao HB, Chu LY, Jaleel CA, Zhao CX (2008) Water-deficit stress-induced anatomical changes in higher plants. C R Biol 331:215–225

Sharbel TF, Haubold B, Mitchell-Olds T (2000) Genetic isolation by distance in *Arabidopsis thaliana* biogeography and postglacialcolonization of Europe. Mol Ecol 9:2109–2118

Sharp RE, Poroyko V, Hejlek LG, Spollen WG, Springer GK, Bohnert HJ, Nguyen HT (2004) Root growth maintenance during water deficits: physiology to functional genomics. J Exp Bot 55:2343–2351

Sheveleva E, Chmara W, Bohnert HJ, Jensen RG (1997) Increased salt and drought tolerance by D-Ononitol production in transgenic *Nicotiana tabacum* L. Plant Physiol 115:1211–1219

Shinozaki K, Yamaguchi-Shinozaki K (2007) Gene networks involved in drought stress response and tolerance. J Exp Bot 58:221–227

Spollen WG, Tao W, Valliyodan B, Chen K, Hejlek LG, Kim JJ, Lenoble ME, Zhu J, Bohnert HJ, Henderson D, Schachtman DP, Davis GE, Springer GK, Sharp RE, Nguyen HT (2008) Spatial distribution of transcript changes in the maize primary root elongation zone at low water potential. BMC Plant Biol 8:32

Sreenivasulu N, Sopory SK, Kavi Kishor PB (2007) Deciphering the regulatory mechanisms of abiotic stress tolerance in plants by genomic approaches. Gene 388:1–13

Stone SL, Williams LA, Farmer LM, Vierstra RD, Callis J (2006) KEEP ON GOING, a RING E3 ligase essential for *Arabidopsis* growth and development, is involved in abscisic acid signaling. Plant Cell 18:3415–3428

Stranger BE, Forrest MS, Dunning M, Ingle CE, Beazley C, Thorne N, Redon R, Bird CP, de Grassi A, Lee C, Tyler-Smith C, Carter N, Scherer SW, Tavaré S, Deloukas P, Hurles ME, Dermitzakis ET (2007) Relative impact of nucleotide and copy number variation on gene expression phenotypes. Science 315:848–853

Sunkar R, Kapoor A, Zhu JK (2006) Posttranscriptional induction of two Cu/Zn superoxide dismutase genes in *Arabidopsis* is mediated by downregulation of miR398 and important for oxidative stress tolerance. Plant Cell 18:2051–2065

Sunkar R, Chinnusamy V, Zhu J, Zhu JK (2007a) Small RNAs as big players in plant abiotic stress responses and nutrient deprivation. Trends Plant Sci 12:301–309

Sunkar R, Chinnusamy V, Zhu J, Zhu J-K (2007b) Small RNAs as big players in plant abiotic stress responses and nutrient deprivation. Trends Plant Sci 7:301–309

Suzuki M, Wang HH, McCarty DR (2007) \ Repression of the LEAFY COTYLEDON 1/B3 regulatory network in plant embryo development by VP1/ABSCISIC ACID INSENSITIVE 3-LIKE B3 genes. Plant Physiol 143:902–11

Szalma SJ, Hostert BM, Ledeaux JR, Stuber CW, Holland JB (2007) QTL mapping with near-isogenic lines in maize. Theor Appl Genet 114:1211–1228

Taiz L, Zeiger E (2006) Plant Physiology, 4th edn. Sinauer Associates, Sunderland, MA, p 705

Taji T, Ohsumi C, Iuchi S, Seki M, Kasuga M, Kobayashi M, Yamaguchi-Shinozaki K, Shinozaki K (2002) Important roles of drought- and cold-inducible genes for galactinol synthase in stress tolerance in *Arabidopsis thaliana*. Plant J 29:417–26

Taji T, Seki M, Satou M, Sakurai T, Kobayashi M, Ishiyama K, Narusaka Y, Narusaka M, Zhu JK, Shinozaki K (2004) Comparative genomics in salt tolerance between Arabidopsis and aRabidopsis-related halophyte salt cress using Arabidopsis microarray. Plant Physiol 135:1697–1709

Talamè V, Ozturk NZ, Bohnert HJ, Tuberosa R (2007) Barley transcript profiles under dehydration shock and drought stress treatments: a comparative analysis. J Exp Bot 58:229–240

Tarczynski MC, Jensen RG, Bohnert HJ (1993) Stress protection of transgenic tobacco by production of the osmolyte mannitol. Science 259:508–510

Tuberosa R, Salvi S (2006) Genomics-based approaches to improve drought tolerance of crops. Trends Plant Sci 11:405–412

Umezawa T, Okamoto M, Kushiro T, Nambara E, Oono Y, Seki M, Kobayashi M, Koshiba T, Kamiya Y, Shinozaki K (2006) CYP707A3, a major ABA 8'-hydroxylase involved in dehydration and rehydration response in *Arabidopsis thaliana*. Plant J 46:171–182

Van Breusegem F, Bailey-Serres J, Mittler R (2008) Unraveling the tapestry of networks involving reactive oxygen species in plants. Plant Physiol 147:978–984

Vasquez-Robinet C, Mane S, Ulanov AV, Watkinson JI, Stromberg VK, DeKoeyer D, Schafleitner R, Willmot DB, Bonierbale M, Bohnert HJ, Grene R (2008) Physiological and molecular adaptations to drought in Andean potato genotypes. J Exp Bot 59:2109–2123

Vazquez F, Vaucheret H, Rajagopalan R, Lepers C, Gasciolli V, Mallory AC et al (2004) Endogeneous trans-acting siRNAs regulate the accumulation of *Arabidopsis* mRNAs. Mol Cell 16:69–79

Verslues PE, Bray EA (2006) Role of abscisic acid (ABA) and *Arabidopsis thaliana* ABA-insensitive loci in low water potential-induced ABA and proline accumulation. J Exp Bot 57:201–212

von Koskull-Doring P, Scharf KD, Nover L (2007) The diversity of plant heat stress transcription factors. Trends Plant Sci 12:452–457

Wang XM, Devalah SP, Zhang WH (2006) Signaling functions of phosphatidic acid. Progr Lipid Res 45:250–278

Wasilewska A, Vlad F, Sirichandra C, Redko Y, Jammes F, Valon C et al (2008) An update on abscisic acid signaling in plants and more. Mol Plant 1:198–217

Watkinson JI, Hendricks L, Sioson AA, Heath LS, Bohnert HJ, Grene R (2008) Tuber development phenotypes in adapted and acclimated, drought-stressed *Solanum tuberosum* ssp. *andigena* have distinct expression profiles of genes associated with carbon metabolism. Plant Physiol Biochem 46:34–45

Wise MJ (2003) LEAping to conclusions: a computational reanalysis of late embryogenesis abundant proteins and their possible roles. BMC Bioinform 4:52–71

Wise MJ, Tunnacliffe A (2004) POPP the question: what do LEA proteins do. Trends Plant Sci 9:1360–1385

Xiao H, Tattersall EAR, Siddiqua MK (2008) CBF4 is a unique member of the CBF transcription factor family of *Vitis vinifera* and *Vitis riparia*. Plant Cell Environ 31:1–10

Yamaguchi-Shinozaki K, Shinozaki K (2006) Transcriptional regulatory networks in cellular responses and tolerance to dehydration and cold stresses. Annu Rev Plant Biol 57:781–803

Yanhui C, Xiaoyuan Y, Kun H, Meihua L, Jigang L, Zhaofeng G, Zhiqiang L, Yunfei Z, Xiaoxiao W, Xiaoming Q, Yunping S, Li Z, Xiaohui D, Jingchu L, Xing-Wang D, Zhangliang C, Hongya G, Li-Jia Q (2006) The MYB transcription factor superfamily of *Arabidopsis*: expression analysis and phylogenetic comparison with the rice MYB family. Plant Mol Biol 60:107–124

Yazaki J, Shimatani Z, Hashimoto A, Nagata Y, Fujii F, Kojima K, Suzuki K, Taya T, Tonouchi M, Nelson C, Nakagawa A, Otomo Y, Murakami K, Matsubara K, Kawai J, Carninci P, Hayashizaki Y, Kikuchi S (2004) Transcriptional profiling of genes responsive to abscisic acid and gibberellin in rice: phenotyping and comparative analysis between rice and *Arabidopsis*. Physiol Genom 17:87–100

Ye Q, Steudle E (2006) Oxidative gating of water channels (aquaporins) in corn roots. Plant Cell Environ 29:459–470

Zhang W, Qin C, Zhao J, Wang X (2004) Phospholipase D {alpha}1-derived phosphatidic acid interacts with ABI1 phosphatase 2C and regulates abscisic acid signaling. Proc Natl Acad Sci USA 101:9508–9513

Zhang X, Garreton V, Chua NH (2005a) The AIP2 E3 ligase acts as a novel negative regulator of ABA signaling by promoting ABI3 degradation. Genes Dev 19:1532–1543

Zhang W, Yu L, Zhang Y, Wang X (2005b) Phospholipase D the signaling networks of plant response to abscisic acid and reactive oxygen species. Biochim Biophys Acta 1736:1–9

Zhang Y, Yang C, Li Y, Zheng N, Chen H, Zhao Q, Gao T, Guo H, Xie Q (2007) SDIR1 is a RING finger E3 ligase that positively regulates stress-responsive abscisic acid signaling in *Arabidopsis*. Plant Cell 19:1912–1929

Zhou N, Robinson SJ, Huebert T, Bate NJ, Parkin IAP (2007) Comparative genome organization reveals a single copy of CBF in the freezing tolerant crucifer *Thlaspi arvense*. Plant Mol Biol 65:693–705

Zhu JK (2001) Cell signaling under salt, water and cold stresses. Curr Opin Plant Biol 4:401–406

Chapter 14
Dehydrins: Molecular Biology, Structure and Function

Sylvia K. Eriksson and Pia Harryson

14.1 Introduction

Most plants experience times of less optimal growth conditions. Fast changes in temperature, drought or growth in saline soils impose stress on growing plants. Sometimes these conditions get severe, even life frightening and the plants need to take action in trying to stay alive. In these cases, plants activate several fast responding support systems that increase the chances of survival. Among the induced stress responses are synthesis of osmolytes and different stress proteins, like LEA proteins to which the dehydrins belong (Ingram and Bartels 1996).

The first reports on the existence of a LEA protein were published more than 25 years ago by Dure, who named this group of proteins (Dure and Chlan 1981; Dure and Galau 1981; Dure et al. 1981). LEA proteins were first found to accumulate in seeds just before desiccation during the late stages of maturation. Since then, extensive research has been carried out on several aspects of LEA proteins and new members are continuously discovered (Tunnacliffe and Wise 2007; Hundertmark and Hincha 2008). Their high cellular accumulation, not only in seeds but also in different plant tissues, in response to different water-related stresses suggests a role in stress survival. Although extensive research efforts have been made to understand their role in stress, the molecular function of most LEA proteins is still unclear.

The large family of LEA proteins is divided into six groups, three major groups 1–3 and three minor groups 4–6. The division into the different groups is based on sequence similarities, often occurring in short conserved sequence motifs (Galau and Hughes 1987; Dure et al. 1989; Dure 1993a, b), which can be repeated several times within the amino acid sequence of a LEA protein. Although the different LEA groups are not similar in sequence, most LEA proteins share some common characteristics: they are very hydrophilic, have a low content of hydrophobic amino acids and a high content of charged amino acids. Because of these characteristics they tend to have either high or low isoelectric points (pIs). Moreover, they lack or have a low content of the amino acids cysteine and tryptophan and some also have a high content of glycine (Tunnacliffe and Wise 2007; Battaglia et al. 2008; Hundertmark and Hincha 2008). The LEA proteins with a high glycine content also fall into the more general group of proteins called hydrophilins (Garay-Arroyo et al. 2000).

U. Lüttge et al. (eds.), *Plant Desiccation Tolerance*, Ecological Studies 215,
DOI 10.1007/978-3-642-19106-0_14,

14.2 Dehydrins (Group 2 LEA Proteins)

The group 2 of the LEA proteins are commonly called dehydrins because of an early proposed function in plants surviving drought stress (Close et al. 1989). Dehydrins are exclusively found in plants, whereas members of the other LEA groups are also present in organisms such as bacteria or nematodes (Goyal et al. 2005; Tunnacliffe and Wise 2007). Not only are dehydrins exclusive to plants, they also seem to be ubiquitous and have been found in all seed plants investigated and also in mosses (Saavedra et al. 2006) and lycopods (Iturriaga et al. 2006). The dehydrins are characterised by the presence of several short conserved amino acid sequences that sometimes are organised in repeats: the 15 amino acid long K-segment (EKKGIMDKIKEKLPG), the 7 amino acid long Y-segment at the N-terminal ((V/T)D(E/Q)YGNP) and the S-segment (SDSSSSSSS), a poly-serine stretch in some dehydrins (Close et al. 1989; Close 1996). The S-segment is sometimes followed by a stretch of highly charged residues that are rich in lysines, aspartic and glutamic acids (Mouillon et al. 2006) (Fig. 14.1). By definition, all dehydrins contain at least one copy of the K-segment with a varied composition of the other segments (Dure 1993a, b; Close 1996). The K-segment is particularly abundant in cold-induced dehydrins where it could be repeated up to 11 times (Close 1997). No correlation to any particular stress can be found for the other motifs. The nomenclature of the dehydrins is written $Y_nS_nK_n$ (Close 1996). Following this, the dehydrins can be divided into five subgroups, according to the presence of the segments: K_n, SK_n, Y_nSK_n, Y_nK_n and K_nS. Typically, one plant species has several

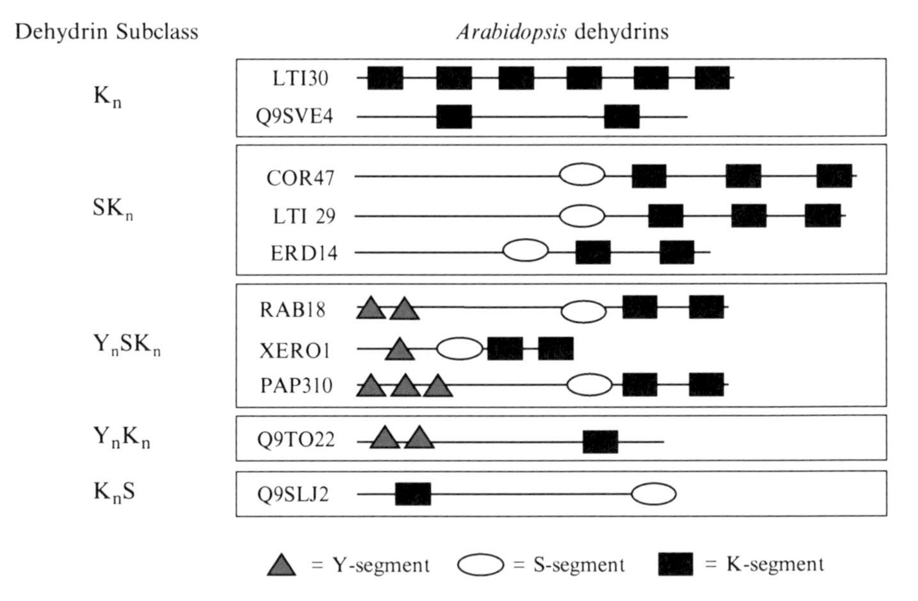

Fig. 14.1 The five different dehydrin subclasses based on the occurrence of the conserved segments and how the ten dehydrins in *Arabidopsis* are represented in different subclasses

dehydrins representing different subgroups. As an example, *Arabidopsis thaliana* has ten different dehydrins divided into the following subgroups: 2 K_n, 3 SK_n, 3 Y_nSK_n, 1 Y_nK_n and 1 K_nS (Fig. 14.1). Since most dehydrins are highly charged proteins, another way to classify dehydrins is by their net charge. Dehydrins with a low pI are called acidic dehydrins and the rest are grouped as basic/neutral dehydrins. The length of dehydrin proteins varies greatly, from approximately 80 to over 600 amino acids.

14.3 The Cellular Localisation of Dehydrin Proteins in Plants

Dehydrins have been found in plant tissues such as seeds, buds, roots, root tips, stems, vascular tissues, flowers and leaves (Rorat et al. 2004). Certain dehydrins seem to have a more specific location for example in guard cells, at the plasmodesmata or in pollen sacs (Nylander et al. 2001). Whereas some dehydrins are only found after stress, a small number seems to be constitutively expressed (Houde et al. 1992; Danyluk et al. 1994). However, as a general rule, most dehydrins show a broad localisation after stress (Rorat et al. 2004). Out of the ten dehydrins found in non-stressed *Arabidopsis*, four occur in vegetative tissues (in seedlings, leaves, roots, buds, stems and flowers), whereas two other dehydrins are restricted to seeds: one is found in roots only and the last three are found both in vegetative tissues and in seeds (Hundertmark and Hincha 2008). The dehydrins are highly soluble proteins and many of them accumulate in the cytoplasm. In addition, dehydrins have also been found in connection with plasma membranes, in the nucleus, in the vacuole, in mitochondria, at proplastid membranes or in protein bodies (Danyluk et al. 1998; Rinne et al. 1999; Borovskii et al. 2002; Heyen et al. 2002; Mehta et al. 2009). Some have been predicted to be in chloroplasts or peroxisomes (Tunnacliffe and Wise 2007).

14.4 Expression of Dehydrins

Dehydrins are expressed at high levels under conditions of low cellular water content, such as cold, drought, high salt concentrations or after ABA application. Plants have often multiple dehydrin genes: *Arabidopsis* has 10, rice has 8, barley has 13 and populus has 3 genes (Choi et al. 1999; Sterky et al. 2004; Wang et al. 2007; Hundertmark and Hincha 2008). Many of these genes are responsive to more than one stress. Out of the ten *Arabidopsis* genes, six are upregulated in response to salt stress, four respond to cold, one to drought and five increase after application of ABA in leaves (Nylander et al. 2001; Hundertmark and Hincha 2008). Some specific *Arabidopsis* examples are Cor47 (SK_3) and Lti29 (SK_3) that respond to salt, cold and ABA, Erd10 (SK_2) that responds to salt and ABA, Lti30 (K_6) that responds to salt and cold and Rab18 (Y_2SK_2) to salt, drought and ABA (Hundertmark and Hincha 2008). Hence, there is no obvious connection between subgroup classification and stress response (Close 1997; Nylander et al. 2001).

In addition to the high expression in response to stress, trees living in temperate zones show a longer and seasonal expression of dehydrins. In most trees, submitted to cold winters, there is a peak of dehydrin expression just before leaf senescence, the levels of dehydrins stay high over the winter in buds and stems to decline in the spring and be very low during the warmer summer months (approximately 2 months) (Sarnighausen et al. 2002; Welling et al. 2004). A similar yearly variation can be seen in woody shrubs such as *Vaccinum* (Muthalif and Rowland 1994). An analogous developmentally programmed expression is also seen in maturing seeds where a peak of expression often is correlated with the late desiccation phase (Han et al. 1997; Ismail et al. 1999; Lee et al. 2006). The level of expression in seeds appears to be higher as compared to any vegetative tissue (Hundertmark and Hincha 2008). Interestingly, ABA treatment of seed and leaves leads to little overlap in expression profiles of dehydrins, which suggests regulation by different signal transduction pathways (Hundertmark and Hincha 2008). Although dehydrin expression profiles in response to different stresses is commonly reported, data on actual protein levels are very limited. The protein concentration is generally assumed to be high during stress, but only a few studies report on the actual cellular protein level (Dure 1993a, b). This complicates the in vitro experimental approach of functional studies where the cellular concentrations only can be approximated.

14.5 Transgenic Plants Overexpressing Dehydrins and Knockout Mutants

The construction of transgenic plants overexpressing dehydrin genes gives valuable insight into their possible role in plants growing under various stresses. In general, most overexpressed dehydrins have a positive effect on the stress under study, but there are also examples where overexpression of dehydrin genes has no effect at all (summarised in Table 14.1). One such example is the overexpression of the *Arabidopsis* Rab18 that had no effect on drought or freezing tolerance. However, when the same gene was co-expressed with the dehydrin Cor47, *Arabidopsis* freezing tolerance was improved, but no effect was seen on survival during drought (Puhakainen et al. 2004).

A powerful method to study the effect of dehydrin expression is by knockout mutants, i.e. by silencing the gene expression. There are only a few reports on knockout mutants for any of the LEA proteins, and the only report on a dehydrin is a knockout mutant of the moss *Physcomitrella patens* dehydrin DHNA. In comparison with the WT moss, the knockout showed no phenotypic change or any change in growth response. However, the ability to recover after salt or osmolyte stress was greatly reduced (Saavedra et al. 2006).

Table 14.1 The effects of overexpressing dehydrin genes in transgenic plants

Transgenic plant	Gene origin	Effect in transgenic plants	References
Rice	Wheat dehydrin	Improved drought resistance	Cheng et al. (2002)
Tobacco	*Citrus* dehydrin	Increased germination and growth at low temperature	Hara et al. (2003)
Tobacco	*Craterostigma* Two dehydrins	No effect on drought tolerance	Iturriaga et al. (1992)
Tobacco	*Brassica* BjDHN2/ BjDHN3	Decreased electrolyte leakage	Xu et al. (2008)
Strawberry	Wheat Wcor410	Improved freeze tolerance of leaves	Houde et al. (2004)
Cucumber	Potato DHN24	Improved chilling tolerance	Yin et al. (2006)
Arabidopsis	Maize DHN1/Rab17	Improved osmotic stress tolerance	Figueras et al. (2004)
Arabidopsis	*Arabidopsis* Rab18/ Cor47	Improved freeze tolerance No effect on drought	Puhakainen et al. (2004)
Arabidopsis	*Arabidopsis* Lti29/ Lti30	Improved freeze tolerance No effect on drought	Puhakainen et al. (2004)
Arabidopsis	Wheat DHN5	Improved osmotic stress and salt tolerance	Brini et al. (2007)

14.6 Structure and Function of Dehydrins: Dehydrins – Intrinsically Disordered Proteins

The dehydrins are probably the structurally most investigated group of the LEA proteins. Their sequence characteristics, i.e. highly hydrophilic and highly charged in combination with a low amount of hydrophobic residues, implicate a highly flexible structure, i.e. the sequence composition does not allow the formation of an extensive hydrophobic core typical for folded proteins. In accordance, several CD (circular dichroism) studies of dehydrins in solution show a high degree of disorder with a minor (less than 2%) inclusion of α-helices (Lisse et al. 1996; Ismail et al. 1999; Soulages et al. 2003; Bokor et al. 2005; Mouillon et al. 2006, 2008). This lack of a fixed structure is also shown by ^{1}H NMR and solid-state NMR studies, both of which confirm a flexible and extended conformation (Lisse et al. 1996; Bokor et al. 2005). Hence, the dehydrins fall into the group of intrinsically disordered proteins, i.e. proteins that lack a fixed 3D structure under physiological conditions (Dunker et al. 2002; Fink 2005; Uversky et al. 2005). This group of proteins is also sometimes called natively unfolded or intrinsically unstructured proteins. As a consequence of not having a structure, the dehydrins stay soluble even after boiling, an attribute taken advantage of in purification procedures of dehydrins (Svensson et al. 2000; Livernois et al. 2009). Further, the lack of structure makes it impossible to crystallise them.

The question is if we can learn something about the dehydrin structure and function by comparing them with disordered proteins in general? The majority of disordered proteins are involved in regulation, signalling or control (Uversky et al. 2005; Dunker et al. 2008). Some functions are rarely or never represented by disordered proteins, for example enzymatic functions. Most disordered proteins bind to cellular targets, either permanently (such as assemblers) or transiently (chaperones or interaction sites) and as they interact with their targets they assume an at least partly folded structure (Dunker et al. 2002; Wright and Dyson 2009). Here, the disorder- and binding-induced conformational changes switch the functions on and off. In this respect, most disordered proteins follow the rule that a fixed structure is necessary for an interaction to take place. It should also be mentioned that a minority of disordered proteins are less prone to fold and have functions such as entropic chains or flexible linkers (Magidovich et al. 2007). The cellular targets of disordered proteins can for example be other proteins, membranes, RNA, DNA or metal ions. Following this, the transition to a functional state of most disordered proteins comes with an increase in structure (Uversky et al. 2005). If this functional switch is monitored experimentally, it means that the random coil content of the CD spectra of disordered proteins decrease with a simultaneous increase in secondary structure (α-helix and β-sheets), i.e. a switch from a coil spectrum to a more helical or beta-like spectrum (Fig. 14.2). It is interesting to note that the conformational changes of dehydrins are sometimes conditional, i.e. the protein structure responds to changes in the environment such as temperature (McNulty et al. 2006), pH (Tornroth-Horsefield et al. 2006), availability of water (Luo and Baldwin 1997) or macromolecular crowding (Minton 2005a, b; Hall 2006).

Following the hypothesis that induced folding might be the functional switch, the structural behaviour of dehydrins has been followed by CD spectroscopy at different temperatures, different pH values or in the presence of molecules that mimic the cellular milieu or targets such as lipid vesicles, metal ions and different sugars. Further, the influence of structure promoting molecules, such as trifluoroethylene (TFE) and sodium dodecyl sulphate (SDS) on the structure, has been monitored. The outcome of these experiments is summarised below.

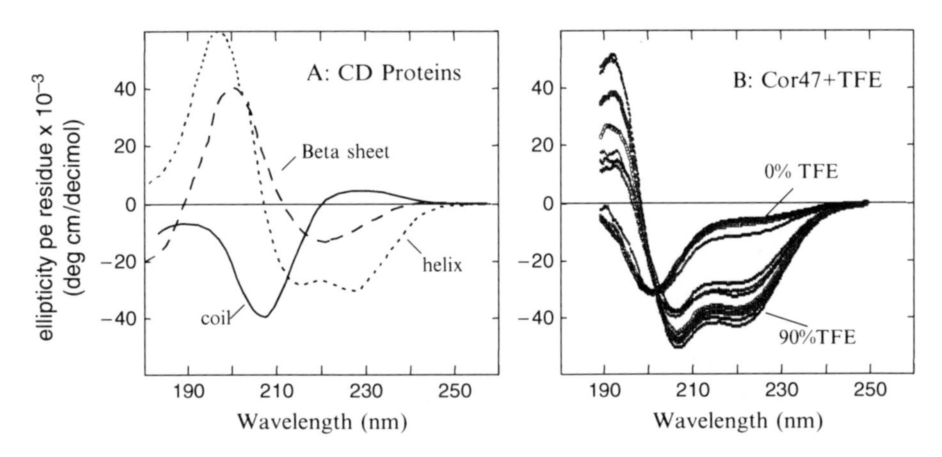

Fig. 14.2 (**a**) Protein CD spectra representing pure α-helix, β-sheet and random coil. (**b**) The dehydrin Cor47 titrated with the strong helical inducer TFE in 10% steps, showing a conformational transition from a coil spectrum at 0% TFE to a more helical spectrum at 90% TFE (Mouillon et al. 2006)

14.7 Structural Responses to TFE

To investigate if dehydrins are able to fold at all, structural transition of dehydrins in the presence of the strong helical promoter TFE has been followed by CD studies. This experimental set-up also gives information on the structural propensity of dehydrins, i.e. how easily any helixes are formed. The native CD spectra of dehydrins in physiological buffer show a very small helical content of only 0.7–5% (Lisse et al. 1996; Koag et al. 2003; Soulages et al. 2003; Mouillon et al. 2006). Upon addition of TFE, dehydrins from soybean, *Craterostigma* and *Arabidopsis* form a small amount of helices. Even when high TFE concentrations are used (up to 90%), the helical content is relatively small and varies for different dehydrins between 2 and 45% (Mouillon et al. 2006). The highly conserved K-segment has been proposed to form an amphiphatic α-helix when it interacts with membranes. It is therefore expected that the K-segment easily would form helixes in TFE. However, the helical content of several of the dehydrins in TFE is too low to account for all K-segments in helical conformation (Soulages et al. 2003; Mouillon et al. 2006). Also, the K-segment, as a short 15 amino acid peptide, is only 10% helical in 90% TFE (Mouillon et al. 2006). The difference in helical content of dehydrins in TFE can therefore not be correlated to the appearance of K-segments or any of the conserved segments, but it is a consequence of the amino acid composition of the whole polypeptide.

14.8 Dehydrins and Background Crowding

As a consequence of the decrease of the water content of the cell, the background concentration of other molecules is enriched. This leads to a geometric restriction of the available space for proteins. Typically, proteins function under conditions where the solute concentration is around 400 g/L (Zimmerman and Minton 1993). In severely dehydrated plant cells, this concentration can be even substantially higher (Bartels and Salamini 2001). Such increased background of crowding molecules is expected to have an important impact on cellular processes (Ellis 2001a, b). By simply competing for volume, macromolecular crowding favours compact states of protein chains and increases the stability of native proteins. Such cellular crowding also greatly slows down diffusion-based reactions (Ellis 2001a, b; Minton 2005a, b, 2006). Crowding also promotes protein assembly by favouring compact conformations over extended ones, and the reduction in bulk water will directly influence membrane topology and dynamics.

If the dehydrin function is correlated with a conformational change, dehydrins must be stable in their disordered state without any structure even against a highly crowded background like in a drought-stressed cell. This means that they must withstand structural collapse due to excluded volume effects.

When the structural behaviour of three dehydrins from *Arabidopsis* was tested against different crowded backgrounds of either naturally occurring sugars or polymers, they stayed mainly disordered (Mouillon et al. 2008). Interestingly, the dehydrins stayed disordered even under severe crowding, under which conditions other disordered proteins that fold upon binding to their targets would tend to collapse (Flaugh and Lumb 2001). It seems as if the dehydrin sequence is highly evolved and adapted to remain disordered also under conditions of severe cellular crowding where unfolded states of other proteins would tend to collapse (Mouillon et al. 2008). In this respect, the dehydrins differ from the class of disordered proteins that rely on folding to become functional and tend to partly fold due to crowding. The function of the dehydrins could therefore lie in the interactions of the conserved segments with their specific biological targets. Accordingly, the disordered appearance of the dehydrins may be converted into active structures provided that the conditions inside a drought-stressed plant cell are sufficiently well reproduced.

14.9 Structural Responses to Temperature

Helical propensity increases with decreased temperature and reversibly high temperature melts out structures such as α-helices. Since some dehydrins are highly expressed at low temperature, the idea that changes in temperature could promote a structural change in dehydrins was suggested. Hence, as temperature decreases, dehydrins would collapse into more helical conformation and possibly also gain function. It seems that the structure of the dehydrins responds opposite to what is expected. In CD studies, the dehydrins respond to elevated temperature by an apparent increase in helical structure (Soulages et al. 2003; Mouillon et al. 2006). However, this apparent increase in dehydrin structure is actually a decrease of the non-classical secondary structure polyproline helix (PII). Both dehydrins from soybean and *Arabidopsis* responded to increased temperature indicative of inclusion of PII conformation. Approximately 10–20% of the dehydrin structures are estimated to be in PII helixes. Notably, the PII content of dehydrins seems not to be controlled by any of the conserved segments (Mouillon et al. 2006). PII is a highly extended helical conformation and as temperature is increased this conformation is melted out in favour of coil conformation. PII helices lack intra-molecular hydrogen bonds and rely on bridging water molecules to keep their extended helical shape and are thus a highly hydrated conformation in comparison with α-helices. It is also this lack of direct intra-molecular hydrogen bonds that makes the PII structure intrinsically hard to detect spectroscopically. This type of expanded secondary structure represents an intriguing example of structures involving water binding combined with high structural adaptability. PII structures have been linked to several specific functions in animal and bacterial systems. For example, in transcription, signal transduction, cell motility and the immune response (Kelly et al. 2001), they are involved in binding to lipids (Kanai et al. 2001) and possibly also to nucleic acids (Hicks and Hsu 2004). The flexible PII structure is believed to allow an induced fit to the target molecules (Bochicchio and Tamburro 2002). It has been suggested that PII conformation is a characteristic feature of most disordered proteins and may also be a characteristic feature of unfolded globular proteins (Whittington et al. 2005; Shi et al. 2006). In agreement with this, other disordered LEA proteins show a similar temperature dependence as the dehydrins, i.e. indicating PII structures (Soulages et al. 2002). This ubiquitous occurrence of the PII structure could thus reflect a biological preference for maintaining unfolded proteins in a well-defined state.

14.10 Interaction to Lipid Vesicles and Sodium Dodecyl Sulphate

An interaction between dehydrins and membranes is suggested to stabilise the membrane topology at stress situations. The compulsory K-segment is the proposed binding site and is suggested to form an amphiphatic α-helix upon membrane binding (Close 1996; Koag et al. 2003). When the available data from dehydrin lipid binding experiments are scrutinised, they are ambiguous, as they point to both in favour of lipid binding and not participating in lipid binding. In favour of a membrane binding is the fact that certain dehydrins are found only at membrane surfaces (Danyluk et al. 1998). Also, an acidic dehydrin from maize (SK_3) has been shown to interact with SUVs (small unilamellar vesicles) composed of negatively charged phospholipids (Koag et al. 2003, 2009). Following the lipid interaction, an increase is detected in the α-helical content of the maize dehydrin that correlates with the content of K-segments. When the three K-segments are taken away one at a time by mutation, the truncated mutated versions of the dehydrin bind less to lipid vesicles and also have a lower content of helices, pointing to an involvement of the K-segments (Koag et al. 2009). On the contrary, a SK_2 dehydrin from soybean does not interact with lipid vesicles and does not show a conformational change in the presence of lipid vesicles (Soulages et al. 2003). The two *Arabidopsis* dehydrins ERD10 and ERD14 were found to interact with phospholipid liposomes (Kovacs et al. 2008). However, the structural response of ERD10 and ERD14 is in line with the response of the soybean dehydrin, i.e. no structural changes could be detected in the presence of liposomes. When four dehydrins from *Arabidopsis* (including ERD10) were tested for interaction with phospholipid vesicles of different compositions, one bound strongly to all tested lipid vesicles, one bound only to vesicles with a high negative charge and two (including ERD10) did not bind to any lipid vesicles at all (Kutzer et al., Dept of Biochemistry, Stockholm University, unpublished results). A factor that somewhat complicates direct comparison of the results is that the methods vary by which lipid binding is studied.

SDS solution is sometimes used to mimic membrane surfaces. Dehydrins from maize, soybean, *Arabidopsis* and the *bona fide* desiccation-tolerant *Craterostigma* increased their helical content in the presence of SDS (Lisse et al. 1996; Koag et al. 2003; Soulages et al. 2003). Interestingly, this is also true for dehydrins that do not interact with lipid vesicles, for example the soybean dehydrin or the two

Arabidopsis dehydrins that do not interact with lipids (Soulages et al. 2003; Eriksson S and Harryson P, Dept of Biochemistry, Stockholm University, unpublished results). In this respect, the response of dehydrins to SDS is similar to that of helical inducers such as TFE.

14.11 Chelating: Metal Binding

As water is withdrawn from the cellular interior, solutes and ions increase in concentration. These ions can cause damage, since they catalyse the production of reactive oxygen species (ROS). Dehydrins have been suggested to chelate metals and thus decrease the ionic concentration. The metal binding can be part of a function as antioxidants, since low ion concentrations would decrease the production of ROS. In line with this, a dehydrin from *Citrus* has been shown to protect liposomes from peroxidation (Hara et al. 2004). In support of a role as "metal sponges", several dehydrins have been demonstrated to bind metal ions such as Fe^{3+}, Fe^{2+}, Cu^{2+}, Mn^{2+} and Zn^{2+} (Svensson et al. 2000; Hara et al. 2004; Xu et al. 2008). The natural high content of charged amino acids and histidine residues is suggested to have a role in the metal coordination since dehydrins lack metal binding motifs (Hara et al. 2005). Phosphorylation of dehydrins in vitro has also been shown to have a positive effect on ion binding since it increases the binding affinity of calcium (Heyen et al. 2002; Alsheikh et al. 2003). Despite this strong binding of metal ions, no structural effects of such metal coordination have been detected for the two dehydrins from *Arabidopsis* (Mouillon et al. 2006).

14.12 Structural Responses to pH Changes

Changes in pH may destabilise folded proteins and hence alter protein structure. The disordered dehydrins could respond to alterations in pH by protonation or deprotonation of side groups of certain amino acids that would change the overall (and local) net charge of the polypeptide. However, a change in net charge due to pH titrations does not give rise to any conformational changes when tested between pH 4 and pH 9 regardless if the dehydrin is originally acidic or neutral/alkaline (Eriksson S and Harryson P, Dept of Biochemistry, Stockholm University, unpublished results).

14.13 Posttranslational Modifications: Phosphorylation

Phosphorylation of proteins is central to signal transduction pathways and plays an important role in the regulation of cellular processes. Dehydrins have been found to be phosphorylated both in vivo and in vitro (Alsheikh et al. 2003; Jiang and Wang 2004; Mouillon et al. 2008). Casein Kinase II (CKII) might be involved in the phosphorylation of the serine segment and has been shown to phosphorylate several dehydrins from subgroups that have the S-segment in vitro (Heyen et al. 2002; Riera et al. 2004; Alsheikh et al. 2005; Mouillon et al. 2008; Mehta et al. 2009). Another conserved motif that is targeted for phosphorylation was recently found; dehydrins containing the sequence LXRXXS are targeted by the SnRK2-10 kinase (Vlad et al. 2008). Examples of phosphorylated dehydrins include Xero1 from *Arabidopsis*, Rab17 from maize and Rab21 from rice. As a result of phosphorylation, the membrane-associated VCaB45 dehydrin from celery and the *Arabidopsis* ERD14 bind significantly more calcium when they are phosphorylated (Heyen et al. 2002; Alsheikh et al. 2003). Further, the cytosolic localisation of AmDHN1 from *Avicennia* and Rab17 from maize is altered upon phosphorylation, and the phosphorylated proteins are also found in the nucleus (Jensen et al. 1998; Mehta et al. 2009). There are no structural changes upon phosphorylation by CKII of the two *Arabidopsis* Cor47 and Lti29 proteins; neither did phosphorylation alter their response to osmolytes or crowding agents (Mouillon et al. 2008).

14.14 Chaperone Activity

A general role as protein stabilisers is often proposed for the dehydrins. It is thought that an interaction with other proteins would prevent them from denaturing or from forming inactive aggregates during stress. Such a function is supported by the notion that intrinsic structural disorder is a prevalent feature of proteins with chaperone activity. The mode of action of disordered chaperones is suggested to be different from the classical chaperones that interact with their targets mainly by hydrophobic attraction. One proposed model is the "entropy transfer" model where the disordered chaperone binds unspecifically to denatured proteins and solubilises (i.e. destabilises) the misfolded and energetically trapped peptides (Tompa and Csermely 2004). Along the same line, a function as "molecular shields" has been suggested where the interaction would have a steric and thus a stabilising effect. In this case, the binding would be guided by electrostatic attractions (Hundertmark and Hincha 2008). Also, the high expression of LEA proteins in combination with a high sugar accumulation would indirectly increase the background crowding in cells during stress and hence have an indirect stabilising effect on folded proteins and aggregation (Hundertmark and Hincha 2008; Mouillon et al. 2008). The capacity of dehydrins to rescue enzymatic activity or prevent protein aggregation during drought, low temperatures or even high temperatures has often been tested experimentally. For example, the activity of lactate dehydrogenase (LDH) or citrate synthase (CS) are commonly used as markers for dehydrin ability to rescue protein function during stress. The addition of dehydrins often protects enzyme functions under stress conditions better than BSA (bovine serum albumin) in in vitro assays. The role of the conserved segments in such chaperon activity is not clear, but it is suggested that they serve as binding sites. Table 14.2 shows some of the results from such experiments.

Table 14.2 Ability of dehydrins to rescue enzymatic activity under different stresses in vitro

Dehydrin	Plant species	Experimental stress	Rescue enzyme activity of	References
CuCor15	*Citrus*	Low temperature and drought	LDH and MDH	Sanchez-Ballesta et al. (2004)
Dehydrin	Birch	Drought	Amylase	Rinne et al. (1999)
ERD10/ LTI29	*Arabidopsis*	High temperature	Lysozyme, ADH, luciferase and CS	Kovacs et al. (2008)
ERD14	*Arabidopsis*	High temperature	Lysozyme, ADH, luciferase and CS	Kovacs et al. (2008)
DHN	Soybean	Low temperature	No stabilising effect on LDH over BSA	Momma et al. (2003)
ERD10/ LTI29	*Arabidopsis*	Low temperature	LDH and MDH	Reyes et al. (2008)
WCS120	Wheat	Low temperature	LDH	Houde et al. (1995)
PCA60	Peach	Low temperature	LDH	Wisniewsk et al. (1999)
Dehydrin	*Citrus*	Low temperature	LDH	Hara et al. (2001)
P-80	Barley	Low temperature	LDH	Bravo et al. (2003)

ADH alcohol dehydrogenase, *LDH* lactate dehydrogenase, *MDH* malate dehydrogenase, *CS* citrate synthase

14.15 Outlook

Since the dehydrin family is divided into several subgroups, based on the occurrence of the conserved segments, one could question if they have one unique function as a group, or if the subgroups have different stress-related functions. It could also be that the function is indeed common and the different segments present conserved interaction sites for different cellular targets. A factor that complicates the general conclusions that can be drawn from the experimental data is that many studies are done on a single dehydrin, representing only one subgroup of dehydrins from one plant investigated under one kind of stress. More comparative functional studies with several dehydrins representing the different subgroups, and also preferably the same subgroup but different plant species, would help to shed light on this complex matter. Taken together, the experimental results suggest that plant dehydrins have multifunctional roles during environmental stress, despite their diffuse structural characteristics encountered *in vitro*. Dehydrins are the most abundant proteins in the proteome of desiccated vegetative tissues in desiccation-tolerant plants and in dry seeds. Therefore, it is postulated that they make an important contribution to the complex phenomenon of desiccation tolerance.

References

Alsheikh MK, Heyen BJ, Randall SK (2003) Ion binding properties of the dehydrin ERD14 are dependent upon phosphorylation. J Biol Chem 278:40882–40889

Alsheikh MK, Svensson JT, Randall SK (2005) Phosphorylation regulated ion-binding is a property shared by the acidic subclass dehydrins. Plant Cell Environ 28:1114–1122

Bartels D, Salamini F (2001) Desiccation tolerance in the resurrection plant *Craterostigma plantagineum*. A contribution to the study of drought tolerance at the molecular level. Plant Physiol 127:1346–1353

Battaglia M, Olvera-Carrillo Y, Garciarrubio A, Campos F, Covarrubias AA (2008) The enigmatic LEA proteins and other hydrophilins. Plant Physiol 148:6–24

Bochicchio B, Tamburro AM (2002) Polyproline II structure in proteins: identification by chiroptical spectroscopies, stability, and functions. Chirality 14:782–792

Bokor M, Csizmok V, Kovacs D, Banki P, Friedrich P, Tompa P, Tompa K (2005) NMR relaxation studies on the hydrate layer of intrinsically unstructured proteins. Biophys J 88: 2030–2037

Borovskii GB, Stupnikova IV, Antipina AI, Vladimirova SV, Voinikov VK (2002) Accumulation of dehydrin-like proteins in the mitochondria of cereals in response to cold, freezing, drought and ABA treatment. BMC Plant Biol 2:5

Bravo L, Gallardo J, Navarrete A, Olave N, Martínez J, Alberdi M, Close T, Corcuera L (2003) Cryoprotective activity of a cold-induced dehydrin purified from barley. Physiol Plant 118: 262–269

Brini F, Hanin M, Lumbreras V, Irar S, Pages M, Masmoudi K (2007) Functional characterization of DHN-5, a dehydrin showing a differential phosphorylation pattern in two *Tunisian durum* wheat (*Triticum durum* Desf.) varieties with marked difference in salt and drought tolerance. Plant Sci 172:20–28

Cheng Z, Targolli J, Huang X, Wu R (2002) Wheat LEA genes, PMA80 and PMA1959 enhance dehydration tolerance of transgenic rice (*Oryza sativa*) L. Mol Breed 10:71–82

Choi D-W, Zhu B, Close T (1999) The barely (*Hordeum vulgare* L.) dehydrins multigene family: sequences, allele types, chromosome assignments and expression characteristics of 11 Dhn genes of cv Dicktoo. Theor Appl Genet 98:1234–1247

Close TJ (1996) Dehydrins: emergence of a biochemical role of a family of plant dehydration proteins. Physiol Plant 97:795–803

Close TJ (1997) Dehydrins: a commonality in the response of plants to dehydration and low temperature. Physiol Plant 100:291–296

Close TJ, Kortt AA, Chandler PM (1989) A cDNA-based comparison of dehydration-induced proteins (dehydrins) in barley and corn. Plant Mol Biol 13:95–108

Danyluk J, Houde M, Rassart E, Sarhan F (1994) Differential expression of a gene encoding an acidic dehydrin in chilling sensitive and freezing tolerant gramineae species. FEBS Lett 344:20–24

Danyluk J, Perron A, Houde M, Limin A, Fowler B, Benhamou N, Sarhan F (1998) Accumulation of an acidic dehydrin in the vicinity of the plasma membrane during cold acclimation of wheat. Plant Cell 10:623–638

Dunker AK, Brown CJ, Lawson JD, Iakoucheva LM, Obradovic Z (2002) Intrinsic disorder and protein function. Biochemistry 41:6573–6582

Dunker AK, Silman I, Uversky VN, Sussman JL (2008) Function and structure of inherently disordered proteins. Curr Opin Struct Biol 18:756–764

Dure L (1993a) Structural motifs in LEA proteins

Dure L 3rd (1993b) A repeating 11-mer amino acid motif and plant desiccation. Plant J 3:363–369

Dure L 3rd, Greenway SC, Galau GA (1981) Developmental biochemistry of cottonseed embryogenesis and germination: changing messenger ribonucleic acid populations as shown by in vitro and in vivo protein synthesis. Biochemistry 20:4162–4168

Dure L, Chlan C (1981) Developmental biochemistry of cottonseed embryogenesis and germination: XII. Purification and properties of principal storage proteins. Plant Physiol 68:180–186

Dure L, Crouch M, Harada J, Ho T-H, Mundy J, Quatrano R, Thomas T, Sung Z (1989) Common amino acid sequence domains among the Lea proteins of higher plants. Plant Mol Biol 12:475–486

Dure L, Galau GA (1981) Developmental biochemistry of cottonseed embryogenesis and germination: XIII. Regulation of biosynthesis of principal storage proteins. Plant Physiol 68: 187–194

Ellis RJ (2001a) Macromolecular crowding: an important but neglected aspect of the intracellular environment. Curr Opin Struct Biol 11:114–119

Ellis RJ (2001b) Macromolecular crowding: obvious but underappreciated. Trends Biochem Sci 26:597–604

Figueras M, Pujal J, Saleh A, Save R, Pages M, Goday A (2004) Maize Rab17 overexpression in Arabidopsis plants promotes osmotic stress tolerance. Ann Appl Biol 144:251–257

Fink AL (2005) Natively unfolded proteins. Curr Opin Struct Biol 15:35–41

Flaugh SL, Lumb KJ (2001) Effects of macromolecular crowding on the intrinsically disordered proteins c-Fos and p27(Kip1). Biomacromolecules 2:538–540

Galau GA, Hughes DW (1987) Coordinate accumulation of homeologous transcripts of seven cotton Lea gene families during embryogenesis and germination. Dev Biol 123:213–221

Garay-Arroyo A, Colmenero-Flores JM, Garciarrubio A, Covarrubias AA (2000) Highly hydrophilic proteins in prokaryotes and eukaryotes are common during conditions of water deficit. J Biol Chem 275:5668–5674

Goyal K, Pinelli C, Maslen SL, Rastogi RK, Stephens E, Tunnacliffe A (2005) Dehydration-regulated processing of late embryogenesis abundant protein in a desiccation-tolerant nematode. FEBS Lett 579:4093–4098

Hall D (2006) Protein self-association in the cell: a mechanism for fine tuning the level of macromolecular crowding? Eur Biophys J 35:276–280

Han B, Hughes DW, Galau GA, Bewley JD, Kermode AR (1997) Changes in late-embryogenesis-abundant (LEA) messenger RNAs and dehydrins during maturation and premature drying of *Ricinus communis* L. seeds. Planta 201:27–35

Hara M, Fujinaga M, Kuboi T (2004) Radical scavenging activity and oxidative modification of citrus dehydrin. Plant Physiol Biochem 42:657–662

Hara M, Fujinaga M, Kuboi T (2005) Metal binding by citrus dehydrin with histidine-rich domains. J Exp Bot 56:2695–2703

Hara M, Terashima S, Fukaya T, Kuboi T (2003) Enhancement of cold tolerance and inhibition of lipid peroxidation by citrus dehydrin in transgenic tobacco. Planta 217:290–298

Hara M, Terashima S, Kuboi T (2001) Charaterization and cryoprotective activity of cold responsive dehydrin from *Citrus unshiu*. J Plant Physiol 158:1333–1339

Heyen BJ, Alsheikh MK, Smith EA, Torvik CF, Seals DF, Randall SK (2002) The calcium-binding activity of a vacuole-associated, dehydrin-like protein is regulated by phosphorylation. Plant Physiol 130:675–687

Hicks JM, Hsu VL (2004) The extended left-handed helix: a simple nucleic acid-binding motif. Proteins 55:330–338

Houde M, Dallaire S, N'Dong D, Sarhan F (2004) Overexpression of the acidic dehydrin WCOR410 improves freezing tolerance in transgenic strawberry leaves. Plant Biotechnol J 2:381–387

Houde M, Daniel C, Lachapelle M, Allard F, Laliberte S, Sarhan F (1995) Immunolocalization of freezing-tolerance-associated proteins in the cytoplasm and nucleoplasm of wheat crown tissues. Plant J 8:583–593

Houde M, Danyluk J, Laliberte JF, Rassart E, Dhindsa RS, Sarhan F (1992) Cloning, characterization, and expression of a cDNA encoding a 50-kilodalton protein specifically induced by cold acclimation in wheat. Plant Physiol 99:1381–1387

Hundertmark M, Hincha DK (2008) LEA (late embryogenesis abundant) proteins and their encoding genes in *Arabidopsis thaliana*. BMC Genomics 9:118

Ingram J, Bartels D (1996) The molecular basis of dehydration tolerance in plants. Annu Rev Plant Physiol Plant Mol Biol 47:377–403

Ismail AM, Hall AE, Close TJ (1999) Purification and partial characterization of a dehydrin involved in chilling tolerance during seedling emergence of cowpea. Plant Physiol 120:237–244

Iturriaga G, Cushman M, JC C (2006) An EST catalogue from the resurrection plant *Selaginella lepidophylla* reveals abiotic stress adaptive genes. Plant Sci 170:1173–1184

Iturriaga G, Schneider K, Salamini F, Bartels D (1992) Expression of desiccation-related proteins from the resurrection plant Craterostigma plantagineum in transgenic tobacco. Plant Mol Biol 20:555–558

Jensen AB, Goday A, Figueras M, Jessop AC, Pages M (1998) Phosphorylation mediates the nuclear targeting of the maize Rab17 protein. Plant J 13:691–697

Jiang X, Wang Y (2004) Beta-elimination coupled with tandem mass spectrometry for the identification of in vivo and in vitro phosphorylation sites in maize dehydrin DHN1 protein. Biochemistry 43:15567–15576

Kelly M, Chellgren B, Rucker A, Troutman J, Fried M, Miller A, Creamer T (2001) Host-guest study of left-handed polyproline II helix formation. Biochemistry 40:14376–14383

Koag MC, Fenton RD, Wilkens S, Close TJ (2003) The binding of maize DHN1 to lipid vesicles. Gain of structure and lipid specificity. Plant Physiol 131:309–316

Koag MC, Wilkens S, Fenton RD, Resnik J, Vo E, Close TJ (2009) The K-segment of maize DHN1 mediates binding to anionic phospholipid vesicles and concomitant structural changes. Plant Physiol 150:1503–1514

Kovacs D, Kalmar E, Torok Z, Tompa P (2008) Chaperone activity of ERD10 and ERD14, two disordered stress-related plant proteins. Plant Physiol 147:381–390

Lee CS, Chien CT, Lin CH, Chiu YY, Yang YS (2006) Protein changes between dormant and dormancy-broken seeds of *Prunus campanulata* Maxim. Proteomics 6:4147–4154

Lisse T, Bartels D, Kalbitzer HR, Jaenicke R (1996) The recombinant dehydrin-like desiccation stress protein from the resurrection plant *Craterostigma plantagineum* displays no defined three-dimensional structure in its native state. Biol Chem 377:555–561

Livernois AM, Hnatchuk DJ, Findlater EE, Graether SP (2009) Obtaining highly purified intrinsically disordered protein by boiling lysis and single step ion exchange. Anal Biochem 392:70–76

Luo P, Baldwin RL (1997) Mechanism of helix induction by trifluoroethanol: a framework for extrapolating the helix-forming properties of peptides from trifluoroethanol/water mixtures back to water. Biochemistry 36:8413–8421

Magidovich E, Orr I, Fass D, Abdu U, Yifrach O (2007) Intrinsic disorder in the C-terminal domain of the Shaker voltage-activated K+ channel modulates its interaction with scaffold proteins. Proc Natl Acad Sci USA 104:13022–13027

McNulty BC, Tripathy A, Young GB, Charlton LM, Orans J, Pielak GJ (2006) Temperature-induced reversible conformational change in the first 100 residues of alpha-synuclein. Protein Sci 15:602–608

Mehta PA, Rebala KC, Venkataraman G, Parida A (2009) A diurnally regulated dehydrin from Avicennia marina that shows nucleo-cytoplasmic localization and is phosphorylated by Casein kinase II in vitro. Plant Physiol Biochem 47:701–709

Minton AP (2005a) Influence of macromolecular crowding upon the stability and state of association of proteins: predictions and observations. J Pharm Sci 94:1668–1675

Minton AP (2005b) Models for excluded volume interaction between an unfolded protein and rigid macromolecular cosolutes: macromolecular crowding and protein stability revisited. Biophys J 88:971–985

Minton AP (2006) Macromolecular crowding. Curr Biol 16:R269–R271

Momma M, Kaneko S, Haraguchi K, Matsukura U (2003) Peptide mapping and assessment of cryoprotective activity of 26/27-kDa dehydrin from soybean seeds. Biosci Biotechnol Biochem 67:1832–1835

Mouillon JM, Eriksson SK, Harryson P (2008) Mimicking the plant cell interior under water stress by macromolecular crowding: disordered dehydrin proteins are highly resistant to structural collapse. Plant Physiol 148:1925–1937

Mouillon JM, Gustafsson P, Harryson P (2006) Structural investigation of disordered stress proteins. Comparison of full-length dehydrins with isolated peptides of their conserved segments. Plant Physiol 141:638–650

Muthalif MM, Rowland LJ (1994) Identification of dehydrin-like proteins responsive to chilling in floral buds of blueberry (Vaccinium, section Cyanococcus). Plant Physiol 104:1439–1447

Nylander M, Svensson J, Palva ET, Welin BV (2001) Stress-induced accumulation and tissue-specific localization of dehydrins in *Arabidopsis thaliana*. Plant Mol Biol 45:263–279

Puhakainen T, Hess MW, Makela P, Svensson J, Heino P, Palva ET (2004) Overexpression of multiple dehydrin genes enhances tolerance to freezing stress in *Arabidopsis*. Plant Mol Biol 54:743–753

Reyes JL, Campos F, Wei H, Arora R, Yang Y, Karlson DT, Covarrubias AA (2008) Functional dissection of hydrophilins during in vitro freeze protection. Plant Cell Environ 31:1781–1790

Riera M, Figueras M, Lopez C, Goday A, Pages M (2004) Protein kinase CK2 modulates developmental functions of the abscisic acid responsive protein Rab17 from maize. Proc Natl Acad Sci USA 101:9879–9884

Rinne PL, Kaikuranta PL, van der Plas LH, van der Schoot C (1999) Dehydrins in cold-acclimated apices of birch (Betula pubescens ehrh.): production, localization and potential role in rescuing enzyme function during dehydration. Planta 209:377–388

Rorat T, Grygorowicz WJ, Irzykowski W, Rey P (2004) Expression of KS-type dehydrins is primarily regulated by factors related to organ type and leaf developmental stage during vegetative growth. Planta 218:878–885

Saavedra L, Svensson J, Carballo V, Izmendi D, Welin B, Vidal S (2006) A dehydrin gene in *Physcomitrella patens* is required for salt and osmotic stress tolerance. Plant J 45:237–249

Sanchez-Ballesta MT, Rodrigo MJ, Lafuente MT, Granell A, Zacarias L (2004) Dehydrin from citrus, which confers in vitro dehydration and freezing protection activity, is constitutive and highly expressed in the flavedo of fruit but responsive to cold and water stress in leaves. J Agric Food Chem 52:1950–1957

Sarnighausen E, Karlson D, Ashworth E (2002) Seasonal regulation of a 24-kDa protein from red-osier dogwood (*Cornus sericea*) xylem. Tree Physiol 22:423–430

Shi Z, Chen K, Liu Z, Kallenbach NR (2006) Conformation of the backbone in unfolded proteins. Chem Rev 106:1877–1897

Soulages JL, Kim K, Arrese EL, Walters C, Cushman JC (2003) Conformation of a group 2 late embryogenesis abundant protein from soybean. Evidence of poly (L-proline)-type II structure. Plant Physiol 131:963–975

Soulages JL, Kim K, Walters C, Cushman JC (2002) Temperature-induced extended helix/random coil transitions in a group 1 late embryogenesis-abundant protein from soybean. Plant Physiol 128:822–832

Sterky F, Bhalerao RR, Unneberg P, Segerman B, Nilsson P, Brunner AM, Charbonnel-Campaa L, Lindvall JJ, Tandre K, Strauss SH, Sundberg B, Gustafsson P, Uhlen M, Bhalerao RP, Nilsson O, Sandberg G, Karlsson J, Lundeberg J, Jansson S (2004) A Populus EST resource for plant functional genomics. Proc Natl Acad Sci USA 101:13951–13956

Svensson J, Palva ET, Welin B (2000) Purification of recombinant *Arabidopsis thaliana* dehydrins by metal ion affinity chromatography. Protein Expr Purif 20:169–178

Tompa P, Csermely P (2004) The role of structural disorder in the function of RNA and protein chaperones. FASEB J 18:1169–1175

Tornroth-Horsefield S, Wang Y, Hedfalk K, Johanson U, Karlsson M, Tajkhorshid E, Neutze R, Kjellbom P (2006) Structural mechanism of plant aquaporin gating. Nature 439:688–694

Tunnacliffe A, Wise MJ (2007) The continuing conundrum of the LEA proteins. Naturwissenschaften 94:791–812

Uversky VN, Oldfield CJ, Dunker AK (2005) Showing your ID: intrinsic disorder as an ID for recognition, regulation and cell signaling. J Mol Recognit 18:343–384

Vlad F, Turk BE, Peynot P, Leung J, Merlot S (2008) A versatile strategy to define the phosphorylation preferences of plant protein kinases and screen for putative substrates. Plant J 55:104–117

Wang X-S, Zhu H-B, Jin G-L, Liu H-L, Wu W-R, Zhu J (2007) Genome scale identification and analysis of LEA genes in rice (*Oryza sativa* L.). Plant Sci 172:414–420

Welling A, Rinne P, Viherä-Aarnio A, Kontunen-Soppela S, Heino P, ET P (2004) Photoperiod and temperature differentially regulate the expression of two dehydrin genes during overwintering of birch (*Betula pubescens* Ehrh.). J Exp Bot 55:507–516

Whittington SJ, Chellgren BW, Hermann VM, Creamer TP (2005) Urea promotes polyproline II helix formation: implications for protein denatured states. Biochemistry 44:6269–6275

Wisniewsk M, Webb R, Balsamo R, Close T, Yu X, Griffith M (1999) Purification, immunolocalization, cryoprotective and antifreeze activity of PCA60: a dehydrin from peach (*Prunus persica*). Physiol Plant 105:600–608

Wright PE, Dyson HJ (2009) Linking folding and binding. Curr Opin Struct Biol 19:31–38

Xu J, Zhang YX, Wei W, Han L, Guan ZQ, Wang Z, Chai TY (2008) BjDHNs confer heavy-metal tolerance in plants. Mol Biotechnol 38:91–98

Yin Z, Rorat T, Szabala B, Ziolkowska A, Malepszy S (2006) Expression of a *Solanum sogarandinum* SK3-type dehydrin enhances cold tolerance in transgenic cucumber seedlings. Plant Sci 170:1164–1172

Zimmerman SB, Minton AP (1993) Macromolecular crowding: biochemical, biophysical, and physiological consequences. Annu Rev Biophys Biomol Struct 22:27–65

Chapter 15
Understanding Vegetative Desiccation Tolerance Using Integrated Functional Genomics Approaches Within a Comparative Evolutionary Framework

John C. Cushman and Melvin J. Oliver

15.1 Introduction

Desiccation tolerance (DT), the ability of cells to survive the air-dried state (equilibration to the water potential of the air which is generally low), is not uncommon in the plant kingdom. The vast majority of plants develop tissues that can withstand desiccation and for the most part these tissues are propagules: spores, pollen, and seeds. Vegetative DT is relatively common in the less complex plants that make up the algae, lichens, bryophytes, and perhaps the hornworts (Alpert 2006; Wood 2007), but is rare in the more complex vascular plants (Porembski and Barthlott 2000; Oliver et al. 2000; Proctor and Pence 2002). Vegetative DT was likely lost during the first steps in the evolution of tracheophytes (Oliver et al. 2000) around 415 million years ago (Soltis et al. 2002). Early mechanisms of vegetative desiccation tolerance have been postulated to be energetically expensive and perhaps structurally demanding (requiring simplicity) and thus were discarded in lieu of increased growth rates, structural and morphological complexity, and mechanisms for increased water conservation and efficient carbon fixation. The distribution of desiccation-tolerant plants is described in detail in Chap. 8.

Mechanistically, one can see trends within the evolutionary pathway for vegetative DT. DT has been postulated to have first evolved in spores of the land plants, and probably their charophyte alga relatives, prior to expression in vegetative tissues. The mechanism for tolerance in the early land plants resembles that seen in the modern day desiccation-tolerant bryophytes: a mechanism based upon constitutive cellular protection coupled with a rehydration-induced recovery process that involves cellular repair (Oliver et al. 2005). The available data suggest that the early reappearance of vegetative DT in vascular plants, using modern day ferns as a model, involved an environmentally inducible cellular protection mechanism as well as a rehydration-induced recovery process, indicating, at least conceptually, a simple progression from the mechanisms employed by bryophytes (Bewley et al. 1993). Seed plants evolved around 340 million years ago (Soltis et al. 2002), and it is at this time that DT appeared as a developmentally controlled program of what appears to be almost exclusively a cellular protection process (Oliver et al. 2000; Berjak et al. 2007; LePrince and Buitink 2007). In the angiosperms, the

U. Lüttge et al. (eds.), *Plant Desiccation Tolerance*, Ecological Studies 215,
DOI 10.1007/978-3-642-19106-0_15,

predominant mechanism for vegetative DT is the environmental induction of a cellular protection process with apparently little need for a rehydration-induced recovery or repair mechanisms (Ingram and Bartels 1996; Oliver et al. 2000; Hoekstra et al. 2001; Moore et al. 2009).

Although we have DT has played in the evolution of the land plants, much is yet to be uncovered. Our understanding of the underlying mechanisms, their components, the genes, gene networks, and the regulatory processes that control them is still in its infancy. The attainment of an understanding of vegetative DT mechanisms has profound importance, for elucidating not only the intricacies of plant evolution, but also the more practical aspects of the plant sciences that lie within the agricultural arena. Unraveling the genetic components that control and establish tolerance to severe dehydration will generate novel strategies for the improvement of drought tolerance in crops for which the toleration of dehydration is a major component. This is proving to be the case as discussed in a recent review by Moore et al. (2009) and as we allude to in this chapter.

The recent advances in systems biology and the tools that have been developed to look at biological processes on a large scale, the "omics" technologies, have not yet been fully employed in the study of DT. These technologies offer the most exciting possibilities for expanding our knowledge of DT in plants. Here, we review the gene discovery and transcriptomic studies that have been performed in resurrection species and discuss how future studies and improvements in these are expected to revolutionize our understanding of DT. Although less commonly performed, we also summarize recent proteomic and metabolite profiling studies using various resurrection plant species.

15.2 Targeted Gene Discovery

Early targeted gene isolation and expression studies in resurrection plants have been limited to a relatively few species including the moss *Tortula ruralis* (Scott and Oliver 1994; O'Mahony and Oliver 1999a, b; Wood et al. 2000; Zeng and Wood 2000; Chen et al. 2002; Zeng et al. 2002), the clubmosses *Selaginella lepidophylla* (Zentella et al. 1999) and *Selaginella tamariscina* (Liu et al. 2008), the dicotyledonous species *Craterostigma plantagineum*, which has received the most attention (Bartels et al. 1990, 1992; Iturriaga et al. 1992, 1996; Michel et al. 1994; Velasco et al. 1994, 1998; Bernacchia et al. 1995, 1996; Chandler and Bartels 1997; Furini et al. 1997; Ingram et al. 1997; Frank et al. 1998; Mariaux et al. 1998; Kleines et al. 1999; Kirch et al. 2001; Deng et al. 2002, 2006; Hilbricht et al. 2002; Phillips et al. 2002a, b; Rodrigo et al. 2004; Ditzer and Bartels 2006), and the monocotyledonous species *Sporobulus stapfianus* (Gaff et al. 1997; Clugston et al. 1998; Blomstedt et al. 1998a, b; O'Mahony and Oliver 1999a, b; Neale et al. 2000; Le et al. 2007), *Xerophyta viscosa* (Mundree et al. 2000; Mowla et al. 2002; Garwe et al. 2003, 2006; Lehner et al. 2008), *Xerophyta humilis* (Collett et al. 2003; Illing et al. 2005; Mulako et al. 2008), and *Xerophyta villosa* (Collett et al. 2004).

Other resurrection plant species have been characterized, but no gene sequence information has yet been reported for them including the moss *Polytrichum formosum* (Proctor et al. 2007a, b), the lichen *Cladonia convoluta* (Tuba et al. 1998), the clubmoss *Selaginella bryopteris* (Deeba et al. 2009), the fern *Polypodium virginianum* (Reynolds and Bewley 1993), the homoiochlorophyllous angiosperms *Haberlea rhodopensis* (Georgieva et al. 2007, 2009), *Boea hygrometrica* (Jiang et al. 2007), and *Boea hygroscopica* (Bochicchio et al. 1998), the angiosperms *Ramonda serbica* (Živković et al. 2005; Degl'Innocenti et al. 2008; Veljovic-Jovanovic et al. 2008), *Craterostigma wilmsii* (Cooper and Farrant 2002; Vicré et al. 2004), and *Lindernia brevidens*, a close relative of *C. plantagineum* (Smith-Espinoza et al. 2007; Phillips et al. 2008), the resurrection grass *Eragrostis nindensis* (Vander Willigen et al. 2003; Illing et al. 2005), and the dicotyledonous, woody, medicinal shrub, *Myrothamnus flabellifolia* (Moore et al. 2006, 2007).

15.3 Gene Discovery Using Expressed Sequence Tags

The most basic approach toward transcriptome analysis has been to collect expressed sequence tags (ESTs) using traditional Sanger sequencing. ESTs are typically automatically curated, single-read sequences of cDNA (complementary DNA molecules derived from reverse-transcribed cellular mRNA populations) (Rudd 2003). The relatively inexpensive nature of random sampling of cDNA libraries has made EST sequencing a very attractive and popular route for sampling transcriptomes (Rudd 2003) and also provides the raw material for the fabrication of cDNA or oligonucleotide microarrays (Alba et al. 2004). In the absence of complete genome sequence data, EST data can provide a low-cost, accessible way to efficiently sample the actively transcribed portions of a genome (Rudd 2003). Comparison of available EST data among diverse taxa of resurrection species or closely related "sister groups" also provides a means to provide novel insights into shared or unique molecular processes required for DT. Once whole-genome sequence information becomes available for a resurrection species, EST data will provide an invaluable resource for genome annotation.

Despite its vast utility, EST sequencing has a number of weaknesses. First, the sampling depth of most traditional EST projects is often limited to only a few hundred or thousand sequenced cDNAs. Although the cDNA library might accurately reflect the relative abundance of a particular transcript, some transcripts, such as low abundance, will be represented poorly in the cDNA library and genes, which are not expressed in the particular organ or tissue from which the library was prepared, will be absent (Rudd 2003). Enrichment strategies, such as normalization and subtraction, can partially overcome inadequate sampling (Bonaldo et al. 1996) and can dramatically improve the sequence diversity within a particular cDNA library by equalizing the relative occurrence of abundant versus rare transcripts. However, such strategies can never fully compensate for inadequate sampling leading to unreliable estimates of relative transcript abundance, particularly of low to moderately expressed mRNAs. Fortunately, more high-throughput sampling technologies have been developed that allow for more cost-effective, in-depth quantitative mRNA expression profiling (see Sect. 15.9).

Second, the sequence quality of cDNA-derived ESTs derived from Sanger sequencing is typically about 97% accurate when compared with genomic reference sequences considering all types of errors including insertions, deletions, and substitutions (Hillier et al. 1996). These errors can be the result of the poor fidelity of the reverse transcriptase and sequencing polymerase (Arezi and Hogrefe 2007) and base-calling accuracy (Li et al. 2004; Prosdocimi et al. 2007). Furthermore, if EST sequences are not carefully cleansed, they can contain xenocontaminants (e.g., vector, polylinker, and primer-adaptor sequences) or sequences from foreign organisms (e.g., *E. coli* and fungi) and abundant structural or regulatory RNAs (e.g., rRNAs and organellar transcripts). Typically, such contaminants can represent 1–3% of all ESTs (Lee and Shin 2009). In general, high-throughput sequencing strategies that provide highly redundant sampling technologies will result in far more accurate transcriptome sequencing data (Sect. 15.9).

Third, EST sequencing will often not provide a complete representation of gene models or full-length cDNAs from which they were derived. To overcome this limitation, full-length cDNA collections have been developed for many important model desiccation-sensitive species including *Arabidopsis* (Seki et al. 2002, 2004), soybean (Umezawa et al. 2008), and the model halophyte, *Thellungiella halophila* (Taji et al. 2008). Such full-length cDNA collections should also be developed as a key component for whole-genome annotation for selected resurrection species. In addition, full-length cDNA collections provide for the efficient exploration of plant gene function in heterologous hosts and plant improvement by using heterologous gene resources (Ichikawa et al. 2006; Kondou et al. 2009).

15.4 Transcriptome Analysis of Nonvascular Resurrection Plants

Targeted gene characterization studies have expanded to become large-scale EST sequencing efforts within a handful of resurrection model species (Table 15.1). Early small-scale EST sequencing (152 ESTs), from a cDNA library from polysomal mRNP fractions of desiccated leaves of the desiccation-tolerant bryophyte, *T. ruralis*, showed that 71% of the ESTs in the library represented novel sequences (Wood et al. 1999). More extensive EST sequencing of cDNA library from rehydrated, rapid-dried *T. ruralis* resulted in the characterization of ~10,368 ESTs representing 5,563 genes of which 2,242 (40.3%) were classified as unknowns, indicating the possibility that this species serves as a genetic reservoir for novel genes involved in stress tolerance (Oliver et al. 2004). Some of the most abundant transcripts in this cDNA library encode late embryogenesis abundant (LEA) proteins, suggesting that these proteins might play a role in the recovery from desiccation upon

Table 15.1 Current status of large-scale expressed sequence tag sequencing projects in resurrection species

Species	cDNA library	Sanger ESTs[a]	SSH ESTs[a]	Roche/454 life sciences ESTs[a]	References
Tortula ruralis	Polysomal mRNA desiccated gametophytes	152	–	–	Wood et al. (1999)
Tortula ruralis	Rehydrated rapid-dried gametophytes	9,074	–	–	Oliver et al. (2004)
Tortula ruralis	Total and polysomal RNA from rapid- and slow-dried and rehydrated gametophytes	–	768	–	Oliver et al. (2009)
Selaginella lepidophylla	Dehydrating fronds	8,355	–	612,206	Iturriaga et al. (2006), Cushman, unpublished
Xerophyta humilis	Dehydrating/ rehydrating roots and leaves	403	–	–	Collett et al. (2004)
Sporobolus stapfianus	Dehydrating leaves	14,515	–	490,144	Oliver, unpublished
Craterostigma plantagineum	Dehydrated, desiccated, rehydrated, and untreated leaves			182 Mb transcript	Rodriguez et al. (2010)

[a]All ESTs reported were cleansed. Additional cDNA sequences may also be present in GenBank database

rehydration (Oliver et al. 2004). Comparison of these *T. ruralis* ESTs to available ESTs from the desiccation-sensitive moss, *Physcomitrella patens*, revealed that while *T. ruralis* is closely related, as both species are bryophytes, there is substantial phylogenetic distance between these two species (Oliver et al. 2004). These studies also led to the identification of desiccation- and rehydration-specific ubiquitin genes (O'Mahony and Oliver 1999a, b).

15.5 Transcriptome Analysis in Vascular Resurrection Plants

Large-scale gene discovery efforts in vascular resurrection plants have been limited to all but a few model species. One of the earliest reports of cloning and sequencing large numbers of cDNA clones from a resurrection species used differential, subtractive, or cold-plaque screening to isolate and characterize 200 cDNA clones from *C. plantagineum* leaves dried for 1 h or to complete dryness (Bockel et al. 1998). Recently, a very comprehensive transcriptome analysis has become available for cDNAs generated from different physiological stages of *C. plantagineum* (Rodriguez et al. 2010). Genes encoding abundant drought-induced genes correlated with DT or low abundance transcripts encoding gene products not previously associated with drought stress in *S. stapfianus* were isolated by differential screening (Blomstedt et al. 1998a) or by "cold-plaque" hybridization procedures (Neale et al. 2000a, b), respectively, suggesting that resurrection plants may possess unique genes and/or regulatory processes that confer DT.

In the new world lycophyte, *S. lepidophylla,* native to Mexico and the southwestern United States, 1,046 ESTs were obtained from a cDNA library constructed from plants undergoing desiccation representing 873 unique transcripts (Iturriaga et al. 2006). Comparison of the *S. lepidophylla* ESTs with 1,301 unigenes from *Selaginella moellendorffii* revealed that 63% of genes were unique to *S. lepidophylla.* In contrast to the 2,181 ESTs from *S. moellendorffii*, the desiccation-tolerant *S. lepidophylla* EST collection (Weng et al. 2005) contained a much greater relative percentage of stress response (i.e., LEA proteins), chaperones, and heat shock protein (HSP) ESTs. More importantly, analysis of the most abundant transcripts sampled by EST sequencing revealed that *S. lepidophylla* preferentially expressed genes whose primary assignable function is in stress response pathways (Iturriaga et al. 2006). Currently, a total of 8,355 ESTs have been sequenced using traditional Sanger sequencing from a cDNA library constructed from plants undergoing dehydration or rehydration at intervals of approximately 10% relative water content (RWC) loss or gain (Table 15.1).

Large-scale EST collections have also been generated from the poikilochlorophyllous, monocotyledonous, *Xerophyta humilis*, a resurrection species native of southern Africa (Collett et al. 2004). Four individually normalized cDNA libraries were generated from dehydrating leaf, dehydrating root, rehydrating leaf, and rehydrating root, respectively, at a range of seven different RWCs. Approximately 100 cDNA clones from each library were sequenced, resulting in an annotated set of 424 cDNAs of which 94% of clones were unique. On the basis of this limited evaluation, the cDNA libraries were judged to be normalized successfully. The libraries obtained from dehydrating root and leaf tissues were also found to be enriched for genes known to be associated with water-deficit, osmotic, cold, and pathogen stress responses, relative to the rehydration libraries (Collett et al. 2004). These same 424 cDNAs were then used to fabricate a printed, cDNA microarray, the results of which were also validated using reverse northern slot blot analysis. Between the two hybridization approaches, a total of 55 cDNA clones (13%) exhibited increased relative transcript abundance upon dehydration at either 26 or 9% RWC. Notably, cDNAs encoding LEA proteins, metallothioneins, an oleosin, and biosynthetic enzymes for the biosynthesis of polyols and raffinose family oligosaccharides were included in this group. A total of 79 cDNA clones (18.6%) showed decreased relative transcript abundance upon dehydration. Of these, 25% encoded cDNAs with functions related to photosynthesis and metabolism consistent with the notion that this poikilochlorophyllous species deactivates photosynthetic functions during dehydration. A follow-on study focused on the analysis of 16 cDNAs representing seven different LEA protein groups derived from this *X. humilis* EST collection (Illing et al. 2005). The relative mRNA expression of 13 of these cDNAs was validated by northern blot analysis and all exhibited similar expression profiles with peak expression occurring during desiccation (<65%

RWC) and stably stored in dry (<6% RWC) leaves. In this same study, a comparison of LEA protein mRNA expression, antioxidant enzyme mRNA expression and activity patterns, and sucrose accumulation patterns among desiccation-sensitive and desiccation-tolerant species revealed discrete commonalities with desiccation-tolerant plants and the acquisition of DT in orthodox seeds. Namely, both DT vegetative tissues and seeds shared sucrose accumulation and the expression of a LEA-6 gene and a 1-cys-peroxiredoxin gene as common DT mechanisms (Illing et al. 2005).

15.6 Subtractive Suppression Hybridization

Most EST sequencing projects are conducted from randomly selected clones from cDNA libraries from a specific tissue or developmental stage. Such a random sampling approach is relatively inefficient because it does not selectively target a subpopulation of transcripts within a particular cDNA library. In contrast, the Subtractive Suppression Hybridization (SSH) approach allows cDNA libraries to be constructed so that they are specifically enriched for a subpopulation of transcripts (Diatchenko et al. 1996). SSH combines normalization and subtraction into a single process that involves the use of two hybridizations that normalize and enrich, respectively, for differentially expressed target genes in "tester" versus "driver" cDNA populations. Differentially expressed target cDNAs become preferentially amplified, whereas amplification of nontarget cDNA is suppressed during the PCR stages of the process. The SSH procedure reportedly achieves a greater than 1,000-fold enrichment of differentially expressed cDNAs (Diatchenko et al. 1996). This enrichment reduces the cloning of abundantly expressed genes common to control (driver) and treatment (tester) samples, a process that increases the probability of cloning differentially expressed genes of interest. Many genes of interest are expressed at relatively low abundance or in specific cell or tissue types or under a particular environmental condition, making selective enrichment by SSH an important and useful strategy.

The SSH approach was first used in the DT lycopod *S. tamariscina* to enrich for and identify genes whose mRNA expression is increased following 2–4 h dehydration of fronds (Liu et al. 2008). Out of more than 300 cDNA clones obtained, reverse northern blot analysis revealed that 96 were differentially expressed with 4 and 92 clones had reduced or increased relative mRNA abundance, respectively, following dehydration stress. The most abundantly expressed cDNA encoded an early light-inducible protein B (ELIPB) closely related to the homologous gene from *T. ruralis* (Zeng et al. 2002). A subset (11) of the increased abundance clones were also verified to be abscisic acid (ABA)-responsive genes adding evidence to the notion that ABA mediates dehydration stress responses in *S. tamariscina* (Liu et al. 2008).

The SSH approach was also used in the DT moss *T. ruralis* to enrich for and identify genes with low relative mRNA abundance that might be involved in slow drying or rehydration (Oliver et al. 2009). A total of 768 cDNAs were sequenced. Of these, 614 (80%) were unique demonstrating the effectiveness of the normalization strategy. Half of these cDNAs (298) were not previously obtained from an earlier EST sequencing effort that generated over 10,000 ESTs (Oliver et al. 2004). Furthermore, 59% of the SSH-derived EST contigs could not be annotated by similarity matches to public sequence databases, which might be expected of less well-expressed transcripts. Interestingly, the dehydration SSH EST collection only contained a single gene encoding a protein with a photosynthetic function. In contrast, functional categorization of EST collections from *S. lepidophylla* and *X. humilis* revealed that up to 17% of ESTs had functions related to photosynthesis. This bias against photosynthetic genes in the *T. ruralis* SSH EST collection might be an indication that the photosynthetic apparatus of this DT moss is relatively undamaged by desiccation and rehydration (Oliver et al. 2009) and that its protection constitutes a major component of DT in bryophytes (Oliver et al. 2005; Proctor et al. 2007a, b) (see Chap. 7). In contrast, the rehydration SSH EST collection was enriched for cDNAs that encode components of the protein synthetic apparatus and the translation process, consistent with the important role played by protein synthesis in the recovery of *T. ruralis* from desiccation following rehydration (Oliver et al. 2005, 2009).

A major motivation for the generation of EST collections is also to provide physical probe collections for cDNA microarrays. cDNA collections derived from two different SSH libraries enriched for sequestered transcripts within slow-dried *T. ruralis* gametophytes or transcripts translated after rehydration were used to develop a printed cDNA microarray containing 768 cDNAs (Oliver et al. 2009). Expression profiles within total RNA derived from hydrated gametophytes compared with rapid-dried rehydrated (RDR) and slow-dried (SD), respectively, or within polyA RNA from the polysomal fraction of hydrated gametophytes compared with the polysomal fractions of SD or RDR gametophytes revealed existence of several novel components of the DT mechanism including jasmonic acid signaling, proteosomal activation, and alternative splicing (Oliver et al. 2009).

15.7 cDNA-Amplified Fragment Length Polymorphism

Quantitative cDNA-Amplified Fragment Length Polymorphism (AFLP) is a powerful gene discovery technique for the identification of differentially expressed genes (Bachem et al. 1996; Breyne et al. 2003). The basic approach involves combinatorial restriction enzyme digestion of cDNA followed by selective amplifications of the resulting fragments to produce less complex subsets of transcript tags, which are then separated by electrophoresis on high-resolution polyacrylamide gels and visualized by autoradiography or an automated LI-COR system (Vuylsteke et al. 2007). The original cDNA-AFLP method has undergone several refinements to screen and visualize a majority of both abundant and weak differentially expressed genes on a genome-wide basis (Breyne et al. 2003). Despite the accessibility of cDNA-AFLP analysis for the typical laboratory, to our knowledge there have been no reports of its use on any resurrection species. However, the related technique of mRNA differential display visualized by silver staining was reported to examine mRNA expression patterns in the dicotyledonous, homoiochlorophyllous resurrection plant *B. hygrometrica* (Deng et al. 1999) or by radioactive labeling to isolate a desiccation-rehydration-responsive small GTP-binding protein gene (O'Mahony and Oliver 1999b).

15.8 Comparative Transcriptome Analysis in Resurrection Plants

The identification of differentially expressed genes is only the beginning of the gene discovery process, especially in relation to understanding the underlying genetic components of DT. Desiccation and subsequent rehydration are extreme environmental insults to plant cells whether or not they are tolerant to them. A large proportion of the differential gene expression responses are likely to result from cellular injury. Injury might trigger the up- or downregulation of specific genes that are not involved in promoting adaptation to dehydration *per se* and are thus misleading when it comes to unraveling mechanisms of tolerance and perhaps of less use in drought improvement strategies. The only means by which the adaptive nature of a gene (and its expression) can be inferred is from a phylogenetic perspective, which entails species-to-species comparisons. There are two basic types of comparisons that one can employ to analyze changes in gene expression across species (Fig. 15.1): (1) ancestor–descendant comparisons where one attempts to reconstruct the evolutionary history of a gene and its association with a trait such as DT, and (2) sister-group contrasts where one compares the transcriptomes and gene expression profiles for two closely related species that differ in a critical phenotype, in this case DT. Direct comparisons between species-specific alterations in gene expression associated with DT have not been made using either strategy. The necessary transcriptomic resources for an ancestor–descendant

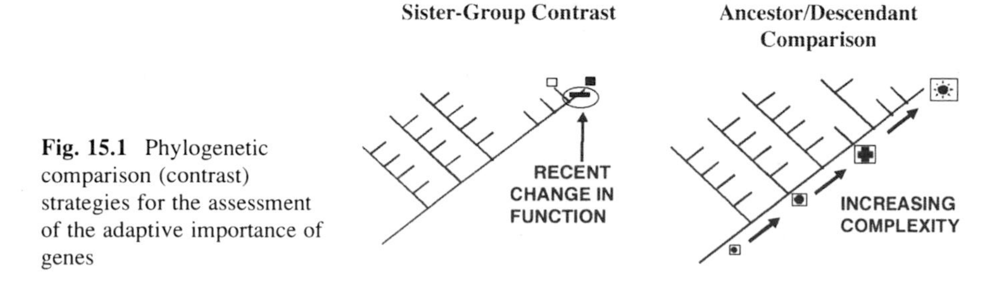

Fig. 15.1 Phylogenetic comparison (contrast) strategies for the assessment of the adaptive importance of genes

comparison for desiccation-tolerant plants are not yet available, but their development is in progress in several labs. There have been some limited efforts to compare tolerant and nontolerant species, but generally they are of limited value because of the evolutionary distance between the two species within the contrast. Illing et al. (2005) compared expression profiles for a number of *X. humilis* genes with published profiles for *Arabidopsis* homologues and was able to infer that there are similarities between the response to desiccation in the resurrection plant to the acquisition of DT in *Arabidopsis* seeds (see below). This is an important observation and adds credence to some of the earlier hypotheses regarding the evolution of DT, but says little about the adaptive nature of the genes investigated. Oliver et al. (2004) compared the transcriptomes of rehydrating *T. ruralis*, a desiccation-tolerant bryophyte, to a publicly available transcriptome for *P. patens*, a desiccation-tolerant bryophyte, and could identify general categories of transcripts that might have indicated adaptive responses to desiccation, in particular those related to maintaining plastid integrity. However, this was a very limited comparison and the two mosses are not particularly closely related. A much closer species-specific transcriptome comparison was reported by Iturriaga et al. (2006), for *S. lepidophylla* and *S. moellendorffii* (see earlier discussion). This comparison requires more direct expression profiling data before any substantial hypotheses can be drawn.

15.9 High-Throughput Sequencing Approaches

Although EST sequencing and its variations described above have been used widely for gene discovery, the depth of coverage is typically limited for digital measurements of gene expression in which the relative abundance of each sequence tag is used to infer the relative abundance of its corresponding transcript present within a particular tissue or condition (Audic and Claverie 1997). Therefore, a number of methods have been developed to provide greater depth of sequencing of short tags, and thereby provide a more accurate, quantitative measure of the relative abundance of a transcript. High-throughput, sequence-based gene expression methods such as Serial Analysis of Gene Expression (SAGE) or Massively Parallel Signature Sequencing (MPSS) can be regarded as complementary to hybridization-based or "closed" transcriptome approaches provided that a sufficient number of biological replicates are performed with each platform (Liu et al. 2007; Vega-Sanchez et al. 2007) and generally provide similar results at high transcript abundance ranges (Nygaard et al. 2008). However, high-throughput sequencing-based methods exhibit increased sensitivity to the detection of low abundance transcripts, increased dynamic range, and have the inherent advantage of measuring new transcripts (Liu et al. 2007). Here, we provide a brief overview of the major transcriptome technologies available currently for genomic and transcriptomic analysis of resurrection plants.

15.9.1 Serial Analysis of Gene Expression

The SAGE method relies on Sanger sequencing of short (10–14 bp), concatenated cDNA tags derived from a cDNA region immediately 3′ to the 3′-most restriction site present in a double-stranded cDNA synthesized on magnetic beads (Velculescu et al. 1995). Although this original method represented an up to 30-fold improvement in tag coverage for the same number of Sanger sequencing reads, such short tags did not always permit unambiguous assignment of tags to corresponding transcripts. Therefore, improved versions of the original method called LongSAGE, which generates 17–19-mer tags (Saha et al. 2002), and SuperSAGE, which generates 26-mer tags (Matsumura et al. 2003, 2008), were developed subsequently. Despite the improved throughput and reduced cost compared to EST sequencing and its widespread use in diverse species (Anisimov 2008), no SAGE-related profiling studies using a resurrection or related species have been reported to date. Furthermore, newer, massively parallel, high-throughput sequencing technologies have recently eclipsed traditional SAGE approaches in both tag size and sampling depth.

15.9.2 Next-Generation Sequencing Technologies

Recent advances in high-throughput sequencing made possible by the commercial introduction of so-called second-generation sequencing instrumentation, which is capable of producing millions of DNA sequence reads in a single run, promise to rapidly transform the pace and scope of functional genomic analyses (Mardis 2008a, b; Simon et al. 2009). Such high-throughput sequencing systems promise to not only drive down the cost of gene discovery by transcriptome or genome sequencing, but also provide genome-wide sequence readouts as endpoints for a wide variety of applications including mutation or polymorphism discovery, comparative genomics, chromatin structural analysis, epigenetic regulation, and discovery of noncoding RNAs (Mardis 2008a, b). Application of these new "open" transcriptome technologies will have a profound influence on our future understanding of resurrection plant gene expression dynamics.

The first so-called next-generation Roche (454) Genome Sequencer (GS) FLX sequencer was commercialized in 2004 by 454 Life Sciences/Roche Applied Science (http://www/454/com) and is based on pyrosequencing (Ronaghi et al. 1998). This system is capable of producing 100–200 Mb during a typical run with an average read length of up to ~500 bp, making this platform well suited for de novo sequencing. (Margulies et al. 2005; Droege and Hill 2008). Given its attractiveness, Roche/454 Life Sciences pyrosequencing has recently been used to characterize a major fraction of the transcriptomes of *S. lepidophylla* and *S. stapfianus* using mixed cDNA libraries prepared from dehydrating and rehydrating tissues in an effort to capture the full repertoire of expressed genes in these species (Table 15.1).

Illumina's SBS Genome Analyzer II system introduced in 2006 uses sequencing-by-synthesis (http://www.illumina.com) technology, which results in shorter (currently ~60 bases) but more abundant ESTs than the Roche/454 Life Sciences platform; however, longer read lengths (>100 bases) are expected to be possible in the near future (Mardis 2008a, b). A typical run will yield 30–50 million sequence tags, making this platform highly desirable for transcriptome profiling, SNP detection (genome resequencing), and genome-wide detection of protein–DNA interactions using ChIP sequencing in organisms with fully sequenced genomes (Bentley 2006; Smith et al. 2008; Simon et al. 2009).

The Applied Biosystems, Inc. Sequencing by Oligo Ligation and Detection (SOLiD) System (http://www.appliedbiosystems.com) generates an adapter-ligated fragment library using emulsion PCR to amplify the fragments on the surface of small magnetic beads similar to the Roche/454 Life Sciences system. However, after amplification, the beads are covalently attached to the surface of a specially treated glass slide, which is placed inside a fluidics cassette inside the sequencer (Mardis 2008a, b). A unique attribute of the SOLiD system is that it contains an inherent quality check feature, called "2 base encoding," to identify miscalled bases from correct base calls during the data analysis step (Mardis 2008a, b; Simon et al. 2009). Each SOLiD run produces 2–3 Gb of sequence data. The major applications for SOLiD include very deep transcriptome profiling, SNP characterization, dissection of chromatin architecture (nucleosome positioning) using ChIP sequencing, and resequencing of bacterial genomes (Simon et al. 2009).

15.10 Protein Expression and Proteomics

Despite the enormous utility of transcriptome analyses, mRNA abundance does not always correlate well with the relative expression of the corresponding protein in eukaryotes (Gygi et al. 1999; Ideker et al. 2001). Posttranslational modifications such as phosphorylation, glycosylation, or N- or C-terminal processing events cannot be assessed using nucleic acid approaches, and small open reading frames are often difficult to confirm without direct sequence information from expressed proteins (Wasinger and Humphery-Smith 1998). Furthermore, proteomic methods are necessary to define the quantity, structure, and function of proteins within the systems biology framework (Phizicky et al. 2003).

The analysis of protein expression changes in resurrection plants has gained increasing attention recently, but also has a long history dating back to the 1980s. One of the earliest studies targeting *S. lepidophylla* investigated changes in the pattern of protein synthesis that occurred during rehydration (Eickmeier 1982). Cytosol-directed protein synthesis occurred within the first 12 h following hydration, whereas organelle-directed protein synthesis remained low until after 12 h of hydration and increased rapidly thereafter coincident with extensive thylakoid membrane proliferation and chloroplast polysome formation within this same time frame (Eickmeier 1982). Alterations in the pattern of protein synthesis were also characterized by comparisons between hydrated and rehydrated gametophytes (2 h of rehydration following rapid desiccation) in *T. ruralis* (Oliver 1991). This study was able to detect the termination or decrease in synthesis of 25 proteins (termed hydrins) and the initiation or substantial increase in the synthesis of 74 others (termed rehydrins). Using a timed labeling strategy, this study was also able to demonstrate that the controls over the reduction or increase in the synthesis of these two groups of proteins are not linked mechanistically. A certain amount of prior water loss was needed to fully activate the synthesis of rehydrins upon rehydration. Discrete protein expression changes have also been documented in the resurrection fern *P. virginianum* undergoing dehydration and rehydration (Reynolds and Bewley 1993). A 22 kDa early-light-induced protein (ELIP) was found to increase in abundance during desiccation and ABA treatment in the thylakoid membranes of *C. plantegineum* chloroplasts and was found to colocalize with the carotenoid zeaxanthin (Alamillo and Bartels 2001). This protein is thought to contribute to protection against photoinhibition caused by dehydration. Analysis of plastidic antenna protein complexes in the resurrection plant *H. rhodopensis* revealed movement of a portion of this complex from PSII to PSI during plant desiccation (Georgieva et al. 2009). This movement was also accompanied by a large increase in zeaxanthin accumulation. The expression of two small heat shock proteins was induced by dehydration and heat stress and exogenous ABA treatment and surprisingly in unstressed vegetative tissues (roots and lower parts of shoots) of *C. plantegineum*, which resembles the expression patterns typically found in desiccation-tolerant zygotic embryos and germinating seeds (Alamillo et al. 1995). Protein expression changes have also been compared in the desiccation-tolerant, resurrection grass *S. stapfianus* and the desiccation-sensitive grass *S. pyramidalis* (Kuang et al. 1995) and the desiccation-tolerant *S. elongatus* versus the desiccation-sensitive *S. pyramidalis* (Ghasempour and Kianian 2007). *In vivo* isotopic labeling of newly synthesized proteins was monitored following 6, 12, and 24 h rehydration in *C. plantagineum* by two-dimensional polyacrylamide gel electrophoresis (2D-PAGE) analysis (Bernacchia et al. 1996). The expression pattern did not change appreciably after 12 h rehydration and none of the proteins were identified.

More recent and more comprehensive proteomic analyses have been performed in several resurrection species. 2D-PAGE analysis of the leaves of the small, dicotyledonous, homoiochlorophyllous resurrection plant *B. hygrometrica* reproducibly identified 223 proteins of which most (60%) were unchanged in abundance in dehydrated versus rehydrated leaves (Jiang et al. 2007). This species differs from many other resurrection species in that detached leaves retain the same ability of DT as that found in intact plants. Analysis of detached leaves avoids potential interference from developmental regulation and long-distance signaling events from other organs. In detached leaves, 35% of the proteins surveyed showed increased abundance upon dehydration, whereas only 5% showed increased abundance following rehydration (Jiang et al. 2007). Of the 14 dehydration-responsive proteins that were analyzed by mass spectrometry (MS), eight were identified as having functional roles in reactive oxygen scavenging, photosynthesis, and metabolism with the remainder being unknown or unidentified.

A more comprehensive proteomic analysis in the monocotyledonous, poikilochlorophyllous, resurrection plant *X. viscosa* surveyed approximately 430 protein spots (Ingle et al. 2007). During dehydration to 65 and 35% RWC, 20 proteins increased in abundance, 13 decreased in abundance, and 21 were specific to dehydration. Of these 54 proteins, 17 were identified by matrix-assisted laser desorption ionization (MALDI)–time of flight (TOF) MS. Proteins with increased abundance during leaf drying included an RNA-binding protein, chloroplast FtsH protease, desiccation-related proteins, and glycolytic and antioxidant enzymes. In contrast, proteins that declined in abundance included four components of PSII, indicating that desiccation involves the dismantling of the thylakoid membranes. In contrast, the abundance of these PSII proteins was largely maintained in the homoiochlorophyllous species *C. plantagineum*, which does not dismantle its thylakoid membranes upon drying (Ingle et al. 2007).

A recent study compared the protein expression profiles in dehydrated and rehydrated *S. bryopteris* fronds, a species in which its detached fronds have the ability to survive desiccation in the same way as the intact plant (Deeba et al. 2009). Analysis of detached fronds avoids potential interference from developmental regulation and long-distance signaling (Jiang et al. 2007). A total of about 250 spots were reproducibly detected by 2D-PAGE with 21 and 27 spots exhibiting increases or decreases in abundance, respectively. Among the 30 most differentially expressed proteins, 9 were identified by LC-MS/MS. Two of the proteins with increased relative abundance included a putative F-box/LRR repeat protein identified as an E3 ubiquitin ligase involved in proteasomal protein degradation and a putative DEAD-box ATP-dependent RNA helicase 5 protein whose expression is known to be increased by salt, dehydration, wounding, and low temperature stresses in *Arabidopsis* (Deeba et al. 2009).

In addition to protein analysis, changes in reversible protein phosphorylation patterns have also been studied in *C. plantagineum* undergoing dehydration stress (Röhrig et al. 2006). Desiccation-induced phosphoproteins were shown to accumulate in desiccated roots and leaves with two phosphoproteins, CDet11-24, a dehydration and ABA-responsive protein and CDeT6-19, a group 2 LEA protein, being especially abundant. Phosphorylation sites were mapped within predicted coiled-coil regions of CDet11-24, indicating that these phosphorylations might influence the stability of coiled-coil interactions with itself or other proteins (Röhrig et al. 2006). In a follow-up study, more than 20 desiccation-induced putative phosphoproteins were enriched using a modified metal oxide affinity chromatography approach, separated by 2D-PAGE, detected by Pro-Q Diamond staining, and identified by MALDI–TOF MS and MS/MS (Röhrig et al. 2008). Of the 20 putative phosphoproteins, 16 were suggested to be *bona fide* phosphoproteins from published evidence. More recently, protocols for the extraction and analysis of nuclear proteins from *X. viscosa* have also been optimized (Abdalla et al. 2009).

15.11 Metabolomics and Fluxomics

In addition to transcriptomic and proteomic analyses, the study of metabolite concentrations and how their abundance changes during the dehydration–rehydration process of resurrection plants is an absolutely essential part of understanding the biochemical and regulator aspects of DT. Quantitative knowledge of the complete set of metabolites and knowledge of which distinct metabolic processes are involved in the production and degradation or flux of discrete sets of metabolites can provide a better representation of the phenotype of an organism than other methods (Wiechert et al. 2007; Cascante and Marin 2008). Although large-scale metabolomics studies have been reported for *Arabidopsis* undergoing dehydration (Urano et al. 2009), similar studies have not yet been performed for any resurrection species; however, such studies are currently in progress for *S. lepidophylla* and *S. stapfianus*.

15.11.1 Sugar Metabolism

Despite the lack of large-scale metabolite studies, many targeted or small-scale investigations into the composition and abundance of key metabolites or changes in the activities of key enzymes have been reported. For example, desiccation-tolerant mosses such as *T. ruralis* maintain a high sucrose content (>100 mg g^{-1} DW) (Smirnoff 1992). Similarly, *Selaginella* species maintain high concentrations of both sucrose and the nonreducing, disaccharide, trehalose (White and Towers 1967; Adams et al. 1990; Iturriaga et al. 2000; Liu et al. 2008), whereas trehalose is only a minor component of accumulated sugars in many other DT species (Ghasempour et al. 1998). Either hydrated or desiccated fronds of *S. lepidophylla* or *S. tamariscina* maintain a high sugar content (>130 mg g^{-1} DW) (Adams et al. 1990; Iturriaga et al. 2000; Liu et al. 2008). In hydrated, growing fronds, trehalose content is high, but then declines 0.8-fold compared with desiccated fronds, whereas sucrose concentrations increase three-fold upon desiccation (Adams et al. 1990). The angiosperm, *C. plantagineum*, maintains a substantial sugar content (>400 mg g^{-1} DW) in both hydrated and desiccated leaves (Bianchi et al. 1991). The monosaccharide 2-octulose is the major soluble sugar in hydrated leaves (430 mg g^{-1} DW), whereas sucrose becomes the dominant sugar upon desiccation (374 mg g^{-1} DW) in *C. plantagineum* and *C. wilmsii* (Cooper and Farrant 2002). A similar, but less dynamic, interconversion of 2-octulose and sucrose has also been observed in *L. brevidens* (Phillips et al. 2008). In comparison, most orthodox seeds accumulate sucrose and large quantities of raffinose series oligosaccharides (RFOs) during the late maturation stages of seed development (Amuti and Pollard 1977). In the intermediate homoiochlorophyllous resurrection grass *S. stapfianus*, hexose sugars (glucose and fructose) increase along with sucrose during dehydration above 50% RWC. However, as RWC declines below 50% during the later stages of dehydration, both hexose sugars decline in abundance, whereas sucrose accumulation continues to increase (Ghasempour et al. 1998; Whittaker et al. 2001). The transient accumulation of hexose sugars might play an osmoregulatory role during the intermediate stages of dehydration (Ghasempour et al. 1998). A similar temporal accumulation pattern is observed in the monocotyledonous, poikilochlorophyllous *X. viscosa*, except that hexose sugars do not accumulate to large concentrations during the dehydration process (Whittaker et al. 2001). In addition to sucrose, RFOs, particularly raffinose, are prominent soluble carbohydrates that accumulate in dehydrated leaves of *X. viscosa* (Peters et al. 2007) as well as other resurrection species including *C. plantagineum* (Norwood et al. 2000), *R. serbica* (Živković et al. 2005), and various resurrection species where they accumulate to about 10% of sucrose content (Ghasempour et al. 1998). The sucrose-to-raffinose mass ratio was very low (1.3:1) in *X. viscosa* compared with other resurrection species where this ratio was much higher (5:1 to 10:1), suggesting that raffinose might also serve a dual role in stress protection and carbon storage (Peters et al. 2007). The *myo*-inositol 1-phosphate synthase (MIPS) gene encoding the enzyme catalyzing the committed step in the formation of *myo*-inositol, which is required for RFOs biosynthesis, has been cloned from *X. viscosa* (Lehner et al. 2008). The MIPS gene and protein increase in abundance following salinity stress and the loss of water to >65% RWC (Lehner et al. 2008). Sucrose accumulates in the drying leaves of the desiccation-tolerant *E. nindensis*, but not in those of the desiccation-sensitive *Eragrostis* species, providing compelling evidence for the importance of sucrose in conferring desiccation tolerance (Illing et al. 2005).

Soluble sugars, such as trehalose, sucrose, and other oligosaccharides, function as compatible solutes in water replacement, preferential exclusion of destabilizing molecules, and vitrification, an immobilized glassy state formed within the cytosol to prevent the occurrence of deleterious reactions in desiccated cells (Buitink et al. 1998; Hoekstra et al. 2001). Sucrose accumulates in all leaf cell types during dehydration (<56% RWC) in *S. stapfianus* as shown by *in situ* staining (Martinelli 2008). Sucrose has also been shown to lower membrane phase transition temperatures of dried membranes *in vitro*, which is beneficial during desiccation (Hoekstra and Golovina 1999). The membranes of desiccation-tolerant species also tend to contain higher unsaturated phospholipid concentrations, which help prevent phase transition of membranes (Hoekstra and Golovina 1999).

15.11.2 Enzyme Activities

Early studies examining the conservation of enzyme activities in dried tissues revealed a mean conservation of 94% in the bryophyte *Acrocladium cuspidatum* (Stewart and Lee 1972), with *T. ruralis* showing 70% conservation for Rubisco along with substantial conservation of several other enzymes (Sen Gupta 1977). Dried *S. lepidophylla* fronds retained an average of 75% activity of ten enzymes (Harten and Eickmeier 1986), whereas desiccation-tolerant angiosperm species, such as *X. viscosa* and *M. flabellifolia*, generally showed lower amounts of enzyme content. The conservation of these enzyme activities is thought to aid in the rapid resumption of metabolic activity in resurrection plants following rehydration.

During dehydration, hexokinase activities increase concomitantly with increased sucrose accumulation and decreased hexose sugar accumulation in both *S. stapfianus* and *X. viscosa* (Whittaker et al. 2001). Hexokinases catalyze the conversion of glucose, and with fructose at a lesser efficiency, to hexose monophosphates with the conversion of ATP to ADP for sucrose production. Leaf Glc-6-P, Fru-6-P, and ATP concentrations also increased in support of the observed increases in hexokinase activity. In contrast, fructokinase activity was unchanged during dehydration. The expression of genes encoding both sucrose-phosphate synthase (SPS) (Ingram et al. 1997) and sucrose synthase (SuSy) (Kleines et al. 1999) have been shown to increase during dehydration in *C. plantagineum*. SPS is considered the major enzyme of sucrose biosynthesis. The increase in SuSy has been proposed to be important for supplying carbon via sucrose catabolism to fuel glycolysis during dehydration and/or following rehydration (Kleines et al. 1999). SPS activity has also shown to increase during dehydration of *S. stapfianus* leaves coincident with sucrose accumulation (Whittaker et al. 2007). In addition to sugar metabolism, nitrogen and amino acid accumulation and metabolism might also play important roles in DT. In *S. stapfianus*, total amino acid content increased during the latter stages of water loss (>80% RWC) likely derived from insoluble protein breakdown (Whittaker et al. 2007). Specifically, the accumulation of large amounts of arginine and asparagine as nitrogen reserves might serve as essential nitrogen and carbon resources useful for successful rehydration (Martinelli et al. 2007).

15.11.3 Reactive Oxygen Scavenging

In addition to sugar metabolism, a critical stress adaptive mechanism for DT involves the detoxification of reactive oxygen intermediates (ROIs), whose production and accumulation increase as a result of environmental stresses including dehydration stress (Mittler 2002). Detoxification can be brought about by free-radical scavenging enzyme systems, which are active only under partially hydrated conditions, and by molecular antioxidants (e.g., ascorbate, glutathione, polyols, carbohydrates, proteins such as peroxiredoxin, and amphiphilic compounds, such as tocopherols, quinones, flavonoids, and phenolics), which can operate under dry conditions to alleviate oxidative stress damage (Vertucci and Farrant 1995). ROI scavenging enzymes such as ascorbate peroxidase (AP), glutathione reductase (GR), and superoxide dismutase (SOD) increased during early or late stages of drying in both *C. wilmsii* and *X. viscosa* (Sherwin and Farrant 1998). GR and SOD activities also increased during the early stages of rehydration, indicating that these enzymes likely afford critical free-radical protection until full rehydration and metabolic recovery has been achieved (Sherwin and Farrant 1998). However, antioxidant enzyme responses can vary depending on the species. GR activity increased upon drying in *S. stapfianus* (Sgherri et al. 1994a, b) and *B. hygroscopica* (Sgherri et al. 1994a, b), whereas AP activity declined or remained constant during dehydration. In *T. ruralis*, AP and catalase activity decreased during drying, whereas SOD activity remained constant (Seel et al. 1992).

In *C. plantagineum*, only AP increased, whereas AP, GR, and SOD activity increased to varying extents following dehydration of *M. flabellifolia* and *X. humilis* (Farrant 2000). Comparison of ROI scavenging enzyme activities (e.g., AP, GR, and SOD) among three *Eragrostis* grass species showed that while these activities were elevated in all species during the early stages of dehydration, they remained elevated only in the DT species (*E. nindensis*) and declined in the closely related desiccation-sensitive species at RWC <70% (Illing et al. 2005). While these ROI scavenging enzymes are not considered unique to DT, their retention in this DT species is likely a consequence of the general enzyme protection mechanisms of resurrection species. However, some enzymes, such as 1-Cys peroxiredoxin, whose expression is induced upon drying in the resurrection plant *X. viscosa* (Mowla et al. 2002) and expressed during rehydration in the DT moss *T. ruralis* (Oliver 1996), might be considered an evolutionary prerequisite for DT (Illing et al. 2005). Polyphenol oxidase (PPO), which catalyzes the oxidation of mono- and *o*-diphenols to *o*-diquinones, showed increased protein abundance and enzyme activity in dehydrating leaves of *B. hygroscopica* (Jiang et al. 2007). PPO activity has also been shown to increase several-fold in leaves of *R. serbica* during dehydration stress (Veljovic-Jovanovic et al. 2008). Polyphenols are powerful detoxifiers of toxic reactive oxygen species (Rice-Evans et al. 1997), and might function as antioxidants during the first few hours of rehydration (Veljovic-Jovanovic et al. 2008).

In addition to enzymes, various metabolites afford protection to resurrection plants in the dried state. Anthocyanidins protect against light stress during desiccation by providing sunscreen to reduce damage to photosynthetic pigments chlorophyll and carotenoids and/or free-radical quenching. Anthocyanin content increased six-fold in the homoiochlorophyllous species *C. wilmsii*, but then declined rapidly following rehydration (Sherwin and Farrant 1998). Anthocyanin content also increases in the related species *C. pumilum* (Hoekstra et al. 2001). The resurrection fern *Polypodium polypodioides* (Muslin and Homann 1992) and the moss *T. ruralis* (Seel et al. 1992) are able to prevent and/or repair photooxidative damage of the photosynthetic apparatus despite the retention of chlorophyll in the dried state. In contrast, the poikilochlorophyllous species *X. viscosa* also displayed a sixfold increase in anthocyanin content upon drying, and this level was retained during rehydration (Sherwin and Farrant 1998). Because poikilochlorophyllous species lose their chlorophyll and dismantle their thylakoid membranes during dehydration, this retention of anthocyanin during rehydration might protect against photooxidation while the photosynthetic apparatus is being reassembled (Sherwin and Farrant 1998). Carotenoid content was retained during drying in *C. wilmsii*, but declined in *X. viscosa* consistent with the dismantling of its photosynthetic apparatus. An increase in reduced glutathione content, consistent with increased abundance of the GR enzyme itself and presumably the result of increased GR activity, has been reported in *B. hygrometrica* undergoing dehydration (Jiang et al. 2007). Increases in reduced glutathione content have also been reported in other resurrection species undergoing dehydration including *B. hygrometrica* (Navari-Izzo et al. 1997) and *R. serbica* (Augusti et al. 2001). A novel member of the vicinal oxygenase chelate (VOC) superfamily of metalloenzymes was cloned and characterized from *X. humilis* (Mulako et al. 2008). This gene is induced during dehydration in both vegetative and seed tissues of *X. humilis* and in mature, dry seeds of *A. thaliana* and is thought to play a role in the detoxification of methylglyoxal during desiccation (Mulako et al. 2008). Methylglyoxal is a cytotoxic by-product of glycolysis that accumulates following a variety of abiotic stresses.

Certain resurrection plants, such as the woody, medicinal plant, *M. flabellifolia*, can serve as sources for unique polyphenols including procyanidins (Anke et al. 2008). The predominant polyphenol in the leaves of this species is 3,4,5 tri-*O*-galloylquinic acid, a compound shown to stabilize artificial membranes (liposomes) against desiccation damage presumably by preserving the liquid crystalline phase of the membrane (Moore et al. 2005). This compound can also protect linoleic acid against free-radical-induced oxidation *in vitro* by itself becoming oxidized (Moore et al. 2005).

15.11.4 Membranes and Lipids

A common response to water deficit is a sharp reduction in lipid content, and similar reductions in lipids have been observed to occur in many resurrection plant species including *R. serbica* (Quartacci et al. 2002), *S. stapfianus* (Quartacci et al. 1997), and *B. hygroscopica* (Navari-Izzo et al. 1995, 2000). Such a reduction might aid membrane integrity. Another consequence of desiccation is a sharp increase in the free-sterol content, particularly cholesterol and cerebrosides, of the membranes in *R. serbica* (Quartacci et al. 2002). Sterol enrichment is thought to increase membrane rigidity, thereby reducing water permeation rates, which might play an important role in stress tolerance. Reductions in acyl chain unsaturation were also observed in *R. serbica* (Quartacci et al. 2002). Decreased fatty acid unsaturation results in decreased membrane fluidity, resulting in a more rigid, tighter lipid bilayer that might reduce solute leakage. However, such reductions in acyl chain unsaturation have not been observed after desiccation in other species including *B. hygroscopica* (Navari-Izzo et al. 1995), *S. stapfianus* (Quartacci et al. 1997), and *S. tamariscina* (Liu et al. 2008), indicating that species-specific adaptive mechanisms are likely to exist.

15.12 Signaling Pathways

The signaling mechanisms that coordinate the constitutive or inducible DT are not well understood (see also Chap. 13). Several studies have investigated the presence and alterations in endogenous concentrations of ABA in response to dehydration as evidence for its involvement in the maintenance or acquisition of DT (see Chaps. 9, 12, and 16). In the moss *T. ruralis*, ABA is not detectable upon drying and exogenous application of ABA does not appear to trigger the synthesis of mRNA or proteins (Bewley et al. 1993). Instead, bryophytes appear to rely on alterations in translational controls to mount a response to desiccation in contrast to the well-established signaling pathways associated with abiotic stress responses in angiosperms. The recent characterization of significantly accumulating transcripts in the polysomal mRNA pools of dehydrated *T. ruralis* invokes the likely participation of gamma-aminobutyric acid (GABA) and jasmonic acid signaling in DT (Oliver et al. 2009).

In vascular resurrection species, however, ABA appears to play a key role in the acquisition of DT. ABA concentrations increase threefold in dehydrated fronds of *S. tamariscina* relative to fully hydrated fronds, indicating that ABA probably plays a role in the acquisition of DT (Liu et al. 2008). A subset (11) of the increased abundance clones were also verified to be ABA-responsive genes verifying that ABA mediates dehydration stress responses in *S. tamariscina* (Liu et al. 2008). Further evidence for the production of ABA in a *Sellaginella* species was provided by the discovery of cDNAs encoding 9-cis-epoxycarotenoid dioxyenase, a key dehydration stress-inducible enzyme of the ABA biosynthetic pathway from *S. lepidophylla* (Iturriaga et al. 2006). In *C. plantagineum*, treatment of desiccation-sensitive callus with ABA renders it DT and induces a set of mRNAs comparable with that activated by whole plant dehydration (Bartels et al. 1990). ABA content increased in *S. stapfianus* leaves in response to dehydration and peaks between 40 and 15% RWC (Gaff and Loveys 1992). Exogenous application of ABA to leaves of *S. stapfianus* also results in the elevated mRNA accumulation of four dehydration-induced cDNA clones (Blomstedt et al. 1998b). However, exogenous ABA application cannot rescue the survival of detached leaves undergoing desiccation, so non-ABA-dependent processes are also likely to be involved.

T-DNA-mediated activation tagging screens of transgenic *C. plantagineum* callus leading to the creation of dominant mutants has led to the discovery of calli, which are capable of surviving desiccation without prior ABA treatment (Furini et al. 1997; Smith-Espinoza et al. 2005). The genes targeted by the T-DNA include two *Craterostigma* desiccation-tolerant (CDT) gene family members that encode naturally occurring siRNA molecules with features of a short interspersed element retrotransposon (SINE) that likely plays important roles in ABA signal transduction, because both activation mutant lines exhibit increased expression of ABA-induced LEA proteins (Phillips et al. 2007; Hilbricht et al. 2008). However, their involvement in ABA-independent DT pathways cannot be excluded (Furini et al. 1997; Smith-Espinoza et al. 2005).

In addition to ABA signaling, phospholipid-based signaling is also likely to play an important role in the acquisition of DT. In *C. plantagineum*, activity of phospholipase D (PLD), which catalyzes the hydrolysis of phosphatidylcholine and other phospholipids to phosphatidic acid (PA), which in turn regulates protein kinases or small GTP-binding proteins, is induced within minutes of dehydration and is not induced by exogenous ABA treatment (Frank et al. 2000). One member of the PLD family (*Cp*PLD-1) in *C. plantagineum* is constitutively expressed, whereas the other (*Cp*PLD-2) is responsive to dehydration and ABA (Frank et al.

2000). Furthermore, upon water-deficit stress in *C. plantagineum*, PLD and diacylglycerol kinase (DAG kinase) activities can lead to the increased accumulation of PA and diacylglycerol pyrophosphate (DGPP), which itself can serve as a second messenger (Munnik et al. 2000). The recent identification of multiple, rehydration-reversible phosphorylation events of a functionally diverse set of *C. plantagineum* proteins triggered by dehydration reinforces the importance of protein kinase and protein phosphatase activity in metabolic regulation during the dehydration/rehydration cycle (Röhrig et al. 2008). Furthermore, the identification of EBP1, a key regulator of cell growth and differentiation in plants as a phosphoprotein (Horváth et al. 2006), or 14-3-3 GF14 omega protein, which undergoes phosphorylation only during rehydration and might be involved in cell cycle checkpoint regulation (Sorrell et al. 2003), implicates their involvement in critical signaling and regulatory events required for DT.

15.13 Developmental Pathways of Seeds and DT Vegetative Tissues

Given the similarities of DT in orthodox seeds and DT in the vegetative tissues of resurrection plants (Illing et al. 2005), one might postulate that DT is simply a reiteration of seed desiccation. The molecular networks that control seed maturation in *Arabidopsis* during ripening, dormancy, and germination have been well characterized (Holdsworth et al. 2007, 2008). Therefore, one might expect that a comparison of the molecular networks in both seed dormancy and DT in a phylogenetic context would reveal evolutionarily conserved control pathways/networks that are common to both networks. Ongoing integrated transcriptomic, proteomic, and metabolomic studies are expected to result in characterizing these molecular networks in *T. ruralis*, *S. lepidophylla*, *C. plantagineum*, and *S. stapfianus* in the near future (see also Chap. 16). However, given that very different developmental and tissue-specific controls likely exist between the two systems, one might also expect fundamental differences to exist. One striking example of these differences is illustrated by a recent study in which a seed-specific transcription factor from sunflower, HaHSFA9, was constitutively expressed in transgenic tobacco (Prieto-Dapena et al. 2008). Overexpression of this transcription factor conferred improved (~40%) survival to 3-week-old seedling to severe (−40 MPa) dehydration. The overexpression of HaHSFA9 was correlated with the ectopic expression of small, seed-specific heat shock proteins, but neither LEA protein expression, nor elevated levels of glucose, sucrose, trehalose, RFOs, or proline. The HSPs are thought to confer the observed severe dehydration tolerance by the expected chaperone functions of preventing stress-induced denaturation of protein structure or aggregation of proteins as well as protecting membrane integrity (Prieto-Dapena et al. 2008). While the levels of dehydration tolerance are not like those exhibited by resurrection plants (e.g., −50 to > −100 MPa), they are better than all previous reports of engineered dehydration tolerance in sensitive species in the literature. Therefore, additional regulators of genes encoding different desiccation protectants, such as LEA proteins, are likely to exist that further contribute to the severe DT of orthodox seeds and resurrection plants. An example of other possible components that contribute to DT in seeds, and possibly in resurrection plants, is group 1 LEA proteins. Loss of these proteins in an *Arabidopsis* T-DNA knockout line caused seeds to dry out more rapidly than wild-type seeds resulting in more rapid acquisition of DT (Manfre et al. 2008). In addition to transcriptional regulators, components of signaling pathways can also be expected to play critical roles in seed development as well as DT. For example, a triple T-DNA knockout of three SNF1 (sucrose nonfermenting 1)-related protein kinases (SnRKs), SnRK2.2, SnRK2.3, and SnRK2.6, resulted in severe defects during seed development and seed dormancy by disruption of ABA signaling pathways via control of gene expression programs through the phosphorylation of ABI5 and other transcription factors (Nakashima et al. 2009).

15.14 Conclusion

The last decade has resulted in a dramatic increase in the volume of gene discovery efforts and large-scale transcriptomic and proteomic studies performed in a variety of model resurrection species. To date, the vast majority of transcript data have been obtained using traditional Sanger sequencing of ESTs from only a very limited number of varieties and species. However, transcriptome analysis in resurrection species is rapidly entering a new phase of discovery that will embrace the use of a variety of next-generation sequencing platforms. In the near future, large amounts of gene sequence information and transcriptome-scale mRNA expression data will be available using one or more of these platforms, resulting in the rapid characterization of key signaling and regulatory components required for the control of the dehydration/rehydration cycle of DT. Integration of this information with other "omics" datasets including proteomics and metabolomics from closely related species that are desiccation-sensitive versus desiccation-tolerant will help to elucidate the discrete differences in gene content, signaling, and regulatory mechanisms that are responsible for the development of DT. Such information will also inform the development of novel strategies to engineer improvements in dehydration tolerance in crops.

Acknowledgments This work was supported by grants from the United States Department of Agriculture (USDA) National Research Initiative (NRI) (CREES-NRI-2007-02007) to MJ and JCC, and the University of Nevada Agricultural Experiment Station. Support for the Nevada Proteomics Center was made possible by NIH Grant Number P20 RR-016464 from the INBRE-BRIN Program of the National Center for Research Resources and the NIH IDeA Network of Biomedical Research Excellence (INBRE, RR-03-008).

References

Abdalla K, Thomson J, Rafudeen M (2009) Protocols for nuclei isolation and nuclear protein extraction from the resurrection plant *Xerophyta viscosa* for proteomic studies. Anal Biochem 384:365–367

Adams R, Kendall E, Kartha K (1990) Comparison of free sugars in growing and desiccated plants of *Selaginella lepidophylla*. Biochem Syst Ecol 18:107–110

Alamillo J, Bartels D (2001) Effects of desiccation on photosynthesis pigments and the ELIP-like dsp 22 protein complexes in the resurrection plant *Craterostigma plantagineum*. Plant Sci 160:1161–1170

Alamillo J, Almoguera C, Bartels D, Jordano J (1995) Constitutive expression of small heat shock proteins in vegetative tissues of the resurrection plant *Craterostigma plantagineum*. Plant Mol Biol 29:1093–1099

Alba R, Fei Z, Payton P, Liu Y, Moore S, Debbie P, Cohn J, D'Ascenzo M, Gordon J, Rose J, Martin G, Tanksley S, Bouzayen M, Jahn M, Giovannoni J (2004) ESTs, cDNA microarrays, and gene expression profiling: tools for dissecting plant physiology and development. Plant J 39:697–714

Alpert P (2006) Constraints of tolerance: why are desiccation-tolerant organisms so small or rare? J Exp Biol 209:1575–1584

Amuti K, Pollard C (1977) Soluble carbohydrates of dry and developing seeds. Phytochemistry 156:529–532

Anisimov S (2008) Serial Analysis of Gene Expression (SAGE): 13 years of application in research. Curr Pharm Biotechnol 9:338–350

Anke J, Petereit F, Engelhardt C, Hensel A (2008) Procyanidins from *Myrothamnus flabellifolia*. Nat Prod Res 22:1243–1254

Arezi B, Hogrefe H (2007) Escherichia coli DNA polymerase III epsilon subunit increases Moloney murine leukemia virus reverse transcriptase fidelity and accuracy of RT-PCR procedures. Anal Biochem 360:84–91

Audic S, Claverie J (1997) The significance of digital gene expression profiles. Genome Res 7:986–995

Augusti A, Scartazza A, Navari-Izzo F, Sgherri C, Stevenovic B, Brugnoli E (2001) Photosystem II photochemical efficiency, zeaxanthin, and antioxidant contents in the poikilohydric *Ramonda serbica* during dehydration and rehydration. Photosyn Res 67:79–88

Bachem C, van der Hoeven R, de Bruijn S, Vreugdenhil D, Zabeau M, Visser R (1996) Visualization of differential gene expression using a novel method of RNA fingerprinting based on AFLP: analysis of gene expression during potato tuber development. Plant J 9:745–753

Bartels D, Schneider K, Terstappen G, Piatkowski D, Salamini F (1990) Molecular cloning of abscisic acid-modulated genes which are induced during desiccation of the resurrection plant *Craterostigma plantagineum*. Planta 181:27–34

Bartels D, Hanke C, Schneider K, Michel D, Salamini F (1992) A desiccation-related Elip-like gene from the resurrection plant *Craterostigma plantagineum* is regulated by light and ABA. EMBO J 11:2771–2778

Bentley D (2006) Whole-genome re-sequencing. Curr Opin Genet Dev 16:545–552

Berjak P, Farrant J, Pammenter N (2007) Plant desiccation tolerance. In: Jenks M, Woods A (eds) Seed desiccation-tolerance mechanisms. Blackwell, Oxford, UK, pp 151–192

Bernacchia G, Schwall G, Lottspeich F, Salamini F, Bartels D (1995) The transketolase gene family of the resurrection plant *Craterostigma plantagineum*: differential expression during the rehydration phase. EMBO J 14:610–618

Bernacchia G, Salamini F, Bartels D (1996) Molecular characterization of the rehydration process in the resurrection plant *Craterostigma plantagineum*. Plant Physiol 111:1043–1050

Bewley J, Reynolds T, Oliver M (1993) Evolving strategies in the adaptation to desiccation. In: Close T, Bray E (eds) Plant responses to cellular dehydration during environmental stress,

current topics in plant physiology, vol 10, Am Soc Plant Physiol Series. American Society of Plant Physiologists, Rockville, MD, pp 193–201
Bianchi G, Gamba A, Murelli C, Salamini F, Bartels D (1991) Novel carbohydrate metabolism in the resurrection plant *Craterostigma plantagineum*. Plant J 1:355–359
Blomstedt C, Gianello R, Gaff D, Hamill J, Neale A (1998a) Differential gene expression in desiccation-tolerant and desiccation-sensitive tissue of the resurrection grass, *Sporobolus stapfianus*. Aust J Plant Physiol 25:937–946
Blomstedt C, Gianello R, Hamill J, Neale A, Gaff D (1998b) Drought-stimulated genes correlated with desiccation tolerance of the resurrection grass *Sporobolus stapfianus*. Plant Growth Regul 24:153–161
Bochicchio A, Vazzana C, Puliga S, Alberti A, Cinganelli S, Vernieri P (1998) Moisture content of the dried leaf is critical to desiccation tolerance in detached leaves of the resurrection plant *Boea hygroscopica*. Plant Growth Regul 24:163–170
Bockel C, Salamini F, Bartels D (1998) Isolation and characterization of genes expressed during early events of the dehydration process in the resurrection plant *Craterostigma plantagineum*. J Plant Physiol 152:158–166
Bonaldo M, Lennon G, Soares M (1996) Normalization and subtraction: two approaches to facilitate gene discovery. Genome Res 6:791–806
Breyne P, Dreesen R, Cannoot B, Rombaut D, Vandepoele K, Rombauts S, Vanderhaeghen R, Inzé D, Zabeau M (2003) Quantitative cDNA-AFLP analysis for genome-wide expression studies. Mol Genet Genomics 269:173–179
Buitink J, Mireilleet M, Claessens M, Marcus A, Hemminga A, Hoekstra F (1998) Influence of water content and temperature on molecular mobility and intracellular glasses in seeds and pollen. Plant Physiol 118:531–541
Cascante M, Marin S (2008) Metabolomics and fluxomics approaches. Essays Biochem 45:67–81
Chandler J, Bartels D (1997) Structure and function of the vp1 gene homologue from the resurrection plant *Craterostigma plantagineum* Hochst. Mol Gen Genet 256:539–546
Chen X, Kanokporn T, Zeng Q, Wilkins T, Wood A (2002) Characterization of the V-type H((+))−ATPase in the resurrection plant Tortula ruralis: accumulation and polysomal recruitment of the proteolipid c subunit in response to salt-stress. J Exp Bot 53:225–232
Clugston S, Daub E, Honek J (1998) Identification of glyoxalase I sequences in Brassica oleracea and *Sporobolus stapfianus*: evidence for gene duplication events. J Mol Evol 47:230–234
Collett H, Butowt R, Smith J, Farrant J, Illing N (2003) Photosynthetic genes are differentially transcribed during the dehydration-rehydration cycle in the resurrection plant, *Xerophyta humilis*. J Exp Bot 54:2593–2595
Collett H, Shen A, Gardner M, Farrant J, Denby K, Illing N (2004) Towards transcript profiling of desiccation tolerance in *Xerophyta humilis*: construction of a normalized 11 k *X. humilis* cDNA set and microarray expression analysis of 424 cDNAs in response to dehydration. Physiol Plant 122:39–53
Cooper K, Farrant J (2002) Recovery of the resurrection plant *Craterostigma wilmsii* from desiccation: protection versus repair. J Exp Bot 53:1805–1813
Deeba F, Pandey V, Pathre U, Kanojiya S (2009) Proteome analysis of detached fronds from a resurrection plant *Selaginella bryopteris* – response to dehydration and rehydration. J Proteomics Bioinform 2:108–116
Degl'Innocenti E, Guidi L, Stevanovic B, Navari F (2008) CO2 fixation and chlorophyll a fluorescence in leaves of *Ramonda serbica* during a dehydration-rehydration cycle. J Plant Physiol 165:723–733
Deng X, Hu Z, Wang H (1999) mRNA differential display visualized by silver staining tested on gene expression in resurrection plant *Boea hygrometrica*. Plant Mol Biol Rep 17:279
Deng X, Phillips J, Meijer A, Salamini F, Bartels D (2002) Characterization of five novel dehydration-responsive homeodomain leucine zipper genes from the resurrection plant *Craterostigma plantagineum*. Plant Mol Biol 49:601–610

Deng X, Phillips J, Bräutigam A, Engström P, Johannesson H, Ouwerkerk P, Ruberti I, Salinas J, Vera P, Iannacone R, Meijer A, Bartels D (2006) A homeodomain leucine zipper gene from *Craterostigma plantagineum* regulates abscisic acid responsive gene expression and physiological responses. Plant Mol Biol 61:469–489

Diatchenko L, Lau Y, Campbell A, Chenchik A, Moqadam F, Huang B, Lukyanov S, Lukyanov K, Gurskaya N, Sverdlov E, Siebert P (1996) Suppression subtractive hybridization: a method for generating differentially regulated or tissue-specific cDNA probes and libraries. Proc Natl Acad Sci USA 93:6025–6030

Ditzer A, Bartels D (2006) Identification of a dehydration and ABA-responsive promoter regulon and isolation of corresponding DNA binding proteins for the group 4 LEA gene CpC2 from *C. plantagineum*. Plant Mol Biol 61:643–663

Droege M, Hill B (2008) The Genome Sequencer FLX System–longer reads, more applications, straight forward bioinformatics and more complete data sets. J Biotechnol 136:3–10

Eickmeier W (1982) Protein synthesis and photosynthetic recovery in the resurrection plant, *Selaginella lepidophylla*. Plant Physiol 69:135–138

Farrant J (2000) A comparison of mechanisms of desiccation tolerance among three angiosperm resurrection plant species. Plant Ecol 151:29–39

Frank W, Phillips J, Salamini F, Bartels D (1998) Two dehydration-inducible transcripts from the resurrection plant *Craterostigma plantagineum* encode interacting homeodomain-leucine zipper proteins. Plant J 15:413–421

Frank W, Munnik T, Kerkmann K, Salamini F, Bartels D (2000) Water deficit triggers phospholipase D activity in the resurrection plant *Craterostigma plantagineum*. Plant Cell 12:111–124

Furini A, Koncz C, Salamini F, Bartels D (1997) High level transcription of a member of a repeated gene family confers dehydration tolerance to callus tissue of *Craterostigma plantagineum*. EMBO J 16:3599–3608

Gaff D, Loveys B (1992) Abscisic acid levels in drying plants of a resurrection grass. Trans Malays Soc Plant Physiol 3:286–287

Gaff D, Bartels D, Gaff J (1997) Changes in gene expression during drying in a desiccation-tolerant grass *Sporobolus stapfianus* and a desiccation-sensitive grass *Sporobolus pyramidalis*. Aust J Plant Physiol 24:617–622

Garwe D, Thomson J, Mundree S (2003) Molecular characterization of XVSAP1, a stress-responsive gene from the resurrection plant *Xerophyta viscosa* Baker. J Exp Bot 54:191–201

Garwe D, Thomson J, Mundree S (2006) XVSAP1 from *Xerophyta viscosa* improves osmotic-, salinity- and high-temperature-stress tolerance in Arabidopsis. Biotechnol J 1:1137–1146

Georgieva K, Szigeti Z, Sarvari E, Gaspar L, Maslenkova L, Peeva V, Peli E, Tuba Z (2007) Photosynthetic activity of homoiochlorophyllous desiccation tolerant plant *Haberlea rhodopensis* during dehydration and rehydration. Planta 225:955–964

Georgieva K, Röding A, Büchel C (2009) Changes in some thylakoid membrane proteins and pigments upon desiccation of the resurrection plant *Haberlea rhodopensis*. J Plant Physiol 166 (14):1520–1528

Ghasempour H, Kianian J (2007) The study of desiccation-tolerance in drying leaves of the desiccation-tolerant grass *Sporobolus elongatus* and the desiccation-sensitive grass *Sporobolus pyramidalis*. Pak J Biol Sci 10:797–801

Ghasempour H, Gaff D, Williams R, BGianellow R (1998) Contents of sugars in leaves of drying desiccation tolerant flowering plants, particularly grasses. Plant Growth Regul 24:185–191

Gygi S, Rochon Y, Fransz B, Aebersold R (1999) Correlation between protein and mRNA abundance in yeast. Mol Cell Biol 19:1720–1730

Harten J, Eickmeier W (1986) Enzyme dynamics of the resurrection plant *Selaginella lepidophylla* (Hook. & Grev.) spring during rehydration. Plant Physiol 82:61–64

Hilbricht T, Salamini F, Bartels D (2002) CpR18, a novel SAP-domain plant transcription factor, binds to a promoter region necessary for ABA mediated expression of the CDeT27-45 gene from the resurrection plant *Craterostigma plantagineum* Hochst. Plant J 31:293–303

Hilbricht T, Varotto S, Sgaramella V, Bartels D, Salamini F, Furini A (2008) Retrotransposons and siRNA have a role in the evolution of desiccation tolerance leading to resurrection of the plant *Craterostigma plantagineum*. New Phytol 179:877–887

Hillier L, Lennon G, Becker M, Bonaldo M, Chiapelli B, Chissoe S, Dietrich N, DuBuque T, Favello A, Gish W, Hawkins M, Hultman M, Kucaba T, Lacy M, Le M, Le N, Mardis E, Moore B, Morris M, Parsons J, Prange C, Rifkin L, Rohlfing T, Schellenberg K, Bento Soares M, Tan F, Thierry-Meg J, Trevaskis E, Underwood K, Wohldman P, Waterston R, Wilson R, Marra M (1996) Generation and analysis of 280,000 human expressed sequence tags. Genome Res 6:807–828

Hoekstra F, Golovina E (1999) Membrane behavior during dehydration: implications for desiccation tolerance. Russ J Plant Physiol 46:295–306

Hoekstra F, Golvina E, Buitink J (2001) Mechanisms of plant desiccation tolerance. Trends Plant Sci 6:431–438

Holdsworth M, Rinch-Savage W, Grappin P, Job D (2007) Post-genomics dissection of seed dormancy and germination. Trends Plant Sci 13:7–13

Holdsworth M, Bentsink L, Soppe W (2008) Molecular networks regulating Arabidopsis seed maturation, after-ripening, dormancy and germination. New Phytol 179:33–54

Horváth B, Magyar Z, Zhang Y, Hamburger A, Bakó L, Visser R, Bachem C, Bögre L (2006) EBP1 regulates organ size through cell growth and proliferation in plants. EMBO J 25:4909–4920

Ichikawa T, Nakazawa M, Kawashima M, Iizumi H, Kuroda H, Kondou Y, Tsuhara Y, Suzuki K, Ishikawa A, Seki M, Fujita M, Motohashi R, Nagata N, Takagi T, Shinozaki K, Matsui M (2006) The FOX hunting system: an alternative gain-of-function gene hunting technique. Plant J 48:974–985

Ideker T, Thorsson V, Ranish J, Christmas R, Buhler J, Eng J, Bumgarner R, Goodlett D, Aebersold R, Hood L (2001) Integrated genomic and proteomic analyses of a systematically perturbed metabolic network. Science 292:929–934

Illing N, Denby K, Collett H, Shen A, Farrant J (2005) The signature of seeds in resurrection plants: a molecular and physiological comparison of desiccation tolerance in seeds and vegetative tissues. Integr Comp Biol 45:771–787

Ingle R, Schmidt U, Farrant J, Thomson J, Mundree S (2007) Proteomic analysis of leaf proteins during dehydration of the resurrection plant *Xerophyta viscosa*. Plant Cell Environ 30:435–446

Ingram J, Bartels D (1996) The molecular basis of dehydration tolerance in plants. Annu Rev Plant Physiol Plant Mol Biol 47:377–403

Ingram J, Chandler J, Gallagher L, Salamini F, Bartels D (1997) Analysis of cDNA clones encoding sucrose-phosphate synthase in relation to sugar interconversions associated with dehydration in the resurrection plant *Craterostigma plantagineum* Hochst. Plant Physiol 115:113–121

Iturriaga G, Schneider K, Salamini F, Bartels D (1992) Expression of desiccation-related proteins from the resurrection plant *Craterostigma plantagineum* in transgenic tobacco. Plant Mol Biol 20:555–558

Iturriaga G, Leyns L, Villegas A, Gharaibeh R, Salamini F, Bartels D (1996) A family of novel myb-related genes from the resurrection plant *Craterostigma plantagineum* are specifically expressed in callus and roots in response to ABA or desiccation. Plant Mol Biol 32:707–716

Iturriaga G, Gaff D, Zentella R (2000) New desiccation-tolerant plants, including a grass, in the central highlands of Mexico, accumulate trehalose. Aust J Bot 48:153–158

Iturriaga G, Cushman M, Cushman J (2006) An EST catalogue from the resurrection plant *Selaginella lepidophylla* reveals abiotic stress-adaptive genes. Plant Biol 170:1173–1184

Jiang G, Wang Z, Shang H, Yang W, Hu Z, Phillips J, Deng X (2007) Proteome analysis of leaves from the resurrection plant *Boea hygrometrica* in response to dehydration and rehydration. Planta 225:1405–1420

Kirch H, Nair A, Bartels D (2001) Novel ABA- and dehydration-inducible aldehyde dehydrogenase genes isolated from the resurrection plant *Craterostigma plantagineum* and *Arabidopsis thaliana*. Plant J 28:555–567

Kleines M, Elster R, Rodrigo M, Blervacq A, Salamini F, Bartels D (1999) Isolation and expression analysis of two stress-responsive sucrose-synthase genes from the resurrection plant *Craterostigma plantagineum* (Hochst.). Planta 209:13–24

Kondou Y, Higuchi M, Takahashi S, Sakurai T, Ichikawa T, Kuroda H, Yoshizumi T, Tsumoto Y, Horii Y, Kawashima M, Hasegawa Y, Kuriyama T, Matsui K, Kusano M, Albinsky D, Takahashi H, Nakamura Y, Suzuki M, Sakakibara H, Kojima M, Akiyama K, Kurotani A, Seki M, Fujita M, Enju A, Yokotani N, Saitou T, Ashidate K, Fujimoto N, Ishikawa Y, Mori Y, Nanba R, Takata K, Uno K, Sugano S, Natsuki J, Dubouzet J, Maeda S, Ohtake M, Mori M, Oda K, Takatsuji H, Hirochika H, Matsui M (2009) Systematic approaches to using the FOX hunting system to identify useful rice genes. Plant J 57(5):883–894

Kuang J, Gaff D, Gianello R, Blomstedt C, Neale A, Hamill J (1995) Desiccation-tolerant grass *Sporobolus stapfianus* and a desiccation-sensitive grass *Sporobolus pyramidalis*. Aust J Plant Physiol 22:1027–1034

Le T, Blomstedt C, Kuang J, Tenlen J, Gaff G, Hamill J, Neale A (2007) Desiccation-tolerance specific gene expression in leaf tissue of the resurrection plant *Sporobolus stapfianus*. Funct Plant Biol 34:589–600

Lee B, Shin G (2009) CleanEST: a database of cleansed EST libraries. Nucleic Acids Res 37: D686–D689

Lehner A, Chopera D, Peters S, Keller F, Mundree S, Thomson J, Farrant J (2008) Protection mechanisms in the resurrection plant *Xerophyta viscosa*: cloning, expression, characterisation and role of XvINO1, a gene coding for a myo-inositol 1-phosphate synthase. Funct Plant Biol 35:26–39

LePrince O, Buitink J (2007) The glassy state in dry seeds and pollen. In: Jenks M, Woods A (eds) Plant desiccation tolerance. Blackwell, Oxford, UK, pp 193–214

Li M, Nordborg M, Li L (2004) Adjust quality scores from alignment and improve sequencing accuracy. Nucleic Acids Res 32:5183–5191

Liu F, Jenssen T, Trimarchi J, Punzo C, Cepko C, Ohno-Machado L, Hovig E, Kuo W (2007) Comparison of hybridization-based and sequencing-based gene expression technologies on biological replicates. BMC Genomics 8:153

Liu M-S, Chien C-T, Lin T-P (2008) Constitutive components and induced gene expression are involved in the desiccation tolerance of *Selaginella tamariscina*. Plant Cell Physiol 49:653–663

Manfre A, LaHatte G, Climer C, Marcotte WJ (2008) Seed dehydration and the establishment of desiccation tolerance during seed maturation is altered in the *Arabidopsis thaliana* mutant atem6-1. Plant Cell Physiol 50:243–253

Mardis E (2008a) Next-generation DNA sequencing methods. Annu Rev Genomics Hum Genet 9:387–402

Mardis E (2008b) The impact of next-generation sequencing technology on genetics. Trends Genet 24:133–141

Margulies M, Egholm M, Altman W, Attiya S, Bader J, Bemben L, Berka J, Braverman M, Chen Y, Chen Z, Dewell S, Du L, Fierro J, Gomes X, Godwin B, He W, Helgesen S, Ho C, Irzyk G, Jando S, Alenquer M, Jarvie T, Jirage K, Kim J, Knight J, Lanza J, Leamon J, Lefkowitz S, Lei M, Li J, Lohman K, Lu H, Makhijani V, McDade K, McKenna M, Myers E, Nickerson E, Nobile J, Plant R, Puc B, Ronan M, Roth G, Sarkis G, Simons J, Simpson J, Srinivasan M, Tartaro K, Tomasz A, Vogt K, Volkmer G, Wang S, Wang Y, Weiner M, Yu P, Begley R, Rothberg J (2005) Genome sequencing in microfabricated high-density picolitre reactors. Nature 437:376–380

Mariaux J, Bockel C, Salamini F, Bartels D (1998) Desiccation- and abscisic acid-responsive genes encoding major intrinsic proteins (MIPs) from the resurrection plant *Craterostigma plantagineum*. Plant Mol Biol 38:1089–1099

Martinelli T (2008) In situ localization of glucose and sucrose in dehydrating leaves of *Sporobolus stapfianus*. J Plant Physiol 165:580–587

Martinelli T, Whittaker A, Bochicchio A, Vazzana C, Suzuki A, Masclaux-Daubresse C (2007) Amino acid pattern and glutamate metabolism during dehydration stress in the 'resurrection' plant *Sporobolus stapfianus*: a comparison between desiccation-sensitive and desiccation-tolerant leaves. J Exp Bot 58:3037–3046

Matsumura H, Reich S, Ito A, Saitoh H, Kamoun S, Winter P, Kahl G, Reuter M, Kruger D, Terauchi R (2003) Gene expression analysis of plant host-pathogen interactions by SuperSAGE. Proc Natl Acad Sci USA 100:15718–15723

Matsumura H, Krüger D, Kahl G, Terauchi R (2008) SuperSAGE: a modern platform for genome-wide quantitative transcript profiling. Curr Pharm Biotechnol 9:368–374

Michel D, Furini A, Salamini F, Bartels D (1994) Structure and regulation of an ABA- and desiccation-responsive gene from the resurrection plant *Craterostigma plantagineum*. Plant Mol Biol 24:549–560

Mittler R (2002) Oxidative stress, antioxidants and stress tolerance. Trends Plant Sci 7:405–410

Moore J, Westall K, Ravenscroft N, Farrant J, Lindsey G, Brandt W (2005) The predominant polyphenol in the leaves of the resurrection plant Myrothamnus flabellifolius, 3, 4, 5 tri-O-galloylquinic acid, protects membranes against desiccation and free radical-induced oxidation. Biochem J 385:301–308

Moore J, Nguema-Ona E, Chevalier L, Lindsey G, Brandt W, Lerouge P, Farrant J, Driouich A (2006) Response of the leaf cell wall to desiccation in the resurrection plant Myrothamnus flabellifolius. Plant Physiol 141:651–662

Moore J, Lindsey G, Farrant J, Brandt W (2007) An overview of the biology of the desiccation-tolerant resurrection plant *Myrothamnus flabellifolia*. Ann Bot Lond 99:211–217

Moore J, Le N, Brandt W, Driouich A, Farrant J (2009) Towards a systems-based understanding of plant desiccation tolerance. Trends Plant Sci 14:110–117

Mowla S, Thomson J, Farrant J, Mundree S (2002) A novel stress-inducible antioxidant enzyme identified from the resurrection plant *Xerophyta viscosa* Baker. Planta 215:716–726

Mulako I, Farrant J, Collett H, Illing N (2008) Expression of Xhdsi-1VOC, a novel member of the vicinal oxygen chelate (VOC) metalloenzyme superfamily, is up-regulated in leaves and roots during desiccation in the resurrection plant *Xerophyta humilis* (Bak) Dur and Schinz. J Exp Bot 59:3885–3901

Mundree S, Whittaker A, Thomson J, Farrant J (2000) An aldose reductase homolog from the resurrection plant *Xerophyta viscosa* Baker. Planta 211:693–700

Munnik T, Meijer H, Ter Riet B, Hirt H, Frank W, Bartels D, Musgrave A (2000) Hyperosmotic stress stimulates phospholipase D activity and elevates the levels of phosphatidic acid and diacylglycerol pyrophosphate. Plant J 22:147–154

Muslin E, Homann P (1992) Light as a hazard for the desiccation-resistant "resurrection" fern *Polypodium polypodioides* L. Plant Cell Enivon 15:81–89

Nakashima K, Fujita Y, Kanamori N, Katagiri T, Umezawa T, Kidokoro S, Maruyama K, Yoshida T, Ishiyama K, Kobayashi M, Shinozaki K, Yamaguchi-Shinozaki K (2009) Three Arabidopsis SnRK2 protein kinases, SRK2D/SnRK2.2, SRK2E/SnRK2.6/OST1 and SRK2I/SnRK2.3, involved in ABA signaling are essential for the control of seed development and dormancy. Plant Cell Physiol 50:1345–1363

Navari-Izzo F, Ricci F, Vazzana C, Quartacci M (1995) Unusual composition of thylakoid membranes of the resurrection plant *Boea hygroscopica*: changes in lipids upon dehydration and rehydration. Physiol Plant 94:135–142

Navari-Izzo F, Meneguzzo S, Loggini B, Vazzana C, Sgherri C (1997) The role of the glutathione system during dehyration of *Boea hygroscopica*. Physiol Plant 99:23–30

Navari-Izzo F, Quartacci M, Pinzino C, Rascio N, Vazzana C, Sgherri C (2000) Protein dynamics in thylakoids of the desiccation-tolerant plant *Boea hygroscopica* during dehydration and rehydration. Plant Physiol 124:1427–1436

Neale A, Blomstedt C, Bronson P, Le T-N, Guthridge K, Evans J, Gaff D, Hamill J (2000) The isolation of genes from the resurrection grass *Sporobolus stapfianus* which are induced during severe drought stress. Plant Cell Enivon 23:265–277

Norwood M, Truesdale M, Richter A, Scott P (2000) Photosynthetic carbohydrate metabolism in the resurrection plant *Craterostigma plantagineum*. J Exp Biol 51:159–165

Nygaard V, Liu F, Holden M, Kuo W, Trimarchi J, Ohno-Machado L, CL C, Frigessi A, Glad I, Wiel M, Hovig E, Lyng H (2008) Validation of oligoarrays for quantitative exploration of the transcriptome. BMC Genomics 9:258

O'Mahony P, Oliver M (1999a) The involvement of ubiquitin in vegetative desiccation tolerance. Plant Mol Biol 41:657–667

O'Mahony P, Oliver M (1999b) Characterization of a desiccation-responsive small GTP-binding protein (Rab2) from the desiccation-tolerant grass *Sporobolus stapfianus*. Plant Mol Biol 39: 809–821

Oliver M (1991) Influence of protoplasmic water loss on the control of protein synthesis in the desiccation-tolerant moss *Tortula ruralis*: ramifications for a repair-based mechanism of desiccation tolerance. Plant Physiol 97:1501–1511

Oliver M (1996) Desiccation tolerance in vegetative plant cells. Physiol Plant 97:779–787

Oliver M, Tuba Z, Mishler B (2000) The evolution of vegetative desiccation tolerance in land plants. Plant Ecol 151:85–100

Oliver M, Dowd S, Zaragoza J, Mauget S, Payton P (2004) The rehydration transcriptome of the desiccation-tolerant bryophyte *Tortula ruralis*: transcript classification and analysis. BMC Genomics 5:89

Oliver M, Velten J, Mishler B (2005) Desiccation tolerance in bryophytes: a reflection of the primitive strategy for plant survival in dehydrating habitats? Integr Comp Biol 45:788–799

Oliver M, Hudgeons J, Dowd S, Payton P (2009) A combined subtractive suppression hybridization and expression profiling strategy to identify novel desiccation response transcripts from *Tortula ruralis* gametophytes. Physiol Plant 136:437–460

Peters S, Mundree S, Thomson J, Farrant J, Keller F (2007) Protection mechanisms in the resurrection plant *Xerophyta viscosa* (Baker): both sucrose and raffinose family oligosaccharides (RFOs) accumulate in leaves in response to water deficit. J Exp Bot 58:1947–1956

Phillips J, Hilbricht T, Salamini F, Bartels D (2002a) A novel abscisic acid- and dehydration-responsive gene family from the resurrection plant *Craterostigma plantagineum* encodes a plastid-targeted protein with DNA-binding activity. Planta 215:258–266

Phillips J, Oliver M, Bartels D (2002b) Molecular genetics of desiccation and tolerant systems. In: Black HWPM (ed) Desiccation and plant survival. CABI, Wallingford, UK

Phillips J, Dalmay T, Bartels D (2007) The role of small RNAs in abiotic stress. FEBS Lett 581: 3592–3597

Phillips J, Fischer E, Baron M, van den Dries N, Facchinelli F, Kutzer M, Rahmanzadeh R, Remus D, Bartels D (2008) *Lindernia brevidens*: a novel desiccation-tolerant vascular plant, endemic to ancient tropical rainforests. Plant J 54:938–948

Phizicky E, Bastiaens P, Zhu H, Snyder M, Fields S (2003) Protein analysis on a proteomic scale. Nature 422:208–215

Porembski S, Barthlott W (2000) Granitic and gneissic outcrops (inselbergs) as center of diversity for desiccation-tolerant vascular plants. Plant Ecol 151:19–28

Prieto-Dapena P, Castaño R, Almoguera C, Jordano J (2008) The ectopic overexpression of a seed-specific transcription factor, HaHSFA9, confers tolerance to severe dehydration in vegetative organs. Plant J 54:1004–1014

Proctor M, Pence V (2002) Vegetative tissues: bryophytes, vascular resurrection plants and vegetative propagules. In: Black M, Pritchard H (eds) Desiccation and survival in plants: drying without dying. CABI, Oxford

Proctor M, Ligrone R, Duckett J (2007a) Desiccation tolerance in the moss *Polytrichum formosum*: physiological and fine-structural changes during desiccation and recovery. Ann Bot Lond 99:75–93

Proctor M, Oliver M, Wood A, Alpert P, Stark L, Cleavitt N, Mishler B (2007b) Desiccation-tolerance in bryophytes: a review. Bryologist 110:595–621

Prosdocimi F, Lopes D, Peixoto F, Mourão M, Pacífico L, Ribeiro R, Ortega J (2007) Effects of sample re-sequencing and trimming on the quality and size of assembled consensus sequences. Genet Mol Res 6:756–765

Quartacci M, Forli M, Rascio N, Dalla Vecchia F, Bochicchio A, Navari-Izzo F (1997) Desiccation-tolerant *Sporobolus stapfianus*: lipid composition and cellular ultrastructure during dehydration and rehydration. J Exp Bot 48:1269–1279

Quartacci M, Glisić O, Stevanović B, Navari-Izzo F (2002) Plasma membrane lipids in the resurrection plant *Ramonda serbica* following dehydration and rehydration. J Exp Bot 53:2159–2166

Reynolds T, Bewley J (1993) Characterization of protein synthetic changes in a desiccation-tolerant fern, *Polypodium virginianum*. Comparison of the effects of drying, rehydration and abscisic acid. J Exp Bot 44:921–928

Rice-Evans C, Miller N, Paganga G (1997) Antioxidant properties of phenolic compounds. Trends Plant Sci 2:152–159

Rodrigo M, Bockel C, Blervacq A, Bartels D (2004) The novel gene CpEdi-9 from the resurrection plant *C. plantagineum* encodes a hydrophilic protein and is expressed in mature seeds as well as in response to dehydration in leaf phloem tissues. Planta 219:579–589

Rodriguez M, Edsgard D, Hussain SS, Alquezar A, Rasmussen M, Gilbert T, Nielsen B, Bartels D, Mundy J (2010) Transcriptomes of the desiccation tolerant resurrection plant *Craterostigma plantagineum*. Plant J 63:212–228

Röhrig H, Schmidt J, Colby T, Bräutigam A, Hufnagel P, Bartels D (2006) Desiccation of the resurrection plant *Craterostigma plantagineum* induces dynamic changes in protein phosphorylation. Plant Cell Environ 29:1606–1617

Röhrig H, Colby T, Schmidt J, Harzen A, Facchinelli F, Bartels D (2008) Analysis of desiccation-induced candidate phosphoproteins from *Craterostigma plantagineum* isolated with a modified metal oxide affinity chromatography procedure. Proteomics 8:3548–3560

Ronaghi M, Uhlén M, Nyrén P (1998) A sequencing method based on real-time pyrophosphate. Science 281:363–365

Rudd S (2003) Expressed sequence tags: alternative or complement to whole genome sequences? Trends Plant Sci 8:321–329

Saha S, Sparks A, Rago C, Akmaev V, Wang C, Vogelstein B, Kinzler K, Velculescu V (2002) Using the transcriptome to annotate the genome. Nat Biotechnol 20:508–512

Scott H, Oliver M (1994) Accumulation and polysomal recruitment of transcripts in response to desiccation and rehydration of the moss *Tortula ruralis*. J Exp Bot 45:577–583

Seel W, Hendry G, Lee J (1992) Effects of desiccation on some activated oxygen processing enzymes and antioxidants in mosses. J Exp Bot 43:1031–1037

Seki M, Narusaka M, Kamiya A, Ishida J, Satou M, Sakurai T, Nakajima M, Enju A, Akiyama K, Oono Y, Muramatsu M, Hayashizaki Y, Kawai J, Carninci P, Itoh M, Ishii Y, Arakawa T, Shibata K, Shinagawa A, Shinozaki K (2002) Functional annotation of a full-length Arabidopsis cDNA collection. Science 296:141–145

Seki M, Satou M, Sakurai T, Akiyama K, Iida K, Ishida J, Nakajima M, Enju A, Narusaka M, Fujita M, Oono Y, Kamei A, Yamaguchi-Shinozaki K, Shinozaki K (2004) RIKEN Arabidopsis full-length (RAFL) cDNA and its applications for expression profiling under abiotic stress conditions. J Exp Bot 55:213–223

Sen Gupta A (1977) Non-autotrophic CO2 fixation by mosses. University of Calgary, Calgary

Sgherri C, Loggini B, Bochicchio A, Navari-Izzo F (1994a) Antioxidant system in *Boea hygroscopica*: changes in response to desiccation and rehydration. Phytochemistry 35:377–381

Sgherri C, Loggini B, Puliga S, Navari-Izzo F (1994b) Antioxidant system in *Sporobolus stapfianus*: changes in response to desiccation and rehydration. Phytochemistry 35:561–565

Sherwin H, Farrant J (1998) Protection mechanisms against excess light in the resurrection plants *Craterostigma wilmsii* and *Xerophyta viscosa*. Plant Growth Regul 24:203–210

Simon S, Zhai J, Nandety R, McCormick K, Zeng J, Mejia D, Meyers B (2009) Short-read sequencing technologies for transcriptional analyses. Annu Rev Plant Biol 60:305–333

Smirnoff N (1992) The carbohydrates of bryophytes in relation to desiccation-tolerance. J Bryol 17:185–191

Smith A, Xuan Z, Zhang M (2008) Using quality scores and longer reads improves accuracy of Solexa read mapping. BMC Bioinform 9:128

Smith-Espinoza C, Phillips J, Salamini F, Bartels D (2005) Identification of further *Craterostigma plantagineum* cdt mutants affected in abscisic acid mediated desiccation tolerance. Mol Genet Genomics 274:364–372

Smith-Espinoza C, Bartels D, Phillips J (2007) Analysis of a LEA gene promoter via Agrobacterium-mediated transformation of the desiccation tolerant plant *Lindernia brevidens*. Plant Cell Rep 26:1681–1688

Soltis P, Soltis D, Savolainen V, Crane P, Barraclough T (2002) Rate heterogeneity among lineages of tracheophytes: integration of molecular and fossil data and evidence for molecular living fossils. Proc Natl Acad Sci USA 99:4430–4435

Sorrell D, Marchbank A, Chrimes D, Dickinson J, Rogers H, Francis D, Grierson C, Halford N (2003) The Arabidopsis 14-3-3 protein, GF14omega, binds to the Schizosaccharomyces pombe Cdc25 phosphatase and rescues checkpoint defects in the rad24- mutant. Planta 218:50–57

Stewart G, Lee J (1972) Desiccation injury in mosses II: the effects of moisture stress on enzyme levels. New Phytol 71:461–466

Taji T, Sakurai T, Mochida K, Ishiwata A, Kurotani A, Totoki Y, Toyoda A, Sakaki Y, Seki M, Ono H, Sakata Y, Tanaka S, Shinozaki K (2008) Large-scale collection and annotation of full-length enriched cDNAs from a model halophyte, *Thellungiella halophila*. BMC Plant Biol 8:115

Tuba Z, Proctor M, Csintalan Z (1998) Ecological responses of homoiochlorophyllous and poikilochlorophyllous desiccation tolerance plants: a comparison and an ecological perspective. Plant Growth Regul 24:211–217

Umezawa T, Sakurai T, Totoki Y, Toyoda A, Seki M, Ishiwata A, Akiyama K, Kurotani A, Yoshida T, Mochida K, Kasuga M, Todaka D, Maruyama K, Nakashima K, Enju A, Mizukado S, Ahmed S, Yoshiwara K, Harada K, Tsubokura Y, Hayashi M, Sato S, Ana T, Ishimoto M, Funatsuki H, Teraishi M, Osaki M, Shinano T, Akashi R, Sakaki Y, Yamaguch-Shinozaki K, Shinozaki K (2008) Sequencing and analysis of approximately 40,000 soybean cDNA clones from a full-length-enriched cDNA library. DNA Res 15:333–346

Urano K, Maruyama K, Ogata Y, Morishita Y, Takeda M, Sakurai N, Suzuki H, Saito K, Shibata D, Kobayashi M, Yamaguchi-Shinozaki K, Shinozaki K (2009) Characterization of the ABA-regulated global responses to dehydration in Arabidopsis by metabolomics. Plant J 57:1065–1078

Vander Willigen C, Pammenter N, Jaffer M, Mundree S, Farrant J (2003) An ultrastructural study using anhydrous fixation of *Eragrostis nindensis*, a resurrection grass with both desiccation-tolerant and -sensitive tissues. Funct Plant Biol 30:281–290

Vega-Sanchez M, Gowda M, Wang G (2007) Tag-based approaches for deep transcriptome analysis in plants. Plant Sci 173:371–380

Velasco R, Salamini F, Bartels D (1994) Dehydration and ABA increase mRNA levels and enzyme activity of cytosolic GAPDH in the resurrection plant *Craterostigma plantagineum*. Plant Mol Biol 26:541–546

Velasco R, Salamini F, Bartels D (1998) Gene structure and expression analysis of the drought- and abscisic acid-responsive CDeT11-24 gene family from the resurrection plant *Craterostigma plantagineum* Hochst. Planta 204:459–471

Velculescu V, Zhang L, Vogelstein B, Kinzler K (1995) Serial analysis of gene expression. Science 270:484–487

Veljovic-Jovanovic S, Kukavica B, Navari-Izzo F (2008) Characterization of polyphenol oxidase changes induced by desiccation of *Ramonda serbica* leaves. Physiol Plant 132:407–416

Vertucci C, Farrant J (1995) Acquisition and loss of desiccation tolerance. In: Kigel J, Galili G (eds) Seed development and germination. Marcel Dekker, New York, pp 237–271

Vicré M, Lerouxel O, Farrant J, Lerouge P, Driouich A (2004) Composition and desiccation-induced alterations of the cell wall in the resurrection plant *Craterostigma wilmsii*. Physiol Plant 120:229–239

Vuylsteke M, Peleman J, van Eijk M (2007) AFLP-based transcript profiling (cDNA-AFLP) for genome-wide expression analysis. Nat Protoc 2:1399–1413
Wasinger V, Humphery-Smith I (1998) Small genes/gene-products in Escherichia coli K-12. FEMS Microbiol Lett 169:375–382
Weng J, Tanurdzic M, Chapple C (2005) Functional analysis and comparative genomics of expressed sequence tags from the lycophyte *Selaginella moellendorffii*. BMC Genomics 6:85–99
White E, Towers G (1967) Comparative biochemisty of the lycopods. Phytochemistry 6:663–667
Whittaker A, Bochicchio A, Vazzana C, Lindsey G, Farrant J (2001) Changes in leaf hexokinase activity and metabolite levels in response to drying in the desiccation-tolerant species *Sporobolus stapfianus* and *Xerophyta viscosa*. J Exp Bot 52:961–969
Whittaker A, Martinelli T, Farrant J, Bochicchio A, Vazzana C (2007) Sucrose phosphate synthase activity and the co-ordination of carbon partitioning during sucrose and amino acid accumulation in desiccation-tolerant leaf material of the C4 resurrection plant *Sporobolus stapfianus* during dehydration. J Exp Bot 58:3775–3787
Wiechert W, Schweissgut O, Takanaga H, Frommer W (2007) Fluxomics: mass spectrometry versus quantitative imaging. Curr Opin Plant Biol 10:323–330
Wood A (2007) New Frontiers in Bryology and Lichenology. The Nature and Distribution of Vegetative Desiccation-tolerance in Hornworts, Liverworts and Mosses. Bryologist 110:163–177
Wood A, Duff R, Oliver M (1999) Expressed sequence Tags (ESTs) from desiccated *Tortula ruralis* identify a large number of novel plant genes. Plant Cell Physiol 40:361–368
Wood A, Joel Duff R, Oliver M (2000) The translational apparatus of *Tortula ruralis*: polysomal retention of transcripts encoding the ribosomal proteins RPS14, RPS16 and RPL23 in desiccated and rehydrated gametophytes. J Exp Bot 51:1655–1662
Zeng Q, Wood A (2000) A cDNA encoding ribosomal protein RPL15 from the desiccation-tolerant bryophyte *Tortula ruralis*: mRNA transcripts are stably maintained in desiccated and rehydrated gametophytes. Biosci Biotechnol Biochem 64:2221–2224
Zeng Q, Chen X, Wood A (2002) Two early light-inducible protein (ELIP) cDNAs from the resurrection plant *Tortula ruralis* are differentially expressed in response to desiccation, rehydration, salinity, and high light. J Exp Bot 53:1197–1205
Zentella R, Mascorro-Gallardo J, Van Dijck P, Folch-Mallol J, Bonini B, Van Vaeck C, Gaxiola R, Covarrubias A, Nieto-Sotelo J, Thevelein J, Iturriaga G (1999) A *Selaginella lepidophylla* trehalose-6-phosphate synthase complements growth and stress-tolerance defects in a yeast tps1 mutant. Plant Physiol 119:1473–1482
Živković T, Quartacci M, Stevanović B, Marinone F, Navari-Izzo F (2005) Low-molecular weight substances in the poikilohydric plant *Ramonda serbica* during dehydration and rehydration. Plant Sci 168:105–111

Chapter 16
Resurrection Plants: Physiology and Molecular Biology

Dorothea Bartels and Syed Sarfraz Hussain

Abbreviations

ABA Abscisic acid
LEA Late embryogenesis abundant
ROS Reactive oxygen species

16.1 Evolution and Geographic Distribution of Desiccation-Tolerant Plants

16.1.1 A Window into Past Research of Desiccation Tolerance

Three centuries ago science of desiccation tolerance began with a lengthy period of discovery and doubt. In 1743, Henry Baker announced to the Royal Society in London that some animals could tolerate desiccation, which means that they could dry to equilibrium with air and resume normal functions upon rehydration (Keilin 1959). Henry Baker's example of desiccation tolerance was the larva of the nematode *Anguillulina tritici*. The discovery that a nematode could lose virtually all its free internal water without dying was remarkable, because most animals and plants die instantly, if their cells equilibrate with even moderately dry air, because water maintains the structure of intracellular macromolecules and membranes. The next step was to identify more organisms that tolerate desiccation. Subsequently, it was also observed in rotifers by van Leeuwenhoek in 1702 (Keilin 1959). Eventually desiccation tolerance was reported in four other phyla of animals, in some algae, fungi and bacteria, in ferns, in most bryophytes, lichens, seeds of flowering plants, and in about 300 unique angiosperm species termed resurrection plants (Alpert 2006). The vegetative tissues of resurrection plants are able to survive protoplastic dehydration of less than 2% relative water content. The majority of flowering plants and also gymnosperms have desiccation-tolerant seeds or pollen. This trait is strictly developmentally regulated, it is acquired during embryogenesis (Bartels et al. 1988) or pollen development, but lost during germination.

Among plants, desiccation tolerance is common in primitive plants such as lichens and bryophytes. This form of anhydrobiosis seems to comprise constitutively

U. Lüttge et al. (eds.), *Plant Desiccation Tolerance*, Ecological Studies 215,
DOI 10.1007/978-3-642-19106-0_16,

expressed cell protection mechanisms associated with inducible repair systems that are activated after rehydration (Oliver and Bewley 1997 and covered in detail in Chap. 4). Also among algae, it is possible to find species able to tolerate desiccation, either in terrestrial or in marine intertidal algae (Trainor and Gladych 1995; Abe et al. 2001).

The taxonomic representation of desiccation tolerance in plants is now fairly well established. Although desiccation tolerance in plants is very uncommon, it occurs in many different taxa except gymnosperms (Alpert 2000). Desiccation-tolerant species are found on all continents and among species of all growth forms except trees, but it remains a mystery why desiccation tolerance is not more widespread.

There have been only few systematic surveys for desiccation tolerance within taxa or habitats. The list of tolerant vascular plants from different regions published by Gaff and co-workers (Gaff 1977, 1986; Gaff and Latz 1978), from rock outcrops (Porembski and Barthlott 2000) and reports by Fischer (1992, 1995) (Table 16.1) are probably the closest approaches to survey higher plants (see also Tables 8.1 and 9.1). The overall pattern one sees is that taxonomic and geographic breadth is contrasted with ecological narrowness. In the 1960s, researchers started to investigate the physiological ecology of desiccation tolerance in plants, and since the 1980s, emphasis has shifted to the biochemistry and molecular biology of desiccation tolerance.

16.1.2 Evolution of Desiccation Tolerance

Many primitive autotrophs, such as cyanobacteria, are desiccation tolerant. This indicates that desiccation tolerance is a very early trait in evolution. According to Oliver et al. (2000), phylogenetic evidence exists that among land plants tolerance to vegetative desiccation was present in the basal clade of bryophytes. This adaptive trait was likely to be critical for colonization by primitive terrestrial forms of the relatively dry land environment (Kappen and Valladares 1999). With the evolution of more complex vascular plants, desiccation tolerance was lost in vegetative tissues but was retained in reproductive tissues (Oliver et al. 2000). The acquisition of desiccation tolerance is part of a maturation programme during seed development in most higher plants (Angelovici et al. 2010). The majority of terrestrial plants are capable of producing desiccation-tolerant structures such as seeds and pollen, which can remain viable in the desiccated state for long periods as demonstrated in the case of the ancient *Nelumbo nucifera* (sacred lotus) seed from China (Shen-Miller et al. 1995). It is hypothesized that desiccation tolerance was crucial for ancestral freshwater autotrophs to live on land. Independent evolution or re-evolution of desiccation tolerance has happened in the Selaginellales, leptosporangiate ferns, and in at least ten families of angiosperms. Evolutionary progress may be evident from the fact that pteridophytes, like bryophytes, are able to

Table 16.1 Desiccation-tolerant species in addition to those listed by Proctor and Pence (2002) (see also Tables 8.1 and 9.1)

Species	Country	References
Aspleniaceae		
Asplenium ceterach L. (syn.: *Ceterach officinarum*)		
Asplenium aureum and at least three further *Asplenium* species (e.g., *A. lolegnamense*) from subg. Ceterach	Canary Islands	Fischer unpublished
Gesneriaceae		
Streptocarpus: ca. 20 species (e.g., *S. bindseilii* Eb. Fisch.)	Africa	Fischer unpublished
15 Species (e.g., *S. ibityiensis* Burtt)	Madagasca	Fischer unpublished
Linderniaceae		
Craterostigma plantagineum Hochst.	Trop. Africa, Arabia, India	Fischer (1992)
Craterostigma pumilum Hochst.	East Africa, Arabia	Fischer (1992)
Craterostigma hirsutum S.Moore	East Africa	Fischer (1992)
Craterostigma purpureum Lebrun and Toussaint	Central Africa	Fischer (1992)
Craterostigma lanceolatum Skan	East Africa, Southern Afri.	Fischer (1992)
Craterostigma longicarpum Hepper	East Africa	Fischer (1992)
Craterostigma alatum Hepper	East Africa	Fischer (1992)
Craterostigma smithii S.Moore	East Africa	Fischer (1992)
Craterostigma wilmsii Diels	South Africa	Fischer (1992)
Chamaegigas intrepidus Dinter ex Heil (monotypic)	Namibia	Fischer (1992)
Lindernia acicularis Eb.Fisch.	East Africa	Fischer (1992)
Lindernia brevidens Skan	East Africa	Fischer (1992)
Lindernia abyssinica Engl.	East Africa	Fischer (1992)
Lindernia angolensis (Skan) Eb.Fisch.	Angola	Fischer (1992)
Lindernia crassifolia (Engl.) Eb.Fisch.	Angola	Fischer (1992)
Lindernia wilmsii (Engl.) Philcox	East and South Africa	Fischer (1992)
Lindernia welwitschii (Engl.) Eb.Fisch.	Angola	Fischer (1992)
Lindernia scapoidea Eb.Fisch.	Angola	Fischer (1992)
Lindernia yaundensis (S.Moore) Eb.Fisch.	Cameroon	Fischer (1992)
Lindernia niamniamensis Eb.Fisch. and Hepper	East Africa	Fischer (1992)
Lindernia philcoxii Eb.Fisch.	East Africa	Fischer (1992)
Lindernia pulchella (Skan) Philcox	East and South Africa	Fischer (1992)
Lindernia sudanica Eb.Fisch. and Hepper	East Africa	Fischer (1992)
Lindernia linearifolia (Engl.) Eb.Fisch.	Angola	Fischer (1992)
Lindernia monroi (S.Moore) Eb.Fisch.	Southern Africa	Fischer (1992)
Lindernia horombensis Eb.Fisch.	Madagascar	Fischer (1995)
Lindernia andringitrae Eb.Fisch.	Madagascar	Fischer (1995)

synthesize rehydrin proteins but can also synthesize dehydrin proteins, which are typical for angiosperms (Oliver et al. 2000).

The genes involved in desiccation tolerance were not lost, but were instead recruited for activation of drying responses in reproductive structures through inductive mechanisms during developmental programs of the plant. It seems that resurrection plants have re-directed gene expression of seed-specific genes to be expressed in vegetative tissues (Illing et al. 2005). Recent phylogenetic evidence suggests that vascular plants gained the ability to withstand desiccation of their vegetative tissues from a mechanism present first in spore-bearing plants, and that this evolution (or re-evolution) has occurred on at least ten independent occasions within the angiosperms (Oliver et al. 2005). The presence of this information in the genomes of vascular plants permitted many species to spread over regions with seasonal rain fall and to re-evolve desiccation tolerance in vegetative tissues based on inducible responses triggered by environmental cues (Rascio and La Rocca 2005). The loss of desiccation tolerance in vegetative tissues can be interpreted as the result of suppression of genes in vegetative tissues (Dickie and Pritchard 2002). There may have been a selection against desiccation tolerance in vegetative tissues in favour of other traits. It can be concluded that the ability to survive dehydration must be basic and ubiquitous. On the other hand, desiccation tolerance must have been acquired at various advanced stages of organismic complexity by evolving metabolic or structural protective mechanisms. These must be very specific, either because of a distinct combination of quantitatively selected ubiquitous "house-keeping" mechanisms (Illing et al. 2005) or because of inventing genetic structures that act species-specific, like the *CDT-1* gene in *Craterostigma plantagineum* (Furini et al. 1997; Bartels and Salamini 2001).

16.1.3 Geographic Distribution and Ecology

Resurrection plants are found in places where rainfall is seasonal and sporadic. They often grow on rocky outcrops at low to moderate elevations in tropical and subtropical climates (Porembski and Barthlott 2000). Under these conditions, they are subjected to frequent cycles of drying and rehydration throughout the year and thus tolerate being dry in a broad range of temperatures (Mundree et al. 2002; Moore et al. 2005, 2007). Most resurrection plants have been reported from the southern hemisphere of Africa, India, Australia, and South America.

Resurrection plants have been identified within the angiosperms among both monocotyledonous and dicotyledonous plants. The dicotyledonous plants are represented mainly in the *Linderniaceae*, *Scrophulariaceae,* and *Myrothamnaceae* families, whereas the monocotyledonous are more scattered among different families. A first phylogenetic analysis among the *Scrophulariaceae* suggests a clustering of desiccation-tolerant plants represented by the genera *Craterostigma* and *Lindernia* (Rahmanzadeh et al. 2005).

It has been suggested that desiccation tolerance is connected with a size limitation, since all examples of desiccation-tolerant flowering plants do not exceed a certain height (Bewley and Krochko 1982), the largest known resurrection plant is the small woody shrub *Myrothamnus flabellifolius*, which can be between 0.5 m and 1.5 m tall (Moore et al. 2007). Most resurrection plants are herbaceous plants.

A list of desiccation-tolerant tracheophytes was compiled by Proctor and Pence (2002) and by Fischer (1992, 1995) (Table 16.1). It shows that there are more than 300 species of vascular plants. This number indicates that less than 0.15% of the total species possess vegetative desiccation tolerance.

The majority of resurrection plants were originally described in the 1970s (Gaff 1971; Gaff and Ellis 1974; Gaff and Churchill 1976; Gaff and Latz 1978). Most of the desiccation-tolerant angiosperms such as *M. flabellifolia* (Child 1960), *Xerophyta* spp., or *Craterostigma* spp. (Gaff 1977) are native to Southern Africa (especially South Africa, Namibia, and Zimbabwe) or to Australia, as it is the case for *Borya nitida* (Gaff and Churchill 1976). Physiological and anatomical characterizations of resurrection plants have been well investigated in species such as *B. nitida* (Gaff et al. 1976; Hetherington et al. 1982), *Craterostigma wilmsii* (Sherwin and Farrant 1996, 1998; Vicré et al. 1999; Cooper and Farrant 2002), *Eragrostis nindensis* (Vander Willigen et al. 2001, 2004), *M. flabellifolius* (Sherwin and Farrant 1996; Farrant and Kruger 2001), or *Reaumuria soongorica* (Liu et al. 2007) and most of the molecular aspects of desiccation tolerance have been studied in *C. plantagineum* (Bartels et al. 1990; Bartels and Salamini 2001; Hilbricht et al. 2002; Phillips et al. 2002), *Xerophyta viscosa* (Mundree et al. 2000; Mowla et al. 2002; Garwe et al. 2003) and *Xerophyta humilis* (Collett et al. 2003) and *Sporobolus* (Neale et al. 2000). A list of the most widely studied resurrection plants is compiled in Table 16.2.

Table 16.2 Best-studied resurrection plants

Name	Family	Mono/ dicot	Origin	Poikilochlorophyllous (P) Homoiochlorophyllous (H)
Xerophyta viscosa	Velloziaceae	Monocot	Southern Africa	P
Xerophyta humilis	Velloziaceae	Monocot	Southern Africa	P
Mysothamnus flabellifolia	Myrothamnaceae	Dicot	Southern Africa	H
Sporobolus stapfianus	Poaceae	Monocot	Southern Africa	
Eragrostis nindensis	Poaceae	Monocot	Southern Africa	
Craterostigma plantagineum	Scrophulariaceae	Dicot	Southern Africa	H
Craterostigma wilmsii	Scrophulariaceae	Dicot	Southern Africa	H
Lindernia brevidens	Linderniaceae	Dicot	East Africa	H
Boea hygrometrica	Gesneriaceae	Dicot	China	H
Bornetella nitida	Dasycladaceae	–	Africa, Australia, Asia	
Selaginella lepidophylla	Selaginellaceae	–	North and South America	
Tortula ruralis	Pottiacaeae	–	North America	

16.1.4 Diversity Within Linderniaceae

Phylogenetic analysis showed that desiccation-tolerant *Craterostigma* species cluster with *Lindernia* species, and the close phylogenetic relationship is supported by microsynteny between desiccation-related genes (Phillips et al. 2008). Interestingly, some members of the Linderniaceae are desiccation tolerant and some desiccation sensitive. This observation is being utilized to understand the evolution of desiccation tolerance. *Lindernia brevidens*, a close relative of *C. plantagineum*, was surprisingly identified as desiccation tolerant. This is remarkable, as *L. brevidens* is endemic to montane rainforests of East Africa, where it never experiences dry periods. *L. brevidens* occurs exclusively in two areas of the ancient Eastern Arc Mountains, which were protected from Pleistocene droughts by the mild temperatures of the Indian Ocean. High levels of species endemism and species diversity have developed in this area because of its protection and isolation. Although *L. brevidens* evolved in rainforest areas, it did not lose desiccation tolerance. Molecular analysis of desiccation-related genes suggests a high level of conservation between *L. brevidens* and *C. plantagineum* including the octulose sucrose conversion during desiccation (Phillips et al. 2008). It appears that *L. brevidens* is a neoendemic species, which has retained desiccation tolerance through genome stability. The question why *L. brevidens* did not lose desiccation tolerance cannot be answered; one hypothesis is that desiccation tolerance is linked to another trait, which is advantageous for *L brevidens* in its ecological niche. Comparative analysis between the closely related species with contrasting desiccation tolerance phenotypes could be a useful approach to understand the acquisition of desiccation tolerance.

16.2 Cellular Aspects

16.2.1 Morphological Adaptations

Leaf folding is one of the most obvious morphological changes in desiccation-tolerant vascular plants (Gaff 1989; Scott 2000; Farrant et al. 2003; Vander Willigen et al. 2003; Farrant et al. 2007). Leaves of *C. wilmsii* or *C. plantagineum*, which are fully expanded when watered, progressively curl inward during drying and become tightly folded so that only the abaxial surfaces of the older leaves in the outer whorl are exposed to the sun (Fig. 16.1 and Sherwin and Farrant 1998). Leaf folding is thought to limit oxidative stress damage from UV radiation and is thus an important morphological adaptation for surviving desiccation. Indeed, *C. wilmsii* plants do not survive desiccation in sunlight, if the leaves are mechanically prevented from folding (Farrant et al. 2003).

The leaf blades of *X. humilis* fold in half along the midrib upon dehydration, leaving only the abaxial surface exposed to the light (Sherwin and Farrant 1998).

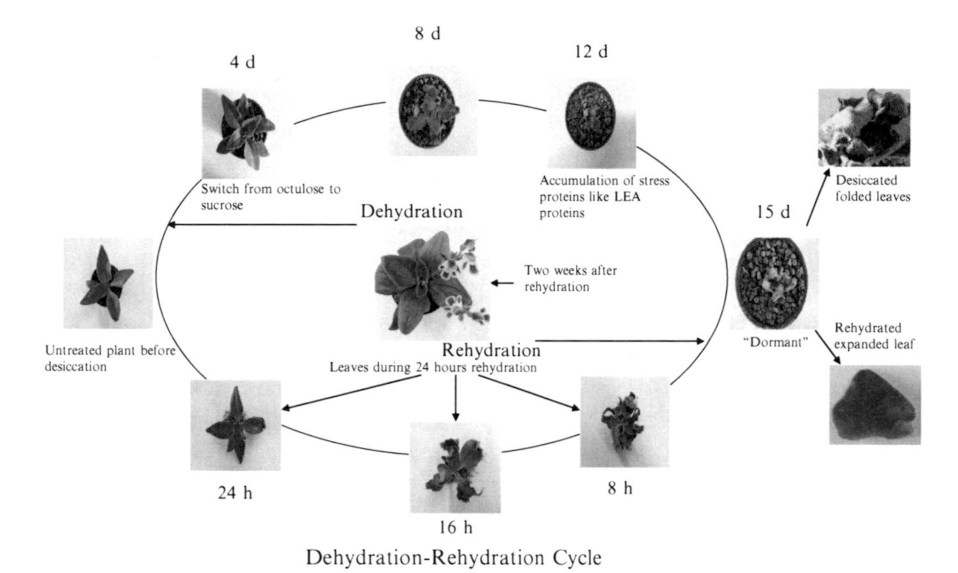

Fig. 16.1 Presentation of a dehydration and rehydration cycle in the resurrection plant *Craterostigma plantagineum*. A single plant was followed throughout the dehydration–rehydration cycle. This plant dehydrates to 2–5% RWC within 15 days and rapidly rehydrates within 24 h and starts photosynthesis. It has been estimated that leaves of this plant shrink to around 15% of the original leaf area. The plant is able to flower, as shown in the photograph in the middle; the photograph was taken 15 days after rehydration

In *Sporobulus stapfianus*, the leaf adaxial side, which is most exposed to sun radiation, is very rich in epicuticular waxes, whose function is, besides decreasing transpiration, to reflect light and to limit irradiation and temperature increase (Dalla Vecchia et al. 1998). Leaf movements occurring during dehydration have been suggested to reduce the effective transpiring surface during early stage dehydration and/or to prevent excessive irradiation of air-dry younger tissue (Gaff 1989; Farrant 2000).

16.2.2 Mechanical Stress: Cell Wall Changes, Vacuole Fragmentation and Water Substitution

One problem that resurrection plants face is the reduction of cell volume that occurs during desiccation. The examples below show that resurrection plants developed different mechanisms to avoid the potential mechanical stress during dehydration. A loss of water from cells leads to shrinkage of the central vacuole and to a drawing inward of the cell contents. This causes tension between the plasmalemma and the cell wall, which generally exhibits limited elasticity. If the cell membrane detaches from the cell wall, plasmodesmatal connections are ruptured, which can be lethal (Cosgrove 2000). Adaptations have acquired to reduce mechanical stress. In some resurrection plants, mainly dicotyledons, the protoplast shrinkage occurring upon dehydration is accompanied by an extensive folding of cell walls, which results in a contraction of the entire cell and avoids the tearing of the plasmalemma from the cell wall (Farrant and Sherwin 1997; Thomson and Platt 1997; Farrant 2000; Vicré et al. 1999, 2004). This phenomenon prevents the development of negative turgor pressure and reduces the potential for irreversible mechanical damage (Vander Willigen et al. 2001).

Recently, a relatively high degree of cell wall flexibility was shown for *Craterostigma* and *M. flabellifolius*. This feature can be attributed to the specific composition of leaf cell walls. In *C. wilmsii*, a significant increase in xyloglucans and unesterified pectins is observed in the cell wall during drying (Vicré et al. 1999). Dehydration also induces a considerable reduction of glucose in the hemicellulose fraction of *C. wilmsii* cell walls (Vicré et al. 2004). These changes are thought to enhance the tensile strength of the cell wall, allowing it to contract and to fold without collapsing in the dried tissue. An increase in expansin activity during desiccation is associated with a rise in cell wall flexibility and folding. This increase in cell wall flexibility is correlated with an increase in expansin transcript levels and activity (Jones and McQueen-Mason 2004; Moore et al. 2008). Expansins are cell wall loosening factors and are thought to act by disrupting the hydrogen bonds between cellulose and hemicellulose polymers in the cell wall (McQueen-Mason and Cosgrove 1995). Thus, the ability of cell walls to fold by changing their chemistry and texture is an adaptive strategy of resurrection plants, but the molecular basis of this alteration is hardly known. Recently, a glycine-rich protein has been correlated with cell wall flexibility in *Boea hygrometrica* (Wang et al. 2009). These exceptional features are remarkable when compared to the rigidification of the cell wall in response to drought in non-resurrection plants (Lu and Neumann 1998; Munns et al. 2000).

Some species, especially monocotyledons, utilize a strategy of water substitution and try to maintain the original cell volume by replacing water with non-aqueous substances such as amino acids, small proteins, and sugars (Farrant 2000; Marais et al. 2004). At the same time, the central vacuole is fragmented into a number of smaller units (Quartacci et al. 1997; Dalla Vecchia et al. 1998; Vander Willigen et al. 2004). This is of advantage during desiccation, as the formation of several small vacuoles increases the surface to volume ratio and thus presents a better mechanical stability (Iljin 1957).

16.2.3 Membrane Fluidity

Membrane fluidity may be another parameter that contributes to cell and tissue flexibility during desiccation. A higher degree in polyunsaturation in membrane phospholipids results in better membrane fluidity. This was demonstrated by changing lipid unsaturation in chloroplasts (Moon et al. 1995). Therefore, membrane composition was analysed in some resurrection plants, but the general picture is still patchy and no general conclusions can be drawn.

Hoekstra (2005) reports a negative correlation between the longevity of desiccation-tolerant tissues and the number of double bonds in the polar lipids of membranes. In desiccation-tolerant vascular plants, dehydration generally causes a decrease in total lipids as well the unsaturation level of individual phospholipids (Quartacci et al. 2002). However, the opposite trend is observed in the resurrection plant *Boea hygroscopica* where increased unsaturation of fatty acids was observed in all lipid classes upon dehydration (Navari-Izzo et al. 1995). In *S. stapfianus* phospholipid content and the level of polyunsaturated lipids within the plasma membrane increased in desiccation-tolerant, dried, attached leaves, but decreased in non-desiccation-tolerant detached leaves during desiccation (Quartacci et al. 1997; Neale et al. 2000).

16.3 Physiology

16.3.1 Photosynthesis

During desiccation, angiosperm resurrection plants shut down photosynthesis. This is correlated with down-regulation of photosynthesis-related gene expression (Bockel et al. 1998; Bernacchia et al. 1996) and with degradation of photosynthetic structures. Two different strategies are distinguished according to which plants are classified as homoiochlorophyllous and poikilochlorophyllous species (see Table 16.2). Homoiochlorophyllous species retain the chlorophyll and thylakoid membranes, although changes in photosynthetic pigment distribution are observed (Alamillo and Bartels 2001). In homoiochlorophyllous plants, chloroplasts do, however, undergo changes during desiccation, becoming roundish, with altered inner membrane organization and stacking. Also changes in the ratios between lipids and proteins as well as between the different lipid classes can occur in thylakoid membranes (Thomson and Platt 1997; Farrant 2000; Navari-Izzo et al. 2000; Quartacci et al. 1997).

In poikilochlorophyllous species, chlorophyll and the photosystem complexes are broken down (Tuba et al. 1998). The degradation of chlorophyll is advantageous, because the plants avoid the accumulation of toxic reactive oxygen species (ROS). Photosynthetic capacities are recovered during the rehydration process. Homoiochlorophyllous plants resume photosynthesis faster than poikilochlorophyllous species, which have to synthesize all components de novo. Ingle et al. (2008) demonstrated that the chloroplast biogenesis during rehydration in *X. humilis* resembles etioplast–chloroplast transition and uses similar molecular mechanisms involved in photomorphogenesis. The authors conclude from their observations that chloroplast degradation and regeneration during desiccation and rehydration do not involve novel genes, but require altered gene regulation of genes existing probably in most higher plants.

As the photosynthetic structures in the homoiochlorophyllous species need to be protected, it is not surprising that different classes of desiccation-induced proteins seem to be involved in the maintenance of chloroplast stability (Schneider et al. 1993; Alamillo and Bartels 1996; Neale et al. 2000). One of these is a chloroplast-localized, desiccation-related protein found in the leaves of *C. plantagineum* (Bartels et al. 1992; Alamillo and Bartels 2001) and *S. stapfianus* (Neale et al. 2000) shows high similarities to early light-inducible proteins (ELIPs). The ELIPs are thylakoid chlorophyll binding proteins that are transiently expressed during greening of etiolated plants (Adamska 1997). However, they are also synthesized in mature leaves exposed to excess of light or other environmental stresses causing the enhancement of free radicals (Montané et al. 1997; Zeng et al. 2002; Provart et al. 2003; Savenstrand et al. 2004). These proteins, similar to the HLIPs (high light-induced proteins) of cyanobacteria and the LHC (light harvesting complex) proteins of photosystems, have a protective function against photo-oxidative damage of the photosynthetic apparatus (Hutin et al. 2003). ELIPs might also bind chlorophylls, thus keeping free pigments at low levels in conditions of high light stress (Hutin et al. 2003). Earlier reports suggested that ELIPs might bind zeaxanthin, taking part in the non-photochemical quenching of light energy (Król et al. 1999).

A novel nuclear gene family expressed in response to dehydration has been identified in the leaves of *C. plantagineum* (Phillips et al. 2002). The genes encode plastid-targeted proteins (CpPTP), which have the ability to interact with plastid DNA, suggesting a role in the down-regulation of chloroplast-encoded genes or a role in protecting DNA during dehydration. In addition, several classes of Lea (=late embryogenesis abundant)-like proteins and detoxifying enzymes such as aldehyde dehydrogenase were shown to accumulate in chloroplasts of *C. plantagineum* during dehydration (Schneider et al. 1993; Kirch et al. 2001).

16.3.2 Antioxidant Systems

Recovery of a resurrection plant correlates with its capacity to establish a number of antioxidant protective mechanisms during dehydration and to maintain these systems upon rehydration (Kranner et al. 2002; Kranner and Birtić 2005). The level of reactive oxgen species (ROS) undergoes a significant rise in water-stressed cells that is dictated largely by chloroplasts as they are the most aerobic compartment in plant cells (Moran et al. 1994; Kranner and Lutzoni 1999). As respiration declines early during dehydration, mitochondria may be less involved in ROS production. In conditions of water deficit, stomata close limiting the supply of CO_2 to chloroplasts. This inhibits carbon fixation and causes overexcitation of chlorophylls, which then transfer their energy to oxygen, giving rise to oxygen singlets. *M. flabellifolius* retains high concentrations of chlorophyll during desiccation. When desiccated *M. flabellifolius* was rehydrated, antioxidants such as ascorbate, glutathione, and α-tocopherol accumulated in different tissues. When, however, the antioxidant pathways were broken down during long exposures to light, the plant did not recover anymore from desiccation (Kranner et al. 2002). This demonstrates that antioxidant systems are essential components of the recovery pathway.

In leaves of *X. viscosa*, a new desiccation-inducible antioxidant enzyme, corresponding to a form of 1-cys peroxiredoxin (Prxs), has been identified (Ndima et al. 2001; Mowla et al. 2002). Peroxiredoxins are a class of conserved thiol-specific antioxidant enzymes (Chae et al. 1994), active on substrates such as hydroperoxides, and are common to all organisms from archaebacteria to plants and mammals (Kozak 1999). The Prx found in leaves of *X. viscosa* (XvPer1) shows more than 70% sequence identity to seed-specific 1-cys peroxiredoxins (Aalen et al. 1994; Haslekas et al. 1998; Lewis et al. 2000), and it is the only 1-cys Prx that has been noticed in vegetative tissues. Its transcript is absent in fully hydrated leaves, but it accumulates upon dehydration and upon other stresses that lead to increased levels of ROS (Mowla et al. 2002). The nuclear location of XvPer1 points to an involvement of this protein in protecting the nuclear compartment from free oxygen radicals (Mowla et al. 2002).

16.3.3 Abscisic Acid Regulates Desiccation Tolerance Pathways

The plant hormone abscisic acid (ABA) is known to be involved in embryo maturation and in the response to osmotic stress in vegetative tissues. The key role of ABA was demonstrated by physiological and genetic experiments as well as measurements of ABA levels. In most resurrection plants, including the aquatic species *Chamaegigas intrepidus*, ABA is involved in attaining desiccation tolerance and in stimulating the synthesis of dehydration-induced proteins (Gaff 1980, 1989; Gaff and Loveys 1984; Bartels et al. 1990; Reynolds and Bewley 1993; Hellwege et al. 1994; Schiller et al. 1997, Chaps. 9 and 12). Leaves of *M. flabellifolia* and *B. nitida* did not survive dehydration, if they were dried so rapidly that ABA could not be accumulated (Gaff and Loveys 1984). A role for ABA in desiccation tolerance has also been demonstrated in *C. plantagineum*. Undifferentiated callus tissue of *C. plantagineum*, although intrinsically not desiccation tolerant, acquires tolerance after it has been cultured on medium containing ABA (Bartels et al. 1990). Molecular analysis suggests that ABA in resurrection plants has a similar function in signalling cascades as has ABA in dehydration responses in *Arabidopsis*. The role of ABA in signalling responses to dehydration in *Arabidopsis* has been intensively analysed through molecular genetics (Seki et al. 2007).

16.4 Gene Expression

Only a few resurrection plants have been studied on the molecular level, the best-studied species include *C. plantagineum*, *X. viscosa*, *X. humilis*, *S. stapfianus*, and *B. hygrometrica* (Table 16.2). Desiccation involves sequential gene expression programmes. Moderate tissue dehydration creates a drought condition also requiring the activation of drought tolerance genes in resurrection plants. More substantial drying, leading to a relative water content of 3% or less, entails the utilization of additional strategies and corresponding gene expression programmes. Thus, initially general mechanisms of water-deficit responses shared with drought-tolerant species are implemented; successively, specific programmes of desiccation tolerance are activated (Neale et al. 2000). It is assumed that gene expression in resurrection plants is geared towards the synthesis of protective molecules. Desiccation-induced genes can be grouped into genes encoding regulatory factors and proteins, which may exert a protective function directly or which may represent enzymes catalysing the synthesis of other protective molecules. Examples for different groups are described below.

To investigate the molecular basis of desiccation tolerance in a vascular resurrection species, Iturriaga et al. (2006) characterized 1,046 ESTs from a cDNA library constructed from *S. lepidophylla* undergoing desiccation for 2.5 h, which represented 873 unique transcripts. Putative functions were assigned to 653 (62.4%) of these clones after comparison with protein databases, whereas 212 (20.2%) sequences having significant similarity to known sequences whose functions are unclear and 181 (17.3%) sequences having no similarity to known sequences. For those ESTs for which functional assignments were made, *S. lepidophylla* had a higher percentage of ESTs within categories of molecular chaperons, i.e., heat shock proteins or late embryogenesis abundant (LEA) proteins. Gene discovery efforts, such as EST collections, constitute an important first step in understanding which genes need to be expressed for desiccation tolerance.

16.4.1 Regulatory Molecules

In view of the fact that upon dehydration resurrection plants synthesize transcripts in vegetative tissues that are similar or even identical to the seed maturation progamme of non-tolerant plants, it seems that evolution has found a way to express genes from other pathways for desiccation tolerance in vegetative tissues. This may also be true for other stress pathways, as exemplified by a heat shock transcription factor, which is expressed in response to dehydration in *C. plantagineum* but in *Arabidopsis thaliana* only in response to high temperature (C. Bockel and D. Bartels, University of Bonn unpublished). Another example for utilizing genes of existing pathways is the formation of chloroplasts during rehydration in *X. viscosa* as discussed above. Therefore, the search for regulatory switches is important for understanding the molecular mechanisms of desiccation tolerance. However, little work on regulatory genes has been reported in resurrection plants. The largest group of characterized genes represents members of different classes of transcription factors, which seem to be very similar in sequence and in function to the corresponding counterparts from non-tolerant genetic model plants. When the transcription factors, isolated from resurrection plants, were tested in e.g., *Arabidopsis*, they seem to function like the endogenous homlogues (for examples see below). Like it was shown for non-tolerant plants, the expression of dehydration-responsive genes in resurrection plants is mediated by ABA independent as well as ABA-dependent signal transduction pathways (Bartels 2005). It seems that the transcription factors have been optimized towards target gene activation, and it can be speculated that adaptation towards a special function in the desiccation pathway may modulate their role in gene activation. This hypothesis is in accordance with the observation that many transcription factors have been evolutionary conserved within a broad range of organisms.

Examples of the desiccation-induced transcription factors isolated thus far from resurrection plants have mainly been reported from *C. plantagineum*. Many of them belong to the Myb, the homeodomain-leucine zipper (HD-Zip), or the bZip (basic leucine zipper) families (Iturriaga et al. 1996; Frank et al. 1998; Ditzer and Bartels 2006). Plant myb-related genes comprise a large family and are likely to participate in a variety of functions (Meissner et al. 1999). Two Myb-related genes, *cpm10* and *cpm7*, show differential expression and regulation in response to desiccation and ABA in *C. plantagineum* (Iturriaga et al. 1996). *Cpm10* is expressed in undifferentiated callus tissue and is up-regulated by ABA, while *cpm7* is induced by dehydration in roots. Transgenic *Arabidopsis* plants overexpressing *cpm10* displayed increased tolerance to drought and salt stress (Villalobos et al. 2004). These plants also showed ABA hypersensitivity and glucose insensitivity, suggesting that *cpm10* is involved in mediating ABA and glucose signalling responses in *Arabidopsis* as well as in the response to drought stress. HD-ZIP genes encode proteins that have only been identified in plants so far and are thought to regulate development and responses to environmental cues (Ramanjulu and Bartels 2002). Two HD-Zip genes, CPHB-1 and CPHB-2, are induced by dehydration in leaves and roots of *C. plantagineum*, but show different responses to exogenously applied ABA (Frank et al. 1998). ABA treatment induces the transcription of CPHB-2, but not that of CPHB-1. Five other HD-Zip genes have been isolated from *C. plantagineum*, two of which were induced by ABA in undifferentiated callus, while the other three were not. This suggests that these HD-Zip genes act in different pathways of the dehydration response; some mediated by ABA, while others are independent of ABA (Deng et al. 2002).

Research related to dehydration- and cold-regulated genes in *Arabidopsis* led to the discovery of the dehydration-responsive elements (DRE) (Yamaguchi-Shinozaki and Shinozaki 1993). Therefore, many genes that are responsive to cold and drought stress contain DRE/CRT (dehydration-responsive element/C-repeat) elements within their 5′ regulatory regions and are regulated by DREB/CBF (DRE-binding protein/C-repeat binding factor) transcription factors (Stockinger et al. 1997; Liu et al. 1998). The DREB/CBF proteins belong to the AP2/EREBP (ethylene-responsive element binding protein) family of transcription factors which are important regulators of flower development and plant responses to abiotic stresses (Riechmann and Meyerowitz 1998; Shigyo et al. 2006). One such AP2/EREBP transcription factor has been reported as a regulator of the drought response pathway in *S. stapfianus* (Le 2005).

Other characterized genes that are involved in the regulatory network are phospholipases, VP1 homologues, or the SAP domain transcription factor CpR18 (Bartels et al. 2006). Like for the nature of the regulatory genes, the recognition promoter elements seem to be conserved, and no desiccation-specific elements have been identified so far (Bartels et al. 2006). This supports the hypothesis that no new mechanisms were invented, but that new combinations of regulatory elements are sufficient for expressing genes in desiccation tolerance.

Mutants would be extremely valuable for the unravelling of the basis of desiccation tolerance, but no genetic system has been developed yet to generate mutants in resurrection plants. The reason for this is that most resurrection plants appear not to be diploid. Over 10 years ago an activation tagging approach, which generates dominant mutations led to the discovery of the unusual *CDT-1* gene in *C. plantagineum* (Furini et al. 1997). No homologues of this gene have so far been identified in species apart from *Craterostigma*. Constitutive overexpression of *cdt-1* leads to desiccation tolerance in *C. plantagineum* callus tissue activating the expression of desiccation-induced transcripts. *CDT-1* transcripts accumulate in response to desiccation, but no corresponding polypeptides are synthesized; thus, *CDT-1* functions as a regulatory, non-protein coding RNA (Hilbricht et al. 2008 (see also Chap. 13.3)). *CDT-1* is a member of a large gene family, of which all members have features of retrotransposons (Smith-Espinoza et al. 2005). Retrotransposon elements may be required for expression of desiccation tolerance, as they are able to activate genes via small RNAs such as CDT-1. Small RNAs as regulators of gene expression are emerging players for essential environmental and developmental switches (Phillips et al. 2007).

16.4.2 Aquaporins

Aquaporin-mediated water transport across membranes provides a fundamental contribution to plant–water relations (Maurel and Chrispeels 2001 and Chap. 10). This makes it obvious to assume that aquaporins are involved in desiccation tolerance. In leaves of *C. plantagineum*, one TIP (tonoplast intrinsic protein) and several PIP (plasma membrane intrinsic protein) have been found, whose genes are differentially expressed during dehydration (Mariaux et al. 1998). A dehydration-responsive TIP has also been isolated from leaves of *S. stapfianus* (Neale et al. 2000). One particularly relevant finding is the identification of a tonoplast aquaporin, namely TIP 3;1, in dried leaves of *E. nindensis* (Vander Willigen et al. 2004). The interest in this desiccation-dependent channel protein is due to the fact that a TIP 3;1 had never been noticed in vegetative tissues. TIP 3;1 proteins (formerly called α-TIPs) are considered to be unique to orthodox seeds (Maurel et al. 1995). It participates in modulating the water permeability of the tonoplast during rehydration and may aid in mobilizing sugars, proteins, and other solutes that have accumulated inside the small vacuoles of bundle sheath cells with which it is associated (Vander Willigen et al. 2004).

16.4.3 Carbohydrates

A common observation in the desiccation process is the accumulation of soluble sugars, and the importance of sugars has been implicated in several studies. During desiccation, starch is rapidly converted into glucose (Crowe et al. 1998). Apart from a role in osmotic adjustment during the dehydration phase, the protective effects of sugars may also include protein stabilization in desiccated cells (Ramanjulu and Bartels 2002). Conversion of starch into sucrose is attributed primarily through activation of sucrose phosphate synthase by reversible protein phosphorylation upon perception of osmotic stress. Sucrose and trehalose have been proposed to prevent protein denaturation and membrane fusions (Crowe et al. 1998; Peters et al. 2007). Both sugars are capable of forming biological glasses (a process called vitrification) within the dried cell. Vitrification of the cytoplasm may not be due to the effects of sugars only, but probably results from the interaction of sugars with other molecules, most likely proteins (Hoekstra 2005). Cells of many desiccation-tolerant plants and animals undergo vitrification upon drying to protect organelles from damage (Buitink and Leprince 2004), and conformational changes of proteins and membrane fusion are prevented (Crowe et al. 1998). Vitrification is thought to limit the production of free radicals by slowing down chemical reactions and molecular diffusion in the cytoplasm (Hoekstra 2005). Thus, vitrification may be an important protective mechanism for resurrection plants.

In fully hydrated *S. stapfianus*, leaves, glucose, fructose, and galactose are present in large amounts. During dehydration, sucrose increases to high levels in air-dry leaves to become the predominant sugar (Ghasempour et al. 1998). Other sugars such as raffinose and trehalose are also detected, and they may complement the role of sucrose as osmotic protectants during drying (Rascio and La Rocca 2005). The changes in sugar composition observed in *S. stapfianus* leaves were quantitatively similar to those reported for other resurrection plants where the common trend was the conversion of monosaccharides in hydrated leaves into sucrose in dried leaves and vice versa in rehydrated leaves (Murelli et al. 1996). In *C. plantagineum*, the unusual eight-carbon sugar, 2-octulose, is the predominant sugar in hydrated leaves, which is converted to sucrose, with the reverse process being observed during rehydration (Bianchi et al. 1991). Sucrose also accumulates in the roots of drying *C. plantagineum*, although the most prevalent sugar stachyose does not undergo big changes (Bianchi et al. 1991; Norwood et al. 2003).

Sugar conversions during the desiccation process have been correlated with corresponding gene expression. Again, it seems that no new genes have evolved, but the conversion processes utilize existing genes of the sugar metabolism. Sucrose biosynthetic genes have been shown to be induced by both desiccation and ABA in the resurrection plant *C. plantagineum* (Kleines et al. 1999). Perhaps gene duplications may happen during acquisition of desiccation tolerance to accommodate desiccation-related metabolism. A possible example for this is the presence of a transketolase localized in the cytoplasm of *C. plantagineum*. It has been suggested that this transketolase is involved in the octulose sucrose conversions during the dehydration/rehydration cycle and has changed its substrate specificity (Willige et al. 2009).

16.4.4 Compatible Solutes

A common phenomenon in drought stress is the accumulation of organic compatible solutes because they are supposed to stabilize proteins and membranes (Levitt 1980; Crowe and Crowe 1992). Upon dehydration, many plants accumulate nontoxic solutes such as proline, mannitol, and glycine betaine (Chen and Murata 2002). The compatible solute accumulation results in an increase in cellular osmolarity, which in turn leads to an influx of water into, or at least a reduced efflux from cells (Hare et al. 1998). The main function of compatible solutes is, however, not so much to increase the water holding capacity of cells, but compatible solutes have been proposed to protect cells through the stabilization of cytoplasmic constituents and ion sequestration (Hare et al. 1998). This is due to the preferential exclusion of these compounds from the surfaces of proteins and membranes, which thus remain preferentially hydrated (Hoekstra et al. 2001).

Several resurrection plants have been analysed for the presence of potential compatible solutes. A diverse spectrum of molecules, which may serve as compatible solutes, has been identified, but no general mechanism is identified. Inositol, present in *X. viscosa*, may be an effective osmoprotectant (Mayee et al. 2005). Application of proline had no effect on detached leaves of *B. nitida* and *M. flabellifolius* (Gaff 1980). In the latter species, polyphenols have been identified that might be relevant to desiccation tolerance, as provenances from Namibia subjected to greater drought stress were genetically different from those in South Africa and contained more and different polyphenols (e.g., 3,4,5-tri-*O*-galloquinic acid) (Moore et al. 2005). *M. flabellifolius* contains a broad spectrum of potential osmotica in particular glucose–glycerol, which has been directly correlated with osmotic tolerance in cyanobacteria (Bianchi et al. 1993).

16.4.5 Protective Proteins: LEA Proteins and Heat Shock Proteins

The abundant accumulation of proteins during desiccation is the most obvious change on the molecular level. The majority of the proteins are LEA proteins. LEA proteins accumulate during seed maturation and during dehydration of vegetative tissues. A protective role for the LEA proteins has been suggested, and it is assumed that they are a major contribution to desiccation tolerance in resurrection plants (Ingram and Bartels 1996; Mtwisha et al. 2006). Despite this fact LEA proteins are not discussed further in this chapter, as there have been excellent recent reviews discussing LEA proteins in detail (Battaglia et al. 2008; Tunnacliffe and Wise 2007 and references within). Besides LEA proteins also heat shock proteins are associated with desiccation. The role of heat shock proteins in this context is summarized.

Proteins commonly indicated as heat shock proteins, being up-regulated by heat stress, also seem to be implicated in desiccation tolerance (Goyal et al. 2005). They are induced by the same stresses as LEA proteins and their synthesis coincides with the acquisition of desiccation tolerance (Wehmeyer et al. 1996; Mtwisha et al. 2006). These proteins function as molecular chaperones and play primary roles in protein biosynthesis, folding, assembly, intracellular localization, secretion, and degradation of other proteins (Feder and Hofmann 1999). Intracellular proteins such as heat shock proteins can compensate the loss of hydrogen bonds to water molecules by hydrogen's bonding to other molecules. Generally, HSPs are able to maintain partner proteins in a folding competent, folded or unfolded, state to minimize the aggregation of non-native proteins or to target non-native or aggregated proteins for degradation or removal from the cell (Feder and Hofmann 1999). This function could be important in maintaining cells in resurrection plants viable.

Small heat shock proteins are expressed in the vegetative tissue of *C. plantagineum* during water stress and are correlated with desiccation-tolerant callus (Alamillo et al. 1995) and in the cysts of the desiccation-tolerant brine shrimp *Artemia franciscana* under desiccation stress (Clegg 2005).

16.5 Rehydration

Most research on resurrection plants has examined strategies that counteract desiccation damage and promote survival in the dry state. Less attention has been paid to the rehydration process. The events occurring during rehydration involve gradual recovery of correct cellular organization, which is accompanied by reactivation of normal metabolic functions. The processes leading to the reestablishment of vital functions and metabolic activities are an integral part of the entire phenomenon of desiccation tolerance. It seems that there are differences in the rehydration process in lower plants such as mosses and in higher vascular plants. It has been discussed whether desiccation tolerance in mosses such as *Tortula ruralis* is constitutive, with recovery essentially a matter of reassembly of components conserved intact through a drying–rehydration cycle, or non-constitutive repair processes. The idea of a repair-based mechanism of desiccation tolerance in bryophytes was introduced by Bewley (Bewley 1979; Bewley and Krochko 1982). It was extended by Oliver and Bewley (1984) to take into account the fine-structural change, and it has since been reiterated in a succession of reviews (Bewley and Oliver 1992; Oliver 1996; Oliver and Bewley 1997; Oliver et al. 1998).

The successful rehydration in *C. plantagineum* requires a discrete period of time (Bartels et al. 1990). Dried *Craterostigma* plants complete the water uptake in 12–15 h when placed in a water-saturated environment. The comparison of protein patterns from dried and rehydrated leaves of *C. plantagineum* revealed that no new mRNAs were detectable during early phases of rehydration

(Bernacchia et al. 1996). Within the first 6–12 h of rewatering, desiccation-related transcripts and their corresponding proteins disappear. Only after 12–15 h, when the leaf tissues reach about 80–90% RWC, changes occur in translatable mRNA populations and protein profiles, mainly related to enhanced expression of genes that were down-regulated upon drying. New photosynthetic pigments are synthesized, and respiration, photosynthesis and other metabolic pathways are reactivated, showing that mRNAs or enzymes necessary to restore cell activities had been accumulated in a stable form during dehydration and were sufficiently protected in the dry state.

Even though the desiccation tolerance of resurrection plants largely depends on strategies of cell protection during drying, these plants also have mechanisms for preventing and/or repairing cell damage upon rehydration (Cooper and Farrant 2002). In rehydrated leaves of several species, the de novo synthesis of some proteins, also produced upon drying, has been observed. Proteins involved in both dehydration and rehydration include the previously mentioned expansins (Cooper and Farrant 2002). Moreover, aquaporins, whose mRNAs are produced and stored during the late stage of drying, also seem to play an important role in rehydration by regulating the rate of water flow into cells (Mariaux et al. 1998).

16.6 Conclusions and Outlook

In this chapter, we have tried to summarize what is known about the physiology and molecular analysis of desiccation tolerance in higher plants. The phenomenon of desiccation tolerance has been well characterized for a few species. However, the mechanism is not understood yet. The obstacles are the complexity of desiccation tolerance and the lack of genetic approaches connected with the complex genomes. The genomes known so far are relatively small in estimated size, but they are not diploid. Recently, DNA sequencing technologies have become available at an economically affordable level. This will allow transcriptome analysis for plant species for which no genome sequence is available and thus a comparative genomic analysis could provide more insight into desiccation tolerance (Rodriguez et al. 2010). It can be expected from such approaches to identify genes that are unique to desiccation-tolerant species and thus could give a clue for the requirements of desiccation tolerance.

In addition to understanding which genes are expressed and which proteins are active, epigenetic mechanisms turn out to be a new dimension of modifying adaptations to responses to environmental stresses. This has not been studied at all in any resurrection plants but needs to be considered for explaining desiccation tolerance mechanisms. Epigenetic mechanisms can be seen as enhancing the complexity of regulation of gene expression and may enhance the flexibility of plant genomes to adapt to diverse environments.

Acknowledgements We thank Prof. E. Fischer for advice on the geographical distribution of desiccation-tolerant species and for providing the information contained in Table 16.1. Work in the laboratory of D.B. was supported by grants from the German Research Council, by the European training network ADONIS and by an ERA-PG project. D.B. is a member of the European COST action INPAS (International Network of Plant Abiotic Stress). We thank C. Marikar for help with the manuscript preparation.

References

Aalen RB, Opsahl-Ferstad HG, Linnestad C, Olsen OA (1994) Transcripts encoding an oleosin and a dormancy-related protein are present in both the aleurone layer and the embryo of developing barley (*Hordeum vulgare* L.) seeds. Plant J 5:385–396

Abe S, Kurashima A, Yokohama Y, Tanaka J (2001) The cellular ability of desiccation tolerance in Japanese intertidal seaweeds. Bot Mar 44:125–131

Adamska I (1997) ELIPs. Light-induced stress proteins. Physiol Plant 100:794–805

Alamillo JM, Bartels D (1996) Light and stage of development influence the expression of desiccation-induced genes in the resurrection plant *Craterostigma plantagineum*. Plant Cell Environ 19:300–310

Alamillo JM, Bartels D (2001) Effects of desiccation on photosynthesis pigments and the ELIP-like dsp 22 protein complexes in the resurrection plant *Craterostigma plantagineum*. Plant Sci 160:1161–1170

Alamillo J, Roncarati R, Heino P, Velasco R, Nelson D, Elster R, Brenacchia G, Furini A, Schwall G, Salamini F, Bartels D (1995) Molecular analysis of desiccation tolerance in barley embryos and in the resurrection plant *Craterostigma plantagineum*. Agronomie 2:161–167

Alpert P (2000) The discovery, scope, and puzzle of desiccation tolerance in plants. Plant Ecol 151:5–17

Alpert P (2006) Constraints of tolerance: why are desiccation-tolerant organisms so small and rare. J Exp Biol 209:1575–1584

Angelovici R, Galili G, Fernie AR, Fait A (2010) Seed desiccation: a bridge between maturation and germination. Trends Plant Sci 15(4):211–218

Bartels D (2005) Desiccation tolerance studied in the resurrection plant *Craterostigma plantagineum*. Integr Comp Biol 45:696–701

Bartels D, Salamini F (2001) Desiccation tolerance in the resurrection plant *Craterostigma plantagineum*. A contribution to the study of drought tolerance at the molecular level. Plant Physiol 127:1346–1353

Bartels D, Singh M, Salamini F (1988) Onset of desiccation tolerance during development of the barley embryo. Planta 175:485–492

Bartels D, Schneider K, Terstappen G, Piatkowski D, Salamini F (1990) Molecular cloning of abscisic acid-modulated genes which are induced during desiccation of the resurrection plant *Craterostigma plantagineum*. Planta 181:27–34

Bartels D, Hanke C, Schneider K, Michel D, Salamini F (1992) A desiccation-related ELIP-like gene from the resurrection plant *Craterostigma plantagineum* is regulated by light and ABA. EMBO J 11:2771–2778

Bartels D, Ditzer A, Furini A (2006) What can we learn from resurrection plants? In: Ribaut JM (ed) Drought adaptation in cereals. New York, Hayworth, pp 599–622

Battaglia M, Olvera-Carrillo Y, Garciarrubio A, Campos F, Covarrubias AA (2008) The enigmatic LEA proteins and other hydrophilins. Plant Physiol 148:6–24

Bernacchia G, Salamini F, Bartels D (1996) Molecular characterization of the rehydration process in the resurrection plant *Craterostigma plantagineum*. Plant Physiol 111:1043–1050

Bewley JD (1979) Physiological aspects of desiccation tolerance. Ann Rev Plant Physiol 30:195–238

Bewley JD, Krochko JE (1982) Desiccation tolerance. In: Pirson A, Zimmermann MH (eds) Encyclopedia of plant physiology, vol 12b, New series. Springer, Heidelberg, pp 325–378

Bewley JD, Oliver MJ (1992) Desiccation-tolerance in vegetative plant tissues and seeds: protein synthesis in relation to desiccation and a potential role for protection and repair mechanisms. In: Osmond CB, Somero G (eds) Water and life: a comparative analysis of water relationship at the organismic, cellular and molecular levels. Springer, Berlin, pp 141–160

Bianchi G, Gamba A, Murelli C, Salamini F, Bartels D (1991) Novel carbohydrate metabolism in the resurrection plant *Craterostigma plantagineum*. Plant J 1(3):355–359

Bianchi G, Gamba A, Limiroli R, Pozzi N, Elste R, Salamini F, Bartels D (1993) The unusual sugar composition in leaves of the resurrection plant *Myrothamnus flabellifolia*. Physiol Plant 87:223–226

Bockel C, Salamini F, Bartels D (1998) Isolation and characterization of genes expressed during early events of the dehydration process in the resurrection plant *Craterostigma plantagineum*. J Plant Physiol 152:158–166

Buitink J, Leprince O (2004) Glass formation in plant anhydrobiotes: survival in the dry state. Cryobiology 48:215–228

Chae HZ, Chung SJ, Rhee SG (1994) Thioredoxin-dependent peroxide reductase from yeast. J Biol Chem 269:27670–27678

Chen THH, Murata N (2002) Enhancement of tolerance of abiotic stress by metabolic engineering of betaines and other compatible solutes. Curr Opin Plant Biol 5:250–257

Child GF (1960) Brief notes on the ecology of the resurrection plant *Myrothamnus flabellifolia* with mention of its water-absorbing abilities. J S Afr Bot 26:1–8

Clegg JS (2005) Desiccation tolerance in encysted embryos of the animal extremophile, *Artemia*. Integr Comp Biol 45:715–724

Collett H, Butowt R, Smith J, Farrant JM, Illing N (2003) Photosynthetic genes are differentially transcripted during the dehydration-rehydration cycle in the resurrection plant, *Xerophyta humilis*. J Exp Bot 54:2593–2595

Cooper K, Farrant JM (2002) Recovery of the resurrection plant *Craterostigma wilmsii* from desiccation: protection versus repair. J Exp Bot 53:1805–1813

Cosgrove DJ (2000) Expansive growth of plant cell walls. Plant Physiol Biochem 38:109–124

Crowe JH, Crowe LM (1992) Membrane integrity in anhydrobiotic organisms: towards a mechanism for stabilizing dry cells. In: Somero GN, Osmond CB, Bolin CL (eds) Water and life. A comparative analysis of water relationships at the organismic, cellular and molecular levels. Springer, Berlin, pp 87–113

Crowe JH, Carpenter JF, Crowe LM (1998) The role of vitrification in anhydrobiosis. Annu Rev Plant Physiol 60:73–103

Dalla Vecchia F, Asmar TE, Calamassi R, Rascio N, Vazzana C (1998) Morphological and ultrastructural aspects of dehydration and rehydration in leaves of *Sporobolus stapfianus*. Plant Growth Reg 24:219–228

Deng X, Phillips J, Meijer AH, Salamini F, Bartels D (2002) Characterization of five novel dehydrationresponsive homeodomain leucine zipper genes from the resurrection plant *Craterostigma plantagineum*. Plant Mol Biol 49:601–610

Dickie JB, Pritchard HW (2002) Systematic and evolutionary aspects of desiccation tolerance in seeds. In: Black M, Pritchard HW (eds) Desiccation and survival of plants – drying without dying. CABI, Wallingford, pp 239–259

Ditzer A, Bartels D (2006) Identification of stress-responsive promoter elements and isolation of corresponding DNA binding proteins for the LEA gene *CpC2* promoter. Plant Mol Biol 61:643–663

Farrant JM (2000) A comparation of mechanisms of desiccation tolerance among three angiosperm resurrection plant species. Plant Ecol 151:29–39

Farrant JM, Kruger LA (2001) Longevity of dry *Myrothamnus flabellifolius* in simulated field conditions. Plant Growth Reg 35:109–120

Farrant JM, Sherwin HW (1997) Mechanisms of desiccation tolerance in seeds and resurrection plants. In: Taylor AG, Huang XL (eds) Progress in seed research. Proceedings of the second international conference on seed science and technology. Communication Services of the New York State Agricultural Experimental Station, Geneva, NY, pp 109–120

Farrant JM, Vander Willigen C, Loffell DA, Bartsch S, Whittaker A (2003) An investigation into the role of light during desiccation of three angiosperm resurrection plants. Plant Cell Environ 26:1275–1286

Farrant JM, Brandt W, Lindsey GG (2007) An overview of mechanisms of desiccation tolerance in selected angiosperm resurrection plants. Plant Stress 1(1):72–84

Feder ME, Hofmann GE (1999) Heat shock proteins, evolutionary and ecological physiology. Ann Rev Physiol 61:243–282

Fischer E (1992) Systematik der afrikanischen Lindernieae (Scrophulariaceae). Tropische und subtropische Pflanzenwelt 81:1–365

Fischer E (1995) Revision of the Lindernieae (Scrophulariaceae) in Madagascar. 1. The genera *Lindernia* Allioni and *Crepidorhopalon* E. FISCHER. Bull Mus Natl Hist Nat Paris 4e sér., 17, section B, Adansonia: 227–257

Frank W, Phillips JR, Salamini F, Bartels D (1998) Two dehydration-inducible transcripts from the resurrection plant *Craterostigma plantagineum* encode interacting homeodomain-leucine zipper proteins. Plant J 15:413–421

Furini A, Koncz C, Salamini F, Bartels D (1997) High level transcription of a member of a repeated gene family confers dehydration tolerance to callus tissue of *Craterostigma plantagineum*. EMBO J16:3599–3608

Gaff DF (1971) Desiccation-tolerant flowering plants in southern Africa. Science 174:1033–1034

Gaff DF (1977) Desiccation tolerant vascular plants of southern Africa. Oecologia 31:95–109

Gaff DF (1980) Protoplasmic tolerance of extreme water stress. In: Turner NC, Kramer PJ (eds) Adaptation of plants to water and high temperature stress. Wiley, New York, pp 207–230

Gaff DF (1986) Desiccation tolerant "resurrection" grasses from Kenya and West Africa. Oecologia 70:118–120

Gaff DF (1989) Responses of desiccation-tolerant "resurrection" plants to water stress. SPB Academic, The Hague

Gaff DF, Churchill DM (1976) *Borya nitida* Labill. – an Australian species in the Liliaceae with desiccation-tolerant leaves. Aust J Bot 24:209–224

Gaff DF, Ellis RP (1974) Southern African grasses with foliage that revives after dehydration. Bothalia 11:305–308

Gaff DF, Latz PK (1978) The occurrence of resurrection plants in the Australian flora. Aust J Bot 26:485–492

Gaff DF, Loveys BR (1984) Abscisic-acid content and effects during dehydration of detached leaves of desiccation tolerant plants. J Exp Bot 35:1350–1358

Gaff DF, Zee S-Y, O'Brien TP (1976) The fine structure of dehydrated and reviving leaves of *Borya nitida* Labill. – a desiccation-tolerant plant. Aust J Bot 24:225–236

Garwe D, Thomson JA, Mundree SG (2003) Molecular characterization of XVSAP1, a stress-responsive gene from the resurrection plant *Xerophyta viscosa* Baker. J Exp Bot 54:191–201

Ghasempour HR, Gaff DF, Williams RPW, Gianello RD (1998) Contents of sugars in leaves of drying desiccation-tolerant flowering plants, particularly grasses. Plant Growth Reg 24:185–191

Goyal K, Walton LJ, Tunnacliffe A (2005) LEA proteins prevent protein aggregation due to water stress. Biochem J 388:151–157

Hare PD, Cress WA, Van Staden J (1998) Dissecting the role of osmolyte accumulation during stress. Plant Cell Environ 21:535–553

Haslekas C, Stacy RAP, Nygaard V, Culianez-Macia FA, Aalen RB (1998) The expression of a peroxiredoxin antioxidant gene, AtPer1, in *Arabidopsis thaliana* is seed-specific and related to dormancy. Plant Mol Biol 36:833–845

Hellwege EM, Dietz KJ, Volk OH, Hartung W (1994) Abscisic acid and the induction of desiccation tolerance in the extremely xerophilic liverwort *Exormotheca holstii*. Planta 194:525–531
Hetherington SE, Smillie RM, Hallam ND (1982) Humidity-sensitive degreening and regreening of leaves of *Borya nitida* Labill. as followed by changes in chlorophyll fluorescence. Aust J Plant Physiol 9:587–599
Hilbricht T, Salamini F, Bartels D (2002) CpR18, a novel SAP-domain plant transcription factor, binds to a promoter region necessary for ABA mediated expression of the CdeT27–45 gene from the resurrection plant *Craterostigma plantagineum* Hochst. Plant J 31:293–303
Hilbricht T, Varotto S, Sgaramella V, Bartels D, Salamini F, Furini A (2008) Retrotransposons and siRNA have a role in the evolution of desiccation tolerance leading to resurrection of the plant *Craterostigma plantagineum*. New Phytol 179:877–887
Hoekstra FA (2005) Differential longevities in desiccated anhydrobiotic plant systems. Integr Comp Biol 45:725–733
Hoekstra FA, Golovina EA, Buitink J (2001) Mechanisms of plant desiccation-tolerance. Trends Plant Sci 6:431–438
Hutin C, Nussaume L, Moise N, Moya I, Kloppstech K, Havaux M (2003) Early light-induced proteins protect Arabidopsis from photooxidative stress. Proc Natl Acad Sci USA 100:4921–4926
Iljin WS (1957) Drought-resistance in plants and physiological processes. Annu Rev Plant Physiol 3:341–363
Illing N, Denby KJ, Collett H, Shen A, Farrant JM (2005) The signature of seeds in resurrection plants: a molecular and physiological comparison of desiccation tolerance in seeds and vegetative tissues. Integr Comp Biol 45:771–787
Ingle RA, Collett H, Cooper K, Takahashi Y, Farrant JM, Illing N (2008) Chloroplast biogenesis during rehydration of the resurrection plant *Xerophyta humilis*: parallels to the etioplast-chloroplast transition. Plant Cell Environ 31:1813–1824
Ingram J, Bartels D (1996) The molecular basis of dehydration-tolerance in plants. Annu Rev Plant Physiol Plant Mol Biol 47:377–403
Iturriaga G, Leyns L, Villegas A, Gharaibeh R, Salamini F, Bartels D (1996) A family of novel myb-related genes from the resurrection plant *Craterostigma plantagineum* are specifically expressed in callus and roots in response to ABA or desiccation. Plant Mol Biol 32:707–716
Iturriaga G, Cushman MAF, Cushman JC (2006) An EST catalogue from the resurrection plant *Selaginella lepidophylla* reveals abiotic stress-adaptive genes. Plant Sci 170:1173–1184
Jones L, McQueen-Mason S (2004) A role for expansins in dehydration and rehydration of the resurrection plant *Craterostigma plantagineum*. FEBS Lett 559:61–65
Kappen L, Valladares F (1999) Opportunistic growth and desiccation tolerance: the ecological success of poikilohydrous autotrophs. In: Pugnaire FI, Valladares F (eds) Handbook of functional plant ecology. Marcel Dekker, New York, pp 10–80
Keilin D (1959) The problem of anabiosis or latent life: history and current concept. Proc R Soc Lond B Biol Sci 150:149–191
Kirch HH, Nair A, Bartels D (2001) Novel ABA- and dehydration-inducible aldehyde dehydrogenase genes isolated from the resurrection plant *Craterostigma plantagineum* and *Arabidopsis thaliana*. Plant J 28(5):555–567
Kleines M, Elster RC, Rodrigo MJ, Blervacq AS, Salamini F, Bartels D (1999) Isolation and expression analysis of two stress-responsive sucrose-synthase genes from the resurrection plant *Craterostigma plantagineum* (Hochst.). Planta 209:13–24
Kozak CA (1999) Genetic mapping of six mouse peroxiredoxin genes and fourteen peroxiredoxin related sequences. Mamm Genome 10:1017–1019
Kranner I, Birtić S (2005) A modulating role for antioxidants in desiccation tolerance. Integr Comp Biol 45:734–740
Kranner I, Lutzoni F (1999) Evolutionary consequences of transition to a lichen symbiotic state and physiological adaptation to oxidative damage associated with poikilohydry. In: Lerner HR (ed) Plant responses to environmental stress. From phytohormones to genome reorganization. Marcel Dekker, New York, pp 591–628

Kranner I, Beckett RP, Wornik S, Zorn M, Pfeinhofer W (2002) Revival of a resurrection plant correlates with its antioxidant status. Plant J 31:13–24

Król M, Ivanov MG, Jansson S, Kloppstech K, Huner NPA (1999) Greening under high light or cold temperature affects the level of xanthophyllcycle pigments, early light-inducible proteins, and light-harvesting polypeptides in wild-type barley and the chlorina f2 mutant. Plant Physiol 120:193–203

Le TN (2005) Genetics of desiccation tolerance in the resurrection plant *Sporobolus stapfianus*. PhD thesis. Monash University, Melbourne, Australia

Levitt J (1980) Responses of plants to environmental stresses, vol 2, Water, radiation, salt and other stresses. Academic, New York

Lewis ML, Miki K, Veda T (2000) FePer1, a gene encoding an evolutionary conserved 1-Cys peroxiredoxin in buckwheat (*Fagopyrum esculentum* Moench), is expressed in a seed-specific manner and induced during seed germination. Gene 246:81–91

Liu Q, Kasuga M, Sakuma Y, Abe H, Miura S, Yamaguchi-Shinozaki K, Shinozaki K (1998) Two transcription factors, DREB1 and DREB2, with an EREBP/AP2 DNA binding domain separate two cellular signal transduction pathways in drought and low temperature responsive gene expression, respectively, in Arabidopsis. Plant Cell 10:1391–1406

Lu ZJ, Neumann PM (1998) Water stressed maize, barley and rice seedlings show species diversity in mechanisms of leaf growth inhibition. J Exp Bot 49:1945–1952

Liu Y, Tengguo Z, Li X, Wang J (2007) Protective mechanisms of desiccation tolerance in *Reaumuria soongorica*. Sci China Ser C-Life Sci 50:15–21

Marais S, Thomson JA, Farrant JM, Mundree SG (2004) X_V VHA-C"-1 a novel stress responsive V-ATPhase Subnit C" homologue isolated from the resurrection plant *Xerophyta viscosa*. Plant Physiol 122:54–64

Mariaux JB, Bockel C, Salamini F, Bartels D (1998) Desiccationand abscisic acid-responsive genes encoding major intrinsic proteins (MIP) from the resurrection plant *Craterostigma plantagineum*. Plant Mol Biol 38:1089–1099

Maurel C, Chrispeels MJ (2001) Aquaporins. A Molecular entry into plant water relations. Plant Physiol 125:135–138

Maurel C, Kado RT, Guern J, Chrispeels MJ (1995) Phosphorylation regulates the water channel activity of the seed-specific aquaporin α-TIP. EMBO J 14:3028–3035

Mayee MB, Mundree SG, Majumder AL (2005) Molecular cloning, bacterial overexpression and characterization of L-myo-inositol 1-Phosphate Synthase from a monocotyledonous resurrection plant, *Xerophyta viscosa*. J Plant Biochem Biotechnol 14:95–98

McQueen-Mason SJ, Cosgrove DJ (1995) Expansin mode of action on cell walls – analysis of wall hydrolysis, stress-relaxation, and binding. Plant Physiol 107:87–100

Meissner RC, Jin H, Cominelli E, Denekamp M, Fuertes A, Greco R, Kranz HD, Penfield S, Petronik K, Urzainqui A, Martin C, Paz-Ares J, Smeekens S, Tonelli C, Weisshaar B, Baumann E, Klimyuk V, Marillonnet S, Patel S, Speulman E, Tissier AF, Bouchez D, Jones JDG, Periera A, Wisman E, Bevan M (1999) Function search in a large transcription factor gene family in Arabidopsis assessing the potential of reverse genetics to identify insertional mutations in R2R3 MYB genes. Plant Cell 11:1827–1840

Montané MH, Dreyer S, Triantaphylides C, Kloppstech K (1997) Early light-inducible proteins during long-term acclimation of barley to photooxidative stress caused by light and cold: high level of accumulation by posttranscriptional regulation. Planta 202:293–302

Moon BY, Higashi S, Gombos Z, Murata N (1995) Unsaturation of the membrane lipids of chloroplasts stabilizes the photosynthetic machinery against low-temperature photoinhibition in transgenic tobacco plants. Proc Natl Acad Sci USA 92(14):6219–6223

Moore JP, Lindsey FJM, GG BWF (2005) The South African and Namibian populations of the resurrection plant *Myrothamnus flabellifolius* are genetically distinct and display variation in their galloquinic acid composition. J Chem Ecol 31:2823–2834

Moore JP, Lindsey GG, Farrant JM, Brandt WG (2007) An overview of the biology of desiccation tolerant resurrection plant *Myrothamnus flabellifolia*. Ann Bot 99:211–217

Moore JP, Vicré-Gibouin M, Farrant JM, Driouich A (2008) Adaptations of higher plant cell walls to water loss: drought vs desiccation. Physiol Plant 134:237–245
Moran JF, Becana M, Iturbe-Ormaetxe I, Frechilla S, Klucas RV, Aparecio-Tejo P (1994) Drought induces oxidative stress in pea plants. Planta 194:346–352
Mowla SB, Thomson JA, Farrant JM, Mundree SG (2002) A novel stress-inducible antioxidant enzyme identified from the resurrection plant *Xerophyta viscosa* Baker. Planta 215:716–726
Mtwisha L, Farrant J, Brandt W, Lindsey GG (2006) Protection mechanisms against water deficit stress: desiccation tolerance in seeds as a study case. In: Ribaut J (ed) Drought adaptation in cereals. Haworth, New York, pp 531–549
Mundree SG, Whittaker A, Thomson JA, Farrant JM (2000) An aldose reductase homolog from the resurrection plant *Xerophyta viscose* Baker. Planta 211:693–700
Mundree SG, Baker B, Mowla S, Peters S, Marais S, Vander Willigen C, Govender K, Maredza A, Muyanga S, Farrant JM, Thomson JA (2002) Physiological and molecular insights into drought tolerance. Afr J Biotechnol 1:28–38
Munns R, Passioura JB, Guo JM, Chazen O, Cramer GR (2000) Water relations and leaf expansion: importance of time scale. J Exp Bot 51:1495–1504
Murelli C, Adamo V, Finzi PV, Albini FM, Bochicchio A, Picco AM (1996) Sugar biotransformations by fungi on leaves of the resurrection plant *Sporobolus stapfianus*. Phytochemistry 43:741–745
Navari-Izzo F, Ricci F, Vazzana C, Quartacci MF (1995) Unusual composition of thylakoid membranes of the resurrection plant Boea hygroscopica: changes in lipids upon dehydration and rehydration. Physiol Plant 94:135–142
Navari-Izzo F, Quartacci MF, Pinzino C, Rascio N, Vazzana C, Sgherri C (2000) Protein dynamics in thylakoids of the desiccation-tolerant plant Boea hygroscopica during dehydration and rehydration. Plant Physiol 124:1427–1436
Ndima T, Farrant JM, Thomson J, Mundree S (2001) Molecular characterization of XVT8, a stress-responsive gene from the resurrection plant *Xerophyta viscosa* Baket. Plant Growth Reg 35:137–145
Neale AD, Blomstedt CK, Bronson P, Le TN, Guthridge K, Evand D, Gaff DF, Hamill JD (2000) The isolation of genes from the resurrection grass *Sporobolus stapfianus* which are induced during severe drought stress. Plant Cell Environ 23:265–277
Norwood M, Toldi O, Richter A, Scott P (2003) Investigation into the ability of roots of the poikilohydric plant *Craterostigma plantagineum* to survive dehydration stress. J Exp Bot 54:2313–2321
Oliver MJ (1996) Desiccation-tolerance in vegetative plant cells. Physiol Plant 97:779–787
Oliver MJ, Bewley JD (1984) Desiccation and ultrastructure in bryophytes. Adv Bryology 2:91–131
Oliver MJ, Bewley JD (1997) Desiccation-tolerance in plant tissues. A mechanistic overview. Hortic Rev 18:171–214
Oliver MJ, Wood AJ, O'Mahony P (1998) "To dryness and beyond" – preparation for the dried state and rehydration in vegetative desiccation tolerant plants. Plant Growth Reg 24:193–201
Oliver MJ, Tuba Z, Mishler BD (2000) The evolution of vegetative desiccation tolerance in land plants. Plant Ecol 151:85–100
Oliver MJ, Velten J, Mishler BD (2005) Desiccation tolerance in bryophytes: a reflection of the primitive strategy for plant survival in dehydrating habitats? Integr Comp Biol 45:788–799
Peters S, Mundree SG, Thomson JA, Farrant JM, Keller F (2007) Protection mechanisms in the resurrection plant *Xerophyta viscosa* (Baker): both sucrose and raffinose family oligosacharides (RFOs) accumulate in leaves in response to water deficit. J Exp Bot 58(8):1947–1956
Phillips JR, Hilbricht T, Salamini F, Bartels D (2002) A novel abscisic acid- and dehydration-responsive gene family from the resurrection plant *Craterostigma plantagineum* encodes a plastid-targeted protein with DNA binding activity. Planta 215:258–266
Phillips JR, Dalmay T, Bartels D (2007) The role of small RNAs in abiotic stress. FEBS Lett 581:3592–3597

Phillips JR, Fischer E, Baron M, van den Dries N, Facchinelli F, Kutzer M, Rahmanzadeh R, Remus D, Bartels D (2008) Lindernia brevidens: a novel desiccation-tolerant vascular plant, endemic to ancient tropical rainforests. Plant J 54:938–948

Porembski S, Barthlott W (2000) Granitic and gneissic outcrops (inselbergs) as centers of diversity for desiccation-tolerant vascular plants. Plant Ecol 151:19–28

Proctor MCF, Pence VC (2002) Vegetative tissues: bryophytes, vascular resurrection plants, and vegetative propogules. In: Black M, Pritchard HW (eds) Desiccation and survival in plants: drying without dying. CABI, Wallingford, Oxon, pp 207–237

Provart NJ, Gil P, Chen W, Han B, Chang HS, Wang X, Zhu T (2003) Gene expression phenotypes of Arabidopsis associated with sensitivity to low temperatures. Plant Physiol 132:893–906

Quartacci MF, Forli M, Rascio N, DallaVecchia F, Bochicchio A, Navari-Izzo F (1997) Desiccation-tolerant *Sporobolus stapfianus*: lipid composition and cellular ultrastructure during dehydration and rehydration. J Exp Bot 48:1269–1279

Quartacci MF, Glisic O, Stevanovic B, Navari-Izzo F (2002) Plasma membrane lipids in the resurrection plant *Ramonda serbica* following dehydration and rehydration. J Exp Bot 53:2159–2166

Rahmanzadeh R, Müller K, Fischer E, Bartels D, Borsch T (2005) The Linderniaceae and Gratiolaceae are further lineages distinct from the Scrophulariaceae (Lamiales). Plant Biol 7:1–12

Ramanjulu S, Bartels D (2002) Drought- and desiccation-induced modulation of gene expression in plants. Plant Cell Environ 25:141–151

Rascio N, La Rocca N (2005) Resurrection plants: the puzzle of surviving extreme vegetative desiccation. Crit Rev Plant Sci 24:209–225

Reynolds TL, Bewley JD (1993) Abasicic acid enhances the ability of the desiccation tolerant fern *Polypodium virginianum* to withstand drying. J Exp Bot 44:1771–1779

Riechmann JL, Meyerowitz EM (1998) The AP2/EREBP family of plant transcription factors. Biol Chem 379:633–646

Rodriguez M, Edsgard D, Hussain SS, Alquezar A, Rasmussen M, Gilbert T, Nielsen B, Bartels D, Mundy J (2010) Transcriptomes of the desiccation tolerant resurrection plant *Craterostigma plantagineum*. Plant J 63:212–228

Savenstrand H, Olofsson M, Samuelsson M, Strid A (2004) Induction of early-inducible protein gene expression in *Pisum sativum* after exposure to low levels of UV-B radiation and other environmental stresses. Plant Cell Rep 22:532–536

Schiller P, Heilmeier H, Hartung W (1997) Absisic acid (ABA) relations in the aquatic resurrection plant *Chamaegigas intrepidus* under naturally fluctuating environmental conditions. New Phytol 136:603–611

Schneider K, Wells B, Schmelzer E, Salamini F, Bartels D (1993) Desiccation leads to the rapid accumulation of both cytosol and chloroplast proteins in the resurrection plant *Craterostigma plantagineum* Hochst. Planta 189:120–131

Scott P (2000) Resurrection plants and the secrets of eternal leaf. Ann Bot 85:159–166

Seki M, Umezawa T, Urano K, Shinozaki K (2007) Regulatory metabolic networks in drought stress responses. Curr Opin Plant Biol 10:296–302

Shen-Miller J, Mudgett MB, Schopf JW, Clarke S, Berger R (1995) Exceptional seed longevity and robust growth – ancient sacred lotus from China. Am J Bot 82:1367–1380

Sherwin HW, Farrant JM (1996) Differences in rehydration of three desiccation tolerant angiosperm species. Ann Bot 78:703–710

Sherwin HW, Farrant JM (1998) Protection mechanisms against excess light in the resurrection plants *Craterostigma wilmsii* and *Xerophyta viscosa*. Plant Growth Reg 24:203–210

Shigyo M, Haseba M, Ito M (2006) Molecular evolution of the AP2 subfamily. Gene 366:256–265

Smith-Espinoza CJ, Phillips JR, Salamini F, Bartels D (2005) Identification of further *Craterostigma plantagineum* cdt mutants affected in abscisic acid mediated desiccation tolerance. Mol Gen Genom 274:364–372

Stockinger EJ, Gilmour SJ, Thomashow MF (1997) Arabidopsis thaliana CBF1 encodes an AP2 domain-containing transcriptional activator that binds to the C-repeat/DRE, a cisacting DNA regulatory element that stimulates transcription in response to low temperature and water deficit. Proc Natl Acad Sci USA 94:1035–1040

Thomson WW, Platt KA (1997) Conservation of cell order in desiccated mesophyll of *Selaginella lepidophylla* ([Hook and Grev.] Spring). Ann Bot 79:439–447

Trainor FR, Gladych R (1995) Survival of algae in a desiccated soil: a 35-year study. Phycologia 34:191–192

Tuba Z, Proctor MCF, Csintalan Z (1998) Ecophysiological responses of homoiochlorophyllous and poikilochlorophyllous desiccation-tolerant plants: a comparison and an ecological perspective. Plant Growth Reg 24:211–217

Tunnacliffe A, Wise MJ (2007) The continuing conundrum of the LEA proteins. Naturwissenschaften 94:791–812

Vander Willigen C, Farrant JM, Pammenter NW (2001) Anomalous pressure volume curves of resurrection plants do not suggest negative turgor. Ann Bot 88:537–543

Vander Willigen C, Pammenter CNW, Jaffer MA, Mundree SG, Farrant JM (2003) An ultrastructural study using anhydrous fixation of *Eragrostis nindensis*, a resurrection grass with both desiccation-tolerant and sensitice tissues. Funct Plant Biol 30:1–10

Vander Willigen C, Pammenter NW, Mundree SG, Farrant JM (2004) Mechanical stabilization of desiccated vegetative tissues of the resurrection grass *Eragrostis nindensis*: does a TIP 3;1 and/or compartimentation of subcellular components and metabolites play a role? J Exp Bot 397:651–661

Vicré M, Sherwin HW, Driouich A, Jaffer MA, Farrant JM (1999) Cell wall characteristics and structure of hydrated and dry leaves of the resurrection plant *Craterostigma wilmsii*, a microscopical study. J Plant Physiol 155:719–726

Vicré M, Lerouxel O, Farrant J, Lerouge P, Driouich A (2004) Composition and desiccation-induced alterations of the cell wall in the resurrection plant *Craterostigma wilmsii*. Physiol Plant 120:229–239

Villalobos MA, Bartels D, Iturriaga G (2004) Stress tolerance and glucose insensitive phenotypes in Arabidopsis overexpressing the CpMYB10 transcription factor gene. Plant Physiol 135:309–324

Wang L, Shang H, Liu X, Wu R, Zheng M, Phillips J, Bartels D, Deng X (2009) A cell wall localised glycine-rich protein plays a role in dehydration tolerance in the resurrection plant *Boea hygrometrica*. Plant Biol 12:1–13

Wehmeyer N, Hernandez LD, Finkelstein RR, Vierling E (1996) Synthesis of small heat shock proteins is part of the developmental program of late seed maturation. Plant Physiol 112:747–757

Willige B, Kutzer M, Tebartz F, Bartels D (2009) Subcellular localization and enzymatic properties of differentially expressed transketolase genes isolated from the desiccation tolerant resurrection plant *Craterostigma plantagineum*. Planta 229:659–666

Yamaguchi-Shinozaki K, Shinozaki K (1993) The plant hormone abscisic acid mediates the drought-induced expression but not the seed-specific expression of rd22, a gene responsive to dehydration stress in *Arabidopsis thaliana*. Mol Gen Genet 238:17–25

Zeng Q, Chen X, Wood AJ (2002) Two early light-inducible protein (ELIP) cDNAs from the resurrection plant *Tortula ruralis* are differentially expressed in response to desiccation, rehydration, salinity, and high light. J Exp Bot 53:1197–1205

Part IV
Synopsis

Chapter 17
Synopsis: Drying Without Dying

Dorothea Bartels, Ulrich Lüttge, and Erwin Beck

This chapter focuses on aspects of desiccation tolerance of plants which connect several chapters addressing general principles and future research goals.

Volume 215 of Ecological Studies demonstrates that many more plant species as commonly assumed are able to survive periodic desiccation (Chaps. 2, 4, 6, 8, 9). This perception relates not only to non-vascular but also to vascular plants (1,300 species, Chap. 8) whose water management in the soil–plant–air continuum comprises several compartments and mutually interdependent steps which are vulnerable to desiccation and, surprisingly, also rehydratation (Chap. 10).

The majority of the poikilohydrous vascular plants are ferns and fern allies or tropical and subtropical epiphytes on the stems and in the canopies of big phorophytes (Chap. 8). Another ecological center of desiccation tolerant vascular plant species are tropical and subtropical inselbergs. Interestingly, in both types of habitats poikilohydrous and homoiohydrous plants occur side-by-side indicating that desiccation tolerance is not a "*conditio sine qua non*" for vascular plants to colonize such ecological niches. However, including also non-vascular plants in the survey results in the surprising perception that many, if not the majority of desiccation tolerant plant species are found in moist, at least periodically wet habitats (however, see also Chaps. 5 and 8 for a theory of niche separation). In that regard one extreme case of poikilohydric plants is the tiny aquatic angiosperm *Chamaegigas intrepidus* from the Scrophulariaceae, growing in ephemeric rock pools of Central Namibia (Chap. 12). Contrary to the general opinion, lichens, which are considered to be the most desiccation tolerant organisms, are more sensitive to dehydration than mosses, and even more sensitive than desiccation tolerant vascular plants – at least if the maximal length of time is considered which the different organisms are known to survive at experimental desiccation (Chap. 6).

Although some poikilohydrous vascular plants from the Cyperaceae and Velloziaceae can become several hundred years old (Chap. 8), physiological stress imposed by losing more than 90% of tissue water and remaining in the desiccated state for months at elevated day-temperatures and high irradiation usually reduces the life-span of the plants. Therefore tall plant statures are rare among the desiccation tolerant terrestrial plants. Nevertheless some of them have been termed shrubs or "arborescent," but the size even of the old pseudostemmed perennials which are – except *Myrothamnus flabellifolius* – all monocots, does not exceed a few meters.

U. Lüttge et al. (eds.), *Plant Desiccation Tolerance*, Ecological Studies 215,
DOI 10.1007/978-3-642-19106-0_17, © Springer-Verlag Berlin Heidelberg 2011

Likewise the size of the leaves of the resurrection plants is usually small and many of them possess anatomical structures which upon drought allow reduction of the transpiring surface, e.g., by folding or rolling in, or getting appressed against the stem (*M. flabellifolius*, Chap. 9). A low surface to volume ratio is also typical for many non-vascular plants including lichens (Chap. 5). In vascular plants folding or rolling in takes place along sclerenchymatous tissue strands which provide a support for the dehydrated thin-walled cells, like in the spongy or palisade tissues.

Nevertheless, loss of the bulk of the tissue water results in an extraordinary osmotic and mechanical stress. Vascular desiccation-tolerant plants can apply two strategies to minimize dehydration stress: water replacement or a reduction in cellular volume (Gaff 1971). The water replacement approach reduces in the first place the mechanical stress on the cellular substructures (Farrant 2000). Upon desiccation, in cells of some resurrection plants the big central vacuole that stabilizes the hydrated cells is replaced by many small vacuoles which contain high concentrations of sucrose, proline and protein (Vander Willigen et al. 2004). The other strategy to prevent lethal mechanical stress is to reduce the volume of the cells by extensive cell wall folding which requires a low elastic modulus of the entire cell wall (see Chap. 9 for biochemical explanation). During extensive folding of the cells the plasma membrane remains attached to the cell walls, a phenomenon that is known as cytorrhysis. Preservation of the contact between the plasma membrane and the cell wall is essential for the recovery during rehydration (Hartung et al. 1998; Vicré et al. 1999). In addition to the potential of reversible folding of the cell walls, shrinkage of the cell volume is commonly reduced by accumulation of high concentrations of the above-mentioned compounds in the central vacuole and the cytoplasm. Comparably little is known about the structural cellular responses to desiccation of poikilohydrous cryptogams, such as algae, lichens and mosses and the cyanobacteria.

Water storage is a strategy which allows many poikilohydrous plants to reduce the duration of the desiccated state upon transient or long-lasting water deficiency. Water content of bryophytes and lichens, as related to their dry weight is usually much higher than that of vascular plant species and can exceed the plant biomass 10- to 20-fold (Chap. 6). A high water content of these thallophytes is especially important, because these organisms cannot extract water from the soil by roots. Several bryophytes have developed special structures for water storage, such as the water sacs of *Frullania* or the utricles of the *Sphagnum* leaves. In addition, water retained by capillary forces in wick-like agglomerated rhizoids, or in dense cushions of plantlets constitutes large exogenous reservoirs. Such water relations have been termed ectohydric while thallophytes, which store water inside the thallus, e.g., in gelatinous sheaths around cyanobacterial filaments in cyanobacterial lichens, are considered as endohydric. Separation of photosynthesizing cells or tissues from structures for water storage is essential because of the slow diffusion of CO_2 in liquid water. Interference of surface water with photosynthesis is a special problem of lichens and liverworts whose thallus surface may be "sealed" by a water film (Chap. 6). Likewise, water soaking of the medullar cavities of lichens is considered noxious because of not only affecting photosynthetic CO_2

uptake by the imbedded algae but also respiratory gas exchange of the lichen fungus. For these thallophytes the maximal is therefore not necessarily the optimal water content.

While the majority of the desiccation-tolerant vascular and cryptogamic plants are homoiochlorophyllous, i.e., maintain their chloroplasts also in the dehydrated state, several of the resurrection plants (from the Cyperaceae, Liliaceae, Poaceae and Velloziaceae), which can survive long periods of drought, degrade their chloroplasts to an inoperable state, termed desiccoplast (Chap. 9). Desiccoplasts are different from geronto- or etioplasts as they can recover to functional chloroplasts. By completely degrading the chlorophylls and at least partially degrading the carotenoids they reduce the potential of damage from light absorption in a state in which they are unable to photosynthesize. Concomitantly the thylakoid structure disappears, whereas the chloroplast envelope is apparently maintained. Likewise, mitochondria change their ultra-structure, reducing the number of tubules (Chap. 9). These cellular changes to attain a long-lasting quasi-dormant state require several days and also the reverse processes, the conversion of desiccoplasts to chloroplasts, the re-establishment of the mitochondrial ultra-structure, and the return to full metabolic activity, takes more days. Enhanced respiration rates (resaturation respiration), however, are observed quickly upon rehydration and already before full turgor is attained. Since the roots of the poikilochlorophyllous plants die back during prolonged drought, rewetting is exclusively via the leaves while new adventitious roots are formed with some delay. Whether such changes are considered as disintegration and repair or as "reassembly" and subsequent revival of metabolism remains a matter of opinion.

The more widespread functional strategy, homoiochlorophylly (although some of the chlorophyll may be also degraded as in *M. flabellifolius*), which is typical of both vascular and non-vascular plants, is interpreted as adaptation to more frequent but shorter periods of drought (Chaps. 5, 6, 8, 9).

For homoiochlorophyllous plants irrespective of whether they are thallophytes or vascular plants, irradiance is a major challenge upon desiccation, when light absorption continues but the biochemical machinery for utilization of the products of linear electron flow becomes inactivated. In the cyanobacteria which do not possess the protective xanthophyll cycle (Chaps. 3 and 7) accumulation of sun-screen pigments which reduce the flux density of damaging irradiance arriving in the cell interior provides an effective though not complete protection. Simultaneously desiccation is delayed by extracellular polysaccharides and high concentrations of compatible solutes in the cytoplasm. A more effective and physiologically elegant mechanism is the combination of an effective xanthophyll cycle in the hydrated or partially hydrated state with an ultra-fast thermal deactivation of excitons in the photosynthetic reaction centers when dry. Upon excess irradiance green plants activate the xanthophyll cycle by protonation of a specific PsbS protein. However, only light energy which has not yet been absorbed by the photosynthetic reaction center can be discharged by that mechanism. While the xanthophyll cycle is inactivated by desiccation the second shield becomes activated which is also working in the cyanobacteria: The conversion of photosynthetic reaction centers into energy dissipation centers. Contrary to the xanthophyll cycle the dissipation centers can discharge also the first singlet state of chlorophyll, in other words can discharge excited reaction centers (Chap. 7). Excitation energy that reaches reaction centers can convert them to dissipation centers. After rehydration, this mechanism of energy dissipation is inactivated and the reaction centers regain their photosynthetic capacity. The protein that may be involved in the reversible conversion of reaction centers appears to be constitutively produced but may require activation by desiccation.

Irrespective of the various mechanisms to avoid damage by radicals formed upon high irradiance, biota depending on an oxidative atmosphere cannot avoid producing reactive oxygen species (ROS) by their own metabolism. Stress upon de- and subsequent rehydration still increases the extent of such reactions. Thus detoxification of ROS is a challenge for all plants during dehydration, in the desiccated state and during rehydration. However, as discussed in Chap. 11 cells require a balance of ROS, as some of these molecules are essential in signaling processes. Similar to activation of stress protective genes the ROS producing and detoxifying systems have been best studied in *Arabidopsis*. When desiccation-tolerant plants have been analyzed, it also appears that novel genes have not evolved, but that detoxification mechanisms, which are present in the non-desiccation-tolerant plants, are very effectively used in desiccation-tolerant organisms. Detoxification of ROS is a major component of desiccation tolerance, as it was nicely demonstrated by Kranner et al. (2002) in field studies of the resurrection plant *Myrothamnus*. Efficiency of ROS detoxification may be achieved by having several different parallel pathways in place leading to detoxification of reactive molecules. Parallel pathways will give a certain buffer and level of robustness to organisms, if one pathway may become less effective through mutation events.

The chapters presented in Part III of the book demonstrate that molecular and biochemical approaches have been successfully implemented in desiccation tolerance studies and have contributed to understanding of desiccation tolerance at the cellular level. The approaches have also changed the type of questions being asked. Whereas studies on the whole-plant level tried to analyze the responses of entire plants to desiccation (Part II), now the focus is on key molecules contributing to desiccation tolerance. The functional role of these molecules is addressed, as exemplified in Chap. 14 for the dehydrin stress proteins. The authors describe some very detailed biochemical and biophysical studies which allow proposing new hypotheses for the function, which in turn have to be tested on the whole-plant level.

As it has often been pointed out not only the desiccation but also the rehydration process is critical for cellular integrity, and it has been assumed that repair processes may take place at this stage. However, approaches to identify rehydration specific gene products were not very successful in angiosperms with a few exceptions. The situation is different in the moss *Tortula ruralis* where a cellular recovery program is activated during rehydration (Oliver et al. 2009). It appears that in higher plants recovery of cellular integrity is being anticipated during desiccation so that a number of gene products accumulate during dehydration and can be active in the very early stages of rehydration.

To detect the functions of the stress-related transcripts a mutational genetic approach is required, but no desiccation-tolerant plant model system has yet been identified which is amenable for generating mutants. In fact, all desiccation-tolerant plants analyzed so far are not diploid with the possible exception of the grass *Oropetium thomae* (Bartels and Mattar 2002). Therefore dehydration tolerance aspects of the genetic model system *Arabidopsis thaliana* have been included in this volume (Chap. 13). The chapter describes the complexity of the signaling pathway starting with sensing water depletion and leading to the activation of a transcriptional network. Transcript analyses during early stages of dehydration in desiccation-tolerant plants have shown that the same regulatory pathways are activated as in *Arabidopsis* (Bartels and Salamini 2001). Therefore, much can be learned about the identity of the molecules involved in response to dehydration from the large data sets available for *Arabidopsis* as presented by Grene et al. in Chap. 13. This reinforces the hypothesis that acquisition of desiccation tolerance involves mainly conserved genes present in most, if not all higher plants. Thus regulation of gene expression and epigenetic modulations must be keys to desiccation tolerance. It is remarkable that also gene products which are synthesized in response to dehydration in all desiccation-tolerant plants studied so far are conserved. One of the most abundant groups of gene products are the late embryogenesis abundant (LEA) proteins which are not only found in plants but also in desiccation-tolerant nematodes like *Aphelenchus avenae* (Browne et al. 2002). The conservation across organisms supports a functional requirement of these very hydrophilic proteins in desiccation tolerance. Comparative analyses across different desiccation-tolerant organisms may help to identify key molecules considering the difficulties of genetic approaches.

A very unique exception is the *CDT1* (= *Craterostigma* desiccation tolerance 1) gene which was discovered in the resurrection plant *Craterostigma*, and which is linked to desiccation tolerance, but so far has not been found in any other plant. It has been proposed that *CDT1* functions as regulatory RNA and may be involved in activating massive transcription of genes encoding stress protective products (Chap. 16). It has only been recognized over the last years that small RNAs are important regulators of gene expression in all processes of plants including stress responses. Studies of small RNAs in desiccation-tolerant plants have not been published yet. When this information will be available in the future, specific regulatory RNA components may be revealed (see Chap. 13.3, page 266). This is a novel research field which is not covered in this volume, but will certainly need attention.

Generation of ROS is a general response to different stressors. This may be an adaptation to the true environmental conditions under which most desiccation-tolerant plants grow. They have not only to endure times without water but in addition they are also exposed to high temperature and/or irradiance. This is to some extent reflected on the molecular level in two ways. Not only the synthesis of drought stress-related LEA proteins is observed but also synthesis of heat shock proteins is reported in desiccated plants (Alamillo et al. 1995).Vice versa LEA proteins have been found in response to temperature stress. Integration of multiple environmental factors is achieved on the promoter level of genes. Promoters of stress responsive genes are made up of functional *cis* elements regulating the gene expression level. There are several examples that the same gene is activated by different stressors such as dehydration, salt or low temperature (Yamaguchi-Shinozaki and Shinozaki 2006). So far promoter analysis has mainly focused on one stress factor but it is the task of future research to decipher transcriptional networks as described in Chaps. 13 and 15.

Chapter 15 emphasizes that new technologies, which allow analyzing the transcripts, proteins and metabolites on a large scale, will lead to a major advance in understanding the mechanistic basis of desiccation tolerance displayed in resurrection plants. So far these technologies have mainly been applied to genetic model plants. New DNA sequencing technologies and comprehensive bioinformatic tools allowed for the first time to get an overall picture of the transcriptome of a desiccation-tolerant plant during the different stages of hydration (Chap. 16; Rodriguez et al. 2010). This could lead to novel gene discoveries when the approaches will be applied to desiccation-tolerant plants. These studies may allow addressing open questions such as how is the cell cycle regulated during desiccation and rehydration, when is normal cell division taken up after desiccation and how is cellular homeostasis re-established. Discovery of so far unknown molecules has to be combined by detailed biochemical and, if possible, genetic studies to reveal the function. The contribution of biochemical studies to understanding functions and properties of molecules can be seen in Chap. 14 for the dehydrin proteins which were the first molecules to be discovered when molecular approaches were started to analyze responses to dehydration in vegetative tissues and acquisition of desiccation tolerance in seeds.

The book illustrates that a lot of data have been generated on the whole-plant level concerning distribution and ecological considerations, on the cellular level and on the molecular level in diverse desiccation-tolerant organisms. These data together with continuously growing genome sequence information will allow new experimental approaches which should give a general overview of the desiccated and rehydrated cell. It should be the aim to determine the transcriptome, proteome, and metabolome and then ask how do the different components interact and what is the hierarchical order of events. In this way, a system biology approach should become feasible to uncover the secret of desiccation survival.

References

Alamillo JM, Almoguera C, Bartels D, Jordano J (1995) Constitutive expression of small heat shock proteins in vegetative tissues of the resurrection plant *Craterostigma plantagineum*. Plant Mol Biol 29:1093–1099

Bartels D, Mattar M (2002) *Oropetium thomaeum*: a resurrection grass with a diploid genome. Maydica 47:185–192

Bartels D, Salamini F (2001) Desiccation tolerance in the resurrection plant *Craterostigma plantagineum*. A contribution to the study of drought tolerance at the molecular level. Plant Physiol 127:1346–1353

Browne J, Tunnacliffe A, Burnell A (2002) Plant desiccation gene found in a nematode. Nature 416:38

Farrant JM (2000) A comparison of mechanisms of desiccation tolerance among three angiosperm resurrection plant species. Plant Ecol 151:29–39

Gaff DF (1971) Desiccation-tolerant flowering plants of Southern Africa. Science 174:1033–1034

Hartung W, Schiller P, Dietz K-J (1998) Physiology of poikilohydric plants. Prog Bot 59:299–327

Kranner I, Beckett RP, Wornik S, Zorn M, Pfeinhofer W (2002) Revival of a resurrection plant correlates with its antioxidant status. Plant J 31:13–24

Oliver MJ, Hudgeons J, Dowd SE, Payton PR (2009) A combined subtractive suppression hybridization and expression profiling strategy to identify novel desiccation response transcripts from *Tortula ruralis* gametophytes. Physiol Plant 136:437–460

Rodriguez M, Edsgard D, Hussain SS, Alquezar A, Rasmussen M, Gilbert T, Nielsen B, Bartels D, Mundy J (2010) Transcriptomes of the desiccation tolerant resurrection plant *Craterostigma plantagineum*. Plant J 63:212–228

Vander Willigen C, Pammenter NW, Mundree SG, Farrant JM (2004) Mechanical stabilization of desiccated vegetative tissues of the resurrection grass *Eragrostis nindensis*: does a TIP 3;1 and/or compartmentalization of subcellular components and metabolites play a role? J Exp Bot 55:651–661

Vicré M, Sherwin HW, Driouich A, Jaffer MA, Farrant JM (1999) Cell water characteristics and structure of hydrated and dry leaves of the resurrection plant *Craterostigma wilmsii*, a microscopical study. J Plant Physiol 155:719–726

Yamaguchi-Shinozaki K, Shinozaki K (2006) Transcriptional regulatory networks in cellular responses and tolerance to dehydration and cold stresses. Annu Rev Plant Biol 57:781–803

Index

U. Lüttge et al. (eds.), *Plant Desiccation Tolerance*, Ecological Studies 215,
DOI 10.1007/978-3-642-19106-0, © Springer-Verlag Berlin Heidelberg 2011

L

M

N

Q